T0183245

Lecture Notes in Computer Science 9982

Commenced Publication in 1973
Founding and Former Series Editors:
Gerhard Goos, Juris Hartmanis, and Jan van Leeuwen

More information about this series at http://www.springer.com/series/7409

Paul Groth · Elena Simperl
Alasdair Gray · Marta Sabou
Markus Krötzsch · Freddy Lecue
Fabian Flöck · Yolanda Gil (Eds.)

The Semantic Web – ISWC 2016

15th International Semantic Web Conference
Kobe, Japan, October 17–21, 2016
Proceedings, Part II

 Springer

Editors

Paul Groth
Elsevier Labs
Amsterdam
The Netherlands

Elena Simperl
University of Southampton
Southampton
UK

Alasdair Gray
Heriot-Watt University
Edinburgh
UK

Marta Sabou
Vienna University of Technology
Vienna
Austria

Markus Krötzsch
Technische Universität Dresden
Dresden
Germany

Freddy Lecue
Accenture Technology Labs
Dublin
Ireland

and

Inria
Sophia Antipolis
France

Fabian Flöck
GESIS-Leibniz Institute for the Social
    Sciences
Cologne
Germany

Yolanda Gil
University of Southern California
Marina del Rey, CA
USA

ISSN 0302-9743   ISSN 1611-3349   (electronic)
Lecture Notes in Computer Science
ISBN 978-3-319-46546-3   ISBN 978-3-319-46547-0   (eBook)
DOI 10.1007/978-3-319-46547-0

Library of Congress Control Number: 2016951984

LNCS Sublibrary: SL3 – Information Systems and Applications, incl. Internet/Web, and HCI

Printed on acid-free paper

This Springer imprint is published by Springer Nature
The registered company is Springer International Publishing AG
The registered company address is: Gewerbestrasse 11, 6330 Cham, Switzerland

Preface

The International Semantic Web Conference (ISWC) continues to be the premier forum for Semantic Web researchers and practitioners to gather and share exciting new findings and experiences. The community has steadily grown in size and scope over the years, covering many aspects of Semantic Web technologies that lie at the intersection of semantic technologies, data, and the Web. Basic research has renewed importance as an engine of scientific understanding and of new ideas. The broad range of applications of Semantic Web technologies in real settings help us appreciate the accomplishments of the field as well as the limitations and challenges ahead. In addition to building on well-established standards, the community is always generating shared resources and infrastructure. There is a palpable excitement as we witness the Web becoming more machine readable every day.

This volume contains the proceedings of ISWC 2016 with all the papers accepted to the main conference tracks. This year, in addition to the traditional ISWC Research Track we solicited submissions to an Applications Track and a new Resources Track. A new Journal Track was introduced to expand the scope of the conference. The main conference call for papers received 326 responses, over 60 more than the total for the 2015 conference.

The Research Track continues to be the most popular category for submissions. This year, the track solicited novel and significant research contributions addressing theoretical, analytical, empirical, and practical aspects of the Semantic Web. In addition to work building on W3C Semantic Web recommendations (e.g., RDF, OWL, SPARQL, etc.), investigations on other approaches to the intersection of semantics and the Web were encouraged. The track received 212 submissions. After a bidding process, each was reviewed by at least four anonymous members of the Program Committee of the track including one senior Program Committee member. Authors were given a chance to respond to the reviews during an author rebuttal period. The senior Program Committee member was responsible for promoting discussion among the reviewers and making a final recommendation to the program chairs. Papers were discussed in a Program Committee meeting, and the chairs made final determinations about acceptance. These proceedings include the 39 papers that were accepted for presentation at the conference.

The Applications Track solicited submissions exploring the benefits and challenges of applying semantic technologies in concrete, practical applications, in contexts ranging from industry to government and science. The track accepted submissions in three categories: (1) in-use applications providing evidence that there is actual, significant use of the proposed application or tool by the target user group, preferably outside the group that conducted the development; (2) industry applications describing a business case or motivation and demonstrating their impact in the respective industry while ideally positioning the value of the tool or system for the Semantic Web community; (3) emerging applications describing early reports on real-world projects, exposing substantial research contributions and lessons learned in terms of semantics requirements, testing of approaches or infrastructure, and evaluations of early proto-types. The track received a total of 43 submissions and accepted 12. Each submission was reviewed by at least three Program Committee members of the track. Authors had the opportunity to submit a rebuttal to the reviews to clarify questions posed by Program Committee members. The program chairs made final decisions about accep-tance: 23 submissions were emerging applications and seven of them were accepted, 15 were in-use applications and four were accepted, and five were industry applications and one was accepted.

The newly introduced Resources Track sought submissions providing a concise and clear description of a resource and its (expected) usage. Traditional resources are considered to be ontologies, vocabularies, datasets, benchmarks and replication studies, services, and software. These resources are important outputs of any scientific work. Sharing these resources with the research community does not only ensure the reproducibility of results, but also has the benefit of supporting other researchers in their own work. Although high-quality shared resources have a key role and an essential impact on the advancement of a research community, the academic acknowledgement for sharing such resources is low. Therefore, many researchers primarily focus on publishing scientific papers and lack the motivation to share their resources. An additional challenge is that resources are often shared without following best practices, for example, at non-permanent URLs that become unavailable within a few months. The Resources Track aimed to encourage resource sharing following best practices within the Semantic Web community. Besides more established types of resources, the track solicited submissions of new types of resources such as ontology design patterns, crowdsourcing task designs, workflows, methodologies, and protocols and measures. The track received 71 submissions. At least three Program Committee members for the track reviewed each paper using a structured review form that focused on best practices for publishing a resource. After an author rebuttal period and sub-sequent discussion among the reviewers, the program chairs decided on the final acceptance of 24 resource papers that are included in these proceedings and were invited to be presented at the conference.

A new Journal Track was introduced this year to invite presentations at the con-ference about recent papers in the main journals where the community publishes. This inaugural track targeted the *Journal of Web Semantics* and the *Semantic Web Journal*. Authors of papers accepted during the past year that were not previously presented at a main Semantic Web conference could self-nominate their paper. From the 49 self-nominations, the editorial boards of the respective journals chose 12. These papers are not included in these proceedings, but we list full citations of the papers that can be found in the journals.

There are 75 papers included in these proceedings for the Research, Applications, and Resources tracks. The substantial amount of papers in the Resources Track attest to the strong culture in the Semantic Web community of disseminating research products and continuing to extend the pool of shared resources, and doing so beyond ontologies and software.

The conference proceedings were meticulously assembled by Fabian Flöck as proceedings chair, who worked with the chairs to compile all the papers from the authors, produce the table of contents and the front matter, and submit everything to the publishers. Silvio Peroni and Christoph Lange served as metadata chairs, organizing structured descriptions of the contents of the proceedings so they can be made available as semantic content in linked open data format. This year we accepted paper submissions in HTML format, but only received one submission in this format.

The conference included a variety of events that are traditional at ISWC and enrich the opportunities for interaction, learning, and mentoring.

The ISWC 2016 program included invited talks from prominent researchers within and outside the field. Christian Bizer from the University of Mannheim talked about "Is the Semantic Web What We Expected? Adoption Patterns and Content-Driven Challenges." Hiroaki Kitano of from Sony Computer Science Labs, the Okinawa Institute of Science and Technology, and the Systems Biology Institute discussed "Artificial Intelligence to Win the Nobel Prize and Beyond: Creating the Engine for Scientific Discovery." Kathleen McKeown of Columbia University, gave a talk titled "At the Intersection of Data Science and Language."

The Posters and Demos session, chaired by Takahiro Kawamura and Heiko Paulheim, included 55 posters and 47 demos selected among 115 total submissions. A Lightning Talks session offered time to those who wanted to take to the stage briefly to offer late-breaking results, discussion topics, and perspectives.

Thanks to our workshop and tutorial chairs, Chiara Guidini and Heiner Stuckenschmidt, the conference started off with very successful focused and highly interactive events. Five tutorials were held on ontology design patterns, RDF-stream processing, link discovery, Semantic Web for Internet/Web of Things, and SPARQL querying benchmarks. Moreover, 15 workshops were also held to foster discussions on specific topics of interest and to catalyze emerging communities. Also prior to the conference there was a discussion to envision the future of the Semantic Web Challenge.

The doctoral consortium chairs, Philippe Cudre-Mauroux, Riichiro Mizoguchi, and Natasha Noy, reviewed submissions from students still working on their PhD, and organized an event that gave them an opportunity to share their research ideas in a critical but supportive environment, to get feedback from mentors who are senior members of the community, to explore issues related to academic and research careers, and to build relationships with other PhD students from around the world. This program was complemented by activities put together by Abraham Bernstein, Daniel Garijo, and Matthew Horridge as student coordinators, who arranged travel awards, a mentoring lunch, and other informal opportunities for students to meet other members of the Semantic Web community.

The organization of a conference goes well beyond putting together a scientific program. There were many volunteers who worked hard to support the large event that ISWC has become, with hundreds of attendees from all over the world. We are very grateful to Hideaki Takeda, who as local arrangements chair led a skilled team to support the hotel accommodations, arrange conference facilities, develop the conference website, and take care of the myriad of details involved in supporting a scientific conference. We thank all of them for making the conference a fun event and for hosting us in the beautiful city of Kobe. The city's diverse surroundings (from the modern Kobe port to the mountainous Arima hot spring) and cultural heritage (from the Ikuta shrine to Nada Sake breweries) inspired all participants to think more broadly and about the longer-term legacy of their work. We are especially thankful to Ikki Ohmukai and Kouji Kozaki as vice chairs of the local committee and Rathachai Chawuthai as the Web master.

Sponsorship is crucial to support the conference. We would like to thank our sponsorship chairs, Makoto Iwayama and Carlos Pedrinaci, for their thorough and tireless work at arranging sponsorship, and to all of our sponsors for their generous contributions. We would also like to thank Amit Sheth for submitting a proposal to the National Science Foundation that helped secure support for student travel to the conference. The continued support from the National Science Foundation is greatly appreciated.

We are also grateful to the Semantic Web Science Association (SWSA), and in particular to its chair, Natasha Noy, and its treasurer, Guus Schreiber, for their sponsorship and for maintaining all the historical records of previous conferences containing precious data and advice. We are also grateful to Steffen Staab, Ulrich Wechselberger, Jeff Heflin, and the rest of the Organizing Committee of ISWC 2015, who were always at hand to answer our questions and provide thoughtful advice.

Last but not least, we would like to thank Miel Vander Sande, our publicity chair, who took all the announcements to mailing lists, social media, and other outlets to ensure dissemination and awareness of all the conference events.

We hope that these proceedings and the events at ISWC 2016 will contribute to a lasting legacy of this conference for many years to come.

October 2016

Paul Groth & Elena Simperl
Program Committee Co-chairs, Research Track
Alasdair Gray & Marta Sabou
Program Committee Co-chairs, Resources Track
Markus Krötzsch & Freddy Lecue
Program Committee Co-chairs, Applications Track
Yolanda Gil
General Chair

Conference Organization

General Chair

Yolanda Gil   University of Southern California, USA

Local Chair

Hideaki Takeda   National Institute of Informatics, Japan

Research Track Chairs

Paul Groth   Elsevier Labs, The Netherlands
Elena Simperl   University of Southampton, UK

Resources Track Chairs

Alasdair Gray   Heriot-Watt University, UK
Marta Sabou   Vienna University of Technology, Austria

Applications Track Chairs

Markus Krötzsch   TU Dresden, Germany
Freddy Lecue   Accenture Technology Labs, Ireland/Inria, France

Workshop and Tutorial Chairs

Chiara Ghidini   Fondazione Bruno Kessler, Italy
Heiner Stuckenschmidt   University of Mannheim, Germany

Posters and Demos Track Chairs

Takahiro Kawamura   Japan Science and Technology Agency, Japan
Heiko Paulheim   University of Mannheim, Germany

Journal Track Chairs

Abraham Bernstein   University of Zurich, Switzerland
Pascal Hitzler   Wright State University, USA
Guus Schreiber   VU University Amsterdam, The Netherlands

Doctoral Consortium Chairs

Philippe Cudré-Mauroux   University of Fribourg, Switzerland
Riichiro Mizoguchi   Japan Advanced Institute of Science and Technology,
                              Japan
Natasha Noy   Google, Inc., USA

Proceedings Chair

Fabian Flöck   GESIS - Leibniz Institute for the Social Sciences,
                              Germany

Sponsorship Chairs

Makoto Iwayama   Hitachi, Ltd., Japan
Carlos Pedrinaci   The Open University, UK

Metadata Chairs

Silvio Peroni   University of Bologna, Italy
Christoph Lange   University of Bonn/Fraunhofer IAIS, Germany

Publicity Chair

Miel Vander Sande   Ghent University – iMinds, Belgium

Student Coordinators

Abraham Bernstein   University of Zurich, Switzerland
Daniel Garijo   Universidad Politécnica de Madrid, Spain
Matthew Horridge   Stanford University, USA

Senior Program Committee - Research Track

Harith Alani   The Open University, UK
Philipp Cimiano   Bielefeld University, Germany
Oscar Corcho   Universidad Politécnica de Madrid, Spain
Philippe Cudré-Mauroux   University of Fribourg, Switzerland
Fabien Gandon   Inria, France
Birte Glimm   University of Ulm, Germany
Lalana Kagal   Massachusetts Institute of Technology, USA
Craig Knoblock   University of Southern California, USA
Vanessa Lopez   IBM Research, Ireland
David Martin   Nuance Communications, USA
Diana Maynard   University of Sheffield, UK

Axel-Cyrille Ngonga Ngomo	University of Leipzig, Germany
Axel Polleres	Vienna University of Economics and Business - WU Wien, Austria
Sebastian Rudolph	Technische Universität Dresden, Germany
Stefan Schlobach	Vrije Universiteit Amsterdam, The Netherlands
Mari Carmen Suárez-Figueroa	Universidad Politénica de Madrid, Spain
Maria Esther Vidal	Universidad Simon Bolivar, Venezuela

Program Committee - Research Track

Maribel Acosta	Karlsruhe Institute of Technology, Germany
Muhammad Intizar Ali	National University of Ireland, Ireland
Faisal Alkhateeb	Yarmouk University, Jordan
Pramod Anantharam	Bosch Research and Technology Center, USA
Manuel Atencia	Université Grenoble Alpes and Inria, France
Medha Atre	Unaffiliated
Nathalie Aussenac-Gilles	IRIT CNRS, France
Jie Bao	Memect, China
Payam Barnaghi	University of Surrey, UK
Christian Bizer	University of Mannheim, Germany
Eva Blomqvist	Linköping University, Sweden
Kalina Bontcheva	University of Sheffield, UK
Paolo Bouquet	University of Trento, OKKAM srl, Italy
Loris Bozzato	Fondazione Bruno Kessler, Italy
Adrian M.P. Brasoveanu	MODUL Technology GmbH, Germany
Carlos Buil Aranda	Universidad Técnica Federico Santa María, Chile
Paul Buitelaar	National University of Ireland, Ireland
Grégoire Burel	The Open University, UK
Andrea Calí	University of London, Birkbeck College, UK
Jean-Paul Calbimonte	HES-SO Valais, EPFL, France
Diego Calvanese	Free University of Bozen-Bolzano, Italy
Amparo E. Cano	Aston University, UK
Iván Cantador	Universidad Autónoma de Madrid, Spain
Irene Celino	CEFRIEL, Italy
Davide Ceolin	University Amsterdam, The Netherlands
Pierre-Antoine Champin	LIRIS, France
Gong Cheng	Nanjing University, China
Michael Compton	CSIRO, Australia
Sam Coppens	Autodesk, Inc., Ireland
Isabel Cruz	University of Illinois at Chicago, USA
Bernardo Cuenca Grau	University of Oxford, UK
Edward Curry	Digital Enterprise Research Institute, Ireland
Claudia d'Amato	University of Bari, Italy
Mathieu D'Aquin	The Open University, UK

Danica Damljanovic	Pure AI, UK
Brian Davis	National University of Ireland, Ireland
Daniele Dell'Aglio	DEIB, Politecnico di Milano, Italy
Emanuele Della Valle	DEIB, Politecnico di Milano, Italy
Elena Demidova	University of Southampton, UK
Stefan Dietze	L3S Research Center, Germany
Ying Ding	Indiana University, USA
John Domingue	The Open University, UK
Jérôme Euzenat	Inria and Université Grenoble Alpes, France
Nicola Fanizzi	Università di Bari, Italy
Anna Fensel	University of Innsbruck, Austria
Miriam Fernandez	The Open University, UK
Lorenz Fischer	Swisscom AG, Switzerland
Achille Fokoue	IBM Research, USA
Adam Funk	University of Sheffield, UK
Olaf Görlitz	Cognizant Technology Solutions, Germany
Aldo Gangemi	Université Paris 13 and CNR-ISTC, France
José María García	University of Seville, Spain
Raúl García-Castro	Universidad Politécnica de Madrid, Spain
Anna Lisa Gentile	University of Mannheim, Germany
Jose Manuel Gomez-Perez	Expert System, Spain
Tudor Groza	The Garvan Institute of Medical Research, Australia
Michael Gruninger	University of Toronto, Canada
Christophe Guéret	BBC, UK
Giancarlo Guizzardi	Federal University of Espirito Santo, Brazil
Claudio Gutierrez	Universidad de Chile, Chile
Peter Haase	metaphacts, Germany
Harry Halpin	World Wide Web Consortium, USA
Andreas Harth	Karlsruhe Institute of Technology, Germany
Olaf Hartig	Linköping University, Sweden
Tom Heath	Open Data Institute, UK
Pascal Hitzler	Wright State University, USA
Rinke Hoekstra	VU Amsterdam/University of Amsterdam, The Netherlands
Aidan Hogan	Universidad de Chile, Chile
Andreas Hotho	University of Würzburg, Germany
Geert-Jan Houben	TU Delft, The Netherlands
Wei Hu	Nanjing University, China
Eero Hyvönen	Aalto University, Finland
Luis-Daniel Ibáñez	University of Southampton, UK
Krzysztof Janowicz	University of California, Santa Barbara, USA
Ernesto Jimenez-Ruiz	University of Oxford, UK
Hanmin Jung	Korea Institute of Science and Technology Information, South Korea
Benedikt Kämpgen	FZI Forschungszentrum Informatik, Germany
Mark Kaminski	University of Oxford, UK

David Karger	Massachusetts Institute of Technology, USA
Hong-Gee Kim	Seoul National University, South Korea
Matthias Klusch	DFKI, Germany
Jacek Kopecky	University of Portsmouth, UK
Manolis Koubarakis	National and Kapodistrian University of Athens, Greece
Tobias Kuhn	Vrije Universiteit Amsterdam, The Netherlands
Werner Kuhn	University of California, Santa Barbara, USA
Steffen Lamparter	Siemens AG, Corporate Technology, Germany
Agnieszka Lawrynowicz	Poznań University of Technology, Poland
Chengkai Li	University of Texas at Arlington, USA
Juanzi Li	Tsinghua University, China
Nuno Lopes	TopQuadrant, Inc., Ireland
Chun Lu	SEPAGE S.A.S, France
Markus Luczak-Roesch	Victoria University of Wellington, New Zealand
Maria Maleshkova	Karlsruhe Institute of Technology, Germany
Erik Mannens	iMinds – Ghent University, Belgium
Robert Meusel	University of Mannheim, Germany
Peter Mika	Yahoo! Research, UK
Riichiro Mizoguchi	Japan Advanced Institute of Science and Technology, Japan
Dunja Mladenic	Jozef Stefan Institute, Slovenia
Luc Moreau	University of Southampton, UK
Boris Motik	University of Oxford, UK
Enrico Motta	The Open University, UK
Nadeschda Nikitina	Oxford University, UK
Andriy Nikolov	fluid Operations AG, Germany
Lyndon Nixon	MODUL University Vienna, Austria
Beng Chin Ooi	National University of Singapore, Singapore
Massimo Paolucci	DoCoMo Euro labs, Germany
Bijan Parsia	University of Manchester, UK
Peter Patel-Schneider	Nuance Communications, USA
Terry Payne	University of Liverpool, UK
Carlos Pedrinaci	The Open University, UK
Silvio Peroni	DASPLab, DISI, University of Bologna, Italy
Dimitris Plexousakis	Institute of Computer Science, FORTH, Greece
Guilin Qi	Southeast University, China
Ganesh Ramakrishnan	IIT Bombay, India
Maya Ramanath	IIT Delhi, India
Achim Rettinger	Karlsruhe Institute of Technology, Germany
Giuseppe Rizzo	ISMB, Italy
Dumitru Roman	SINTEF/University of Oslo, Norway
Riccardo Rosati	Sapienza Università di Roma, Italy
Marco Rospocher	Fondazione Bruno Kessler, Italy
Matthew Rowe	Lancaster University, UK

Harald Sack	Hasso Plattner Institute/University of Potsdam, Germany
Hassan Saif	The Open University, UK
Francois Scharffe	3Top, USA
Ansgar Scherp	Kiel University and ZBW – Leibniz Information Center for Economics, Germany
Marco Schorlemmer	IIIA-CSIC, Spain
Daniel Schwabe	PUC-Rio, Brazil
Juan Sequeda	Capsenta, USA
Estefania Serral	KU Leuven, Belgium
Michael Sintek	DFKI GmbH, Germany
Monika Solanki	University of Oxford, UK
Dezhao Song	Thomson Reuters, USA
Steffen Staab	University of Koblenz-Landau, Germany and University of Southampton, UK
Markus Strohmaier	University of Koblenz-Landau, Germany
Gerd Stumme	University of Kassel, Germany
Jing Sun	The University of Auckland, New Zealand
Vojtěch Svátek	University of Economics, Czech Republic
Pedro Szekely	University of Southern California, USA
Annette Ten Teije	Vrije Universiteit Amsterdam, The Netherlands
Matthias Thimm	University of Koblenz-Landau, Germany
Krishnaprasad Thirunarayan	Wright State University, USA
Nicolas Torzec	Yahoo, USA
Sebastian Tramp	eccenca, Germany
Raphaël Troncy	EURECOM, France
Anni-Yasmin Turhan	Technische Universität Dresden, Germany
Jacopo Urbani	Vrije Universiteit Amsterdam, The Netherlands
Victoria Uren	Aston University, UK
Marieke van Erp	Vrije Universiteit Amsterdam, The Netherlands
Frank van Harmelen	Vrije Universiteit Amsterdam, The Netherlands
Jacco van Ossenbruggen	CWI VU Amsterdam, The Netherlands
Ruben Verborgh	Ghent University – iMinds, Belgium
Natalia Villanueva-Rosales	University of Texas at El Paso, USA
Haofen Wang	East China University of Science and Technology, China
Kewen Wang	Griffith University, Australia
Zhichun Wang	Beijing Normal University, China
Yong Yu	Shanghai Jiao Tong University, China
Fouad Zablith	American University of Beirut, Lebanon
Sergej Zerr	University of Southampton, UK
Qingpeng Zhang	Rensselaer Polytechnic Institute, USA

Additional Reviewers - Research Track

Nitish Aggarwal
Albin Ahmeti
Saud Aljaloud
Martin Becker
Aurélien Bénel
David Berry
Georgeta Bordea
Stefano Bortoli
Stefano Braghin
Janez Brank
Michel Buffa
Elena Cabrio
Xuezhi Cao
Claudia Carapelle
David Carral
Roberto Confalonieri
Olivier Corby
Julien Corman
Minh Dao-Tran
Jérôme David
Ronald Denaux
Djellel Eddine Difallah
Anastasia Dimou
Jiwei Ding
Fajar J. Ekaputra
Basil Ell
Michael Färber
Daniel Faria
Yasmin Fathy
Jean-Philippe Fauconnier
Javier D. Fernández
Mariano Fernández-López
Giorgos Flouris
Marvin Frommhold
Irini Fundulaki
Andrés García-Silva
Camilo Garrido
Alain Giboin
Kalpa Gunaratna
Karl Hammar
Tom Hanika

Matthias Hartung
Sona Hasani
Naeemul Hassan
Daniel Hernandez
Yingjie Hu
Robert Isele
Daniel Janke
Stéphane Jean
Soufian Jebbara
Nazifa Karima
Mario Karlovcec
Yevgeny Kazakov
Felix Leif Keppmann
Robin Keskisärkkä
Christoph Kling
Magnus Knuth
Haris Kondylakis
Patrick Koopmann
Aljaz Kosmerlj
Nenad Krdzavac
Adila A. Krisnadhi
Sumant Kulkarni
Sungin Lee
Oliver Lehmberg
Tatiana Lesnikova
Mohamed Nadjib Mami
Nandana
Mihindukulasooriya
Maja Milicic Brandt
Seyed Iman Mirrezaei
Aditya Mogadala
Alexandre Monnin
Jose Mora
Raghava Mutharaju
Thomas Niebler
Charalampos Nikolaou
Nikolay Nikolov
Chifumi Nishioka
Francesco Osborne
Yaser Oulabi
Guillermo Palma

Panayiotis Papadakos
Sujan Perera
Minh Pham
Patrick Philipp
Julien Plu
María Poveda-Villalón
Freddy Priyatna
Alessandro Provetti
Lin Qiu
Kan Ren
Martin Ringsquandl
Dominique Ritze
Markus Rokicki
Marvin Schiller
Lukas Schmelzeisen
Andreas Schmidt
Xin Shuai
Hala Skaf-Molli
Cinzia Incoronata Spina
Simon Steyskal
Dina Sukhobok
Steffen Thoma
Veronika Thost
Pierpaolo Tommasi
Riccardo Tommasini
Serena Villata
Joerg Waitelonis
Simon Walk
Sebastian Walter
Wei Wang
Zhe Wang
Sanjaya Wijeratne
Ian Wood
Tianxing Wu
Yuxin Ye
Ran Yu
Veruska Zamborlini
Gensheng Zhang
Lei Zhang
Linhong Zhu

Program Committee - Resources Track

Maribel Acosta	Karlsruhe Institute of Technology, Germany
Ahmet Aker	Sheffield University, UK
Diana Bental	Heriot-Watt University, UK
Abraham Bernstein	University of Zurich, Switzerland
Eva Blomqvist	Linköping University, Sweden
Sarven Capadisli	University of Bonn, Germany
Tim Clark	Massachusetts General Hospital/Harvard Medical School, USA
Oscar Corcho	Universidad Politécnica de Madrid, Spain
Philippe Cudre-Mauroux	University of Fribourg, Switzerland
Claudia d'Amato	University of Bari, Italy
Mathieu D'Aquin	The Open University, UK
Victor de Boer	VU Amsterdam, The Netherlands
Daniele Dell'Aglio	BEIB, Politecnico di Milano, Italy
Gianluca Demartini	University of Sheffield, UK
Stefan Dietze	L3S Research Center, Germany
Ying Ding	Indiana University, USA
Michel Dumontier	Stanford University, USA
Fajar J. Ekaputra	Technische Universität Wien, Austria
Miriam Fernandez	Knowledge Media Institute, UK
Aldo Gangemi	Université Paris 13 and CNR-ISTC, France
Jose Manuel Gomez-Perez	Expert System, Spain
Alejandra Gonzalez-Beltran	University of Oxford, UK
Pascal Hitzler	Wright State University, USA
Robert Hoehndorf	King Abdullah University of Science and Technology, Saudi Arabia
Rinke Hoekstra	University of Amsterdam, The Netherlands
Matthew Horridge	Stanford University, USA
Andreas Hotho	University of Würzburg, Germany
Antoine Isaac	Europeana and Vrije Universiteit Amsterdam, The Netherlands
Simon Jupp	European Bioinformatics Institute, UK
Elmar Kiesling	TU Wien, Austria
Olga Kovalenko	TU Wien, Austria
Tobias Kuhn	VU University Amsterdam, The Netherlands
Markus Luczak-Roesch	University of Southampton, UK
Maria Maleshkova	Karlsruhe Institute of Technology, Germany
Fiona McNeill	Heriot-Watt University, UK
Chris Mungall	Lawrence Berkeley National Laboratory, USA
Vinh Nguyen	Wright State University, USA
Bijan Parsia	University of Manchester, UK
Heiko Paulheim	University of Mannheim, Germany
Silvio Peroni	University of Bologna, Italy
Valentina Presutti	STLab (ISTC-CNR), Italy

Giuseppe Rizzo	ISMB, Italy
Mariano Rodríguez Muro	IBM Research, USA
Satya Sahoo	Case Western Reserve University, USA
Cristina Sarasua	University Koblenz - Landau, Germany
Stefan Schlobach	Vrije Universiteit Amsterdam, The Netherlands
Jodi Schneider	University of Pittsburgh, USA
Stian Soiland-Reyes	University of Manchester, UK
Valentina Tamma	University of Liverpool, UK
Krishnaprasad Thirunarayan	Wright State University, USA
Ramine Tinati	University of Southampton, UK
Raphaël Troncy	EURECOM, France
Natalia Villanueva-Rosales	University of Texas at El Paso, USA
Simon Walk	Graz University of Technology, Austria
Peter Wetz	TU Wien, Austria
Marcia Zeng	Kent State University, USA
Jun Zhao	University of Oxford, UK

Additional Reviewers - Resources Track

Reihaneh Amini	Kalpa Gunaratna	Laura Rettig
Sebastian Bader	Karl Hammar	Daniel Schlör
David Carral	Robin Keskisärkkä	Saeedeh Shekarpour
Lu Chen	Adila Krisnadhi	Ruben Verborgh
Maxime Déraspe	Sarasi Lalithsena	Ran Yu
Djellel Difallah	María Poveda-Villalón	Amrapali Zaveri
Jhonatan Garcia	Sambhawa Priya	
Nuria García-Santa	José Redondo-García	

Program Committee - Applications Track

Anupriya Ankolekar	Hewlett Packard Labs, USA
Sören Auer	University of Bonn and Fraunhofer IAIS, Germany
Christian Bizer	University of Mannheim, Germany
Oscar Corcho	Universidad Politécnica de Madrid, Spain
Olivier Curé	Université Paris-Est LIGM, France
Mathieu D'Aquin	The Open University, UK
John Davies	BT, UK
Chiara Del Vescovo	British Broadcasting Corporation, UK
Mauro Dragoni	Fondazione Bruno Kessler - FBK-IRST, Italy
Fabien Gandon	Inria, France
Peter Haase	metaphacts, Germany
Aidan Hogan	Universidad de Chile, Chile
Krzysztof Janowicz	University of California, Santa Barbara, USA

Pavel Klinov	Complexible Inc., Germany
Matthias Klusch	DFKI, Germany
Craig Knoblock	University of Southern California, USA
Danh Le Phuoc	Technische Universität Berlin, Germany
Vanessa Lopez	IBM Research, Ireland
Despoina Magka	Facebook, USA
Maria Maleshkova	Karlsruhe Institute of Technology, Germany
Jeff Z. Pan	University of Aberdeen, UK
Peter Patel-Schneider	Nuance Communications, USA
Carlos Pedrinaci	The Open University, UK
Juan F. Sequeda	Capsenta Labs, USA
Dezhao Song	Thomson Reuters, USA
Kavitha Srinivas	IBM Research, USA
Federico Ulliana	Université Montpellier 2, France
Jürgen Umbrich	Vienna University of Economy and Business (WU), Austria
Jacopo Urbani	Vrije Universiteit Amsterdam, The Netherlands
Denny Vrandečić	Google, USA
Peter Yeh	Nuance Communications, USA

Additional Reviewers - Applications Track

Valerio Basile	Ali Khalili	Diego Reforgiato
Michel Buffa	Gregor Leban	Simon Scerri
Alistair Duke	Lionel Médini	Jason Slepicka
Majid Ghasemi-Gol	Michael Mrissa	Charese Smiley
Irlan Grangel	Niklas Petersen	Tobias Weller

Sponsors

Platinum	**Gold Plus**	**Gold**	**Silver**
IBM	Oracle	Recruit	Yahoo Japan
Semantic Software	IOS Press	Technologies	Google
		Fujitsu	
		NTT Resonant	
		SYSTAP/	
		Metaphacts	
		Hitachi	
		Rakuten	

IBM®

semantic software
LIBERATE YOUR DATA

ORACLE®

IOS
P r e s s

RECRUIT
Recruit Technologies Co.,Ltd.

FUJITSU

goo

bla**z**egraph™   metaphacts

HITACHI
Inspire the Next

®Rakuten
Institute of Technology

YAHOO!
JAPAN

Google

Contents – Part II

Resources

Ontological Representation of Audio Features. 3
  Alo Allik, György Fazekas, and Mark Sandler

Abstract Meaning Representations as Linked Data. 12
  Gully A. Burns, Ulf Hermjakob, and José Luis Ambite

Interoperability for Smart Appliances in the IoT World 21
  Laura Daniele, Monika Solanki, Frank den Hartog, and Jasper Roes

An Ontology of Soil Properties and Processes 30
  Heshan Du, Vania Dimitrova, Derek Magee, Ross Stirling,
  Giulio Curioni, Helen Reeves, Barry Clarke, and Anthony Cohn

LODStats: The Data Web Census Dataset 38
  Ivan Ermilov, Jens Lehmann, Michael Martin, and Sören Auer

Zhishi.lemon: On Publishing Zhishi.me as Linguistic Linked Open Data 47
  Zhijia Fang, Haofen Wang, Jorge Gracia, Julia Bosque-Gil,
  and Tong Ruan

Linked Disambiguated Distributional Semantic Networks. 56
  Stefano Faralli, Alexander Panchenko, Chris Biemann,
  and Simone P. Ponzetto

BESDUI: A Benchmark for End-User Structured Data User Interfaces. 65
  Roberto García, Rosa Gil, Juan Manuel Gimeno, Eirik Bakke,
  and David R. Karger

SPARQLGX: Efficient Distributed Evaluation of SPARQL
with Apache Spark 80
  Damien Graux, Louis Jachiet, Pierre Genevès, and Nabil Layaïda

Querying Wikidata: Comparing SPARQL, Relational and Graph Databases . . . 88
  Daniel Hernández, Aidan Hogan, Cristian Riveros, Carlos Rojas,
  and Enzo Zerega

Clinga: Bringing Chinese Physical and Human Geography in Linked
Open Data 104
  Wei Hu, Haoxuan Li, Zequn Sun, Xinqi Qian, Lingkun Xue, Ermei Cao,
  and Yuzhong Qu

LinkGen: Multipurpose Linked Data Generator   113
   Amit Krishna Joshi, Pascal Hitzler, and Guozhu Dong

OntoBench: Generating Custom OWL 2 Benchmark Ontologies   122
   Vincent Link, Steffen Lohmann, and Florian Haag

Linked Data (in Low-Resource) Platforms: A Mapping for Constrained
Application Protocol   131
   Giuseppe Loseto, Saverio Ieva, Filippo Gramegna, Michele Ruta,
   Floriano Scioscia, and Eugenio Di Sciascio

TripleWave: Spreading RDF Streams on the Web   140
   Andrea Mauri, Jean-Paul Calbimonte, Daniele Dell'Aglio,
   Marco Balduini, Marco Brambilla, Emanuele Della Valle,
   and Karl Aberer

Conference Linked Data: The ScholarlyData Project   150
   Andrea Giovanni Nuzzolese, Anna Lisa Gentile, Valentina Presutti,
   and Aldo Gangemi

The OWL Reasoner Evaluation (ORE) 2015 Resources   159
   Bijan Parsia, Nicolas Matentzoglu, Rafael S. Gonçalves, Birte Glimm,
   and Andreas Steigmiller

FOOD: FOod in Open Data   168
   Silvio Peroni, Giorgia Lodi, Luigi Asprino, Aldo Gangemi,
   and Valentina Presutti

YAGO: A Multilingual Knowledge Base from Wikipedia, Wordnet,
and Geonames   177
   Thomas Rebele, Fabian Suchanek, Johannes Hoffart, Joanna Biega,
   Erdal Kuzey, and Gerhard Weikum

A Collection of Benchmark Datasets for Systematic Evaluations
of Machine Learning on the Semantic Web   186
   Petar Ristoski, Gerben Klaas Dirk de Vries, and Heiko Paulheim

Enabling Combined Software and Data Engineering at Web-Scale:
The ALIGNED Suite of Ontologies   195
   Monika Solanki, Bojan Božić, Markus Freudenberg,
   Dimitris Kontokostas, Christian Dirschl, and Rob Brennan

A Replication Study of the Top Performing Systems in SemEval Twitter
Sentiment Analysis   204
   Efstratios Sygkounas, Giuseppe Rizzo, and Raphaël Troncy

VoldemortKG: Mapping schema.org and Web Entities to Linked
Open Data   220
    *Alberto Tonon, Victor Felder, Djellel Eddine Difallah,
    and Philippe Cudré-Mauroux*

AUFX-O: Novel Methods for the Representation of Audio
Processing Workflows   229
    Thomas Wilmering, György Fazekas, and Mark B. Sandler

Applications

Translating Ontologies in Real-World Settings   241
    Mihael Arcan, Mauro Dragoni, and Paul Buitelaar

EnergyUse - A Collective Semantic Platform for Monitoring
and Discussing Energy Consumption   257
    Grégoire Burel, Lara S.G. Piccolo, and Harith Alani

Extracting Semantic Information for e-Commerce   273
    Bruno Charron, Yu Hirate, David Purcell, and Martin Rezk

Building Urban LOD for Solving Illegally Parked Bicycles in Tokyo   291
    Shusaku Egami, Takahiro Kawamura, and Akihiko Ohsuga

Ontology-Based Design of Space Systems   308
    *Christian Hennig, Alexander Viehl, Benedikt Kämpgen,
    and Harald Eisenmann*

Capturing Industrial Information Models with Ontologies and Constraints . . .   325
    *Evgeny Kharlamov, Bernardo Cuenca Grau, Ernesto Jiménez-Ruiz,
    Steffen Lamparter, Gulnar Mehdi, Martin Ringsquandl, Yavor Nenov,
    Stephan Grimm, Mikhail Roshchin, and Ian Horrocks*

Towards Analytics Aware Ontology Based Access to Static
and Streaming Data   344
    *Evgeny Kharlamov, Yannis Kotidis, Theofilos Mailis, Christian Neuenstadt,
    Charalampos Nikolaou, Özgür Özçep, Christoforos Svingos,
    Dmitriy Zheleznyakov, Sebastian Brandt, Ian Horrocks, Yannis Ioannidis,
    Steffen Lamparter, and Ralf Möller*

QuerioDALI: Question Answering Over Dynamic and Linked
Knowledge Graphs   363
    Vanessa Lopez, Pierpaolo Tommasi, Spyros Kotoulas, and Jiewen Wu

Automatic Classification of Springer Nature Proceedings with Smart
Topic Miner   383
    *Francesco Osborne, Angelo Salatino, Aliaksandr Birukou,
    and Enrico Motta*

Semantic Technologies for Data Analysis in Health Care   400
  Robert Piro, Yavor Nenov, Boris Motik, Ian Horrocks, Peter Hendler,
  Scott Kimberly, and Michael Rossman

Building and Exploring an Enterprise Knowledge Graph
for Investment Analysis   418
  Tong Ruan, Lijuan Xue, Haofen Wang, Fanghuai Hu, Liang Zhao,
  and Jun Ding

Extending SPARQL for Data Analytic Tasks   437
  Julian Dolby, Achille Fokoue, Mariano Rodriguez Muro,
  Kavitha Srinivas, and Wen Sun

Author Index   453

Contents – Part I

Research

Structuring Linked Data Search Results Using Probabilistic Soft Logic 3
Duhai Alshukaili, Alvaro A.A. Fernandes, and Norman W. Paton

The Multiset Semantics of SPARQL Patterns 20
Renzo Angles and Claudio Gutierrez

Ontop of Geospatial Databases 37
Konstantina Bereta and Manolis Koubarakis

Expressive Multi-level Modeling for the Semantic Web 53
Freddy Brasileiro, João Paulo A. Almeida, Victorio A. Carvalho,
and Giancarlo Guizzardi

A Practical Acyclicity Notion for Query Answering Over *Horn-*
SRIQ Ontologies 70
David Carral, Cristina Feier, and Pascal Hitzler

Containment of Expressive SPARQL Navigational Queries 86
Melisachew Wudage Chekol and Giuseppe Pirrò

WebBrain: Joint Neural Learning of Large-Scale
Commonsense Knowledge 102
Jiaqiang Chen, Niket Tandon, Charles Darwis Hariman,
and Gerard de Melo

Efficient Algorithms for Association Finding and Frequent Association
Pattern Mining 119
Gong Cheng, Daxin Liu, and Yuzhong Qu

A Reuse-Based Annotation Approach for Medical Documents 135
Victor Christen, Anika Groß, and Erhard Rahm

Knowledge Representation on the Web Revisited: The Case for Prototypes . . . 151
Michael Cochez, Stefan Decker, and Eric Prud'hommeaux

Updating DL-Lite Ontologies Through First-Order Queries 167
Giuseppe De Giacomo, Xavier Oriol, Riccardo Rosati,
and Domenico Fabio Savo

Are Names Meaningful? Quantifying Social Meaning on the Semantic Web. . .   184
  Steven de Rooij, Wouter Beek, Peter Bloem, Frank van Harmelen,
  and Stefan Schlobach

User Validation in Ontology Alignment.   200
  Zlatan Dragisic, Valentina Ivanova, Patrick Lambrix, Daniel Faria,
  Ernesto Jiménez-Ruiz, and Catia Pesquita

Seed, an End-User Text Composition Tool for the Semantic Web.   218
  Bahaa Eldesouky, Menna Bakry, Heiko Maus, and Andreas Dengel

Exception-Enriched Rule Learning from Knowledge Graphs.   234
  Mohamed H. Gad-Elrab, Daria Stepanova, Jacopo Urbani,
  and Gerhard Weikum

Planning Ahead: Stream-Driven Linked-Data Access Under
Update-Budget Constraints.   252
  Shen Gao, Daniele Dell'Aglio, Soheila Dehghanzadeh,
  Abraham Bernstein, Emanuele Della Valle, and Alessandra Mileo

Explicit Query Interpretation and Diversification for Context-Driven
Concept Search Across Ontologies   271
  Chetana Gavankar, Yuan-Fang Li, and Ganesh Ramakrishnan

Predicting Energy Consumption of Ontology Reasoning
over Mobile Devices.   289
  Isa Guclu, Yuan-Fang Li, Jeff Z. Pan, and Martin J. Kollingbaum

Walking Without a Map: Ranking-Based Traversal for Querying
Linked Data.   305
  Olaf Hartig and M. Tamer Özsu

CubeQA—Question Answering on RDF Data Cubes.   325
  Konrad Höffner, Jens Lehmann, and Ricardo Usbeck

Optimizing Aggregate SPARQL Queries Using Materialized RDF Views. . . .   341
  Dilshod Ibragimov, Katja Hose, Torben Bach Pedersen,
  and Esteban Zimányi

Algebraic Calculi for Weighted Ontology Alignments   360
  Armen Inants, Manuel Atencia, and Jérôme Euzenat

Ontologies for Knowledge Graphs: Breaking the Rules   376
  Markus Krötzsch and Veronika Thost

An Extensible Linear Approach for Holistic Ontology Matching.   393
  Imen Megdiche, Olivier Teste, and Cassia Trojahn

Semantic Sensitive Simultaneous Tensor Factorization   411
  Makoto Nakatsuji

Multi-level Semantic Labelling of Numerical Values   428
  Sebastian Neumaier, Jürgen Umbrich, Josiane Xavier Parreira,
  and Axel Polleres

Semantic Labeling: A Domain-Independent Approach   446
  Minh Pham, Suresh Alse, Craig A. Knoblock, and Pedro Szekely

Exploiting Emergent Schemas to Make RDF Systems More Efficient   463
  Minh-Duc Pham and Peter Boncz

Distributed RDF Query Answering with Dynamic Data Exchange   480
  Anthony Potter, Boris Motik, Yavor Nenov, and Ian Horrocks

RDF2Vec: RDF Graph Embeddings for Data Mining   498
  Petar Ristoski and Heiko Paulheim

SPARQL-to-SQL on Internet of Things Databases and Streams   515
  Eugene Siow, Thanassis Tiropanis, and Wendy Hall

Can You Imagine... A Language for Combinatorial Creativity?   532
  Fabian M. Suchanek, Colette Menard, Meghyn Bienvenu,
  and Cyril Chapellier

Leveraging Linked Data to Discover Semantic Relations Within
Data Sources   549
  Mohsen Taheriyan, Craig A. Knoblock, Pedro Szekely,
  and José Luis Ambite

Integrating Medical Scientific Knowledge with the Semantically
Quantified Self   566
  Allan Third, George Gkotsis, Eleni Kaldoudi, George Drosatos,
  Nick Portokallidis, Stefanos Roumeliotis, Kalliopi Pafili,
  and John Domingue

Learning to Assess Linked Data Relationships Using Genetic Programming . . .   581
  Ilaria Tiddi, Mathieu d'Aquin, and Enrico Motta

A Probabilistic Model for Time-Aware Entity Recommendation   598
  Lei Zhang, Achim Rettinger, and Ji Zhang

A Knowledge Base Approach to Cross-Lingual Keyword
Query Interpretation   615
  Lei Zhang, Achim Rettinger, and Ji Zhang

Context-Free Path Queries on RDF Graphs 632
  Xiaowang Zhang, Zhiyong Feng, Xin Wang, Guozheng Rao,
  and Wenrui Wu

Unsupervised Entity Resolution on Multi-type Graphs 649
  Linhong Zhu, Majid Ghasemi-Gol, Pedro Szekely, Aram Galstyan,
  and Craig A. Knoblock

Author Index 669

Resources

Ontological Representation of Audio Features

Alo Allik[✉], György Fazekas, and Mark Sandler

Queen Mary University of London, London, UK
{a.allik,g.fazekas,mark.sandler}@qmul.ac.uk

Abstract. Feature extraction algorithms in Music Informatics aim at deriving statistical and semantic information directly from audio signals. These may be ranging from energies in several frequency bands to musical information such as key, chords or rhythm. There is an increasing diversity and complexity of features and algorithms in this domain and applications call for a common structured representation to facilitate interoperability, reproducibility and machine interpretability. We propose a solution relying on Semantic Web technologies that is designed to serve a dual purpose (1) to represent computational workflows of audio features and (2) to provide a common structure for feature data to enable the use of Open Linked Data principles and technologies in Music Informatics. The Audio Feature Ontology is based on the analysis of existing tools and music informatics literature, which was instrumental in guiding the ontology engineering process. The ontology provides a descriptive framework for expressing different conceptualisations of the audio feature extraction domain and enables designing linked data formats for representing feature data. In this paper, we discuss important modelling decisions and introduce a harmonised ontology library consisting of modular interlinked ontologies that describe the different entities and activities involved in music creation, production and publishing.

Keywords: Semantic audio analysis · Music Information Retrieval · Linked open data · Semantic Web technologies

1 Introduction

The availability of unprecedented amounts of music in digital formats is dramatically changing the way casual and professional users interact with large music collections on the Web. Using textual editorial metadata is no longer sufficient and reliable as the principal means of finding the desired content. Statistical and musical information extracted from digital audio is becoming an increasingly valuable ingredient in strategies for searching, discovering and browsing music in large collections. These strategies are a result of intensive research and development in the Music Information Retrieval (MIR) community with active participation stemming from both academic and commercial interests. Consequently, there is a growing diversity of audio feature extraction algorithms combined with a profusion of audio feature datasets available for research communities and commercial developers. However, it is not always clear what certain

© Springer International Publishing AG 2016
P. Groth et al. (Eds.): ISWC 2016, Part II, LNCS 9982, pp. 3–11, 2016.
DOI: 10.1007/978-3-319-46547-0_1

feature data represents or why two extraction algorithms, identified as identical by their developers, may produce strikingly dissimilar results when applied to the same audio signal. The situation is exacerbated by the lack of common terminology or structuring principles in existing data interchange formats that often have a narrow scope to satisfy tool or task specific requirements. There is a need for more meaningful representation of feature data that would facilitate linking or comparing features produced in different data sources, as well as for generalised descriptions of audio features that would allow easier identification and comparison of audio feature algorithms that produce the data.

We propose a modular approach using Semantic Web ontologies for the representation of audio features. The Audio Feature Ontology framework consists of two separate components, *(i)* a core ontology and *(ii)* a separately maintained extensible vocabulary. This is motivated by the need for mediation between several tool and task specific conceptualisations that exist in this diverse domain. The Audio Feature Vocabulary includes existing audio features and captures computational workflows, providing the terms for specific ontologies without attempting to organise the features hierarchically. The Audio Feature Ontology represents entities in the feature extraction process on different levels of abstraction, modelling the underlying activities involved in problem solving through phases of conceptualisation, modelling and implementation.

2   Background

The need for an ontological representation of audio features was already recognised during the development of the Music Ontology framework [7]. This framework consists of a harmonised library of modular music-related ontologies [1] including a feature ontology. The early version of this ontology was primarily designed to provide terms for the Vamp plugin system[1], an extensible collection of feature extraction algorithms that accept audio signals as input and produce structured feature data as output, including formats prescribed by the original ontology. The plugins are executed in host applications such as the command line Sonic Annotator[2] tool and Sonic Visualiser[3], a desktop application designed to provide visualisations of audio feature data. A number of MIR libraries also release their feature extractors as Vamp plugins. This system has so far been the only solution enabling a shared ontologically structured data representation of audio features. However, the initial ontology does not provide a comprehensive vocabulary of audio features or computational feature extraction workflows. It also lacks concepts to support development of more specific feature extraction ontologies, while structurally it conflates musicological and computational concepts in a way that makes it inflexible for certain modelling requirements [2].

Other existing feature extraction frameworks provide data exchange formats designed for particular workflows or specific tools, providing interoperability on

---

[1] http://www.vamp-plugins.org.

[2] http://www.vamp-plugins.org/sonic-annotator/.

[3] http://www.sonicvisualiser.org.

the syntactic level. However, there is no common structuring principle shared by these different tools and libraries. The motley of output formats is well demonstrated in the representations category of a recent evaluation of feature extraction toolboxes [6]. For example, the popular MATLAB MIR Toolbox[4] export function outputs delimited files as well as Weka Attribute-Relation File Format (ARFF), while Essentia[5] provides YAML and JSON and the YAAFE library outputs CSV and HDF5. The MPEG-7 standard, used as benchmarks for other extraction tools, provides an XML schema for a set of low-level descriptors. The most recent developments in audio feature data formats predominantly employ JavaScript Object Notation (JSON), which is rapidly becoming a ubiquitous data interchange mechanism in a wide range of systems regardless of domain. It is evident that the simplicity of JSON combined with its structuring capabilities make it an attractive option, particularly compared to preceding alternatives including YAML, XML, ARFF, the Sound Description Interchange Format (SDIF) and various delimited formats.

While existing RDF-based solutions face criticism by some domain experts [3], suggesting they are non-obvious, verbose or confusing, we believe this should be addressed in the ontology engineering process. The potential of interoperable representation of audio features on the semantic rather than the syntactic level and the ability to link features with other music related information provides a more sustainable platform for researchers and commercial developers alike. This is in stark contrast with solutions that do not support linking through unique identification of entities existing at different conceptual levels, and do not publish their schema using standardised languages that allow formalising relations, not only concept hierarchies, in this complex domain.

3   Core Ontology Model

In order to address the issues of domain structuring and data representation, we propose a modular framework for the Audio Feature Ontology, separating abstract ontological concepts from more specific vocabulary terminology. The framework also provides for describing extraction workflows and increases flexibility for modelling task and tool specific ontologies. The core structure of the framework separates the underlying classes that represent abstract concepts in the domain from specific named entities. This results in the two main components of the framework defined as:

- Audio Feature Ontology (AFO): https://w3id.org/afo/onto/1.1#
- Audio Feature Vocabulary (AFV): https://w3id.org/afo/vocab/1.1#

The ontology component is structured to reflect different conceptual levels of abstraction of audio features. These layers represent the design process from *(i)* conceptualisation of a feature, through *(ii)* modelling an algorithmic workflow, to

---

[4] http://bit.ly/1rCwJOt.
[5] http://essentia.upf.edu.

(iii) implementation and *(iv)* instantiation in a specific computational context. For example, the abstract concept of Chromagram[6] is separate from its algorithmic model, which involves a sequence of computational operations like cutting an audio signal into frames, calculating the Discrete Fourier Transform for each frame, etc. (see Sect. 4 for a more detailed example). The computational workflow can be implemented in different ways and in various programming languages as components of feature extraction libraries. The implementation layer enables distinguishing a Chromagram written as a Vamp plugin from a Chromagram extractor in the MIR Toolbox. The most concrete layer represents the feature extraction instance in a specific execution context, for example, to reflect the differences of operating systems or hardware on which the extraction occurred. Our proposed layered model is shown in Fig. 1.

Fig. 1. The Audio Feature Ontology core model with four levels of abstraction

The core model of the ontology retains original attributes to distinguish audio features by temporal characteristics and data density. It relies on the Event[7] and Timeline[8] ontologies to provide primary structuring concepts for feature data representation. Temporal characteristics classify feature data either into instantaneous points in time - e.g. event onsets or tonal change moments - or events with known time duration. Data density attributes allow describing how a feature relates to the extent of an audio file: whether it is scattered and occurs irregularly over the course of the audio signal (for example, segmentation or onset features), or the feature is calculated at regular intervals and fixed duration (e.g. signal-like features with regular sampling rate). Figure 2 illustrates how audio features are linked with terms in the Music Ontology and thereby other music-related metadata on the Web. Specific named audio feature entities, such as **afv:Onset**, **afv:Key**, and **afv:MFCC** are subclasses of **afo:AudioFeature**, which, in turn, is a subclass of **event:Event** from the Event Ontology. This way the feature data can be directly linked to time points on the audio signal timeline using the **event:time** property. Listing 1.1 shows a Turtle/RDF example of such linking.

Since there are many different ways to structure audio features depending on a specific task or theoretically motivated organising principle, a common representation would have to account for multiple conceptualisations of the domain and facilitate diverging representations of common features. For example, Mel

---

[6] A feature representing energies of harmonically related frequencies calculated in discrete steps over time. Different musical temperaments yield different chromagrams.

[7] http://motools.sourceforge.net/event/.

[8] http://motools.sourceforge.net/timeline/.

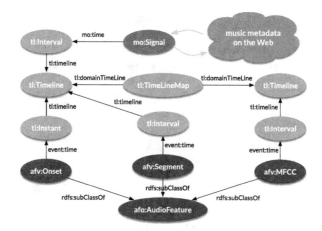

Fig. 2. Framework model showing how feature data representation is linked with music metadata resources on the Web using temporal entities defined in the Timeline ontology

Frequency Cepstral Coefficients (MFCC), that measure rates of energy change in different frequency bands and are widely calculated in many tools and workflows, can be categorised as a "timbral" feature in the psychoacoustic or musicological sense (as in MIR Toolbox for instance), while from the computational point of view, MFCCs could be labelled as a "cepstral" (e.g. in [5]) or "spectral" representation (as in the Essentia library). Collated audio features gathered from relevant literature and extraction software are defined as subclasses in the AFV. Another role of the vocabulary is to define computational extraction workflow descriptions, so that features can be more easily identified and compared by their respective computational signatures. This is discussed in the following section in more detail.

4   Algorithmic Workflow Representation

AFV defines terms that may be subsumed in specific ontologies and implements the *model* layer of the ontology framework. It is a clean version of the catalogue which lists the features without their properties. Many duplications of terms are consolidated. This enables the definition of tool and task specific feature implementations and leaves any categorisation or taxonomic organisation to be specified in the implementation layer.

The vocabulary also specifies computational workflow models for some of the features which lower-level ontologies can be link to. The computational workflow models are based on feature signatures as described in [5]. The signatures represent mathematical operations employed in the feature extraction process with each operation assigned a lexical symbol. It offers a compact description of each feature and facilitates the comparison of features by their computation

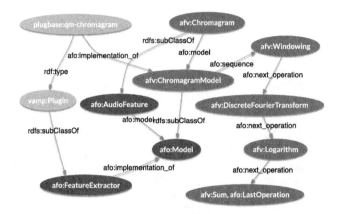

Fig. 3. Computational workflow of the Chromagram feature model linked to the extractor algorithm implemented in a Vamp plugin

workflows. The ontological representation of signatures involves defining a set of OWL classes that describe the representation and sequential nature of the calculations. The operations are implemented as sub-classes of three general classes: transformations, filters and aggregations. For each abstract feature, we define a model property. The OWL range of the model property is a ComputationalModel class in the Audio Feature Ontology namespace. The operation sequence can be defined through this object's operation sequence property. For example, the signature of the Chromagram feature is defined in [5] as "f F l Σ", which designates a sequence of (1) windowing (f), (2) Discrete Fourier Transform (F), (3) logarithm (l) and (4) sum (Σ). Figure 3 shows the resulting graph of the workflow.

5   Audio Content Description

Besides representing the computational steps involved in the extraction process, the framework supports identifying an extracted audio feature by linking it to a corresponding term in the Audio Feature Vocabulary, describing the temporal structure and density of the output data, associating feature data as intervals or instants on the audio signal timeline and associating the output data with feature extraction tools used in the extraction process. It also provides terms to represent inputs and parameters to the feature extraction functions to provide support for development of ontologies specific to a software library.

AFO can facilitate the development of other data formats beside RDF/Turtle that are aligned with linked data principles, such as JSON-LD [4]. JSON-LD is a linked data extension to the standard JSON format that provides an entity-centric representation of RDF/OWL semantics and a means to define a linked data context with URI connections to external ontologies and resources. It has the potential to simplify feature representations while maintaining ontological structuring of the data.

Content-based analyses are becoming crucial in recommendation systems to tackle problems of rarely accessed content for which listening data supporting collaborative filtering is unavailable. These archives are important part of the Web and should be better represented and made accessible on the Semantic Web. The ontology is also a candidate to provide linked data representation for AcousticBrainz[9], which currently includes content-based metadata for over 2 million audio tracks. Adaptation in this context will facilitate significant deployment of musical metadata as linked data, where the feature identification and provenance data describing algorithms, computational tools and services are crucial for interoperability and wider utilisation of such data. The ontology has also been used in large-scale feature extraction projects such as Digital Music Lab[10] and Computational Analysis of the Live Music Archive[11]. The ontology can be deployed to describe large content-based music archives in libraries, music labels and open archives such as the Internet Archive Live Music Archive.

```
:track_1 a mo:Track ;
    dc:title"Afterlife";
    foaf:maker [ a mo:MusicArtist; foaf:name"Desimal"] ;
    mo:available_as <file:///snd/Afterlife.mp3> .

<file:///snd/Afterlife.mp3> a mo:AudioFile ;
    mo:encodes :signal_f6261475.

:signal_f6261475 a mo:Signal ;
    mo:time [
        a tl:Interval ;
        tl:onTimeLine :timeline_aec1cb82
    ] .

:timeline_aec1cb82 a tl:Timeline .

:event_1 a afv:Onset ;
    event:time [
        a tl:Instant ;
        tl:onTimeLine :timeline_aec1cb82 ;
        tl:at"PT1.98S"^^xsd:duration ;
    ] ;

:feature_1 a afv:MFCC ;
    mo:time [
        a tl:Interval ;
        tl:onTimeLine :timeline_aec1cb82 ;
    ] ;
    afo:value ( -26.9344 0.188319 0.106938 ..) .
```

Listing 1.1. An abbreviated example of linking onsets and MFCC features using AFV to the Music Ontology

Beyond representing audio feature data in research workflows, there are many other practical applications for the ontology framework. One of the test cases is providing data services for an adaptive music player that uses audio features to enrich user experience and enables novel ways to search or browse large music collections. The data is used by Semantic Web entities called Dynamic Music

---

[9] http://acousticbrainz.org.

[10] http://dml.city.ac.uk.

[11] http://etree.linkedmusic.org/about/calma.html.

Objects (dymos) [8] that control the audio mixing functionality of the player. Dymos make song selections and determine tempo alignment for cross-fading based on features.

6   Conclusions

The Audio Feature Ontology and Vocabulary provide a framework for representing the semantics of audio features providing interoperability on the conceptual rather than the syntactic level. It provides terminology to facilitate task and tool specific ontology development and serves as a descriptive framework for audio feature extraction. The proposed framework is a significant update to the existing ontology that addresses shortcomings of the original model, which have been identified as barriers to wider adoption in the community. The updates to the original ontology for audio features strive to simplify feature representations and make them more flexible while maintaining ontological structuring and linking capabilities. We produced example ontologies for existing tools including MIR Toolbox, Essentia, and Marsyas. Existing feature extraction tools, including the Sonic Visualiser and Sonic Annotator have been updated to produce RDF/Turtle as well as JSON-LD output. More examples of feature data representation, case studies of use of the ontology framework in emerging applications, and suggestions for best practices are available online: https://w3id.org/afo/onto/1.1#.

Acknowledgments. This work was supported by EPSRC Grant EP/ L019981/1, "Fusing Audio and Semantic Technologies for Intelligent Music Production and Consumption" and the European Commission H2020 research and innovation grant Audio-Commons (688382). Sandler acknowledges the support of the Royal Society as a recipient of a Wolfson Research Merit Award.

References

1. Fazekas, G., Raimond, Y., Jakobson, K., Sandler, M.: An overview of semantic web activities in the OMRAS2 project. J. New Music Res. (JNMR) **39**(4), 295–311 (2010)
2. Fields, B., Page, K., De Roure, D., Crawford, T.: The segment ontology: bridging music-generic and domain-specific. In: Proceedings of the IEEE International Conference on Multimedia and Expo, 11–15 July 2011, Barcelona, Spain (2011)
3. Humphrey, E.J., Salamon, J., Nieto, O., Forsyth, J., Bittner, R., Bello, J.P.: JAMS: a JSON annotated music specification for reproducible MIR research. In: Proceedings of the 15th International Society for Music Information Retrieval Conference, Taipei, Taiwan (2014)
4. Lanthaler, M., Gütl, C.: On using JSON-LD to create evolvable RESTful services. In: Proceedings of the 3rd International Workshop on RESTful Design at WWW 2012 (2012)
5. Mitrovic, D., Zeppelzauer, M., Breiteneder, C.: Features for content-based audio retrieval. In: Advances in Computers, vol. 78, pp. 71–150 (2010)

6. Moffat, D., Ronan, D., Reiss, J.D.: An evaluation of audio feature extraction tool-boxes. In: Proceedings of the 18th International Conference on Digital Audio Effects (DAFx-15), Trondheim, Norway (2015)
7. Raimond, Y., Abdallah, S., Sandler, M., Giasson, F.: The music ontology. In: Proceedings of the 8th International Conference on Music Information Retrieval (ISMIR 2007), 23–27 September, Vienna, Austria (2007)
8. Thalmann, F., Carillo, A.P., Fazekas, G., Wiggins, G.A., Sandler, M.: The mobile audio ontology, experiencing dynamic music objects on mobile devices. In: Proceedings of the 10th IEEE International Conference on Semantic Computing, Laguna Hills, CA, USA (2016)

Abstract Meaning Representations as Linked Data

Gully A. Burns[✉], Ulf Hermjakob, and José Luis Ambite

Information Sciences Institute, University of Southern California,
Marina del Rey, CA 90292, USA
burns@isi.edu

Abstract. The complex relationship between natural language and formal semantic representations can be investigated by the development of large, semantically-annotated corpora. The "Abstract Meaning Representation" (AMR) formulation describes the semantics of a whole sentence as a rooted, labeled graph, where nodes represent concepts/entities (such as PropBank frames and named entities) and edges represent relations between concepts (such as verb roles). AMRs have been used to annotate corpora of classic books, newstext and biomedical literature. Research on semantic parsers that generate AMRs from text is progressing rapidly. In this paper, we describe an AMR corpus as Linked Data (AMR-LD) and the techniques used to generate it (including an open-source implementation). We discuss the benefits of AMR-LD, including convenient analysis using SPARQL queries and ontology inferences enabled by embedding into the web of Linked Data, as well as the impact of semantic web representations directly derived from natural language.

Keywords: Linked linguistic data · Abstract Meaning Representation · AMR · Sembank · Biological pathways

1 Introduction

The Abstract Meaning Representation (AMR) formulation follows the success of using high-quality parallel corpora for machine translation and using Penn Treebank for statistical parsing. Such a "sembank of simple, whole sentence semantic structures" [1] seeks to accelerate natural language understanding research. AMRs abstract away from syntactic variation (so that different ways of saying the same thing map to the same AMR). Consider the text *"SerpinE2 is over-expressed in intestinal epithelial cells transformed by activated MEK1 and onco-genic RAS and BRAF"* [2]). Figure 1 shows this sentence's AMR, and Fig. 2 shows its translation into RDF (AMR-LD). Using AMR-LD with an AMR parser permits semantic representations to be generated directly from text. AMR semantic parsing is an active area of research (e.g., [3–6]).

A significant corpus of AMRs have been generated in the Natural Language Processing (NLP) community. Initially, AMRs were developed as annotations of

© Springer International Publishing AG 2016
P. Groth et al. (Eds.): ISWC 2016, Part II, LNCS 9982, pp. 12–20, 2016.
DOI: 10.1007/978-3-319-46547-0_2

```
# ::id a_pmid_2094_2929.39 ::amr-annotator SDL-AMR-09 ::preferred
# ::tok SerpinE2 is overexpressed in intestinal epithelial cells
#       transformed by activated MEK1 and oncogenic RAS and BRAF
(o / overexpress-01
    :ARG2 (p / protein
          :name (n2 / name :op1 "serpinE2"))
    :ARG3 (c / cell
          :mod (e2 / epithelium)
          :part-of (i / intestine)
          :ARG1-of (t / transform-01
              :ARG0 (a / and
                  :op1 (e3 / enzyme
                        :name (n3 / name :op1 "MEK1")
                        :ARG1-of (a2 / activate-01))
                  :op2 (e4 / enzyme
                        :name (n4 / name :op1 "RAS")
                        :ARG0-of (c2 / cause-01
                            :ARG1 (c3 / cancer)))
                  :op3 (e / enzyme
                        :name (n5 / name :op1 "BRAF")
                        :ARG0-of c2)))))).
```

(a) AMR in its native s-expression syntax

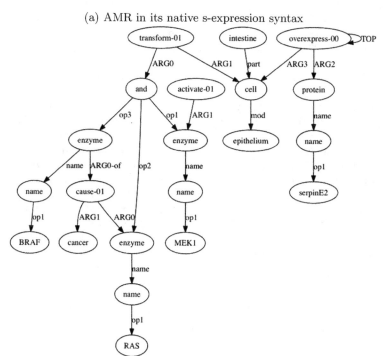

(b) ARMICA graphical rendering of the AMR

Fig. 1. Abstract meaning representation (AMR): biological pathway example

```
@prefix rdf: <http://www.w3.org/1999/02/22-rdf-syntax-ns#> .
@prefix rdfs: <http://www.w3.org/2000/01/rdf-schema#> .
@prefix xsd: <http://www.w3.org/2001/XMLSchema#> .
@prefix ac: <http://amr.isi.edu/rdf/core-amr#> .
@prefix pb:
        <http://verbs.colorado.edu/propbank/framesets-english-aliases#> .
@prefix ae: <http://amr.isi.edu/entity-types#> .
@prefix at: <http://amr.isi.edu/rdf/amr-terms#> .
@prefix up: <http://www.uniprot.org/uniprot/> .
@prefix pfam: <http://pfam.xfam.org/family/> .
@prefix pm: <http://www.ncbi.nlm.nih.gov/pubmed/>
@base : <http://amr.isi.edu/amr_data/> .
@prefix : <a_pmid_2094_2929.39#> .


<a_pmid_2094_2929.39> rdf:type ac:AMR .
<a_pmid_2094_2929.39> ac:has-tokens "SerpinE2 is overexpressed in ..." .
<a_pmid_2094_2929.39> ac:has-id "pmid_1177_7939.53" .
<a_pmid_2094_2929.39> ac:has-annotator "SDL-AMR-09" .
<a_pmid_2094_2929.39> ac:is-preferred "true"^^xsd:boolean .
<a_pmid_2094_2929.39> ac:has-file "alignment-release-bio.txt" .
<a_pmid_2094_2929.39> ac:in-document pm:20942929 .
<a_pmid_2094_2929.39> amr:has-root :o .


:o rdf:type pb:overexpress-01 ;          ac:op3 :e .
   pb:overexpress-01.ARG2 :p ;        :e3 rdf:type ae:enzyme ;
   pb:overexpress-01.ARG3 :c .           rdfs:label "MEK1" ;
:p rdf:type ae:protein ;                 ac:xref up:MP2K1_HUMAN ;
   rdfs:label "serpinE2" ;               ac:xref up:Q02750 ;
   ac:xref up:P07093 ;                   pb:activate-01.ARG1-of :a2 .
   ac:xref up:GDN_HUMAN .             :a2 rdfs:type pb:activate-01 .
:c rdf:type ae:cell ;                 :e4 rdf:type ae:enzyme ;
   ac:mod :e2 ;                          rdfs:label "RAS" ;
   ac:part-of :i ;                       ac:xref pfam:PF00071 ;
   pb:ARG1-of :t .                       pb:cause-01.ARG0-of :c2 .
:e2 rdf:type at:epithelium .          :c2 rdf:type pb:cause-01 ;
:i rdf:type at:intestine .               pb:cause-01.ARG1 :c3 .
:t rdf:type pb:transform-01 ;         :c3 rdf:type at:cancer .
   pb:transform-01.ARG0 :a .          :e rdf:type ae:enzyme ;
:a rdf:type ac:and ;                     rdfs:label "BRAF" ;
   ac:op1 :e3 ;                          pb:cause-01.ARG0-of c2 .
   ac:op2 :e4 ;
```

Fig. 2. AMR-LD example in Turtle syntax

the short novel "The Little Prince". Subsequent releases of core AMR data were derived from newswire and discussion forum sources. In August 2015, the AMR group released 19,572 AMRs through the Linguistic Data Consortium (LDC) licensing process (LDC2015E86: DEFT Phase 2 AMR Annotation R1). Detailed annotation guidelines and active curation tools are available from [7].

In this paper we focus on an *open* AMR corpus produced by the R2L2K2 project at USC/ISI within DARPA's Big Mechanism program [8], which is machine reading the literature on biological pathways in cancer research. In March 2016, the R2L2K2 team (with contributions from USC/ISI, University of Colorado, and SDL) released version 0.8 of the BioAMR corpus consisting of 6,452 sentences. We present this corpus as AMR-LD and describe our approach to convert AMRs into linked data, in effect, linking a rapidly expanding area in natural language research to the semantic web.

2   Linking Abstract Meaning Representation

Converting AMR resources to linked data proceeds in two steps. First, we identify and/or define ontologies that capture the semantics of the original AMR concepts and relations (including the corresponding namespaces). Second, we use entity linkage techniques [9] to map to well-known web resources.

2.1   Representing and Linking AMR Concepts and Relations

The semantics of AMRs is primarily defined by the types of nodes in the graph, which are PropBank frames [10] and AMR entity types. Each PropBank frame defines the applicable roles and their specific meaning. For example, in Fig. 1 the type of the node labeled t is the PropBank frame transform-01 (defined as to "change, cause a change in state"), which has four roles: ARG0, the causer of transformation (Agent); ARG1, the thing changing (Patient); ARG2, the end state (Result); and ARG3, the start state (Material). In our example, the agent of the transformation ARG0 is a set of enzymes (BRAF, MEK1, RAS).

We use a simple meta-model consisting of (1) a main concept class AMR-Concept with subclasses (AMR-PropBank-Frame, AMR-Entity-Type, and AMR-Term) and (2) a main relation AMR-Role with a subrelation (AMR-PropBank-Role). We also define namespaces to make these distinctions explicit, as follows:

AMR-Core (ac) includes the core constructs of the AMR specification: the AMR concept, metadata associated with AMRs (e.g., has-annotator, in-document), and AMR-specific roles (e.g., mod, part-of). We also define a xref role to link to external entities (cf. Sect. 2.2).

AMR-PropBank-Frame (pb) includes PropBank frames (e.g. transform-01, activate-01), the roles used by PropBank Frames (e.g., ARG0, ARG1, ...), and their inverses (e.g., ARG0-of, ARG1-of, ...) with corresponding inverse role assertions (e.g., pb:ARG0-of owl:inverseOf pb:ARG0).

AMR-Entity-Type (ae) includes all named entity types corresponding to common concepts in a domain, (`person`, `organization`, and `location` in general news text, or `enzyme` or `cell` in biomedical text).

AMR-Term (at). AMR parsing tools and human curators are free to create additional entity types, even if those types are not predefined in the AMR-Entity-Type namespace. For example, concepts, such as `cancer` and `intestine` in Fig. 2, are not registered as AMR entity types.

In our translation of AMRs to AMR-LD we closely follow the AMR design. One representational structure we deliberately altered was the naming convention used in the core AMR formulation. This involves the use of a `:name` role to create an instance of a `name` concept containing one ore more `:opN` roles that contain the string tokens of the name, for example, *e.g.*, `:name (n3 / name :op1"MEK1")`. In AMR-LD, we replace this with an standard `rdfs:label` property.

A feature of the PropBank roles `:ARG0`, `:ARG1`, `:ARG2`, `:ARG3` is that their precise semantics may change from frame to frame. Generally, `:ARG0` is the Agent, and `:ARG1` is the Patient. However, this is not always the case. The semantics of `:ARG2`, `:ARG3`, etc., is even more variable. In our presentation in the paper we will show only properties using the `:ARGN` roles. However the tool we describe in Sect. 2.3 can also generate frame-specific roles, like `transform-01.ARG0`. The rationale is to attach precise semantics to roles of different frames, as needed. For example, stating that `transform-01.ARG0` is a subproperty of `vnrole:26.6.1--Agent` role (while not all `ARG0`'s may be agents).

2.2   Representing and Linking AMR Entities

A crucial feature of the AMR-LD representation is that it explicitly links to well-known entities in the Semantic Web using the `xref` property. For example, in Figs. 1 and 2 the AMR node p labeled "serpinE2" corresponds to the UniProtKB protein `GDN_HUMAN` and its synonymous identifier P07093. Similarly the entity e4 labeled "RAS" corresponds to entity PF00071 in the protein family ontology [11]. We could have used `owl:sameAs` to indicate linkages. However, given the strong semantics of `owl:sameAs` and the difficulty of accurately performing entity linkage, we decided to use a more relaxed property like `ac:xref`. The `ac:in-document` property also provides links into the literature. For example, the AMR `<a_pmid_2094_2929.39>` comes from the PubMed article `pm:20942929`. These linkages embed AMR data into the Semantic Web and can significantly enhance the value of AMR corpora by leveraging existing ontologies, as well as provide an entry point into linguistic resources for semantic web applications.

We developed an entity linkage algorithm for common bioentities based on their labels. First we collected protein and chemical names from existing databases, specifically the UniProt knowledge base [12], proteins appearing in pathways in Pathway Commons [13], and chemicals from NCBI's PubChem. Then, we mapped entities appearing in BioAMRs to these resources. For short (protein) names, like "BRAF", we use a combination of string similarity metrics, such as edit distance, and Jaccard similarity over n-grams. For efficiency we include a blocking algorithm based on prefixes of the protein names. Our implementation used the FRIL [14] record linkage system. For long (protein, chemical) names, such as "Cbl E3 ubiquitin ligase", we used traditional information retrieval techniques, such as TF-IDF cosine similarity.

2.3 AMR-LD Open-Source Conversion Software

We developed a Python library for translating the original AMR representation to RDF. The library is hosted on GitHub [15]. The tool provides extensions to connect to different record linkage algorithms. In our development of the bio AMR-LD corpus, we used the L2K2R2 project bioentity mapping web service [16]. We applied this system to the Bio-AMR v0.8 data [7] to generate the publicly available AMR-LD resource at [17]. The conversion proceeds as follows:

1. Generate URLs for AMR elements, qualified by appropriate namespaces.
2. Add RDFS classes to represent AMR Concepts, Entities, Frames and Roles.
3. Convert entity names to standard `rdfs:label` elements.
4. Define elements from the AMR base language, AMR named entity vocabulary, and PropBank frame repository.
5. Link to well-known semantic web entities using `xref` properties.

3 Querying and Reasoning with AMR-LD

An advantage of using linked data standards is that it facilitates data analysis by using existing query and reasoning engines. For example, consider the SPARQL query in Fig. 3 to identify sentences in papers that contain activated entities and their types. Note that we take advantage of path-following queries of SPARQL 1.1 using over the ac:Role predicate which is a super-property of the AMR properties. Similar queries can point to sentences containing specific proteins or chemicals. Leveraging the ontology of AMR types either directly, or through linkage to external ontologies, enables more sophisticated queries. For example, if our ontology includes an axiom that states that `ae:enzyme rdfs:subClassOf ae:protein`, then a query (similar to the one in Fig. 3(a) for sentences) can retrieve proteins of any kind. Finally, although our focus has been on Linked Data, having AMR data represented as an RDF graph can facilitate using Graph databases (e.g., neo4j) that implement more complex graph algorithms, such as shortest paths, or centrality measures.

4 Related Work

Cimiano et al. 2014 [18] describe how semantic web methods can be used in natural language interpretation systems as an implementation of discourse relation theory (DRT) [19]. This "meaning representation" captures the semantics of simple sentences and phrases. This work is not driven by the goal of building a sembank and is driven by data generated from syntactic parse trees. In

```
select ?pmid ?sentence ?activated ?entityType
where {?amr rdf:type ac:AMR .         ?amr :has-pmid ?pmid .
       ?amr :has-tokens ?sentence .   ?amr :root ?aroot .
       ?aroot ac:Role* ?actFrame .    ?actFrame rdf:type pb:activate-01 .
       ?actFrame pb:ARG1 ?actEntity . ?actEntity rdfs:label ?activated .
       ?actEntity rdf:type ?entityType }
```

pmid	sentence	activated	entityType
14656721	As was the case for p38 , ERK1 @/@ 2 was both rapidly and persistently activated in neutrophils exposed to H2O2 (Fig . 2A) .	ERK12	ae:enzyme
15156153	Expression of either Hes1 or Hes5 significantly activated the STAT reporter gene construct in both E13 neuroepithelial cells and MNS @-@ 70 cells ,	STAT	ae:gene

Fig. 3. SPARQL query over AMR-LD to identify sentences in papers that contain activated entities and their types, and some results

other work, the FRED system provides a live system for Semantic Web Machine Reading with a live interface [20]. FRED uses Boxer [21] to generate a RDF representation derived from DRT. Additional components within the FRED architecture include semantic representations for sentiment, citation, and type definitions. Another important framework for semantic representation is Hobb's logical form methodology driven by abductive reasoning [22], which also makes use of the Boxer system.

5 Discussion

The development of AMR as a semantic representation of complete English sentences is blossoming. Current curation efforts provide training data for automatic AMR parsing systems and metrics for evaluating the quality of AMRs ('Smatch' scores [23], which measure overlap between AMR graphs, seen as triples, normalized to 100). Current performance of AMR parsing is $\tilde{6}7.1$ smatch, with human inter-annotator in the 79–83 range [24]. Researchers are using AMRs to build knowledge bases of biological pathways [25]. Additional linkage approaches for AMRs over news text are available [26]. Our dual goal in providing AMR as Linked Data is to empower NLP researchers with a representation naturally suited for inference and embedding in the web knowledge graph, and to provide Semantic Web researchers connections to vast text/linguistic resources.

In future work we plan to investigate improved algorithms for linking AMR data to both entities and concepts from external ontologies, leveraging our work on semantic similarity [27], and on mapping ontologies of Linked Data [28].

Acknowledgments. This work was supported by grant W911NF-14-1-0364.

References

1. Banarescu, L., Bonial, C., Cai, S., Georgescu, M., Griffitt, K., Hermjakob, U., Knight, K., Koehn, P., Palmer, M., Schneider, N.: Abstract meaning representation for sembanking. In: Proceedings of the 7th Linguistic Annotation Workshop and Interoperability with Discourse, pp. 178–186, Sofia, Bulgaria, Assoc. Computational Linguistics (2013)
2. Bergeron, S., et al.: The serine protease inhibitor serpinE2 is a novel target of ERK signaling involved in human colorectal tumorigenesis. Mol. Cancer **9**, 271 (2010)
3. ISI software page. http://www.isi.edu/natural-language/software/
4. Vanderwende, L., et al.: An AMR parser for English, French, German, Spanish and Japanese and a new AMR-annotated corpus. In: NAACL Demonstrations, pp. 26–30. ACL (2015). http://www.aclweb.org/anthology/N15-3006
5. Flanigan, J., et al.: Generation from abstract meaning representation using tree transducers. In: NAACL: Human Language Technologies, pp. 731–739. ACL (2016). http://www.aclweb.org/anthology/N16-1087
6. Rao, S., Vyas, Y., Daume, H., Resnick, P.: Parser for abstract meaning representation using learning to search. In: Proceedings of SemEval 2016 (2016)
7. AMR project website. http://amr.isi.edu/
8. Cohen, P.R.: DARPA's big mechanism program. Phys. Biol. **12**(4), 045008 (2015)
9. Naumann, F., Herschel, M.: An Introduction to Duplicate Detection. Morgan and Claypool Publishers, New York (2010)
10. Palmer, M., Gildea, D., Kingsbury, P.: The proposition bank: an annotated corpus of semantic roles. Comput. Linguist. **31**(1), 71–106 (2005)
11. Pfam: home page. http://pfam.xfam.org
12. UniProt home. http://www.uniprot.org/
13. Pathway commons homepage. http://www.pathwaycommons.org/
14. Jurczyk, P., Lu, J.J., Xiong, L., Cragan, J.D., Correa, A.: FRIL: a tool for comparative record linkage. AMIA Ann. Symp. Proc. **2008**, 440–444 (2008)
15. AMR-linked data github repository. https://github.com/BMKEG/amr-ld/
16. L2K2R2 bioentity mapping web service. http://dna.isi.edu:7080/grounding/
17. Burns, G., Ambite, J.L., Hermjakob, U., The AMR Development Team: Biomedical abstract meaning representation as linked data (v0.8.1). figshare (2016). https://dx.doi.org/10.6084/m9.figshare.3206062.v1
18. Cimiano, P., Unger, C., McCrae, J.P.: Ontology-Based Interpretation of Natural Language. Synthesis Lectures on Human Language Technologies. Morgan & Claypool Publishers, New York (2014)
19. Kamp, H.: A theory of truth and semantic representation. In: Groenendijk, J., Janssen, T., Stokhof, M. (eds.) Formal Methods in the Study of Language. Mathematical Centre, Amsterdam (1981)
20. Presutti, V., Draicchio, F., Gangemi, A.: Knowledge extraction based on discourse representation theory and linguistic frames. In: Teije, A., et al. (eds.) EKAW 2012. LNCS, vol. 7603, pp. 114–129. Springer, Heidelberg (2012). doi:10.1007/978-3-642-33876-2_12. http://wit.istc.cnr.it/stlab-tools/fred/
21. Bos, J.: Wide-coverage semantic analysis with boxer. In: Bos, J., Delmonte, R. (eds.) Semantics in Text Processing (STEP), pp. 277–286. College Publications, London (2008)
22. Hobbs, J.R., Stickel, M.E., Appelt, D.E., Martin, P.A.: Interpretation as abduction. Artif. Intell. **63**(1–2), 69–142 (1993)

23. Cai, S., Knight, K.: Smatch: an evaluation metric for semantic feature structures. In: Proceedings 51st Annual Meeting of the Association for Computational Linguistics, vol. 2, Short Papers, Sofia, Bulgaria, pp. 748–752 (2013)

24. Pust, M., Hermjakob, U., Knight, K., Marcu, D., May, J.: Parsing English into abstract meaning representation using syntax-based machine translation. In: Proceedings of the Conference on Empirical Methods in Natural Language Processing, Lisbon, Portugal, pp. 1143–1154. Association for Computational Linguistics, September 2015

25. Garg, S., Galstyan, A., Hermjakob, U., Marcu, D.: Extracting biomolecular interactions using semantic parsing of biomedical text. In: Proceedings of AAAI (2016)

26. Pan, X., Cassidy, T., Hermjakob, U., Ji, H., Knight, K.: Unsupervised entity linking with abstract meaning representation. In: Proceedings of North American Chapter Association for Computational Linguistics, Denver, Colorado, pp. 1130–1139 (2015)

27. Ashish, N., Dewan, P., Ambite, J.-L., Toga, A.W.: GEM: the GAAIN entity mapper. In: Ashish, N., Ambite, J.-L. (eds.) DILS 2015. LNCS, vol. 9162, pp. 13–27. Springer, Heidelberg (2015). doi:10.1007/978-3-319-21843-4_2

28. Parundekar, R., Knoblock, C.A., Ambite, J.L.: Discovering concept coverings in ontologies of linked data sources. In: Cudré-Mauroux, P., et al. (eds.) ISWC 2012. LNCS, vol. 7649, pp. 427–443. Springer, Heidelberg (2012). doi:10.1007/978-3-642-35176-1_27

Interoperability for Smart Appliances in the IoT World

Laura Daniele[1(✉)], Monika Solanki[2], Frank den Hartog[1], and Jasper Roes[1]

[1] TNO - Netherlands Organization for Applied Scientific Research, The Hague, The Netherlands
{Laura.Daniele,Frank.denHartog,Jasper.Roes}@tno.nl
[2] Department of Computer Science, University of Oxford, Oxford, UK
Monika.Solanki@cs.ox.ac.uk

Abstract. Household appliances are set to become highly intelligent, smart and networked devices in the near future. Systematically deployed on the Internet of Things (IoT), they would be able to form complete energy consuming, producing, and managing ecosystems. Smart systems are technically very heterogeneous, and standardized interfaces on a sensor and device level are therefore needed. However, standardization in IoT has largely focused at the technical communication level, leading to a large number of different solutions based on various standards and protocols, with limited attention to the common semantics contained in the message data structures exchanged at the technical level. The Smart Appliance REFerence ontology (SAREF) is a shared model of consensus developed in close interaction with the industry and with the support of the European Commission. It is published as a technical specification by ETSI and provides an important contribution to achieve semantic interoperability for smart appliances. This paper builds on the success achieved in standardizing SAREF and presents SAREF4EE, an extension of SAREF created in collaboration with the EEBus and Energy@-Home industry associations to interconnect their (different) data models. By using SAREF4EE, smart appliances from different manufacturers that support the EEBus or Energy@Home standards can easily communicate with each other using any energy management system at home or in the cloud.

Keywords: Ontology · Standardization · Semantic interoperability · Internet of things · Smart appliances

Resource type: Ontology
Permanent URL: https://w3id.org/saref and https://w3id.org/saref4ee

1   Introduction

In the past years, standards organizations, alliances and consortia in the Internet of Things (IoT) have focused on enabling interoperability at the technical communication level, creating a large number of different solutions based on various data models, specifications and protocols that are characterized by non-interoperable concepts [1].

© Springer International Publishing AG 2016
P. Groth et al. (Eds.): ISWC 2016, Part II, LNCS 9982, pp. 21–29, 2016.
DOI: 10.1007/978-3-319-46547-0_3

To overcome this fragmentation, it became necessary to move towards a common, harmonized solution that could enable interoperability at the semantic level, i.e., to understand the concepts contained in the message data structures that are exchanged at the underlying technical level, regardless of the specifics of the underlying communication protocols [2]. The need to address the semantics of standards has been acknowledged as an important action in the upcoming IoT standardization activities towards an interoperable and scalable solution across a global IoT ecosystem [3].

In 2015, we created SAREF (https://w3id.org/saref), a reference ontology for smart appliances that focuses on the smart home environment, and provides an important contribution to enable semantic interoperability in the IoT [4]. As explained in our earlier work [5–7], SAREF was originally created for the energy management domain, and was developed in close interaction with the industry with the support of the European Commission (EC). SAREF has been subsequently adopted by ETSI as a Technical Specification [8]. As confirmed in [3], SAREF is a first ontology standard in the Internet of Things (IoT) ecosystem, and sets a template and a base for the development of similar standards for the other verticals to unlock the full potential of IoT. Standards organizations and alliances, such as CENELEC and the Alliance for Internet of Things Innovation (AIOTI), have acknowledged and adopted SAREF in their standardization activities [2, 9].

This paper presents SAREF4EE (https://w3id.org/saref4ee), an extension of SAREF that we created in collaboration with EEBus (http://www.eebus.org/en) and Energy@Home (http://www.energy-home.it), the major Germany- and Italy-based industry associations, to enable the interconnection of their (different) data models. EEBus is an important initiative in the area of the Internet of Things, which has its roots in the sector of smart and renewable energy, and involves partners such as BSH Bosch/Siemens, Miele, Intel and Deutsche Telekom. EEBus developed a standardized and consensus-oriented smart grid and smart home networking concept. The EEBus data model is described in [10]. The Energy@Home association, abbreviated in the rest of the paper as E@H, includes members such as Electrolux, Enel, Indesit, Whirlpool and Telecom Italia. E@H aims at developing and promoting technologies and services for energy efficiency in smart homes, based upon the interaction between user devices and the energy infrastructure. The E@H data model is described in [11]. By using SAREF4EE, smart appliances from manufacturers that support the EEBus or E@H data models will easily communicate with each other using any energy management system at home or in the cloud.

SAREF4EE was created in a 5 months project in collaboration with the EEBus and E@H stakeholders. The result is an OWL-DL ontology that extends SAREF with 115 classes, 31 object properties and 51 data type properties. SAREF4EE is a step forward towards the missing interoperability in the fragmented market of proprietary solutions developed by different consortia in the smart home domain. Towards this aim, SAREF4EE should be used to annotate (or generate) a neutral (protocol-independent) set of messages to be directly adopted by the various manufactures, or mapped to their domain specific solutions of choice, such as ZigBee, Z-Wave, KNX, OCF, AllJoin and so forth. We envision that in the immediate future we will able to demonstrate how appliances that belong to different manufactures from the EEBus and E@H associations can practically communicate with the CEM in the home using a shared semantic standard (SAREF4EE) that abstracts from the specifics of the underlying communication protocols. This paper is structured as follows. The next section presents some background information about SAREF from which SAREF4EE was derived. We then describe the main classes and properties of SAREF4EE, followed by an evaluation according to the desired criteria, and our conclusions.

2 Background

SAREF is a reference ontology that describes the core concepts for the stakeholders in the smart appliances domain and provides a mechanism to map different existing solutions (i.e., data models, protocols, standards) to one another. SAREF is intended to be used as basis for the creation of more specialized ontologies, such as SAREF4EE presented in this paper. The main advantage is that specialized ontologies based on a reference ontology are (semantically) interoperable by construction with other ontologies that reuse the same reference ontology.

SAREF focuses on the concept of "device", which is a tangible object designed to accomplish a particular task in households, common public buildings or offices. In order to accomplish this task, a device performs one or more "functions". For example, a washing machine is designed to wash, and to accomplish this task it performs, among others, a start and stop function. When connected to a network, a device offers a "service", which is a representation of a function to a network that makes the function discoverable, registerable and remotely controllable by other devices in the network. A device is also characterized by an "energy profile" and a "power profile" that can be used to optimize the energy efficiency in a home or office that are part of the building. SAREF is expressed in OWL-DL and contains 110 classes, 31 object properties and 11 datatype properties. A detailed description of the main classes and properties of SAREF can be found in our earlier work [5].

The study to create SAREF was performed in a very transparent manner to allow all stakeholders to provide input and follow the work. We took a multi-channel approach to solicit for review comments: within the span of a year we organized four workshops for stakeholders together with the EC and ETSI, in which we presented the deliverables and work done, and provided an opportunity for stakeholders to provide us with feedback. Besides this quarterly interactive heartbeat of face-to-face gatherings, we had continuous interaction with all involved parties by email, LinkedIn, project's website (https://sites.google.com/site/smartappliancesproject), and by attending additional related events such as ETSI and Home Gateway Initiative (HGI) meetings. This interactive and iterative approach was a key factor of success in the creation of SAREF and its subsequent adoption by the industrial stakeholders. Although this approach created a lot of overhead work, it also guaranteed a higher practical quality of the outcome, reflecting the wishes and intentions of the community in an optimal way, and above all creating the necessary trust into our work, increasing the likelihood of a strong acceptance of the results. In this line, TNO was asked by EEBus and E@H to create the SAREF4EE extension.

3 SAREF4EE

SAREF4EE extends SAREF with 115 classes, 31 object properties and 51 data type properties. Its purpose is to demonstrate the interoperability between EEBus and E@H devices in demand response use cases. In these use cases, customers can offer flexibility to the Smart Grid to manage their smart home devices by means of a Customer Energy Manager (CEM). The CEM is a logical function for optimizing energy consumption and/or production that can reside either in the home gateway or in the cloud. These use cases can be described as follows:

- configuration of devices that want to connect to each other in the home network, for example, to register a new dishwasher to the list of devices managed by the CEM;
- (re-)scheduling of appliances in certain modes and preferred times using power profiles to optimize energy efficiency and accommodate the customer's preferences;
- monitoring and control of the start and status of the appliances;
- reaction to special requests from the Smart Grid, for example, incentives to consume more or less depending on current energy availability, or emergency situations that require temporary reduction of the power consumption.

These use cases are associated with the user stories described in [12], which include, among others, the following examples:

- User wants to do basic settings of their devices;
- User wants to know when the washing machine has finished working;
- User wants their washing done by 5:00 pm. with least electrical power costs;
- User likes to limit own energy consumption up to a defined limit;
- User allows the CEM to reduce the energy consumption of their freezer in a defined range for a specific time, if the grid recognizes (severe) stability issues;
- Grid related emergency situations (blackout prevention).

Figure 1 shows an excerpt of SAREF4EE that represents the concepts of "power profile", "power sequence", "alternative" and "slot" that are used by a device to communicate its energy flexibility to the CEM according to the consumer's preferences and needs. We distinguish between SAREF and SAREF4EE using the prefixes saref: and s4ee:, respectively.

A s4ee:PowerProfile inherits the properties of the more general saref:Profile, extending it with additional properties that are specific for SAREF4EE. The s4ee:PowerProfile is used by a s4ee:Device to expose the power sequences that are potentially relevant for the CEM, for example, a washing machine that wants to communicate its expected energy consumption for a certain day. A s4ee:Device can expose at most one s4ee:PowerProfile, which consists of one or more alternative plans (s4ee:Alternative class). For example, the washing machine mentioned above can offer two alternative plans, a "cheapest" alternative in which the CEM should try to minimize the user's energy bill, and a "greenest" alternative in which the CEM should try to optimize the configuration towards the maximum availability of renewable energy.

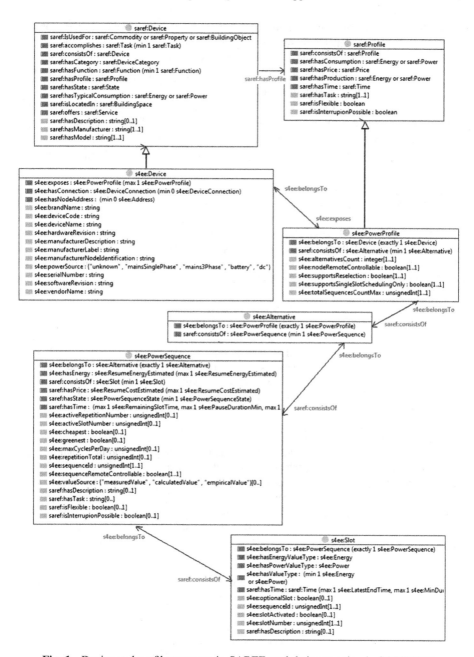

Fig. 1. Device and profile concepts in SAREF, and their extension in SAREF4EE

A `s4ee:Alternative` consists of one or more power sequences (`s4ee: PowerSequence` class). For example, in the "cheapest" alternative mentioned above, the washing machine can ask to allocate two power sequences during the night, while for the "greenest" alternative it can ask to allocate one power sequence in the morning and one in the afternoon. Of these power sequences, one is allocated for the task of washing and cannot be stopped once it started, while the other power sequence is allocated for the task of tumble dry and has the flexibility to be paused by the CEM as long as it finishes within a specified latest end time.

A s4ee:PowerSequence consists of one or more slots (s4ee:Slot class) that represent different phases of consumption (or production) and their values. In the power sequence allocated for washing, for example, various slots can represent the consumption during the different phases of washing, such as heating the water, washing and rinsing.

4   Evaluation

An evaluation of SAREF and SAREF4EE is proposed in Table 1 in accordance to the desired criteria [13], which include ontology-specific criteria and more generic criteria that aim at demonstrating the value, potential for being reused, technical quality and availability of ontologies and other resources in the Web.

Table 1. Evaluation.

Criteria	Evaluation
Design suitability	Based on requirements specifically set by the European Commission [14] and industrial collaborators. Intermediate and final designs interactively reviewed and approved by all involved parties. High practical quality, reflecting the wishes and intentions of the smart appliances community in an optimal way
Elegance & Quality	Design based on Gruber's principles of clarity, coherence, extendibility, minimal encoding bias, and minimum ontological commitment [15]
Logical correctness	Verified using DL reasoners for satisfiability, incoherency and inconsistencies
Reuse	SAREF4EE specializes the energy-related concepts defined in SAREF, and in turn SAREF reuses the following resources:
	• based on 20 domain-specific ontologies, e.g., W3C SSN ontology (https://sites.google.com/site/smartappliancesproject/ontologies); • reuse of W3C WGS84 geo positioning vocabulary and W3C Time ontology via direct import; • reuse of Ontology of units of Measure (OM) individuals
Documentation	Ontologies self-descriptive using labels and definitions. Human readable documentation available with major ontology elements and explanatory diagrams:
	• SAREF4EE documentation - reviewed and approved by EEBus and E@H experts - elaborates on the use cases using some figures (http://ontology.tno.nl/SAREF4EE_Documentation_v0.1.pdf); • SAREF standardized by ETSI as technical specification TS 103 264 [8]

(*Continued*)

Table 1. (*Continued*)

Criteria	Evaluation
Additional value	There is widespread fragmentation in current market and technology. SAREF(4EE) is the first ontology standard in the IoT with focus on understanding the concepts contained in the message data structures that are exchanged at the technical level, regardless of the specifics of the underlying communication protocols
Societal interest	Currently in-home electrical devices from different manufacturers are connected to the Internet, but this connectivity is hardly ever used for devices communicating with each other on an application level. SAREF and SAREF4EE aim to bridge this gap
Impact	SAREF acknowledged as first ontology standard in IoT [3, 4]. SAREF4EE adopted by industrial associations EEbus and E@H to interconnect their data models. Core of ETSI Special Task Force 513 for Maintenance & Evolution of SAREF Reference Ontology
Usage	Used by a wider community beyond the resource creators, such as: • industrial stakeholders, e.g., from the EEBus and E@H associations, whose members include appliances manufacturers such as BSH Bosch/Siemens, Miele, Indesit, Electrolux, and Whirlpool; • oneM2M Standards for M2M and the Internet of Things for their TS-0012 oneM2M Base Ontology, built after and based on SAREF (http://www.onem2m.org/technical/onem2m-ontologies); • AIOTI Alliance for Internet of Things Innovation in their WG03 • IoT standardization activities; • CENELEC in their Smart Home standardization activities [9]; • W3C Linked Building Data community group
Ease to (re)use	Well documented using standard metadata specifications such as Dublin Core. Modular building blocks to allow recombination of different parts to accommodate different needs and points of view. For example, lists of functions, commands and states that can be combined to create complex functionality in a single device
Persistent URI	w3id URIs: https://w3id.org/saref and https://w3id.org/saref4ee
Associated citation	Various citations associated to the resource [5–7].
License	CreativeCommons 3.0 http://creativecommons.org/licenses/by/3.0/
Community registry	Registered in Linked Open Vocabularies (LOV)
Sustainability/Maintenance plan	TNO is involved in a Special Task Force funded by ETSI for the maintenance, evolution and extension of SAREF, which also includes the definition of a sustainable plan for the governance of the SAREF(4EE) network of ontologies in the future

5   Conclusions

This paper presented SAREF4EE, an extension of SAREF that was created in collaboration with the EEBus and Energy@Home industry associations to interconnect their different data models, enabling interoperability between appliances produced by different manufacturers. In this paper, we further evaluated SAREF and SAREF4EE according to some criteria that take into account the value, potential for being reused, technical quality and availability of an ontology. We conclude that SAREF and SAREF4EE satisfy these criteria. Moreover, we conclude that reference ontologies, such as SAREF, and the network of extensions possibly sprouting from them, such as SAREF4EE, play an essential role in the current standardization activities in the IoT domain, and should be used more systematically to enable better interoperability in the current fragmented market of various, often proprietary solutions and technologies. In particular, SAREF and its network of ontologies should be evolved to cover other verticals that are relevant to the IoT and the Smart Home, such as e-health, security and entertainment. To have an actual impact, however, the development of these ontologies should be driven by a real, strong industry demand, and carried out in close collaboration with the stakeholders that are willing to adopt them in specific applications, instead of as an abstract promise for interoperability that is unilaterally pushed to the market. We finally observe that the endorsement by policy makers and standardization bodies (such as the EC and ETSI) is an essential factor to promote the adoption and usage of ontologies by the industry.

Acknowledgments.   The authors would like to thank Daniele Pala and Marco Signa from Energy@Home, and Andy Westermann and Josef Baumeister from EEBus for their insight and expertise that greatly assisted in the creation of SAREF4EE.

References

1. AIOTI WG03- IoT Standardisation: IoT LSP Standard Framework Concept 2.0 (2015)
2. AIOTI WG03- IoT Standardisation: IoT Semantic Interoperability 2.0 (2015)
3. European Commission, Directorate-General for Internal Market, Industry, Entrepreneurship and SMEs: GROW - Rolling Plan for ICT Standardisation (2016). http://ec.europa.eu/DocsRoom/documents/14681/attachments/1/translations/en/renditions/native
4. Viola, R.: New Standard for Smart Appliances in the Smart Home. Digital Agenda for Europe, December 2015. https://ec.europa.eu/digital-agenda/en/blog/new-standard-smart-appliances-smart-home
5. Daniele, L., den Hartog, F., Roes, J.: Created in close interaction with the industry: the smart appliances reference (SAREF) ontology. In: Cuel, R., Young, R. (eds.) FOMI 2015. LNBIP, vol. 225, pp. 100–112. Springer, Heidelberg (2015)
6. Daniele, L., den Hartog, F., Roes, J.: How to keep a reference ontology relevant to the industry: a case study from the smart home. In: Tamma, V., Dragoni, M., Gonçalves, R., Lawrynowicz, A. (eds.) OWLED 2015. LNCS, vol. 9557, pp. 117–123. Springer, Heidelberg (2016). doi:10.1007/978-3-319-33245-1_12

7. den Hartog, F., Daniele, L., Roes, J.: Toward semantic interoperability of energy using and producing appliances in residential environments. In: 12th Annual IEEE Consumer Communications & Networking Conference (CCNC 2015). IEEE Press, January 2015

8. SAREF TS 103 264 V1.1.1, ETSI, November 2015. http://www.etsi.org/deliver/etsi_ts/103200_103299/103264/01.01.01_60/ts_103264v010101p.pdf

9. CENELEC TC59x WG7, prEN50631-x (2016)

10. EEBus initiative: SPINE (Smart Premises Interoperable Neutral-message Exchange) Specification V1.0.0, April 2016. https://www.eebus.org/en/specifications

11. Energy@home project: Energy@home Data Model, v2.1, October 2015. http://www.energy-home.it/Documents/Technical%20Specifications

12. IEC TR 62746-2 Systems interface between customer energy management system and the power management system - Part 2: Use cases and requirements (2015)

13. Gray, A.J.G., Sabou, M.: ISWC2016 Resources Track: Author and Reviewer Instructions. figshare, December 2015. http://dx.doi.org/10.6084/m9.figshare.2016852

14. European Commission: Invitation to tender - Study on the available semantics assets for the interoperability of Smart Appliances. Mapping into a common ontology as a M2 M application layer semantics - SMART 2013/007 (2013)

15. Gruber, T.: Toward principles for the design of ontologies used for knowledge sharing. Int. J. Hum. Comput. Stud. **43**(5–6), 907–992 (1995). Elsevier

An Ontology of Soil Properties and Processes

Heshan Du[1]([✉]), Vania Dimitrova[1], Derek Magee[1], Ross Stirling[2],
Giulio Curioni[3], Helen Reeves[4], Barry Clarke[1], and Anthony Cohn[1]

[1] University of Leeds, Leeds, UK
H.Du@leeds.ac.uk
[2] Newcastle University, Newcastle upon Tyne, UK
[3] University of Birmingham, Birmingham, UK
[4] British Geological Survey, Nottingham, UK

Abstract. Assessing the Underworld (ATU) is a large interdisciplinary
UK research project, which addresses challenges in integrated inter-asset
maintenance. As assets on the surface of the ground (e.g. roads or pavements) and those buried under it (e.g. pipes and cables) are supported
by the ground, the properties and processes of soil affect the performance
of these assets to a significant degree. In order to make integrated decisions, it is necessary to combine the knowledge and expertise in multiple
areas, such as roads, soil, buried assets, sensing, etc. This requires an
underpinning knowledge model, in the form of an ontology. Within this
context, we present a new ontology for describing soil properties (e.g.
soil strength) and processes (e.g. soil compaction), as well as how they
affect each other. This ontology can be used to express how the ground
affects and is affected by assets buried under the ground or on the ground
surface. The ontology is written in OWL 2 and openly available from the
University of Leeds data repository: http://doi.org/10.5518/54.

Keywords: OWL ontology · Soil property/process · Asset maintenance

1 Introduction

Assessing the Underworld (ATU) project is a large interdisciplinary UK research
project, which addresses challenges in asset maintenance, especially how to
reduce the economic, social and environmental costs or impacts of streetworks
required in the maintenance of roads and buried assets (e.g. pipes and cables).
Existing asset management systems (e.g. UK Pavement Management Systems)
can help local authorities or utility companies with financial reporting or assessing the economic costs of the construction, repairing and replacement of assets,
but provide less support in assessing the environmental/social impacts and the
impacts on other assets, which are also important for enabling better informed
decisions in asset maintenance. To establish the total cost (social and environmental impacts along with economic costs) of asset maintenance activities, it
is essential to understand how assets affect each other and how they affect or
are affected by the natural environment and human activities. Within the ATU

© Springer International Publishing AG 2016
P. Groth et al. (Eds.): ISWC 2016, Part II, LNCS 9982, pp. 30–37, 2016.
DOI: 10.1007/978-3-319-46547-0_4

project, we are developing a series of ontologies based on the knowledge and expertise in multiple areas for describing how buried assets, soil, roads, the natural environment and human activities affect each other. The ontology of soil properties and processes (OSP) is a central ontology in this series.

The condition of an asset can affect and be affected by assets close to it. For example, a water pipe burst increases the water content of the soil (including man-made fills) surrounding it. As a consequence, the strength of soil decreases, which has negative effects on the function of soil for supporting buried pipes and cables nearby, as well as the roads or pavements above. In many interactions between assets, as shown in the example above, soil plays a role as a medium. In other words, the conditions of roads and buried assets affect each other through soil, and soil directly affect and is affected by the conditions of roads and buried assets. Therefore, when assessing the impacts of decisions (e.g. whether to fix a water pipe leakage immediately) for surface or buried asset maintenance, it is important to have a basic knowledge of soil properties and processes.

Soil has many properties and is involved in various processes, as described in [15, 16]. The relationships between soil properties and processes are complicated. Whilst it is often easy to know what a process or property affects directly, it is more difficult for people to answer questions like 'what are the factors affecting or affected (directly or indirectly) by a particular property or process of soil?'. To answer such questions, it is necessary to handle information about soil properties, processes and their relationships automatically. This requires a proper ontological model and automated reasoning.

There exist several ontologies [1, 2, 7–9, 12, 17, 18, 20], where soil is defined. Some of them [1, 2, 7, 17] are general environmental or agricultural ontologies, whilst others are specialized for describing soil. These soil ontologies [8, 9, 12, 18, 20] are not publicly available. Most of the existing soil ontologies [8, 9, 18, 20] describe classifications of soil or different types of soil but do not elaborate the various soil properties and processes. The ontologies described in [2, 12] define some soil physical/chemical properties, but limited to those relevant to farming or agricultural applications. None of the existing ontological models provides a systematic and comprehensive description of soil properties and processes, and none of them defines how soil properties and processes affect each other.

The paper addresses this gap. We present a new ontology for describing soil properties and processes, as well as how they affect each other. It reuses and specifies high-level classes in NASA's Semantic Web for Earth and Environmental Terminology (the SWEET ontology) [17], which is widely adopted and extended. The ontology is developed using the NeOn methodology [19]. It is written in OWL 2 Web Ontology Language Manchester Syntax [14], which is based on description logic (DL) [6]. The DL expressivity of the ontology is $\mathcal{SRI}$, allowing transitive relations and inverse relations[1]. The ontology contains 592 concepts and 2243 relation statements (OWL logical axioms), which are based

---

[1] To avoid confusion, we call 'OWL object properties' relations.

on the knowledge of soil experts[2], the SWEET ontology [5,17], English dictionaries [3,4] and a textbook on soil physics [16]. The ontology, together with a tutorial of viewing and querying it, a translation of its main relation statements in natural language and a feedback form, is publicly available [11] from the University of Leeds data repository: http://doi.org/10.5518/54, under the license Creative Commons Attribution 4.0 International (CC BY 4.0)[3].

The rest of the paper is structured as follows. Section 2 describes how the main concepts and relations are defined in the ontology. Section 3 explains how to reason with and query the ontology. Section 4 illustrates the extensibility of the ontology. Section 5 discusses its applications. Section 6 concludes the paper.

2   Defining Soil Properties and Processes

The ontology of soil properties and processes (OSP) defines two main high-level classes or categories: *SoilProperty* and *SoilProcess*. The classes *SoilProperty* and *SoilProcess* are specifications of the classes *Property* and *Process* in the SWEET ontology [5,17] for soil. A property is an attribute, quality, or characteristic of something [4]. A soil property is a property of soil. A process refers to a series of changes that happen naturally over time [3,4]. A soil process is a process involving soil. Following the style of the SWEET ontology, we define different kinds of soil properties and processes as classes in the OSP ontology and classify them into physical, chemical and spatial/biological categories. Table 1 summarizes the number of subclasses of *SoilProperty* and *SoilProcess* in different categories. Though the OSP ontology covers classes in all these categories, it mainly describes soil physical properties and processes.

Table 1. Number of subclasses of *SoilProperty* and *SoilProcess* in the OSP ontology

SoilProperty	Subclasses	SoilProcess	Subclasses
SoilPhysicalProperty	176	SoilPhysicalProcess	111
SoilChemicalProperty	16	SoilChemicalProcess	29
SoilSpatialProperty	4	SoilBiologicalProcess	8

In addition to classifying soil properties and processes into different categories as the SWEET ontology does, we also define soil properties and processes regarding how they affect each other. The main types of relations defined in the OSP ontology are *hasImpactOn* and its inverse *influencedBy*, meaning 'affects or changes' and 'is affected or changed by' respectively. The relations *hasImpactOn* and *influencedBy* are both defined as transitive. A process *q hasImpactOn* a

---

[2] Soil experts were involved in the development and evaluation of the OSP ontology. They checked 1407 relation statements for describing 193 main classes.

[3] https://creativecommons.org/licenses/by/4.0/.

property p, if q causes a change in p. A property p *hasImpactOn* a process q, if a change in p changes how the process q goes. A property p_1 *hasImpactOn* a property p_2, if a change in p_1 causes a change in p_2. A process q_1 *hasImpactOn* a process q_2, if q_1 changes how the process q_2 goes. Other relations defined in the OSP ontology include *hasPossibleCause* and its inverse *hasPossibleEffect*, meaning 'has a possible reason' and 'has a possible consequence' respectively. The word 'possible' means the cause/effect exists in some situation, but may exist or not exist in a particular real world situation considered.

In the OSP ontology, we define relationships between soil properties and processes at the concept level rather than at the individual level. For example, we express 'soil water content *hasImpactOn* soil strength' in description logic as *SoilWaterContent* $\sqsubseteq \exists hasImpactOn.SoilStrength$. Such relation statements are not defined for a particular type of soil at a certain location and depth, but generally applicable to any soil. The statement 'soil water content *hasImpactOn* soil strength' means that for any soil, a change in its water content causes a change in its strength[4]. Similarly, the statement 'soil compaction *hasImpactOn* soil air content' means for any soil, a compaction applied to it changes the air content of it. The statement 'soil compaction *hasImpactOn* soil air movement' means for any soil, a compaction applied to it changes how the air in it moves. The statement 'soil air movement is *influencedBy* soil porosity' means for any soil, a change in its porosity changes how the air in it moves. Note that inverse relations are not always reciprocally asserted in the manually-written ontology.

We rank subclasses of *SoilProperty* and *SoilProcess* based on their usages or how many statements a class is involved. If a class is involved in a statement, then any class equivalent to it is also considered to be involved in the same statement. Table 2 lists the top 20 classes in this ranking, which illustrates the main soil properties and processes defined in the OSP ontology well. For each set of equivalent classes, only one class in it is included in the list to represent the whole set. When naming a soil property/process class, 'Soil' is added to the front to avoid potential name conflicts. A class name indicates its intended meaning. For instance, the class name *SoilWaterInfiltration* is intended to represent the concept of 'infiltration of water into soil'.

3   Querying the OSP Ontology

Description logic (DL) reasoners can be used to reason about statements in the OSP ontology. A basic question a reasoner can answer is whether a statement (e.g. '*SoilWaterEvaporation hasImpactOn SoilStrength*') can be inferred from statements in a given ontology. A justification based explanation framework [13] can be used together with a DL reasoner to explain how a new statement is inferred from existing statements. By reasoning with the OSP ontology, more

---

[4] According to DL/OWL semantics, it means 'every soil water content *hasImpactOn* some soil strength'. The interpretation adopted here is more strict, however, this makes more sense for soil experts and the standard DL inference rules are still sound. In practice, we only expect the model to be applied to a local spatial context.

Table 2. Ranking the subclasses of *SoilProperty* and *SoilProcess* in the OSP ontology by the number of statements involved (usages) — the Top 20

	SoilProperty	Usages	SoilProcess	Usages
1	SoilMoistureContent	93	SoilAeration	70
2	SoilStructure	93	SoilCompaction	59
3	SoilTexture	67	SoilWaterInfiltration	40
4	SoilPorosity	62	SoilCrusting	40
5	SoilTemperature	59	SoilWaterEvaporation	39
6	SoilStrength	52	SoilAggregation	39
7	SoilClayMineral	49	SoilWaterMovement	32
8	SoilBulkDensity	40	SoilErosion	29
9	SoilOrganicMatterContent	39	SoilDrying	29
10	SoilHydraulicConductivity	36	SoilShrinkage	28
11	SoilParticleSpecificSurfaceArea	31	SoilWaterRetention	28
12	SoilClayContent	30	SoilSoluteTransport	26
13	SoilPoreSizeDistribution	28	SoilMicrobialActivity	21
14	SoilCohesion	23	SoilHardsetting	20
15	SoilPlasticity	22	SoilWaterTransmission	19
16	SoilMoisturePotential	18	SoilFreezing	19
17	SoilParticleArrangement	18	SoilDeformation	18
18	SoilBearingCapacity	18	SoilSwelling	18
19	SoilOxygenConcentration	18	SoilDispersion	18
20	SoilAirContent	17	SoilGaseousEmission	17

advanced questions can be answered, such as, for given a class C, list all the classes that C *hasImpactOn* or *influencedBy*.

A simple way to reason with and query the OSP ontology is to use an ontology editor Protégé[5] and its reasoner plugins. Within the DL query tab of Protégé, we can execute queries like getting all subclasses of the class expression '(*hasImpactOn* some *SoilStrength*) and *SoilProperty*'. For each class listed in the query results, its explanations can be obtained easily. We provide a tutorial [10] illustrating (with figures and examples) how to query the OSP ontology. It is available at: http://doi.org/10.5518/54, the same DOI of the OSP ontology.

4 Extensibility of the OSP Ontology

The OSP ontology extends three other top-level classes, *Phenomena*, *HumanActivity* and *Substance*, of the SWEET ontology [5,17]. A phenomenon

---

[5] http://protege.stanford.edu.

refers to a fact/situation that exists and can be observed [3,4]. A human activity refers to a thing that a person/group does or has done [4]. The concept *Substance* [5] covers living entities and non-living entities, such as animals, plants, material things, etc. By including these concepts, the OSP ontology can express how soil properties and processes affect/are affected by other environmental factors (such as water, air, weather and trees) and human activities.

For each of the five top-level classes, its next-level classes in the OSP ontology are shown in Fig. 1. Some of the next-level classes (e.g. *PlanetaryPhenomena*, and *MaterialThing*) are inherited from the SWEET ontology, whilst others (e.g. *PlantProcess* and *SoilSubstance*) are created by specifying high-level concepts in the SWEET ontology for a particular type of objects. All the high-level categories can be extended and enriched further easily by reusing concepts in the SWEET ontology and following similar ways as defining soil properties and processes in Sect. 2. In addition, more high-level categories can be added easily and linked to the existing classes. We are developing ontologies for describing pipe/road properties, processes and phenomena (e.g. defects or failures) in a similar way and linking them to the OSP ontology.

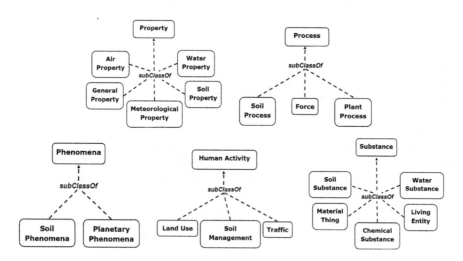

Fig. 1. Class hierarchies of the five top-level classes in the OSP ontology

Note that the OSP ontology mainly describes soil properties and processes from a soil science perspective, but also captures main terminologies (e.g. soil compaction) used in engineering applications. We are aware that terminological differences exist between soil scientists and geotechnical engineers. We follow the concepts defined in soil science mainly because they are more understandable by non-specialists. The OSP ontology can be extended and enriched with parameters[6] (e.g. soil shear strength) and calculations defined in soil mechanics [15].

---

[6] The OSP ontology describes several parameters used for assessing soil structure, under a class called *SoilStructureIndex*.

5   Application

The OSP ontology is intended to be used in a software system that supports inter-asset management/maintenance decisions (ATU decision support system). When making decisions for maintaining a single asset, people (e.g. local authorities and utility companies) often consider the economic cost but may not be aware or take into account the impacts of their decisions or actions on other assets, the natural environment and our daily life. The ATU decision support system will prompt people into asking a series of questions, the answers to which will lead to a complete understanding of the economic, social and economic impacts of different options for dealing with an asset defect/failure. By developing the OSP ontology, we establish a general framework, which extends the SWEET ontology, for expressing how properties, processes, phenomena, human activities and substances affect each other. In this way, many environmental and social impacts can be represented in an ontology or inferred from it using automatic reasoning. The OSP ontology can help users understand the environmental impacts of a process or a human activity related to soil. For example, knowing that a water pipe leakage affects soil water content, we may ask the OSP ontology about the impacts of a change in soil water content. By executing a DL query of getting subclasses of *'influencedBy* some *SoilWaterContent'* in Protégé, a list of 264 classes is returned[7], including various soil properties and processes, as well as factors like building stability, plant growth, air quality, water quality, etc. For classes listed, explanations are provided, which are useful for identifying and understanding indirect or hidden impacts.

The soil properties and processes described in the OSP ontology have many agricultural, engineering and environmental applications (see the subclasses of *SoilFunction*). For example, the OSP ontology describes how the growth of a plant is affected by soil and other environmental factors. The OSP ontology can be applied to a range of contexts that require descriptions of soil properties or processes. For example, the EU project NeTTUN[8] utilises soil properties and processes to provide contextual factors for tunnel construction and maintenance. By extending the OSP ontology and linking it to ontologies for describing the environment, infrastructure assets, human activities and economic cost models, we envisage that the OSP ontology will make a useful contribution to ontological models for environmental, social and economic sustainability.

6   Conclusion

We present a new ontology OSP for describing soil properties, processes and how they affect each other. It is created using reliable knowledge sources and extends the SWEET ontology. It can be reasoned with automatically using DL reasoners and queried using Protégé. The OSP ontology helps people to understand indirect and complicated relationships between soil properties and processes, as well

---

[7] Different types of factors can be filtered out easily using DL query.

[8] http://nettun.org.

as many environmental impacts of a process or a human activity related to soil. The OSP ontology is easy to (re)use, easily extensible, generally applicable, publicly available and findable. As the ATU decision support system develops, we will maintain and enhance this ontology, making new versions publicly available.

Acknowledgments. This research is supported by EPSRC under grant no. EP/K021699/1 which we gratefully acknowledge.

References

1. Eionet GEMET Thesaurus (2015). http://www.eionet.europa.eu/gemet
2. AGROVOC Multilingual agricultural thesaurus (2016). http://aims.fao.org/standards/agrovoc
3. Cambridge Dictionaries Online (2016). http://dictionary.cambridge.org
4. Oxford Dictionaries (2016). http://www.oxforddictionaries.com
5. SWEET: Semantic Web for Earth and Environmental Terminology (2016). https://sweet.jpl.nasa.gov
6. Baader, F., Calvanese, D., McGuinness, D.L., Nardi, D., Patel-Schneider, P.F. (eds.): The Description Logic Handbook. Cambridge University Press, New York (2007)
7. Buttigieg, P.L., Morrison, N., Smith, B., Mungall, C.J., Lewis, S.E.: The environment ontology: contextualising biological and biomedical entities. J. Biomed. Semant. **4**, 43 (2013)
8. Deb, C., Marwaha, S., Malhotra, P., Wahi, S., Pandey, R.: Strengthening soil taxonomy ontology software for description and classification of USDA soil taxonomy up to soil series. In: Proceedings of the 2nd International Conference on Computing for Sustainable Global Development, pp. 1180–1184 (2015)
9. dos Santos Aparício, A., de Farias, O.L.M., dos Santos, N.: Integration of heterogeneous databases and ontologies. Cadernos do IME-Série Inf. **21**, 4–10 (2006)
10. Du, H., Cohn, A.: A Tutorial of Viewing and Querying the Ontology of Soil Properties and Processes. Technical report, University of Leeds (2016)
11. Du, H., Cohn, A.: An Ontology of Soil Properties and Processes. [Dataset]. University of Leeds (2016). http://doi.org/10.5518/54
12. Heeptaisong, T., Shivihok, A.: Soil knowledge-based systems using ontology. In: Proceedings of the International MultiConference of Engineers and Computer Scientists, pp. 1–5 (2012)
13. Horridge, M.: Justification Based Explanation in Ontologies. Ph.D. thesis, University of Manchester (2011)
14. Horridge, M., Patel-Schneider, P.F.: OWL 2 Web Ontology Language Manchester Syntax (2012). https://www.w3.org/TR/owl2-manchester-syntax
15. Knappett, J., Craig, R.: Craig's Soil Mechanics. CRC Press, Boca Raton (2012)
16. Lal, R., Shukla, M.K.: Principles of Soil Physics. CRC Press, New York (2004)
17. Raskin, R.G., Pan, M.J.: Knowledge representation in the semantic web for Earth and environmental terminology (SWEET). Comput. Geosci. **31**(9), 1119–1125 (2005)
18. Shivananda, P., Kumar, P.S.: Building rules based soil classification ontology. Int. J. Comput. Sci. Inf. Technol. Secur. **3**(2), 27 (2013)
19. Suárez-Figueroa, M.C., Gómez-Pérez, A., Fernández-López, M.: The NeOn methodology for ontology engineering. In: Ontology Engineering in a Networked World, pp. 9–34 (2012)
20. Zhao, M., Zhao, Q., Tian, D., Qian, P., Zhang, X.: Ontology-based intelligent retrieval system for soil knowledge. WSEAS Trans. Inf. Sci. Appl. **6**(7), 1196–1205 (2009)

LODStats: The Data Web Census Dataset

Ivan Ermilov[1(✉)], Jens Lehmann[2], Michael Martin[1], and Sören Auer[2]

[1] AKSW, Institute of Computer Science, University of Leipzig, Leipzig, Germany
{iermilov,martin}@informatik.uni-leipzig.de
[2] University of Bonn and Fraunhofer IAIS, Bonn, Germany
{jens.lehmann,auer}@cs.uni-bonn.de

Abstract. Over the past years, the size of the Data Web has increased significantly, which makes obtaining general insights into its growth and structure both more challenging and more desirable. The lack of such insights hinders important data management tasks such as quality, privacy and coverage analysis. In this paper, we present the LODStats dataset, which provides a comprehensive picture of the current state of a significant part of the Data Web. LODStats is based on RDF datasets from *data.gov*, *publicdata.eu* and *datahub.io* data catalogs and at the time of writing lists over 9000 RDF datasets. For each RDF dataset, LODStats collects comprehensive statistics and makes these available in adhering to the *LDSO* vocabulary. This analysis has been regularly published and enhanced over the past five years at the public platform lodstats.aksw.org. We give a comprehensive overview over the resulting dataset.

Resource type: Dataset
Permanent URL: https://datahub.io/dataset/lodstats

1 Introduction

Over the past years, the size of the Data Web has increased significantly, which makes obtaining general insights into its growth and structure both more challenging and more desirable. The expansion of the Data Web can be to a large extent attributed to the efforts in the Semantic Web and Open Government communities. Both communities have a common goal: to provide 5-star[1] RDF datasets to end-users. To achieve this goal, the Semantic Web community introduced a number of requirements for datasets, which should be fulfilled to be included into the *LOD Cloud*[2]. The Semantic Web community has a main dataset registry hub: the datahub[3] data catalog, while Open Government initiatives usually distribute RDF datasets through their own data catalogs (e.g. *data.gov*, *publicdata.eu* and *open.canada.ca*).

---

[1] According to 5-star data model available at http://5stardata.info.
[2] http://linkeddatacatalog.dws.informatik.uni-mannheim.de/state/.
[3] http://datahub.io/.

© Springer International Publishing AG 2016
P. Groth et al. (Eds.): ISWC 2016, Part II, LNCS 9982, pp. 38–46, 2016.
DOI: 10.1007/978-3-319-46547-0_5

All of the mentioned data catalogs utilize CKAN, an open-source data portal platform, which is a de-facto standard for Open Data. CKAN provides a solid framework to organize datasets and to expose metadata about them in various formats, including RDF. However, CKAN does not provide analytics over the registered datasets and highly depends on the user input. Moreover, no single aggregation point exists. These factors limit the possibility to obtain general insights into the Data Web. The lack of such insights hinders important data management tasks such as quality, privacy and coverage analysis.

For this reason, attempts to analyze the Data Web were made previously. *SPARQL Endpoint Status*[4] (SPARQLES) [3] addresses the problem of the availability of SPARQL endpoints over time. SPARQLES aggregates 553 SPARQL endpoints and exposes information on the availability and their features (e.g. support for SPARQL 1.0/1.1, availability of VoID/Service descriptions). *Linked Open Vocabularies*[5] (LOV) [6] is a project for building an RDF vocabulary ecosystem, which can support reuse of vocabulary terms. LOV aggregates the vocabularies from various publishers and establish relationships between them using the VOAF vocabulary. The project collected 548 vocabularies (e.g. DCMI Metadata Terms, Friend of a Friend and others) and enabled vocabulary search by utilizing metrics derived from the analysis of the vocabularies and their relationships. The *vocab.cc* project attempted to fill the gap of vocabulary usage statistics. Being based on the Billion Triples Challenge (BTC) in 2012, vocab.cc introduced four metrics to evaluate the BTC dataset. However, the project has a limited scope (i.e. being restricted to the BTC dataset) and was a one-shot evaluation, and therefore does not provide sustainable statistics over time.

In this paper, we address the above-described gap in the Data Web analysis. We present the *LODStats* dataset, which provides a comprehensive picture of the current state of a significant part of the Data Web. At the time of writing, LODStats aggregates 9960 RDF datasets from the data.gov, publicdata.eu and datahub.io data catalogs. For each RDF dataset, LODStats collects comprehensive statistics adhering to the RDF data model. This analysis has been regularly published and enhanced over the past five years at http://lodstats. aksw.org. We extend our previous work [4,5] as follows: (i) we include data.gov and publicdata.eu data catalogs, which account for 45 % of the RDF datasets (ii) we publish the *LDSO* vocabulary, describing the LODStats data schema and (iii) we enrich the dataset with CKAN metadata. Overall, our contributions are as follows:

- We provide a 5-star RDF dataset containing statistical facts about the Data Web, which is interlinked with CKAN metadata.
- We showcase the usage of the dataset via five use case descriptions.
- We describe insights in the Data Web gained from the analysis of LODStats dataset.
- We maintain LODStats over the past five years, delivering sustainable solution to the Semantic Web community.

---

[4] http://sparqles.ai.wu.ac.at/.
[5] http://lov.okfn.org/.

The rest of the paper is structured as follows: in Sect. 2 we introduce the LODStats web application, Sect. 3 outlines the design of the LODStats dataset, in Sect. 4 we describe use cases supported by the dataset, Sect. 5 exhibits the interfaces to access the dataset, we discuss the insights of the Data Web analysis in Sect. 6, and finally conclude and outline future work in Sect. 7.

2   LODStats: Web Scale RDF Data Analytics

In this section, we briefly outline the inner workings of the LODStats application and show the evolution of the technical solution.

The general overview of the LODStats architecture is depicted in Fig. 1. The *LODStats Statistics Evaluation* (LSE) module performs the execution of the statistical metrics on a dataset and is described in more detail in previous work [4,5].[6] In this paper, we introduce the following new modules. To aggregate the datasets from the data catalogs we implemented the *CKAN Aggregator*[7]. The *Messaging Broker*[8] allows to schedule processing and scale it horizontally (i.e. to distribute datasets processing between LSE modules running in parallel).

We provide interfaces both for human users and machine agents. The *RDB2RDF*[9] module provides virtual RDF views accessible through the *LODStats SPARQL Endpoint* for the consumption of machine agents. For human users, a web front-end is available at http://lodstats.aksw.org.

Moreover, we provide Docker image of the whole system publicly.[10] With LODStats Docker image, the application can be deployed on any Docker-enabled host with one command, namely `docker-compose up -d`.

3   Dataset Modelling

In this section, we describe the LODStats DataSet vOcabulary (LDSO)[11], depicted in Fig. 2. We designed LDSO as an extension of the *Data Catalog Vocabulary* (DCAT) [7] and *Vocabulary of Interlinked Datasets* (VoID) [1] according to the best practices of the vocabulary design, preservation and governance described in [2,6]. In the following, we describe the structure of the vocabulary.

The **ldso:Dataset** class is a representation of a dataset from a CKAN data catalog. Thus, to model **ldso:Dataset** we extend **dcat:Dataset** by adding the *ldso:active* property and reusing general metadata properties such as *dc:identitfier* and *dc:modified*. *ldso:active* is a boolean property, which separates up-to-date (i.e. existing in the CKAN data catalog) and out-dated datasets. **ldso:Dataset** connects to the data.gov, publicdata.eu and datahub.io data

---

[6] For SPARQL endpoints LSE can only infer number of triples.

[7] https://github.com/aksw/ckan-aggregator-py.

[8] We use rabbitmq as a messaging broker https://www.rabbitmq.com/.

[9] For RDB2RDF transformation we utilize Sparqlify http://sparqlify.org/.

[10] https://github.com/aksw/lodstats.docker.

[11] LDSO is published at http://lodstats.aksw.org/ontology/ldso.owl.

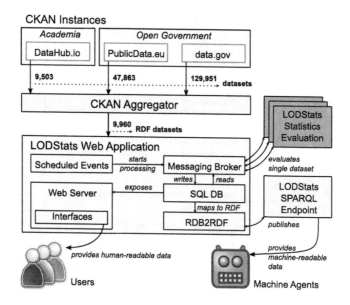

Fig. 1. LODStats architecture overview.

portals (**ldso:CkanCatalog**) via the *dc:isPartOf* property. Also, we interlink instances of **ldso:Dataset** to the corresponding RDF representations in the data portals using *owl:sameAs*. To process a **ldso:Dataset**, the LODStats application utilizes the value of the *dcat:downloadURL* property to retrieve dumps. Subsequently, a **ldso:Dataset** is linked directly to the last evaluation result via *ldso:currentStats*. The modelling of **ldso:Dataset** instances, for example, supports the following queries: (i) *How many RDF datasets are in a particular CKAN data catalog?*, (ii) *What is the ratio between out-dated and up-to-date datasets?*, (iii) *Who is the dataset maintainer and what is her email address?*

A **ldso:StatResult** represents a single evaluation result for a **ldso:Dataset**. **ldso:StatResult** extends **void:Dataset** by adding set of statistical metrics in the LDSO namespace such as *ldso:literals, ldso:blanks, ldso:subclasses*. We connect **ldso:StatResult** to **ldso:Dataset** using the *foaf:primaryTopic* property. The VoID vocabulary introduces the concept of property and class partitions, which represent the subsets of a dataset utilizing particular properties/classes. We extend this design pattern by introducing new partitions, based on datatypes, vocabularies and languages. We interlink **ldso:StatResult** instances to the VoID description of the datasets, generated automatically on dataset evaluation. The modelling of **ldso:StatResult** allows, for example, the following queries: (i) *How many triples (literals, blanks, subclasses) are contained in the dataset?*, (ii) *How many triples in the dataset are adhering to the particular vocabulary (language, datatype)?*, (iii) *What is the size of the dataset dump (in bytes)?*

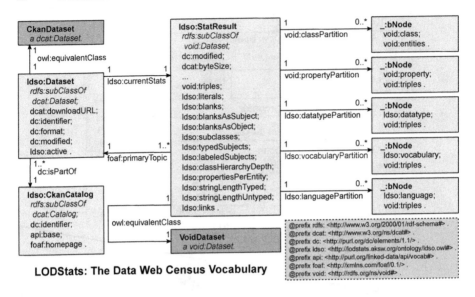

Fig. 2. LODStats vocabulary schema.

4   Relevance of the Dataset

Obtaining *comprehensive* statistical analysis about datasets made available on the Web of Data facilitates a number of important use cases (UC) and provides crucial benefits. These include:

Vocabulary Reuse (UC1). One of the advantages of semantic technologies is to simplify data integration via common vocabularies. However, it is often difficult to identify relevant vocabulary elements. The LODStats web interface stores the usage frequency of vocabulary elements (e.g. property usage count in [4]) and provides search functionality. This allows knowledge engineers to find the most frequent schema elements, which can be used to model the task at hand. Having this functionality encourages reuse of schema elements and, therefore, simplifies data integration, which is one of the central advantages of semantic technologies. LODStats also provides a webservice for this functionality, such that third party tools can easily integrate search for similar classes and properties. For instance, Linked Open Vocabularies (LOV)[12] utilizes vocabulary usage frequency as an indicator showing the users popularity of specific vocabulary inside the Linked Open Vocabularies catalogue.

Quality analysis (UC2). A major problem when using Web Data is quality. However, the quality of the datasets itself is not so much a problem as assessing and evaluating the expected quality and deciding whether it is sufficient for a certain application. Also, on the traditional Web we have very varying quality, but means were established (e.g. page rank) to assess the quality of information on the document web. In order to establish similar measures on the Web of Data

---

[12] http://lov.okfn.org.

it is crucial to assess datasets with regard to incoming and outgoing links, but also regarding the used vocabularies, properties, adherence to property range restrictions, their values etc. The links can be directly used for data quality (e.g. more links – better). The other metrics, for example, can be compared over the time between the datasets. Hence, a statistical analysis of datasets can provide important insights with regard to the expectable quality.

Coverage analysis (UC3). Similarly important as quality is the coverage a certain dataset provides. LODStats can be used to compute several coverage dimensions. For instance, the most frequent properties for a particular dataset can be computed and allow to get an overview over instance data, e.g. whether it contains address information (i.e. vcard:adr usage count > 0). Furthermore, the frequency of namespaces may also be an indicator for the domain of a dataset. The ranges of properties can give insights on whether spatial or temporal information is present in the dataset. In the case of spatial data, for example, we would like to know the region the dataset covers, which can be easily derived from minimum, maximum and average of longitude and latitude properties.

Privacy analysis (UC4). For quickly deciding whether a dataset potentially containing personal information can be published on the Data Web, we need to get a swift overview on the information contained in the dataset without looking at every individual data record (e.g. dataset uses vcard vocabulary). An analysis and summary of all the properties and classes used in a dataset can quickly reveal the type of information and thus prevent the violation of privacy rules.

Link target identification (UC5). Establishing links between datasets is a fundamental requirement for many Linked Data applications (e.g. data integration and fusion). However, as we learned the Web of Linked Data currently still lacks coherence (with less than 10 % of the entities actually being linked). Meanwhile, there are a number of tools available which support the automatic generation of links (e.g. [8,9]). An obstacle for the broad use of these tools is, however, the difficulty to identify suitable link targets on the Data Web. By attaching proper statistics about the internal structure of a dataset (in particular about the used vocabularies, properties etc.) it will be dramatically simplified to quickly identify suitable target datasets for linking. For example, the use of longitude and latitude properties in a dataset indicates that this dataset might be a good candidate for linking spatial objects. If we additionally know the minimum, maximum and average values for these properties, we can even identify datasets which are suitable link targets for a certain region.

5 Availability, Interfaces and Sustainability

In this section, we describe the interfaces to access the dataset as well as how we support sustainability. We publish our dataset on datahub.io data catalog[13]. The datahub.io entry for LODStats includes:

---

[13] Available at http://datahub.io/dataset/lodstats.

– **VoID description.** Machine readable description of the dataset.
– **LDSO vocabulary.** LODStats Dataset Vocabulary.
– **LODStats SPARQL endpoint.** SPARQL endpoint for the application.
– **LODStats RDF dump.** The RDF dump of LODStats dataset (April 2016).
– **VoID descriptions RDF dump.** Automatically generated VoID descriptions from the LODStats application (April 2016).
– **Data.gov, PublicData.eu, Datahub.io RDF dumps.** RDF dumps of the crawled data catalogs (April 2015).

The SPARQL endpoint serves the last output of RDB2RDF module and exposes up-to-date data. We announce the LODStats dataset using public Semantic Web lists and create a Web forum[14] to support community feedback. The sustainability of LODStats is demonstrated through: (i) the LODStats project being running for over the last five years, (ii) a state of the Data Web evaluation being performed every 2 months or at least once per half a year during this period, (iii) the last evaluation was performed just recently.

6   Data Web Statistics Summary

In this section we provide brief overview of the insights into the Data Web, based on the statistics collected over the past five years for the RDF dumps. The general current statistics such as number of triples, entities, literals etc. are available on the LODStats web portal[15].

Over the past five years, the number of the datasets has increased from 422 in 2011 to 9644 in 2015. The burst of the datasets number has occurred in 2015, when we included data catalogs from the Open Governments in LODStats. However, only a small part of the overall amount of triples: 1 % for the PublicData.eu and 3 % for the Data.gov portals, can be attributed to the governmental data catalogs. It can be explained by the fact, that Open Governments publish short documents such as monthly energy consumption or salary rates for the governmental facilities. The connectedness of the Data Web has increased to 40 % since 2011, when only 3 % of the overall amount of triples were links between different datasets.

The further Web Data statistics can be accessed from the LODStats SPARQL endpoint[16]. For instance, the datasets in 2011 can be requested as follows:

```
PREFIX xsd: <http://www.w3.org/2001/XMLSchema#>
PREFIX ldso: <http://lodstats.aksw.org/ontology/ldso.owl#>
PREFIX dc: <http://purl.org/dc/terms/>
PREFIX foaf: <http://xmlns.com/foaf/0.1/>

SELECT ?datasetName ?evaluationDate ?ckanCatalogName {
```

---

[14] https://groups.google.com/forum/#!forum/lodstats.
[15] Statistics can be accessed at http://lodstats.aksw.org/stats.
[16] http://lodstats.aksw.org/sparql.

```
?statResult    a                        ldso:StatResult.
?statResult    dc:modified              ?evaluationDate.
?statResult    foaf:primaryTopic        ?dataset.
?dataset       dc:identifier            ?datasetName.
?dataset       dc:isPartOf              ?ckanCatalog.
?ckanCatalog   dc:identifier            ?ckanCatalogName.
FILTER (?evaluationDate <"2012-01-01T00:00:00"^^xsd:dateTime
&& ?evaluationDate >"2011-01-01T00:00:00"^^xsd:dateTime)}
```

7 Conclusion and Future Work

We presented LODStats – The Data Web Census Dataset, which exposes statistics about the Data Web over the last five years. We exposed the dataset using SPARQL endpoint and as an RDF dump, providing the one point of access at the DataHub.io data catalog. We created a mailing list to collect the feedback from the community and announced the dataset on the major mailing lists.

In the future we will be processing very large datasets with more than hundreds of millions triples, which are expensive to process on a single machine. Additionally, we plan to include metrics for data streams (standing queries and observing their change over time) as well as extending the metrics to compute complex graph properties, and properties related to inference. The timestamps of all individual measurements are available as RDF data via SPARQL endpoint, which we plan to use for providing the timeline views for the different statistics available via LODStats.

Acknowledgments. This work was partly supported by the German Federal Ministry of Education and Research (BMBF) for the LEDS Project (GA no. 03WKCG11C) and by grant from the European Union's Horizon 2020 research Europe flag and innovation programme for the project Big Data Europe (GA no. 644564).

References

1. Alexander, K., Cyganiak, R., Hausenblas, M., Zhao, J.: Describing linked datasets. In: LDOW (2009)
2. Allemang, D., Hendler, J.: Semantic Web for the Working Ontologist: Effective Modeling in RDFS and OWL. Morgan Kaufmann Publishers Inc., San Francisco (2011). ISBN: 9780123859662
3. Buil-Aranda, C., Hogan, A., Umbrich, J., Vandenbussche, P.-Y.: SPARQL web-querying infrastructure: ready for action? In: Alani, H., et al. (eds.) ISWC 2013. LNCS, vol. 8219, pp. 277–293. Springer, Heidelberg (2013). doi:10.1007/978-3-642-41338-4_18
4. Auer, S., Demter, J., Martin, M., Lehmann, J.: LODStats – an extensible framework for high-performance dataset analytics. In: Teije, A., Völker, J., Handschuh, S., Stuckenschmidt, H., d'Acquin, M., Nikolov, A., Aussenac-Gilles, N., Hernandez, N. (eds.) EKAW 2012. LNCS (LNAI), vol. 7603, pp. 353–362. Springer, Heidelberg (2012). doi:10.1007/978-3-642-33876-2_31

5. Ermilov, I., Martin, M., Lehmann, J., Auer, S.: Linked open data statistics: collection and exploitation. In: Klinov, P., Mouromtsev, D. (eds.) KESW 2013. CCIS, vol. 394, pp. 242–249. Springer, Heidelberg (2013). doi:10.1007/978-3-642-41360-5_19
6. Greenberg, E.M.R., Bueno, J.G., de la Fuente, T., Baker, P.-Y.V., Vatant, B.: Requirements for vocabulary preservation and governance. Libr. Hi Tech **31**(4), 657–668 (2013)
7. Maali, F., Erickson, J., Archer, P.: Data catalog vocabulary (dcat). In: W3C Recommendation (2014)
8. Ngonga Ngomo, A.-C., Auer, S.: Limes - a time-efficient approach for large-scale link discovery on the web of data. In: Proceedings of IJCAI (2011)
9. Volz, J., Bizer, C., Gaedke, M., Kobilarov, G.: Discovering and maintaining links on the web of data. In: Bernstein, A., Karger, D.R., Heath, T., Feigenbaum, L., Maynard, D., Motta, E., Thirunarayan, K. (eds.) ISWC 2009. LNCS, vol. 5823, pp. 650–665. Springer, Heidelberg (2009). doi:10.1007/978-3-642-04930-9_41

Zhishi.lemon: On Publishing Zhishi.me as Linguistic Linked Open Data

Zhijia Fang[1]([✉]), Haofen Wang[1], Jorge Gracia[2],
Julia Bosque-Gil[2], and Tong Ruan[1]

[1] East China University of Science and Technology, Shanghai 200237, China
fzjacky@mail.ecust.edu.cn, {whfcarter,ruantong}@ecust.edu.cn
[2] Ontology Engineering Group, Universidad Politécnica de Madrid,
Campus de Montegancedo, Boadilla del Monte, 28660 Madrid, Spain
{jgracia,jbosque}@fi.upm.es

Abstract. Recently, a growing number of linguistic resources in different languages have been published and interlinked as part of the Linguistic Linked Open Data (LLOD) cloud. However, in comparison to English and other prominent languages, the presence of Chinese in such a cloud is still limited, despite the fact that Chinese is the most spoken language worldwide. Publishing more Chinese language resources in the LLOD cloud can benefit both academia and industry to better understand the language itself and to further build multilingual applications that will improve the flow of data and services across countries. In this paper we describe Zhishi.lemon, a newly developed dataset based on the *lemon* model that constitutes the lexical realization of Zhishi.me, one of the largest Chinese datasets in the Linked Open Data (LOD) cloud. Zhishi.lemon combines the *lemon* core with the *lemon* translation module in order to build a linked data lexicon in Chinese with translations into Spanish and English. Links to BabelNet (a vast multilingual encyclopedic resource) have been provided as well. We also present a showcase of this module along with the technical details of transforming Zhishi.me to Zhishi.lemon. The dataset is accessible on the Web for both humans (via a Web interface) and software agents (with a SPARQL endpoint).

Keywords: Linked data · Translation · Multilingualism

1 Introduction

With the development of the Semantic Web, a growing number of structured data in the form of RDF triples have been published and interlinked together as Linked Open Data (LOD) on the Web. Among them, Zhishi.me [10] constitutes

This work was supported by the 863 plan of China Ministry of Science and Technology (No: 2015AA020107), the National Science Foundation of China (No: 61402173), Software and Integrated Circuit Industry Development Special Funds of Shanghai Economic and Information Commission (No: 140304), the LIDER FP7 EU project, and the Spanish Government through the Juan de la Cierva and the FPU programs.

© Springer International Publishing AG 2016
P. Groth et al. (Eds.): ISWC 2016, Part II, LNCS 9982, pp. 47–55, 2016.
DOI: 10.1007/978-3-319-46547-0_6

the first effort to publish Chinese knowledge into the LOD cloud at a large scale. It gathers RDF triples from the three largest Chinese encyclopedic Web sites: Baidu Baike[1], Hudong Baike[2] and Chinese Wikipedia[3].

Recently, there has been a growing trend in publishing language resources (LRs) and interlinking them as part of the Linguistic Linked Open Data (LLOD) cloud[4]. The motivation is to have richer linguistic information accessible on the Web of Data, so that it can be consumed by a new generation of linked data-aware natural language processing (NLP) tools and services. In this context, the *lemon* model (LExicon Model for ONtologies) [9] was designed to bridge the gap between lexical and conceptual information, being now a de-facto standard for representing and publishing lexical resources as linked data on the Web. The *lemon* model has been used to expose bilingual and multilingual dictionaries on the Web of Data [2,5], such as the bilingual from the Apertium initiative [4]. Another example is the RDF version of WordNet [8], created and structured according to *lemon*. Also BabelNet [3], a huge multilingual lexical and encyclopedic resource, has been modeled in *lemon* and published as linked data.

Compared to English and other prevalent languages in the LLOD cloud, resources in Chinese are scarce. In this paper, we move a step towards increasing the presence of lexical information in Chinese in the LLOD cloud while linking it to data in other languages (Spanish and English in particular). Concretely, we have built a new dataset, *Zhishi.lemon*, which constitutes the lexical realization of Zhishi.me. The work closest to ours is Chinese WordNet (CWN) [7]. The main difference between the two efforts is that we focus on entity level word translation while CWN emphasizes conceptual word alignment. Further, Zhishi.me contains a larger number of entities than CWN, usually local denominations, making it a suitable candidate to enrich the LLOD cloud with additional Chinese entries.

In our approach, DBpedia[5] and BabelNet[6] are used as a bridge to help identify correspondences between lexical entries in different languages. We combine the *lemon* core with its translation module [6] to build linked data lexicons in Chinese with translations into Spanish and English. Additional descriptions and lexical relations can be obtained by querying these linked data LRs to enrich the information in Zhishi.me. One advantage is that all the lexical information and translations are external to the original resource (Zhishi.me) so that there is no need to modify it whenever new lexical information is added into Zhishi.lemon. This is consistent with the "semantics by reference" principle followed in *lemon*.

The rest of this paper is organized as follows. Section 2 introduces technical details about linking Zhishi.me to the resources in the LLOD cloud. Section 3 gives an overview of the ontology we designed. Section 4 shows the access mechanisms and experiment results and we conclude the paper in Sect. 5.

---

[1] http://baike.baidu.com/.

[2] http://www.baike.com/.

[3] https://zh.wikipedia.org/.

[4] http://linguistic-lod.org/llod-cloud.

[5] http://dbpedia.org/.

[6] http://babelnet.org/.

2 Linking Zhishi.me to DBpedia and BabelNet

In this section, we introduce the approach used to link Zhishi.me to two wide-spread resources in the LLOD cloud, namely DBpedia (its Spanish and English portions in particular) and BabelNet. Detecting equivalences among them can help identify translations between entity labels[7] expressed in different languages (Chinese, Spanish and English). However, it is impossible to manually align these three large datasets: not only it requires experts who are proficient in all the three languages, but it also needs great human labor due to the numerous resources to be aligned. We present an automatic way to tackle this problem.

DBpedia is making a major impact on the LLOD. In order to link Chinese resources in Zhishi.me to Spanish and English DBpedia, we turn to the cross-lingual equivalence relations in DBpedia to retrieve the corresponding translations. Since Zhishi.me includes Chinese Wikipedia as one of its sources and interlinks its resources to the equivalent ones in the other two sources (Baidu Baike and Hudong Baike), Chinese Wikipedia serves as a bridge to help detect links from Chinese resources to both their Spanish and English equivalences.

The highly multilingual nature of BabelNet can also be exploited to discover additional equivalences between resources in different languages. However, more than one "Babel synset" can be found for every ambiguous Chinese term. In order to identify the correct "Babel synset", we use the category labels in both BabelNet and Zhishi.me to find the disambiguation result for each Chinese term. BabelNet extracts categories of Wikipedia pages and maps them to WordNet. In [11], we publish the Chinese Linked Open Schema, which has further refined the existing categories in Zhishi.me, so that categories in both Zhishi.me and BabelNet are well-organized and of good quality. Therefore, we can collect a set of category labels from both sources for a given term in Zhishi.me. The larger the overlap between the category set of the term in Zhishi.me and that of the "Babel synset" in BabelNet, the higher the probability that the term can be mapped to that "Babel synset". The "Babel synset" with the largest overlap is then selected as the disambiguation result.

3 Ontology Overview

We create a new lexical dataset, *Zhishi.lemon*, which constitutes the lexical realization of Zhishi.me and contains its translations into other languages. In order to explain the dataset concretely, we divide it into two parts (namely the Chinese lexicalization module and the multilingual translation module) and provide a detailed analysis for each of them in the following two subsections.

3.1 Chinese Lexicalization Module

According to the design of the *lemon* core, each entity label in Zhishi.me is modeled as a *lemon: LexicalEntry* whose *lemon: LexicalSense* points to the appropri-

---

[7] In the following sections, we are using the expressions "labels" of entities and "terms" in an interchangeable way.

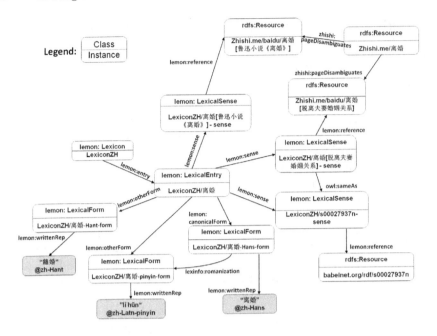

Fig. 1. Chinese lexicalization module

ate ontology reference. Since Zhishi.lemon includes translations among Chinese, Spanish and English, we create a monolingual lexicon for each language which will gradually grow when more resources of the same language are added into Zhishi.lemon. As shown in Fig. 1, we create an instance of *lemon:Lexicon* called "LexiconZH" to gather all Chinese lexical entries.

The *lemon* model assumes that a lexical entry is not semantically disambiguated until an ontology reference provides the semantics of the entry. In Zhishi.me, the title of an article in an encyclopedia site is used as the label of its corresponding entity, and this is the label that we model as a lexical entry. However, labels may be ambiguous, i.e., they can be linked to more than one possible entity. In order to deal with these ambiguities, Zhishi.me uses *zhishi:pageDisambiguates* to represent that a single label refers to more than one entity. As shown in the example presented in Fig. 1, "离婚" can refer to a Chinese novel as well as to a Chinese word with the same sense as "divorce (the legal dissolution of a marriage)" in English. Such word sense disambiguation has been captured in Zhishi.me, so that the label of the subject in a triple describing the *zhishi:pageDisambiguates* relation can be used to create the *lemon:LexicalEntry*, while the object of the triple is its ontology reference.

Given the fact that we are not modifying either Zhishi.me or the other sources (BabelNet, DBpedia), equivalent ontology entities retrieved from DBpedia and BabelNet will be declared at the lexico-semantic layer that Zhishi.lemon describes. In particular, we link the lexical senses that associate a common lexical entry with two semantically equivalent ontology descriptions by using a *owl:sameAs* relation. Figure 1 exemplifies that the Chinese lexical entry " 离婚 " has three possible lexical senses, two of which are describing the same meaning. Accordingly, we treat both of them as equivalent to one another.

We also integrate some features of Chinese itself into our lexicalization model. First of all, both simplified and traditional Chinese characters are standard character sets of contemporary written Chinese. Therefore, it is necessary to model the two different Chinese script variants in Zhishi.lemon. To that end, we propose the use of different language codes in order to distinguish these two different written forms, according to the guidelines provided by W3C[8]: "zh-Hans" to represent simplified Chinese characters while "zh-Hant" for traditional ones.

On the other hand, romanization is another interesting phenomenon presented in the Chinese language. It is a process of transcribing a language into the Latin script. Today, most Chinese use "Hanyu Pinyin" (simply as "pinyin") as a common romanization standard. Since the "pinyin" forms should be included in the model as well, we propose the code "zh-Latn-pinyin" as the language code, which follows the W3C internationalization standards[9].

3.2 Multilingual Translation Module

In Sect. 2, we enrich the multilingual information by discovering alignments with DBpedia and BabelNet. The aligned entity pairs have been used to construct the Zhishi.lemon translation module. Concretely, translation relations can be inferred between terms in different languages when they refer to the same ontology entity. Those lexical senses with an equivalent ontology reference have been regarded as a translation pair to be modeled.

To support the representation of such multilingual information, we use the classes *Translation* and *TranslationSet* in the *lemon* translation module to describe the translation relation. Figure 2 shows our proposed diagram. Lexical entries and their associated properties are used to account for the lexical information, which has been discussed in details in Sect. 3.1. The *TranslationSense* puts the lexical entries from different languages in connection through their lexical senses. *TranslationSet* is designed to group a set of translations, which facilitates querying. For instance, if someone wants to retrieve the Spanish terms of a certain Chinese term, he only needs to query the translation set *tranSet/ES-ZH* instead of searching through the whole dataset.

4 Zhishi.lemon Publication

In this section, we first introduce the online Web access and then give a data statistics of Zhishi.lemon. The data dump is also available via datahub[10].

---

[8] http://www.w3schools.com/tags/ref_language_codes.asp.
[9] https://www.w3.org/International/questions/qa-choosing-language-tags.
[10] https://datahub.io/dataset/zhishi-lemon.

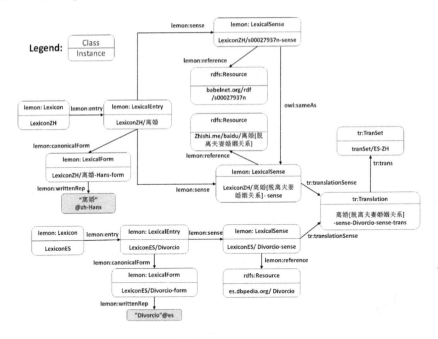

Fig. 2. Multilingual translation module

4.1 IRI Naming Strategy

According to the Linked Data principles, Zhishi.lemon creates IRIs (Internationalized Resource Identifiers) for all resources and provides sufficient information when someone looks up an IRI via the HTTP protocol. Table 1 gives a general view of designed IRI patterns of the dataset. We have followed well established recommendations for this activity [1]. Since Zhishi.lemon consists of a series of linked data lexicons in Chinese, Spanish and English, with translations among them, we use "lexicon" and "tranSet" to indicate the nature of the different resources. In the pattern, [Lang] refers to three possible language marks [ZH], [ES] and [EN] while [LangTag] distinguishes the different character sets used in Chinese. In order to construct the IRIs of the rest of lexical elements, we preserved the labels of the original data in Zhishi.me, denoted as [label], whenever possible, propagating them into the RDF representation. In addition, some other suffixes have been added for the sake of readability: "-form" for lexical forms, "-sense" for lexical senses, and "-trans" for translations. In addition, we have made all the generated information accessible on the Web[11] for both humans (via a Web interface) and software agents (with a SPARQL endpoint).

---

[11] http://lemon.zhishi.me/.

Table 1. IRI patterns

Class	IRI patterns
Lexicon	http://zhishi.me/id/lemon/lexicon[Lang]
Lexical entry	http://zhishi.me/id/lemon/lexicon[Lang]/[label]
Lexical sense	http://zhishi.me/id/lemon/lexicon[Lang]/[label]-sense
Lexical form	http://zhishi.me/id/lemon/lexicon[Lang]/[label]-[LangTag]-form
Translation	http://zhishi.me/id/lemon/tranSet[Lang1]-[Lang2]/[label1]-sense-[label2]-sense-trans
Translation set	http://zhishi.me/id/lemon/tranSet[Lang1]-[Lang2]

4.2 Data Statistics

We first provide a general statistics about Zhishi.lemon. As shown in Table 2, the whole dataset contains 364,765 translations and after its conversion into the *lemon* representation model, 7,036,338 RDF triples were created. Among them, 229,606 resources in Zhishi.me have been found at least one cross-lingual target, in which 218,654 resources come from the English DBpedia, 77,392 from the Spanish DBpedia as well as 16,424 from BabelNet. Since we use a precision-oriented approach for link discovery between Zhishi.lemon and BabelNet, it results in a small number of links. That is to say, the overlap of categories is a hard constraint, so it may filter possible equivalent resources. The precision, however, remains relatively high, which will be evaluated in the next paragraph.

Then, we analyze the quality of the links established between Zhishi.me and BabelNet (with regard to the links to DBpedia, they have been inferred from the multilingual re-directions that Wikipedia contains, so we assume a high quality of such data). Among the 16,424 retrieved translation pairs between Zhishi.me and BabelNet, we select a random subset of 8,536 pairs and ask three students in our laboratory to manually check the quality. Since BabelNet has already provided "sameAs" links between different LRs, we only need to verify whether the Chinese terms in Zhishi.me could be aligned to the corresponding ones in BabelNet, which does not need the annotators to be proficient in all three languages. The experiment shows positive results with an extremely high precision

Table 2. Data statistics

Items	Value
links:BabelNet	16,424
links:DBpedia-en	218,654
links:DBpedia-es	77,392
links:Zhishi.me	229,606
Total translations	364,765
Total triples	7,036,338

Table 3. Comparison with CWN

	CWN	Zhishi.lemon
Word/Lexical Entry	12,726	**215,608**
Sense/Lexical Sense	34,358	**523,585**
Lexical Relation/Translation	47,250	**364,765**

(more than 0.98). After analyzing negative cases, we find that some resources in Zhishi.me may be associated with more than one lexical sense. Here, different lexical senses are included in one resource in Zhishi.me, while its corresponding senses are separated in BabelNet, which leads to a mismatch.

Finally, we compared Zhishi.lemon and CWN. Results are shown in Table 3. First, Zhishi.lemon focuses on translations at entity level instead of conceptual word alignment, as opposed to CWN. We believe that cross-lingual alignment among real-world objects will better benefit the LLOD community. Also, Zhishi.lemon achieves a larger scale in all the three types of elements, namely lexical entries, lexical senses and translations. In this sense, we are convinced that it will greatly help fill the gap between Chinese LRs and the LLOD cloud.

5   Conclusion and Future Directions

In this paper, we introduced Zhishi.lemon, a newly developed dataset that constitutes the lexical realization of Zhishi.me. On the basis of the *lemon* core and its translation module, we built a linked data lexicon in Chinese, with translations into Spanish and English. Links to both DBpedia and BabelNet have also been created. Experiments showed the high quality of Zhishi.lemon, which makes it a promising starting point to respond to the lack of Chinese lexical resources in the LLOD cloud. In the future, we plan to transform more Chinese resources and integrate them into Zhishi.lemon. Identifying new translations to other prevalent languages would be another possible direction. Furthermore, we plan to leverage Zhishi.lemon to build more real-world multilingual applications.

References

1. Archer, P., Goedertier, S., Loutas, N.: Study on persistent URIs. Technical report, December 2012
2. Bosque-Gil, J., Gracia, J., Montiel-Ponsoda, E., Aguado-de Cea, G., Modelling multilingual lexicographic resources for the web of data: the k dictionaries case. In: Proceedings of GLOBALEX'16 workshop at LREC'15, Portoroz, Slovenia, May 2016
3. Ehrmann, M., Cecconi, F., Vannella, D., McCrae, J.P., Cimiano, P., Navigli, R., Representing multilingual data as linked data: the case of BabelNet 2.0. In: Proceedings of LREC 2014, Reykjavik, Iceland, ELRA, May 2014
4. Forcada, M.L., Ginestí-Rosell, M., Nordfalk, J., O'Regan, J., Ortiz-Rojas, S., Pérez-Ortiz, J.A., Sánchez-Martínez, F., Ramírez-Sánchez, G., Tyers, F.M.: Apertium: a free/open-source platform for rule-based machine translation. Mach. Transl. **25**(2), 127–144 (2011)
5. Gracia, J.: Multilingual dictionaries and the web of data. Kernerman Dictionaries News **23**, 1–4 (2015)
6. Gracia, J., Montiel Ponsoda, E., Vila Suero, D., Aguado de Cea, G.: Enabling language resources to expose translations as linked data on the web (2014)
7. Lee, C.-Y., Hsieh, S.-K.: Linguistic linked data in chinese: The case of chinese wordnet. ACL-IJCNLP 2015, p. 70 (2015)

8. McCrae, J., Fellbaum, C., Cimiano, P.: Publishing and linking wordnet using lemon and rdf. In: Proceedings of LDL 2014 (2014)
9. McCrae, J.P., Aguado de Cea, G., Buitelaar, P., Cimiano, P., Declerck, T., Gómez-Pérez, A., Gracia, J., Hollink, L., E. Montiel - Ponsoda, D. Spohr, T. Wunner. Interchanging lexical resources on the semantic web. LREJ **46**(4), 701–719 (2012)
10. Niu, X., Sun, X., Wang, H., Rong, S., Qi, G., Yu, Y.: Zhishi.me - weaving chinese linking open data. In: Aroyo, L., Welty, C., Alani, H., Taylor, J., Bernstein, A., Kagal, L., Noy, N., Blomqvist, E. (eds.) ISWC 2011. LNCS, vol. 7032, pp. 205–220. Springer, Heidelberg (2011). doi:10.1007/978-3-642-25093-4_14
11. Wang, H., Wu, T., Qi, G., Ruan, T.: On publishing chinese linked open schema. In: Mika, P., Tudorache, T., Bernstein, A., Welty, C., Knoblock, C., Vrandečić, D., Groth, P., Noy, N., Janowicz, K., Goble, C. (eds.) ISWC 2014. LNCS, vol. 8796, pp. 293–308. Springer, Heidelberg (2014). doi:10.1007/978-3-319-11964-9_19

Linked Disambiguated Distributional Semantic Networks

Stefano Faralli[1]([✉]), Alexander Panchenko[2],
Chris Biemann[2], and Simone P. Ponzetto[1]

[1] Data and Web Science Group, University of Mannheim, Mannheim, Germany
{stefano,simone}@informatik.uni-mannheim.de
[2] Language Technology Group, TU Darmstadt, Darmstadt, Germany
{panchenko,biem}@lt.informatik.tu-darmstadt.de

Abstract. We present a new hybrid lexical knowledge base that combines the contextual information of distributional models with the conciseness and precision of manually constructed lexical networks. The computation of our count-based distributional model includes the induction of word senses for single-word and multi-word terms, the disambiguation of word similarity lists, taxonomic relations extracted by patterns and context clues for disambiguation in context. In contrast to dense vector representations, our resource is human readable and interpretable, and thus can be easily embedded within the Semantic Web ecosystem.

Resource Type: Lexical Knowledge Base
Permanent URL: https://madata.bib.uni-mannheim.de/id/eprint/171

1 Introduction

Recent years have witnessed an impressive amount of work on the automatic construction of wide-coverage knowledge resources from Wikipedia [3,13] and the Web [7]. Complementary to this, a plethora of works in Natural Language Processing (NLP) has recently focused on combining knowledge bases with distributional information from text. These include approaches that modify Word2Vec [15] to learn sense embeddings [5], methods to enrich WordNet with embeddings for synsets and lexemes [21], acquire continuous word representations by combining random walks over knowledge bases and neural language models [11], or produce joint lexical and semantic vectors for sense representation from text and knowledge bases [4].

In this paper, we follow this line of research and take it one step forward by producing a hybrid knowledge resource, which combines symbolic and statistical meaning representations while (i) staying purely on the lexical-symbolic level, (ii) explicitly distinguishing word senses, and (iii) being human readable. Far from being technicalities, such properties are crucial to be able to embed a resource of this kind into the Semantic Web ecosystem, where human-readable

© Springer International Publishing AG 2016
P. Groth et al. (Eds.): ISWC 2016, Part II, LNCS 9982, pp. 56–64, 2016.
DOI: 10.1007/978-3-319-46547-0_7

distributional representations are explicitly linked to URIfied semantic resources. To this end, we develop a methodology to automatically induce distributionally-based semantic representations from large amounts of text, and link them to a reference knowledge base. This results in a new knowledge resource that we refer to as a *hybrid aligned resource*.

2   Building a Hybrid Aligned Resource

Our resource is built in three main phases (Fig. 1):

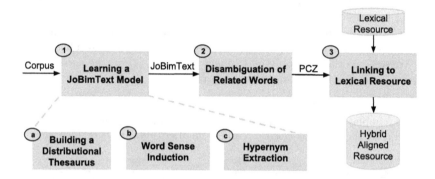

Fig. 1. Overview of our method for constructing a hybrid aligned resource.

(1) **Learning a JoBimText model:** initially, we automatically create a sense inventory from a large text collection using the pipeline of the JoBim-Text project [2,22][1]. The resulting structure contains disambiguated proto-concepts (i.e. senses), their similar and related terms, as well as aggregated contextual clues per proto-concept (Table 1a). This is a distributionally-based conceptualization with some degree of taxonomic information only. Hence the two subsequent phases, which, together with the final resource, represent the novel contribution of this paper.

(2) **Disambiguation of related terms:** we fully disambiguate all lexical information associated with a proto-concept (i.e. similar terms and hypernyms) based on the partial disambiguation from step 1). The result is a proto-conceptualization (PCZ). In contrast to a term-based distributional thesaurus (DT), a PCZ consists of sense-disambiguated entries, i.e. all terms have a sense identifier (Table 1b).

(3) **Linking to a lexical resource:** we align the PCZ with an existing lexical resource (LR). That is, we create a mapping between the two sense inventories and then combine them into a new extended sense inventory, our *hybrid aligned resource*.

---

[1] http://www.jobimtext.org.

2.1 Learning a JoBimText Model

Following [2], we apply a holing operation where each observation in the text is split into a term and its context. The 1000 most significant contexts per term, as determined by the LMI significance measure [8], serve as a representation for the term, and term similarity is defined as the number of common contexts. This procedure induces a weighted similarity graph over terms, also known as Distributional Thesaurus (DT), where each entry of the DT consists of the most similar 200 terms for a given term.

In DTs, entries of polysemous terms t are mixed, i.e. they contain similar terms stemming from several senses respectively usages of the term. Since terms that belong to the same sense are more similar to each other than to terms belonging to a different sense, we can employ graph clustering to partition the open neighbourhood of t in the DT (i.e., terms similar to t and their similarities, without t) to arrive at sense representations for t, characterized by a list of similar terms. We achieve this by applying Chinese Whispers [1] on the ego-network of the term t, as defined by its similar terms as nodes.

Table 1. Examples of entries for "mouse" and "keyboard" from the *news-p1.6* dataset before and after the semantic closure. Trailing numbers indicate sense identifiers.

entry	similar terms	hypernyms	context clues
mouse:NN:0	rat:NN, rodent:NN, monkey:NN, ...	animal:NN, species:NN, ...	rat:NN:conj_and, white-footed:JJ:amod, ...
mouse:NN:1	keyboard:NN, computer:NN, printer:NN ...	device:NN, equipment:NN, ...	click:NN:-prep_of, click:NN:-nn, ...
keyboard:NN:0	piano:NN, synthesizer:NN, organ:NN ...	instrument:NN, device:NN, ...	play:VB:-dobj, electric:JJ:amod, ..
keyboard:NN:1	keypad:NN, mouse:NN, screen:NN ...	device:NN, technology:NN ...	computer:NN:nn, qwerty:JJ:amod ...

(a) JoBimText model entries

entry	similar terms	hypernyms	context clues
mouse:NN:0	rat:NN:0, rodent:NN:0, monkey:NN:0, ...	animal:NN:0, species:NN:1, ...	rat:NN:conj_and, white-footed:JJ:amod, ...
mouse:NN:1	keyboard:NN:1, computer:NN:0, printer:NN:0 ...	device:NN:1, equipment:NN:3, ...	click:NN:-prep_of, click:NN:-nn, ...
keyboard:NN:0	piano:NN:1, synthesizer:NN:2, organ:NN:0 ...	instrument:NN:2, device:NN:3, ...	play:VB:-dobj, electric:JJ:amod, ..
keyboard:NN:1	keypad:NN:0, mouse:NN:1, screen:NN:1 ...	device:NN:1, technology:NN:0 ...	computer:NN:nn, qwerty:JJ:amod ...

(b) Proto-conceptualization entries

Further, we run Hearst patterns [12] over the corpus to extract IS-A (hypernym) relations between terms. We add these hypernyms to senses by aggregating IS-A relations over the list of similar terms for the given sense into a weighted list of hypernyms. Additionally, we aggregate the significant contexts of similar terms per sense to arrive at weighted aggregated context clues. The resulting structure is called the JoBimText model [22] of the corpus. A JoBimText entry consists of a distributionally-induced word sense, a ranked list of similar terms for this sense, a list of superordinate terms and a list of aggregated context clues (note that only unstructured text is required). Table 1(a) shows some JoBimText entries for the polysemous terms "mouse" and "keyboard".

2.2   Disambiguation of Related Terms

While JoBimText models contain sense distinctions, they are not fully disambiguated: the list of similar and hypernyms terms of each sense does not carry sense information. In our example (Table 1a) the sense of "mouse:NN" for the entry "keyboard:NN:1" could either be the "animal" or the "electronic device" one. Consequently, we next apply a semantic closure procedure to arrive at a PCZ in which *all* terms get assigned a unique, best-fitting sense identifier (Table 1b).

At its heart, our method assigns each target word w to disambiguate – namely, a similar and superordinate term from each sense of the JoBimText model – the sense $\hat{s}$ whose context (i.e., the list of of similar or superordinate terms) has the maximal similarity with the target word's context (i.e., the other words in the target word's list of similar or superordinate items) – we use cosine as similarity metric:

$$\hat{s} = \operatorname*{argmax}_{s \in Senses_{JoBim}(w)} cos(ctx(w), ctx(s)). \tag{1}$$

This way we are able to link, for instance, *keyboard:NN* in the list of similar terms for *mouse:NN:1* to its 'device' sense (*keyboard:NN:1*), since *mouse:NN:1* and *keyboard:NN:1* share a large amount of terms from the IT domain.

The structure of a PCZ resembles that of a lexical semantic resource: each term has a list of proto-concepts, and proto-concepts are linked via relations, such as similarity and taxonomic links. Sense distinctions and distributions are dependent on the underlying corpus, which causes the PCZ to naturally adapt to the domain of the corpus. A large difference to manually created resources, however, is the availability of aggregated context clues that allow to disambiguate polysemous terms in text with respect to their sense distinctions. Table 1(b) shows example proto-concepts for the terms "mouse:NN" and "keyboard:NN", taken from our *news-p1.6* PCZ (see Sect. 3.1).

2.3   Linking to a Lexical Resource

We next link each sense in our proto-conceptualization (PCZ) to the most suitable sense (if any) of a Lexical Resource (LR, see Fig. 1 step 3). Our method takes as input:

1. a PCZ $T = \{(j_i, R_{j_i}, H_{j_i})\}$ where j_i is a sense identifier (i.e. *mouse:NN:1*), R_{j_i} the set of its semantically related senses (i.e. $R_{j_i} = \{keyboard:NN:1,$ *computer:NN:0*, ...} and H_{j_i} the set of its hypernym senses (i.e. $H_{j_i} = \{equipment:NN:3, ... \}$;
2. a LR W: we experiment with: WordNet [10], a lexical database for English and BabelNet [16], a very large multilingual "encyclopaedic dictionary";
3. a threshold th over the similarity between pairs of concepts and a number m of iterations as stopping criterion;

and outputs a mapping M, which consists of a set of pairs of the kind (*source, target*) where *source* $\in$ *T.senses* is a sense of the input PCZ T and

$target \in W.senses \cup source$ is the most suitable sense of W or *source* when no such sense is available. At its heart, the mapping algorithm compares the senses across resources with the following similarity function:

$$sim(j, c, M) = \frac{|T.BoW(j, M, W) \cap W.BoW(c)|}{|T.BoW(j, M, W)|}, \text{ where:} \tag{2}$$

1. $T.BoW(j, M, W)$ is the set of words containing all the terms extracted from related/hypernym senses of j and all the terms extracted from the related/hypernym (i.e., already linked in M) synsets in W. For each synset we use all synonyms and content words of the gloss.
2. $W.BoW(c)$ contains the synonyms and the gloss content words for the synset c and all the related synsets of c.

A new link pair (j, c) is then added to M if the similarity score between j and c is greater than or equal to a threshold th. Finally, all unlinked j of T, i.e. proto-concepts that have no corresponding LR sense, are added to the mapping M. We follow the guidelines from McCrae et al. [14] and create an RDF representation to share the mapping between our PCZs and lexical knowledge graphs (i.e., WordNet and BabelNet) in the Linked Open Data Cloud [6].

3 Experiments

3.1 Datasets for the Extraction of the Proto-Conceptualizations (PCZs)

We experiment with two different large corpora, namely a 100 million sentence news corpus (*news*) from Gigaword [17] and LCC [19], and with a 35 million sentence Wikipedia corpus (*wiki*) and different parametrizations of the sense induction algorithm to obtain five proto-conceptualizations (PCZ) with different average sense granularities. Further, we use the method described in [20] to compute a dataset that includes automatically extracted multiword terms. In Table 2, we present figures for our five datasets. For each dataset, Columns 3, 4 and 5 count the overall number of words, including monosemous words and polysemous ones, respectively. For each PCZ we report the cardinality (Column 6), the average polysemy (Column 7) and the maximum polysemy (Column 8). Finally, we report the overall and the average number of related senses and hypernyms (Column 9–12).

3.2 Experiment 1: Disambiguation of the Distributional Thesaurus Entries

Experimental setting. In order to disambiguate a related or superordinate term t in a word sense entry s in the JoBimText model, we compare the related words of s with the related words of each of the senses t_s for the target term t. Similarly, we evaluate the quality of the disambiguation of the JoBim-Text models by judging the compatibility of the similar words for s and t_s.

Table 2. Structural analysis of our five proto-conceptualizations (PCZs).

dataset	n	words			senses	polysemy		rel. senses		hypernyms	
		#	monosemous	polysemous	#	avg.	max	#	avg.	#	avg.
news-p1.6	200	207 k	137 k	69 k	332k	1.6	18	234 k	63.9	15 k	6.9
news-p2.3	50	200 k	99 k	101 k	461 k	2.3	17	298 k	44.3	15 k	5.8
wiki-p1.8	200	206 k	120 k	86 k	368 k	1.8	15	300 k	59.3	15 k	4.4
wiki-p6.0	30	258 k	44 k	213 k	1.5 M	6.0	36	811 k	16.9	52 k	1.7
wiki-mw-p1.6	200	465 k	288 k	176 k	765 k	1.6	13	662 k	46.6	30 k	3.2

For instance, the similar term *mouse:NN*, in the JoBimText model entry for keyboard:NN:1, namely "keypad:NN, *mouse:NN*, screen:NN, ..." is compatible with the related words "keyboard:NN, computer:NN, printer:NN ..." (i.e., those of sense mouse:NN:1) and is not compatible with the related words "cat:NN, rodent:NN, monkey:NN, ..." of mouse:NN:0 (see Table 1).

Our experimental setting is based on three steps: (1) we manually select 17 highly ambiguous *target words*; (2) we collect 19,774 disambiguated entries of the *wiki-p1.6* JoBimText model where the target words appear and randomly sample 15 % of these entries to make annotation feasible; (3) we manually judge entries in the sample on whether the related words of the target word fits the sense assigned or not[2]. Finally, we compute performance by means of standard accuracy – i.e., the proportion of cases in which the similar or hypernym terms from the JoBimText model are correctly disambiguated.

Results and discussion. Our method achieves an accuracy of 0.84 across all parts of speech, including accuracy of 0.94 for nouns, 0.85 for proper nouns, 0.76 for verbs, and 0.63 for adjectives. Different results across parts of speech are due to the different quality of the respective DT clusters. This is because this first experiment also indirectly measures the quality of the senses from the JoBimText model: indeed, a match between two word sense entries is only possible if both of them are interpretable.

To better understand the amount of spurious items in our sense clusters, we performed an additional manual evaluation where, for a sample of 100 randomly sampled noun PCZ items, we counted the ratio between wrong (e.g., *rat* for the computer sense of *mouse*) and correct (*keyboard, computer*, etc.) related concepts that were found within the PCZs. We obtained a macro average of 0.0495 and a micro average of 0.0385 wrong related concepts within the PCZs. Moreover, 83 % of the above sample has no unrelated concepts, and only 2 % has only one unrelated concept with a macro average ratio between the wrong and corrects related PCZ of 0.067. This indicates that, overall, the amount of spurious concepts within clusters is indeed small, thus providing a high-quality context for an accurate disambiguation of noun DT clusters.

---

[2] The target words and annotations can be found at https://goo.gl/jjdhl4.

3.3   Experiment 2: Linking to Lexical Knowledge Bases

Experimental setting. Next, we evaluate the performance of our linking component (Sect. 2.3). For this, we choose two lexical-semantic networks: WordNet [10], which has a high coverage on English common nouns, verbs and adjectives, and BabelNet [16], which also includes a large amount of proper nouns as well as senses gathered from multiple other sources, including Wikipedia.

We follow standard practices (e.g., [16]) and create five evaluation test sets, one for each dataset from Sect. 3.1, by randomly selecting a subset of 300 proto-concepts for each dataset, and manually establishing a mapping from these senses to WordNet and BabelNet concepts (proto-concepts that cannot be mapped are labeled as such in the gold standard). The quality and correctness of the mapping is estimated as accuracy on the ground-truth judgments, namely the amount of true mapping decisions among the total number of (potentially, empty) mappings in the gold standard. We also evaluate our mapping by quantifying Coverage (the percentage of senses of the knowledge base sense inventory covered by the mapping M) and ExtraCoverage (the ratio of concepts in M not linked to the knowledge base sense inventory over the total number of knowledge base senses). The latter is a measure of novelty to quantify the amount of senses discovered in T and not represented by the knowledge base: it indicates the amount of 'added' knowledge we gain with our resource based on the amount of proto-concepts that cannot be mapped and are thus included as novel senses.

Table 3. Results on linking to lexical knowledge bases: number of linked proto-concepts (⟰), Coverage, ExtraCoverage and Accuracy for our five datasets.

Dataset	WordNet				BabelNet			
	⟰ senses	Cov.	ExtraCov.	Accuracy	⟰ senses	Cov.	ExtraCov.	Accuracy
news-p1.6	88 k	34.5 %	206.0 %	86.9 %	164 k	1.3 %	2.9 %	81.8 %
news-p2.3	145 k	38.2 %	267.0 %	93.3 %	236 k	1.4 %	3.9 %	85.1 %
wiki-p1.8	91 k	35.9 %	234.7 %	94.8 %	232 k	1.9 %	2.4 %	86.4 %
wiki-p6.0	400 k	49.9 %	919.9 %	93.5 %	737 k	2.8 %	1.3 %	82.2 %
wiki-mw-p1.6	81 k	30.7 %	581.2 %	95.3 %	589 k	4.7 %	1.8 %	83.8 %

Results and discussion. In Table 3 we present the results using the optimal parameter values (i.e. $th = 0.0$ and $m = 5$)[3]. For all datasets the number of linked senses, Coverage and ExtraCoverage are directly proportional to the number of entries in the dataset – i.e., the finer the concept granularity, as given by a lower sense clustering n parameter, the lower the number of mapped senses, Coverage and ExtraCoverage.

---

[3] To find optimal value for m, we prototyped our approach on a dev set consisting of a random sample of 300 proto-concepts, and studied the curves for the number of linked proto-concepts to WordNet resp. BabelNet. The th value was then selected as to maximize the accuracy.

In general, we report rather low coverage figures: the coverage in WordNet is always lower than 50 % (30 % in one setting) and in BabelNet is in all settings lower than 5 %. Low coverage is due to different levels of granularities between the source and target resource. Our target knowledge bases, in fact, have very fine-grained sense inventories. For instance, BabelNet lists 17 senses of the word "python" including two (arguably obscure ones) referring to particular roller coasters. In contrast, word senses induced from text corpora tend to be coarse and corpus-specific. Consequently, the low coverage comes from the fact that we connect a coarse and a fine-grained sense inventory – cf. also previous work [9] showing comparable proportions between coverage and extra-coverage of automatically acquired knowledge (i.e., glosses) from corpora.

Finally, our results indicate differences between the order of magnitude of the Coverage and ExtraCoverage when linking to WordNet and BabelNet. This high difference depends on the cardinality of the two sense inventories, where Babel-Net has millions of senses while WordNet more than a hundred of thousands – many of them not covered in our corpora. Please note that an ExtraCoverage of about 3 % in BabelNet corresponds to about 300k novel senses. Overall, we take our results to be promising in that, despite the relative simplicity of our approach (i.e., almost parameter-free unsupervised linking), we are able to reach high accuracy figures in the range of around 87 %–95 % for WordNet and accuracies above 80 % for BabelNet.

4   Conclusions

We presented an automatically-constructed hybrid aligned resource that combines distributional semantic representations with lexical knowledge graphs. To the best of our knowledge, we are the first to present such a large-scale, fully URIfied hybrid aligned resource with high alignment quality. As future work, we will explore ways to couple focused crawling [18] with domain-specific PCZs to extend our resource to many domains. Moreover, we aim at using our linguistically-grounded hybrid resource to provide generalizations beyond concepts, such as, for instance, hybrid symbolic and distributional representations of actions and events.

Acknowledgments. We acknowledge the support of the Deutsche Forschungsgemein-schaft (DFG) under the JOIN-T project.

References

1. Biemann, C.: Chinese Whispers: an efficient graph clustering algorithm and its application to natural language processing problems. In: Proceedings of the TextGraphs, pp. 73–80 (2006)
2. Biemann, C., Riedl, M.: Text: now in 2D! a framework for lexical expansion with contextual similarity. J. Lang. Model. **1**(1), 55–95 (2013)
3. Bizer, C., Lehmann, J., Kobilarov, G., Auer, S., Becker, C., Cyganiak, R., Hellmann, S.: DBpedia - a crystallization point for the web of data. JWS **7**(3), 154–165 (2009)

4. Camacho-Collados, J., Pilehvar, M.T., Navigli, R.: NASARI: a novel approach to a semantically-aware representation of items. In: Proceedings of the NAACL-HLT, pp. 567–577 (2015)

5. Chen, X., Liu, Z., Sun, M.: A unified model for word sense representation and disambiguation. In: Proceedings of the EMNLP, pp. 1025–1035 (2014)

6. Chiarcos, C., Hellmann, S., Nordhoff, S.: Linking linguistic resources: examples from the open linguistics working group. In: Chiarcos, C., Nordhoff, S., Hellmann, S. (eds.) Linked Data in Linguistics - Representing and Connecting Language Data and Language Metadata, pp. 201–216. Springer, Heidelberg (2012)

7. Dong, X., Gabrilovich, E., Heitz, G., Horn, W., Lao, N., Murphy, K., Strohmann, T., Sun, S., Zhang, W.: Knowledge vault: a web-scale approach to probabilistic knowledge fusion. In: Proceedings of the KDD, pp. 601–610 (2014)

8. Evert, S.: The Statistics of Word Cooccurrences: Word Pairs and Collocations. Ph.D. thesis, Institut für maschinelle Sprachverarbeitung, University of Stuttgart (2005)

9. Faralli, S., Navigli, R.: Growing multi-domain glossaries from a few seeds using probabilistic topic models. In: Proceedings of the EMNLP, pp. 170–181 (2013)

10. Fellbaum, C. (ed.): WordNet: An Electronic Database. MIT Press, Cambridge (1998)

11. Goikoetxea, J., Soroa, A., Agirre, E.: Random walks and neural network language models on knowledge bases. In: Proceedings of the NAACL HLT, pp. 1434–1439 (2015)

12. Hearst, M.A.: Automatic acquisition of hyponyms from large text corpora. In: Proceedings of the COLING, pp. 539–545 (1992)

13. Hoffart, J., Suchanek, F.M., Berberich, K., Weikum, G.: YAGO2: a spatially and temporally enhanced knowledge base from Wikipedia. ArtInt, pp. 28–61 (2013)

14. McCrae, J.P., Fellbaum, C., Cimiano, P.: Publishing and linking WordNet using lemon and RDF. In: Proceedings of the 3rd Workshop on Linked Data in Linguistics (2014)

15. Mikolov, T., Sutskever, I., Chen, K., Corrado, G.S., Dean, J.: Distributed representations of words and phrases and their compositionality. In: Proceedings of the NIPS, pp. 3111–3119 (2013)

16. Navigli, R., Ponzetto, S.P.: BabelNet: the automatic construction, evaluation and application of a wide-coverage multilingual semantic network. ArtInt. **193**, 217–250 (2012)

17. Parker, R., Graff, D., Kong, J., Chen, K., Maeda, K.: English Gigaword, 5th edn. Linguistic Data Consortium, Philadelphia (2011)

18. Remus, S., Biemann, C.: Domain-specific corpus expansion with focused webcrawling. In: Proceedings of the LREC (2016)

19. Richter, M., Quasthoff, U., Hallsteinsdóttir, E., Biemann, C.: Exploiting the Leipzig corpora collection. In: Proceedings of the IS-LTC (2006)

20. Riedl, M., Biemann, C.: A single word is not enough: ranking multiword expressions using distributional semantics. In: Proceedings of the EMNLP, pp. 2430–2440 (2015)

21. Rothe, S., Schütze, H.: AutoExtend: extending word embeddings to embeddings for synsets and lexemes. In: Proceedings of the ACL, pp. 1793–1803 (2015)

22. Ruppert, E., Kaufmann, M., Riedl, M., Biemann, C.: JoBimViz: a web-based visualization for graph-based distributional semantic models. In: Proceedings of the ACL-IJCNLP System Demonstrations, pp. 103–108 (2015)

BESDUI: A Benchmark for End-User Structured Data User Interfaces

Roberto García[1(⊠)], Rosa Gil[1], Juan Manuel Gimeno[1], Eirik Bakke[2], and David R. Karger[2]

[1] Computer Science and Engineering Department, Universitat de Lleida,
Jaume II 69, 25001 Lleida, Spain
{rgarcia, rgil, jmgimeno}@diei.udl.cat
[2] Computer Science and Artificial Intelligence Laboratory, MIT,
32 Vassar Street, Cambridge, MA, USA
{ebakke, karger}@mit.edu

Abstract. The Semantic Web Community has invested significant research efforts in developing systems for Semantic Web search and exploration. But while it has been easy to assess the systems' computational efficiency, it has been much harder to assess how well different semantic systems' *user interfaces* help their users. In this article, we propose and demonstrate the use of a benchmark for evaluating such user interfaces, similar to the TREC benchmark for evaluating traditional search engines. Our benchmark includes a set of typical user tasks and a well-defined procedure for assigning a measure of performance on those tasks to a semantic system. We demonstrate its application to two such system, Virtuoso and Rhizomer. We intend for this work to initiate a community conversation that will lead to a generally accepted framework for comparing systems and for measuring, and thus encouraging, progress towards better semantic search and exploration tools.

Keywords: Benchmark · User experience · Usability · Semantic data · Exploration · Relational data

Resource type: Benchmark
Permanent URL: http://w3id.org/BESDUI

1 Introduction

One of the main barriers alleged when justifying the lack of the uptake of the Semantic Web is that it has not reached end-users [1]. The amount of semantic data is growing, through open data initiatives like the Linked Open Data Cloud [2] or motivated by SEO benefits like those provided by major search engines for web pages annotated using schema.org [3]. However, this has not noticeably impacted user applications, for instance by the long sought Killer App for the Semantic Web [4].

It might be argued that this is in fact the desired outcome, that client applications should hide the complexities of semantic technologies and that the benefits should just

© Springer International Publishing AG 2016
P. Groth et al. (Eds.): ISWC 2016, Part II, LNCS 9982, pp. 65–79, 2016.
DOI: 10.1007/978-3-319-46547-0_8

be evident server side. For instance, search engines like Google provide better results thanks to semantic annotations that users never see. This, in fact, should usually be the desired outcome when trying to satisfy specific user needs: the user should be provided the simplest user experience possible [5].

For known tasks, such as managing a music collection or an address book, the simplest possible experience is often provided by a task-specific application with a task-specific interface. In this case, any Semantic Web nature of the underlying data will be hidden behind the familiar interface. But there will be other cases where no familiar application exists to camouflage the underlying semantic web data. A user may need to explore a data collection that is too rarely used to have motivated an application—perhaps because they are the only ones managing data in that particular schema. Or they may be seeking to learn something by combining multiple data collections that are not often combined.

A general example of this is *semantic search*, where a user is presented with some arbitrary semantic web data and seeks to find resources that fit some query. Semantic search tools that must work with arbitrary schema cannot hard-code any particular schema into their interfaces. For this task, tech-savvy users can rely on standards like SPARQL to query available data. But this is beyond the capabilities of most users. And even SPARQL-aware developers have trouble querying unfamiliar data collections because it is hard to get a clear idea about what is available from a semantic dataset [6]. Consequently, we focus on more user-friendly visual query tools.

All kinds of users can benefit from tools that make it possible to visually explore semantic data, showing all its richness while provided a smooth user experience. In this particular scenario we might find the Semantic Web killer app that makes all the power of Web-wide connected data available to common users, so they can discover unforeseen connections in it.

Proposals are very disparate [7], ranging from Linked Data browsers [8] to Controlled Natural Language query engines [9] or faceted browsers [10]. This makes them difficult to compare, especially from the user perspective, for instance what ways of exploring the data they provide and how efficient they are from a Quality in Use perspective [11, 12].

To enable comparing proposals in this domain, a reference framework for benchmarking is clearly required, as discussions in this research domain have already highlighted [13]. Moreover, it has also been shown that benchmarks help organizing and strengthening research efforts in a particular research area [14]. An example is the Text REtrieval Conference (TREC) benchmarks [15] which have become the de facto standard for evaluating any text document retrieval system.

In the context of semantic data exploration there have been already some efforts in specific areas. These include the Intelligent Exploration of Semantic Data Challenge[1] and the Biomedical Semantic Indexing and Question Answering one[2]. However, none of them target the general user task of semantic data exploration, nor provide a complete benchmark that facilitates comparability and competition in this research topic.

---

[1] IESD Challenge, https://iesd2015.wordpress.com/iesd-challenge-2015.

[2] BioASQ, http://www.bioasq.org.

On the other hand, there are many benchmarks for performance evaluation from a system perspective, like the Berlin SPARQL Benchmark (BSBM) [16] to evaluate SPARQL query engines, but they do not take into account the end-user perspective.

In this paper, we present a benchmark for semantic data *(graphical) user interfaces* with a set of user tasks to be completed and metrics to measure the performance of the analyzed interfaces at different levels of granularity. We provide a benchmark not just for semantic-web data exploration, but for structured data more generally. This makes it possible to also compare tools available in more mature domains like relational data-bases [17]. It is well known that semantic web data can be squeezed into a traditional relational (SQL) database, and vice versa. Since the GUI systems we consider are aimed at end users, they generally isolate the user from details of the underlying storage mechanism or query engine. Thus, these interfaces can in theory operate over either type of data (modulo some simple-matter-of-programming data transformations). We also hope to further motivate research in semantic data exploration that goes beyond what is possible with other less rich data models.

In Sect. 2, we present our approach to providing a benchmark for structured data exploration. Then, in Sect. 3, we present the benchmark, which is then put into practice with a couple of faceted browsers in Sect. 4. Finally, the conclusions are presented in Sect. 5 and future work in Sect. 6.

2 Approach

Defining a benchmark requires two main decisions. First, we need to choose the *tasks* that will be benchmarked. Second, we need to decide *what to measure* about the systems as they are used for the chosen tasks. In both parts, our choices influence the fidelity of our benchmark. First, our chosen tasks should be representative of the tasks we expect users to perform. They should cover the common cases, and be neither too hard nor too easy. Second, our performance metrics should provide some suggestion of what real users will experience using the system. At the same time they will be easier to adopt if at least some measurement can be done analytically, without actual expensive user studies. These two choices are the "axioms" of our benchmark system; they cannot be proven correct but must instead be justified by experience and argumentation. We will discuss both in detail in the following two sections. For tasks we begin with (then augment) the *Berlin SPARQL Benchmark*, a set of queries initially intended to serve as a benchmark of computational performance. Our performance measure consider basic user operations such as mouse movements and keyboard clicks under the so-called *Keystroke Level Model* [18] of user interaction.

In choosing tasks, we want to avoid introducing bias from an a priori conception of the problem or experience developing our own tools. Consequently, we have looked outward to find sets of typical end-user tasks related to structured data exploration.

Although our main interest is semantic technologies, we prefer a benchmark that can also be applied to relational-database tools, so we can compare them with semantic tools and highlight pros and cons between them. As discussed in the introduction, visual query tools will insulate the user from details of the underlying storage repre-sentation, meaning RDF or relational databases could equally be used as back-ends.

From existing benchmarks with user tasks a clear candidate emerged: the Berlin SPARQL Benchmark (BSBM). Although this benchmark is intended for measuring the *computational* performance of semantic and relational database query engines, it is based on a set of realistic queries inspired by common information needs in these domains. We can therefore leverage the same queries to measure the *user-interaction* performance of visual query systems. Moreover, it is based on a synthetic dataset and a tool that facilitates its generation for a given target size, facilitating thus the distribution of the benchmark. And the data can be generated as SQL or RDF.

All the user tasks are accompanied by both the SPARQL and SQL query to satisfy them. Though from the perspective of a user experience benchmark these queries are technological details that might not be relevant because users can satisfy the tasks by generating different queries, they might be helpful to verify the outcomes of users' tasks and check they are actually getting the intended result.

Therefore, we adopted the proposed user tasks that motivate the actual SPARQL and SQL queries that conform the Berlin SPARQL Benchmark. The tasks are contextualized in an e-commerce scenario, where different vendors offer a set of products and different consumers have posted reviews about these products.

In fact, there are three different sets of tasks in the BSBM depending on task types. The BSBM Explore set of tasks are directly connected to the proposed benchmark aim. There is a second set of Business Intelligence tasks, which are too complex to be considered in the context of data exploration tasks for the moment. Finally, there are Update tasks, which in the future we hope to use to define a benchmark for users *editing*, rather than *searching* semantic data.

Consequently, the data exploration tasks in BSBM have been used as the starting point for the proposed structured data exploration benchmark from a user experience perspective. These are 12 tasks that illustrate the user experience of a user looking for a product. The tasks are presented in the following subsection.

Note that our goal is not to evaluate ecommerce tools specifically. The intended targets are search tools for arbitrary structured data, so cannot have any e-commerce features hard coded into them. Indeed, this domain is so common there are likely to be domain-specific interfaces for tasks in it. However, ecommerce provides a convenient and intuitive domain in which to define queries we expect users to want to carry out. We are interested in general operations, such as combining two constraints, but for concreteness we provide tasks in our benchmark in e-commerce language.

Our benchmark does not aim to assess discoverability/learnability. We posit a user who is already familiar with the tool being evaluated who knows where to access available operations and how to invoke them. To conclude this section, and before starting to describe each task in detail, it is important to note that the SPARQL and SQL queries associated to each task are not included in this paper due to space constraints but are available from the benchmark repository [19].

3   Structured Data Exploration Benchmark

The proposed benchmark currently consists of 12 end-user tasks to be completed with the evaluated tool, listed in Sect. 3.1. For each task we detail the *information need* and provide some context. Then, we give a sample query based on the sample dataset accompanying the benchmark together with the expected outcome.

The proposed benchmark also includes a set of metrics to measure the effectiveness and efficiency of the evaluated tool when performing each of the proposed tasks. These metrics yield numbers that can be used to compare the performance of structured data exploration tools, as detailed in Sect. 3.2.

3.1   End-User Tasks

The following subsections describe each of the 12 end-user tasks. All but one of them are directly adopted from the Berlin SPARQL Benchmark (BSBM). One additional task, Task 2, has been added as a variation of Task 1 to cover a gap in the original benchmark, OR versus AND operations for combining subqueries.

Although the BSBM presents a particular e-commerce schema, we hold that a true semantic web query tool *cannot* make assumptions about the schema of the data it is to query. It should operate equally well on any data schema it encounters. A tool that hard-wires the BSBM schema into its interface will be useless on a different data set and thus is not a true semantic web tool. The BSBM instantiates one arbitrary schema to let us talk about our queries concretely, but the tool being analyzed should not be permitted advance knowledge of this particular instantiation.

Task 1. Find products for a given set of combined features
A consumer seeks a product that present a specific set of features. The corresponding information need for the benchmark dataset specifies a product type from the product hierarchy (one level above leaf level), two different product features that correspond to the chosen product type and that should be present simultaneously and a number between 1 and 500 for a numeric property. For instance:

> *"Look for products of type **sheeny** with product features **stroboscopes** AND **gadgeteers**, and a **productPropertyNumeric1** greater than **450**".*

For the previous query, and considering the sample BSBM 1000 Products dataset[3], the product labels the user should obtain are:

> *"auditoriums reducing pappies" and "driveled".*

Task 2. Find products for a given set of alternative features
A consumer is seeking a product with a general idea about some alternative features of what he wants. This task has been added beyond those provided by BSBM. It makes Task 1 to less specific by considering feature alternatives; the user is interested in any

---

[3] https://github.com/rhizomik/BESDUI/blob/master/Datasets/bsbm-1000products.ttl.tgz.

product that presents at least one of them. This benchmarks how exploration tools lets users define OR operations. A sample query for this task might be:

> *"List products of type **sheeny** with product features **stroboscopes** OR **gadgeteers**, and a productPropertyNumeric1 greater than 450".*

For the previous query, and considering the sample dataset, the product labels the user should obtain if restricted to the first 5 ordered alphabetically are:

> *"aliter tiredest", "auditoriums reducing pappies", "boozed", "byplay", "closely jerries".*

Task 3. Retrieve basic information about a specific product for display purposes
The consumer wants to view basic information about a specific product. For instance:

> *"Get details about product **boozed**".*

From the entry page, and considering the synthetic dataset generated using the BSMB tool, the response should include the following properties for the selected product with their corresponding values, which are omitted due to space restrictions but available from the benchmark repository[4]:

> *"label", "comment", "producer", "productFeature", "propertyTextual1", "propertyTextual2", "propertyTextual3", "propertyNumeric1", "propertyNumeric2", "propertyTextual4", "propertyTextual5", "propertyNumeric4".*

Task 4. Find products having some specific features and not having one feature
After looking at information about some products, the consumer has a more specific idea what she wants, features the products should have and others that should not. The main feature of this task is the use of negation. A sample query for this task is:

> *"Look for products of type **sheeny** with product features **stroboscopes** but **NOT gadgeteers**, and **productPropertyNumeric1** value greater than **300** and **productPropertyNumeric3** smaller than **400**".*

For this query and the BSBM 1000 dataset, the user should obtain:

> *"boozed", "elatedly fidelis release" and "learnable onomatopoeically".*

Task 5. Find products matching two different sets of features
After looking at information about some products, the consumer has a more specific idea what he wants. Therefore, he asks for products matching either one set of features or another set. The complexity in this case is the union of the sets of products selected by two different patterns. For instance:

> *"Look for products of type **sheeny** with product features **stroboscopes** and **gadgeteers** and a **productPropertyNumeric1** value greater than **300** plus those of the same product type with product features **stroboscopes** and **rotifers** and a **productPropertyNumeric2** greater than **400**".*

For the previous query, and the sample dataset, the product labels the user should obtain if restricted to the first 5 ordered alphabetically are:

---

[4] Task 2: https://github.com/rhizomik/BESDUI/blob/master/Benchmarks/3.md.

"auditoriums reducing pappies", "boozed", "driveled", "elatedly fidelis release", "zellations".

Task 6. Find product that are similar to a given product

The consumer has found a product that fulfills his requirements. She now wants to find products with similar features. The corresponding query starts from a product and looks for all other products with at least one common feature and a wider range of values for two of its numeric properties. For instance:

*"Look for products similar to **boozed**, with at least one feature in common, and a **productPropertyNumeric1** value **between 427 and 627** and a **productPropertyNumeric2** value **between 595 and 895** (150 more or less than its value for boozed, 745)".*

For the previous query, and considering the sample dataset, the product labels the user should obtain if restricted to the first 5 ordered alphabetically are:

"debouches orangs unethically", "dirk professionalize", "grappled", "imposed", "pepperiness gothically shiner".

Task 7. Find products having a name that contains some text

The consumer remembers parts of a product name from former searches. She wants to find the product again by searching for the parts of the name that she remembers. The corresponding query is just one of the words from the list of words[5] that were used during dataset generation by the BSBM Data Generator[6]. For instance:

*"Search products whose name contains **ales**".*

For the previous query, and considering the sample dataset, the product labels the user should obtain if restricted to the first 5 ordered alphabetically are:

"cogitations centralest recasting", "overapprehensively dales ventless", "skidooed finales noisemaker" and "unwed convalescents".

Task 8. Retrieve in-depth information about a specific product including offers and reviews

The consumer has found a product which fulfills his requirements. Now he wants in-depth information about this product including offers from German vendors and product reviews if existent. The corresponding query refers to a selected product and defines a current date within the "valid from" and "valid to" range of the offers. Compared to previous tasks, this one introduces being able to pose restrictions to different model entities that are interrelated, in this case vendors and reviews that are interrelated with products and offers. For instance:

*"For the product **waterskiing sharpness horseshoes** list details for all its offers by German vendors and still valid by **2008-05-28** plus details for all reviews for this product, including values for **rating1** and **rating2** if available".*

---

[5] https://github.com/rhizomik/BESDUI/blob/master/Datasets/titlewords.txt.

[6] http://wifo5-03.informatik.uni-mannheim.de/bizer/berlinsparqlbenchmark/spec/BenchmarkRules/index.html#datagenerator.

Considering the benchmark sample dataset, the user should get access to the details about the following offers and reviews:

"Offer10801", "Offer5335", "Offer10597", "Review5481", "Review7546", "Review2669", "Review5731", "Review8494".

Task 9. Give me recent reviews in English for a specific product

The consumer wants to read the 20 most recent English language reviews about a specific product. The corresponding query refers to a selected product. This task required being able to filter literals language and ordering by date. For instance:

*"For the product **waterskiing sharpness horseshoes** list the 20 more recent reviews in English".*

Given the sample dataset, the user should obtain the details for the following reviews in the order they are listed:

"Review5481", "Review8494" and "Review2669".

Task 10. Get Information about a reviewer

In order to decide whether to trust a review, the consumer asks for any kind of information that is available about the reviewer. The corresponding query refers to a selected product. This tasks explore how easy it is to reach the information about a resource from a related one. For instance:

*"Get all available information about **Reviewer11**".*

For the sample dataset, the user obtains all the details about the following reviewer:

"Reviewer1".

Task 11. Get offers for a given product which fulfill specific requirements

The consumer wants to buy from a vendor in the United States that is able to deliver within 3 days and is looking for the cheapest offer that fulfills these requirements. The corresponding query refers to a selected product and defines a current date within the "valid from" and "valid to" range of the offers. For instance:

*"Look for the cheapest and still valid by **2008-06-01** offer for the product **waterskiing sharpness horseshoes** by a **US** vendor that is able to deliver within 3 days".*

Considering the sample dataset, the user interface should get as a response the following offers:

"Offer3499", "Offer11865" and "Offer15103".

Task 12. Export the chosen offer into another information system which uses a different schema

After deciding on a specific offer, the consumer wants to save information about this offer on his local machine using a different schema. The corresponding query refers to a selected offer, or the one considered by the previous task.

*"Save in the local computer the information about the vendor for **Offer3499**, this is **half** the task. To complete it, restrict the output to just label, homepage and country and **map** them to schema.org terms: name, url and nationality".*

3.2   Metrics

Our benchmark gives a number of generic yet typical information-seeking tasks to be measured. We now ask the following three increasingly detailed questions to measure the effectiveness and efficiency of the tool on these tasks:

1. **Capability** (effectiveness). Is performing the task *possible* with the given system?
2. **Operation Count** (efficiency). *How many basic steps* (mouse clicks, keyboard entry, scrolling) must be performed to carry out the given task?
3. **Time** (efficiency). *How quickly* can these steps be executed to carry out the task?

Presumably, a system can be judged superior to another if it can be used to perform more of the tasks, with fewer basic steps that take less time. Our general target is graphical user interfaces for querying structured data. For contrast, if we consider for example a SPARQL command line, a suitably trained user would be able to perform all benchmark tasks with just a single primitive operation (typing the SPARQL query) in a very small amount of time (leaving out designing and debugging the SPARQL query). But most users don't have the training or understanding necessary to use such a tool. Instead, some type of GUI is the norm, and it is such systems we aim to evaluate.

The first two questions, of capability and operation count, can be answered entirely analytically. They simply require identifying and counting up a sequence of operations that complete each task. Ideally, the third question would be answered by a timed user study. However, conducing user studies is a very time consuming activity, especially because it involves recruiting users. To facilitate the application of the benchmark, our proposed metric relies on past HCI research that offers a way to answer the time question analytically as well, by applying known, analytic timing models for primitive actions (keyboard and mouse operations) in the identified sequence.

In particular, the Keystroke Level Model (KLM) [18] gave experimentally derived timings for basic operations such as typing a key, pointing on the screen with the mouse, moving hands back to the keyboard, and so forth. Given a sequence of these basic operations, we can total up their timings to yield an overall predicted execution time for the task.

Our proposal is to use the main interaction operators proposed by KLM and their mapping to time to define the Operation Count and Time metrics. The first one does not distinguish among operations so it is computed as the sum of the counts of all operations. The considered operators and their mappings to time in seconds to compute the Time metric are shown in Table 1.

Table 1. Mapping from KLM Operators to time

KLM Operator	Time (seconds)
K: button press or keystroke, (keys not characters, so shift-C is two)	0.2
P: pointing to a target on a display e.g. with a mouse. Time differs depending on target distance and size, but is held constant for simplicity.	1.1
H: homing the hand(s) on the keyboard or other device, this includes movement between any two devices.	0.4

4 Benchmark Evaluation with Rhizomer and Virtuoso

To facilitate the adoption of the benchmark, and to evaluate its applicability, we have tested it with two of the most sophisticated faceted browsers for semantic data, which also feature pivoting: Rhizomer [10] and Virtuoso Facets [20]. This way we provide a couple of samples that illustrate how the benchmark works. Due to space restrictions, we provide the results for the first 3 tasks.

Moreover, we have set a GitHub repository[7] for the benchmark that can be forked to contribute results for additional tools, which can be then incorporated into the reference repository through a pull request. Additional details about how to contribute to the benchmark are available from the repository. The whole set of results for both Virtuoso and Rhizomer, and other tools, are available from the repository. For instance, the benchmark has been already applied to SIEUFERD [21], a query construction tool through direct manipulation of nested relational results.

Task 1. Results
Rhizomer does not support this kind of query because when defining the values for a particular facet, like **stroboscopes** and **gadgeteers** for **feature**, it is not possible to specify that both should be available for the same product simultaneously. The Capability metric value is then 0 %.

Virtuoso, as it is shown in Fig. 1, can complete this task and the outcome is the expected considering the sample dataset, the products "driveled" and "auditoriums reducing pappies". To complete this task, the interaction steps and KLM Operators are listed in Table 2, while the Operator Count and Time metrics are shown in Table 4.

Task 2. Results
Rhizomer, as shown in Fig. 2, supports this task because its facets can be used to select more than one of their values as alternatives. The interaction steps and KLM Operators are presented in Table 3 and the Operation Count and Time in Table 4.

Virtuoso also supports this task in a very similar way to Task 1, though in this case alternative feature values are defined using the "Add condition: IN" feature. The results are available from Table 4.

Task 3. Results
Rhizomer supports this task. The user clicks the "Quick search…" input field and types "boozed", then selects the entry for the product from the autocomplete.

Virtuoso also supports this task. The user types "boozed" in the entry page search box then clicks the entry for the requested product in the results listing to get the details.

The outcomes for the rest of the tasks cannot be included in this paper due to space restrictions but are available from the benchmark repository (see Footnote 7). From the operation count and their classification following the KLM model detailed in, it is possible to compute the numbers that measure their performance using the metrics

---

[7] https://github.com/rhizomik/BESDUI.

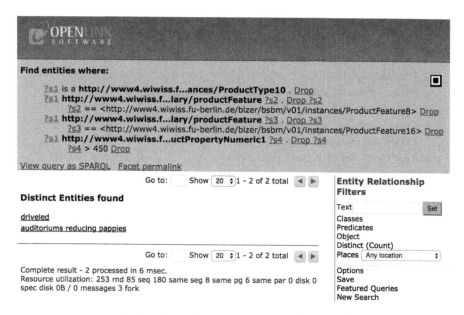

Fig. 1. Using Virtuoso Facets to complete Task 1.

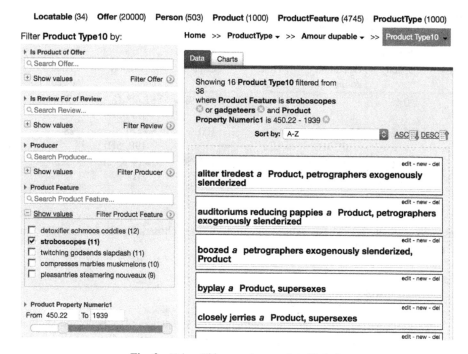

Fig. 2. Using Rhizomer to complete Task 2.

Table 2. Interaction steps and corresponding KLM operators to complete Task 1 using Virtuoso

1. Type "sheeny" and "Enter", then click "ProductType10"	9 K, 2P, 3H
2. Click "Go" for "Start New Facet", then click "Options"	2 K, 2P
3. For "Interence Rule" Click and Select rules graph then "Apply"	2 K, 2P
4. Click "Attributes", then "productFeature" and "stroboscopes"	3 K, 3P
5. Click "Attributes", then "productFeature" and "gadgeteers"	3 K, 3P
6. Click "Attributes" and "productPropertyNumeric1"	2 K, 2P
7. Click "Add condition: None" and select ">"	2 K, 2P
8. Type "450" and click "Set Condition"	5 K, 2P, 2H

Table 3. Interaction steps and KLM operators to complete Task 2 with Rhizomer

1. Click menu "ProductType" and then "Sheeny" submenu	2 K, 2P, 1H
2. Click "Show values" for facet "Product Feature"	1 K, 1P
3. Click facet value "stroboscopes"	1 K, 1P
4. Type in input "Search Product Feature" "gad..."	4 K, 1P, 1H
5. Select "gadgeteers" from autocomplete	1 K, 1P, 1H
6. Set left side of "Product Property Numeric1"slider to "450"	1 K, 2P

Table 4. Summary of benchmark results for Rhizomer and Virtuoso for Tasks 1 to 3, more results available from the BESDUI Repository (see Footnote 7)

Task	Rhizomer			Virtuoso		
	Capacity	Operation count	Time (seconds)	Capacity	Operation count	Time (seconds)
1	0 %	–	–	100 %	**51** (28 K, 18P, 5H)	27.4
2	100 %	**21** (10 K, 8P, 3H)	12.0	100 %	**53** (29 K, 19P, 5H)	28.7
3	100 %	**13** (8 K, 2P, 3H)	5.0	100 %	**14** (9 K, 2P, 3H)	5.2

presented in Sect. 3.2. These values constitute the benchmark evaluation results and are presented in Table 4.

5   Conclusions

As already shown in other research domains, the existence of a benchmarks that facilitate comparing contributions related to a specific research challenge helps foster efforts in that particular domain and clarify the scope of the contributions. In the domain of semantic data exploration and search, there are many proposed tools and surveys but there is not a benchmark to compare them from a user experience perspective.

Our proposal is based on a set of user tasks, most of them borrowed from the Berlin SPARQL Benchmark (BSBM), to be completed using the evaluated tools. Though this tasks are originally conceived to test SPARQL engines' performance, they are very well contextualized in the e-commerce domain, cover a wide range of information needs and are accompanied by a synthetic dataset generator that facilitates the distribution of the benchmark and its deployment, even for very big testing datasets.

Though the dataset is synthetic and, for instance, many resources present funny names like "waterskiing sharpness horseshoes", it is important to note that this does not introduce any significant effect from the user experience perspective when measured using the KLM-based metric. Moreover, it is also important to note that, as testing with real users is very costly and time consuming, the benchmark is based on analytical methods and therefore require only the involvement of a researcher experienced in semantic data exploration tools.

The metrics are Capability, an effectiveness metric that measures if a task can be completed or not using an evaluated tool, Operation Count, an efficiency metric counting how many Keystroke Level Model Operators are required to complete the task, and Time, another efficiency metric that translates the KLM Operators required to complete the task into an approximate amount of time.

Based on the 12 proposed tasks and the 3 metrics, the application of the benchmark consists of trying to complete each of the tasks using the evaluated tool. This can be done without having to recruit test users, an experienced user capable of using the tool is enough. The user record the interaction steps and then translates them to KLM Operators: K for keystrokes or button press, P for pointing to a target with the mouse and H for homing the hands on the keyboard or other device. The amount of operators of each kind should be counted to be able to compute the efficiency metrics. Operator Count is just the total amount of operators needed while Time is the conversion of the operators to a time measure, where K amounts 0.2 s, P 1.1 and H 0.4 s.

The benchmark has been tested with two of the most sophisticate semantic data faceted browsers, Virtuoso Facets and Rhizomer. Just the results for the first three tasks including all the metrics are included in this paper, the rest are available from a GitHub repository intended for maintaining the benchmark, keeping track of evaluations and organizing contributions.

Another interesting effect of using BSBM is that it also provides the benchmark dataset in a format suitable for relational databases so it is possible to use the benchmark to compare semantic-based and relational-based structured data exploration tools. These benchmark results are also available online (see Footnote 7).

6   Future Work

The main objective of this contribution is to foster the formation of a community around the evaluation and comparison of tools for structured data exploration. Consequently, we have prepared a GitHub repository where all the required elements to conduct an evaluation are available. This includes a sample dataset, the descriptions of the tasks, reference SPARQL and SQL queries to test expected responses, descriptions of the metrics and templates to report results.

We plan to add contributing instructions based on common practice in GitHub that encourage forking the repository, making contributions like new evaluation results based on the templates and then doing a pull request to incorporate them in the reference repository. We also expect contributions like additional tasks or metrics, which will be also considered for inclusion. We plan to trigger this community building process by proposing the benchmark as the way to evaluate submission to the IESD Workshop Challenge[1], which some of the authors co-organize.

In addition to this expected community building efforts and results, our plans also include concrete tools to be evaluated, metrics and tasks to consider. First of all, we are currently exploring the tasks also proposed in the BSBM in the Business Intelligence scenario. Though these tasks are much more complex than the Explore tasks, some of them might be interesting to test with tools featuring visualizations.

We also plan to test many more tools using the benchmark, ranging from semantic data tools providing direct manipulation (SParallax, Explorator, Tabulator,…) to tools that facilitate building queries interactively (YASGUI, iSPARQL, AutoSPARQL,…) or relational data exploration tools (Cipher, BrioQuery,…).

Our performance metrics currently emphasize low-level basic operations such as keystrokes and mouse clicks. These can be refined. For example, Fitts's law has related the time to execute a mouse operation to the target regions size and from it's the mouse starting point; this can be incorporated into our timing analysis. At a higher level, our benchmark currently does not capture user effort. An interface that requires the user to think hard about which operation to perform next (and how to do it) will be more taxing and take more time. As discussed above, and extreme model of this is the SPARQL command line, which is extremely efficient in the KLM because all the work is mental, figuring out what SPARQL query to time. Similarly, our benchmark favors complicated UI layouts where all actions are "one click away", neglecting the fact that Fitts' Law indicates that actually *selecting* these actions becomes slower. The KLM model does not capture this, but there are so-called GOMS [22] models that begin to.

To fully evaluate the usefulness of the proposed efficiency metrics, we will also test the systems using user experience evaluation techniques that involve end-users and include measuring the real time users need to complete the tasks.

References

1. Shadbolt, N., Hall, W., Berners-Lee, T.: The semantic web revisited. Intell. Syst. **21**, 96–101 (2006)
2. Cyganiak, R., Jentzsch, A.: The Linking Open Data cloud diagram. http://lod-cloud.net
3. Guha, R.: Introducing schema.org: Search engines come together for a richer web (2011). https://googleblog.blogspot.com.es/2011/06/introducing-schemaorg-search-engines.html
4. Alani, H., Kalfoglou, Y., O'Hara, K., Shadbolt, N.R.: Towards a killer app for the semantic web. In: Gil, Y., Motta, E., Benjamins, V., Musen, M.A. (eds.) ISWC 2005. LNCS, vol. 3729, pp. 829–843. Springer, Heidelberg (2005)

5. Krug, S., Black, R.: Don't Make Me Think! A Common Sense Approach to Web Usability. New Riders Publishing, Indianapolis (2000)
6. Freitas, A., Curry, E., Oliveira, J.G., O'Riain, S.: Querying heterogeneous datasets on the linked data web: challenges, approaches, and trends. IEEE Internet Comput. **16**, 24–33 (2012)
7. Dadzie, A.-S., Rowe, M.: Approaches to visualising linked data: a survey. Semant. Web. **2**, 89–124 (2011)
8. Berners-Lee, T., Chen, Y., Chilton, L., Connolly, D., Dhanaraj, R., Hollenbach, J., Lerer, A., Sheets, D.: Tabulator: exploring and analyzing linked data. In: Proceedings of the 3rd Semantic Web and User Interaction Workshop (SWUI 2006), Athens, Georgia (2006)
9. Kaufmann, E., Bernstein, A.: Evaluating the usability of natural language query languages and interfaces to Semantic Web knowledge bases. Web Semant. Sci. Serv. Agents World Wide Web **8**, 377–393 (2010)
10. Brunetti, J.M., García, R., Auer, S.: From overview to facets and pivoting for interactive exploration of semantic web data. Int. J. Semant. Web Inf. Syst. **9**, 1–20 (2013)
11. Bevan, N.: Extending quality in use to provide a framework for usability measurement. In: Kurosu, M. (ed.) HCD 2009. LNCS, vol. 5619, pp. 13–22. Springer, Heidelberg (2009)
12. González-Sánchez, J.L., García, R., Brunetti, J.M., Gil, R., Gimeno, J.M.: Using SWET-QUM to compare the quality in use of semantic web exploration tools. J. Univ. Comput. Sci. **19**, 1025–1045 (2013)
13. García-Castro, R.: Benchmarking Semantic Web Technology. IOS Press (2009)
14. Sim, S.E., Easterbrook, S., Holt, R.C.: Using benchmarking to advance research: a challenge to software engineering. In: Proceedings of the 25th International Conference on Software Engineering, pp. 74–83. IEEE Computer Society, Washington (2003)
15. Voorhees, E.M., Harman, D.K.: TREC: Experiment and Evaluation in Information Retrieval (Digital Libraries and Electronic Publishing). The MIT Press (2005)
16. Bizer, C., Schultz, A.: The Berlin SPARQL benchmark. Int. J. Semant. Web Inf. Syst. (IJSWIS) **5**, 1–24 (2009)
17. Catarci, T., Costabile, M.F., Levialdi, S., Batini, C.: Visual query systems for databases: a survey. J. Vis. Lang. Comput. **8**, 215–260 (1997)
18. Card, S.K., Moran, T.P., Newell, A.: The keystroke-level model for user performance time with interactive systems. Commun. ACM **23**, 396–410 (1980)
19. García, R., Gil, R., Gimeno, J.M., Bakke, E., Karger, D.R.: BESDUI: A Benchmark for End-User Structured Data User Interfaces. http://w3id.org/BESDUI
20. Erling, O., Mikhailov, I.: RDF support in the virtuoso DBMS. In: Pellegrini, T., Auer, S., Tochtermann, K., Schaffert, S. (eds.) Networked Knowledge - Networked Media. SCI, vol. 221, pp. 7–24. Springer, Heidelberg (2009)
21. Bakke, E., Karger, D.R.: Expressive query construction through direct manipulation of nested relational results. In: Proceedings of the 2016 International Conference on Management of Data (SIGMOD 2016), pp. 1377–1392, ACM, New York (2016)
22. John, B.E., Kieras, D.E.: The GOMS family of user interface analysis techniques: comparison and contrast. ACM Trans. Comput. Hum. Interact. **3**(4), 320–351 (1996)

SPARQLGX: Efficient Distributed Evaluation of SPARQL with Apache Spark

Damien Graux[1,2,3(✉)], Louis Jachiet[1,2,3], Pierre Genevès[1,2,3],
and Nabil Layaïda[1,2,3]

[1] Inria, Paris, France
{damien.graux,louis.jachiet,nabil.layaida}@inria.fr
[2] CNRS, LIG, Grenoble, France
pierre.geneves@cnrs.fr
[3] Université Grenoble Alpes, Grenoble, France

Abstract. SPARQL is the W3C standard query language for querying data expressed in the Resource Description Framework (RDF). The increasing amounts of RDF data available raise a major need and research interest in building efficient and scalable distributed SPARQL query evaluators. In this context, we propose SPARQLGX: our implementation of a distributed RDF datastore based on Apache Spark. SPARQLGX is designed to leverage existing Hadoop infrastructures for evaluating SPARQL queries. SPARQLGX relies on a translation of SPARQL queries into executable Spark code that adopts evaluation strategies according to (1) the storage method used and (2) statistics on data. We show that SPARQLGX makes it possible to evaluate SPARQL queries on billions of triples distributed across multiple nodes, while providing attractive performance figures. We report on experiments which show how SPARQLGX compares to related state-of-the-art implementations and we show that our approach scales better than these systems in terms of supported dataset size. With its simple design, SPARQLGX represents an interesting alternative in several scenarios.

Keywords: RDF system · Distributed SPARQL evaluation

1 Introduction

SPARQL is the standard query language for retrieving and manipulating data represented in the Resource Description Framework (RDF) [11]. SPARQL constitutes one key technology of the semantic web and has become very popular since it became an official W3C recommendation [1].

The construction of efficient SPARQL query evaluators faces several challenges. First, RDF datasets are increasingly large, with some already containing more than a billion triples. To handle efficiently this growing amount of data, we need systems to be distributed and to scale. Furthermore, semantic data often have the characteristic of being dynamic (frequently updated). Thus being able to answer quickly after a change in the input data constitutes a very desirable property for

© Springer International Publishing AG 2016
P. Groth et al. (Eds.): ISWC 2016, Part II, LNCS 9982, pp. 80–87, 2016.
DOI: 10.1007/978-3-319-46547-0_9

a SPARQL evaluator. In this context, we propose SPARQLGX: an engine designed to evaluate SPARQL queries based on Apache Spark [21]: it relies on a compiler of SPARQL conjunctive queries which generates Scala code that is executed by the Spark infrastructure. The source code of our system is available online from the following URL: https://github.com/tyrex-team/sparqlgx.

The paper is organized as follows: we first introduce the technologies that we consider in Sect. 2. Then, in Sect. 3, we describe SPARQLGX and present additional available tools in Sect. 4. Section 5 reports on our experimental validation to compare our implementation with other open source HDFS-based RDF systems. Finally, we review related works in Sect. 6 and conclude in Sect. 7.

2   Background

The Resource Description Framework (RDF) is a language standardized by W3C to express structured information on the Web as graphs [11]. It models knowledge about arbitrary resources using Unique Resource Identifiers (URIs), Blank Nodes and Literals. RDF data is structured in sentences –or triples– written $(s\ p\ o)$, each one having a subject s, a predicate p and an object o.

The SPARQL standard language has been studied under various forms and fragments. We focus on the problem of evaluating the Basic Graph Pattern (BGP) fragment over a dataset of RDF triples. The BGP fragment is composed of conjunctions of triple patterns (TPs). A candidate solution is a mapping from the variables of the query towards values, a candidate solution satisfies a TP when the replacement of the variables of the TP with their value corresponds to a triple that appears in the RDF data. A query solution is a candidate solution that satisfies all the TPs of the query.

Apache Hadoop[1] is a framework for distributed systems based on the Map-Reduce paradigm. The Hadoop Distributed File System (HDFS) is a popular distributed file system handling the distribution of data across a cluster and its replication [19].

Apache Spark [21] is a MapReduce-like data-parallel framework designed for large-scale data processing running on top of the JVM. Spark can be set up to use HDFS.

3   SPARQLGX: General Architecture

In this Section, we explain how we translate queries from our SPARQL fragment into lower-level Scala code [14] which is directly executable with the Spark API. To this end, after presenting the chosen data storage model, we give a translation into a sequence of Spark-compliant Scala-commands for each operator of the considered fragment.

---

[1] Apache Hadoop: http://hadoop.apache.org.

3.1   Data Storage Model

In order to process RDF datasets with Apache Spark, we first have to adopt a convenient storage model on the HDFS. From a "raw" storage (*e.g.* a file in the N-Triple standard which is a simple list of all triples) to complex schemes (*e.g.* involving indexes or B-trees), there are many ways to store RDF data. Any storage choice is a compromise between (1) the time required for converting origin data into the target format, (2) the total disk-space needed, (3) the possible response time improvement induced.

RDF triples have very specific semantics. In a RDF triple $(s\ p\ o)$, the predicate p represents the "semantic relationship" between the subject s and the object o. Thus, there are often relatively few distinct predicates compared to the number of distinct subjects or objects. The vertically partitioned architecture introduced by Abadi *et al.* in [2] takes advantage of this observation by storing the triple $(s\ p\ o)$ in a file named p whose contents keeps only s and o entries.

(1) Converting RDF data into a vertically partitioned dataset does not involve complex computation: each triple is read once and the pair (subject, object) is appended to the predicate file.
(2) For large datasets with only a few predicates, two URIs are stored instead of three which reduce the memory footprint compared with the input dataset.
(3) Having vertically partitioned data reduces evaluation time of triple patterns whose predicate is a constant (*i.e.* not a variable): searches are limited to the relevant files. In practice, one can observe that most SPARQL queries have triple patterns with a constant predicate. [7] showed that graph patterns where all predicates are constant represent 77.81 % of the queries asked to DBpedia and 98.08 % of the ones asked to SWDF.

We believe that vertical partitionning is very well suited for RDF: it implies a pass over the data but with only simple computation, reduces the size of the dataset and provides an indexation.

3.2   SPARQL Fragment Translation

We compute the solution of a conjunction of TPs recursively. Given a conjunction of n TPs we recursively compute the set of solution for the $n-1$ first TPs and then we combine this set with the solutions of the last TP by joining them on their common variables.

To compute the solutions for a unique TP: when the predicate is a constant, we open the relevant HDFS file using `textFile`; otherwise, we have to open all predicate files. Then, using the constants of the TP, we use a `filter` to keep only the matching elements. Finally, we use the variables names appearing in the TP for variables. For instance, the following TP {?s age 21 .} matching people that are 21 years old is translated into:

```
val tp=sc.textFile("age.txt")
        .filter{case(s,obj)=>obj==21}
```

In order to translate a conjunction of TPs (*i.e.* a BGP), the TPs are joined. Two set of partial solutions are joined using their common variables as a key: keyBy in Spark. Joining TPs is then realized with `join` in Spark. For example the following TPs {?s age 21 . ?s gender ?g .} are translated into:

```
val tp1=sc.textFile("age.txt")
          .filter{case(s,obj)=>obj==21}
          .keyBy{case(s,obj)=>s}
val tp2=sc.textFile("gender.txt")
          .keyBy{case(s,g)=>s}
val bgp=tp2.join(tp1).values
```

A join with no common variables corresponds to a cross product (therefore a cartesian in Spark). For a conjunction of n TPs we perform $(n-1)$ joins.

The obtained translation (the Scala-code) thus depends on the initial order of TPs since the joins will be perfomed in the same order. This allows us to develop optimizations based on join commutativity such as the ones presented in Sect. 4.1.

3.3   SPARQL Fragment Extension

Once the TPs are translated, we use a map to retain only the desired fields (*i.e.* the distinguished variables) of the query. At that stage, we can also modify results according to the SPARQL solution modifiers [1] (*e.g.* removing duplicates with distinct, sorting with sortByKey, returning only few lines with take, etc.)

Furthermore, we also easily translate two additional SPARQL keywords: UNION and OPTIONAL, provided they are located at top-level in the WHERE clauses. Indeed, Spark allows to aggregate sets having similar structures with union and is also able to add data if possible with leftOuterJoin. Thus SPARQLGX natively supports a slight extension (UNIONs and OPTIONALs at top level) of the extensively studied SPARQL fragment made of conjunctions of triple patterns.

4   Additional Features

4.1   Optimized Join Order with Statistics

The evaluation process (using Spark) first evaluates TPs and then joins these subsets according to their common variables; thus, minimizing the intermediate set sizes involved in the join process reduces evaluation time (since communication between workers is then faster). Thereby, statistics on data and information on intermediate results sizes provide useful information that we exploit for optimisation purposes.

Given an RDF dataset $\mathcal{D}$ having T triples, and given a place in an RDF sentence $k \in \{subj, pred, obj\}$, we define the selectivity in $\mathcal{D}$ of an element e located at k as: (1) the occurrence number of e as k in $\mathcal{D}$ if e is a constant; (2) T if e is a variable. We note it $sel_{\mathcal{D}}^{k}(e)$. Similarly, we define the selectivity of a TP $(a\ b\ c\ .)$ over an RDF dataset $\mathcal{D}$ as: $SEL_{\mathcal{D}}(a, b, c) = \min(sel_{\mathcal{D}}^{subj}(a),\ sel_{\mathcal{D}}^{pred}(b),\ sel_{\mathcal{D}}^{obj}(c))$.

Thereby, to rank each TP, we compute statistics on datasets counting all the distinct subjects, predicates and objects. This is implemented in a compile-time module that sorts TPs in ascending order of their selectivities before they are translated.

Finally, we also want to avoid cartesian products. Given an ordered list l of TPs we compute a new list l' by repeating the following procedure: remove from l and append to l' the first TP that shares a variable with a TP of l'. If no such TP exists, we take the first.

4.2 SDE: SPARQLGX as a Direct Evaluator

Our tool evaluates SPARQL queries using Apache Spark after preprocessing RDF data. However, in certain situations, data might be dynamic (*e.g.* subject to updates) and/or users might only need to evaluate a single query (*e.g.* when evaluation is integrated into a pipeline of transformations). In such cases, it is interesting to limit as much as possible both the preprocessing time and the query evaluation time.

To take the original triple file as source, we only have to modify in our translation process the way we treat TPs to change our storage model. Instead of searching in predicate files, we directly use the initial file; and the rest of the translation process remains the same. We call this variant of our evaluator the "direct evaluator" or SDE.

5   Experimental Results

In this Section, we present an excerpt of our empirical comparison of our approach with other open source HDFS-based RDF systems. RYA [16] relies on key-value tables using Apache Accumulo[2]. CliqueSquare [8] converts queries in a Hadoop list of instructions. S2RDF [18] is a recent tool that allow to load RDF data according to a novel scheme called ExtVP and then to query the relational tables using Apache SparkSQL [4]. Finally, PigSPARQL [17] just translates SPARQL queries into an executable PigLatin [15] instruction sequence; and RDFHive[3] is a straightforward tool we made to evaluate SPARQL conjunctive queries directly on Apache Hive [20] after a naive translation of SPARQL into Hive-QL.

All experiments are performed on a cluster of 10 Virtual Machines (VM) distributed on two physical machines (each one running 5 of them). The operating system is CentOS-X64 6.6-final. Each VM has 17 GB of memory and 4 cores at 2.1 GHz. We kept the default setting with which HDFS is resilient to the loss of two nodes and we do not consider the data import on the HDFS as part of the preprocessing phase.

We compare the presented systems using two popular benchmarks: LUBM [9] and Watdiv [3]. Table 1 presents characteristics of the considered datasets.

---

[2] Apache Accumulo: accumulo.apache.org.

[3] RDFHive Sources: http://tyrex.inria.fr/rdfhive/home.html.

Table 1. General information about used datasets.

Dataset	Number of Triples	Original File Size on HDFS
Watdiv-100M	109 million	46.8 GB
Lubm-1k	134 million	72.0 GB
Lubm-10k	1.38 billion	747 GB

We rely on three metrics to discuss results (Table 2): query execution times, preprocessing times (for systems that need to preprocess data), and disk footprints. For space reasons, Table 2 presents three Lubm queries: Q1 because it bears large input and high selectivity, Q2 since it has large intermediate results while involving a triangular pattern and Q14 for its simplicity. Moreover, we aggregate Watdiv queries by the categories proposed in the Watdiv paper [3]: 3 complex (QC), 5 snowflake-shaped (QF), 5 linear (QL) and 7 star queries (QS). In Table 2 we indicate "TIMEOUT" whenever the process did not complete within a certain amount of time[4]. We indicate "FAIL" whenever the system crashed before this timeout delay. This regroups several kinds of failure such as unability of evaluating queries and also unability of preprocessing the datasets. We indicate "N/A" whenever the task could not be accomplished because of a failure during the preprocessing phase.

Table 2 shows that SPARQLGX always answer all tested queries on all tested datasets whereas this is not the case with other conventional RDF datastores which either timeout or fail at some point.

Table 2. Compared system performance.

		Conventional RDF Datastores				Direct Evaluators		
		RYA	CliqueSquare	S2RDF	SPARQLGX	SDE	RDFHive	PigSPARQL
Watdiv-100M	Preprocessing (minutes)	35	288	718	24	0	0	0
	Footprint (GB)	11.0	30.2	15.2	23.6	46.8	46.8	46.8
	QC (seconds)	TIMEOUT	333	504	**118**	278	1174	6973
	QF (seconds)	12071	FAIL	771	**182**	355	1640	9904
	QL (seconds)	5895	**94**	490	119	199	1175	5670
	QS (seconds)	1892	FAIL	805	**210**	363	1053	2460
Lubm-1k	Preprocessing (minutes)	34	157	408	55	0	0	0
	Footprint (GB)	16.2	55.8	13.1	39.1	72.0	72.0	72.0
	Q1 (seconds)	192	461	118	**22**	96	281	226
	Q2 (seconds)	TIMEOUT	**105**	1599	320	8917	TIMEOUT	1239
	Q14 (seconds)	66	22	86	**9**	149	274	212
Lubm-10k	Preprocessing (minutes)	410	TIMEOUT	FAIL	672	0	0	0
	Footprint (GB)	177	403	N/A	407	747	747	747
	Q1 (seconds)	1799	524	N/A	**305**	904	1631	2272
	Q2 (seconds)	TIMEOUT	22093	N/A	**19158**	TIMEOUT	TIMEOUT	18029
	Q14 (seconds)	571	731	N/A	**541**	1500	2937	2525

---

[4] We set the timeout delay to 10 hours for the query evaluation stage and to 24 hours for the dataset preprocessing stage.

In addition, SPARQLGX outperforms several implementations in many cases (also as shown on Table 2), while implementing a simple architecture exclusively built on top of open source and publicly available technologies. Furthermore, the SDE variant of our implementation, which does not require any preprocessing phase offers performances similar to the ones obtained with state-of-the-art implementations that require preprocessing.

6 Related Work

In recent years, many RDF systems capable of evaluating SPARQL queries have been developed [12]. These stores can be divided in two categories: centralized systems (*e.g.* RDF-3X [13] or Virtuoso [5]) and distributed ones, that we further review. Distributed RDF stores can in turn be divided into three categories. (1) The *ad-hoc* systems that are specially designed for RDF data and that distribute and store data across the nodes according to custom ad-hoc methods (*e.g.* 4store [10]). (2) Other systems use a communication layer between centralized systems deployed across the cluster and then evaluate sub-queries on each node such as Partout with RDF-3X [6]. (3) Lastly, some RDF systems [8,16–18] are built on top of distributed Cloud platforms such as Apache Hadoop. One major interest of such platforms relies on their common file systems (*e.g.*, HDFS): indeed various applications can access data at the same time and the distribution/replication issues are transparent. These systems [8,16–18], then evaluate SPARQL conjunctive queries using various tools as presented in Sect. 5 (*e.g.* Accumulo, Hive, Spark, etc.). To set up appropriate tools for pipeline applications, we choose to distribute data with a Cloud platform (HDFS) and evaluate queries using Spark. We compared the performances of SPARQLGX with the most closely related implementations in Sect. 5.

Finally, it is worthwhile to notice that SPARQL is a very expressive language which offers a rich set of features and operators. Most evaluators based on Cloud platforms focus on the restricted SPARQL fragment composed of conjunctive queries. SPARQLGX also natively supports a slight extension of this fragment with UNION and OPTIONAL operators at top level.

7 Conclusion

We proposed SPARQLGX: a tool for the efficient evaluation of SPARQL queries on distributed RDF datasets. SPARQL queries are translated into Spark executable code, that attempts to leverage the advantages of the Spark platform in the specific setting of RDF data. SPARQLGX also comes with a direct evaluator based on the same SPARQL translation process and called SDE, for situations where preprocessing time matters as much as query evaluation time. We report on practical experiments with our systems that outperform several state-of-the-art Hadoop-reliant systems, while implementing a simple architecture that is easily deployable across a cluster.

References

1. SPARQL 1.1 overview, March 2013. http://www.w3.org/TR/sparql11-overview/
2. Abadi, D.J., Marcus, A., Madden, S.R., Hollenbach, K.: Scalable semantic web data management using vertical partitioning. In: Proceedings of the 33rd International Conference on Very Large Data Bases, pp. 411–422. VLDB Endowment (2007)
3. Aluç, G., Hartig, O., Özsu, M.T., Daudjee, K.: Diversified stress testing of RDF data management systems. In: Mika, P., et al. (eds.) ISWC 2014. LNCS, vol. 8796, pp. 197–212. Springer, Heidelberg (2014). doi:10.1007/978-3-319-11964-9_13
4. Armbrust, M., Xin, R.S., Lian, C., Huai, Y., Liu, D., Bradley, J.K., Meng, X., Kaftan, T., Franklin, M.J., Ghodsi, A., et al.: Spark SQL: Relational data processing in spark. In: SIGMOD, pp. 1383–1394. ACM (2015)
5. Erling, O., Mikhailov, L.: Virtuoso: RDF support in a native RDBMS. In: de Virgilio, R., Giunchiglia, F., Tanca, L. (eds.) Semantic Web Information Management, pp. 501–519. Springer, Heidelberg (2010)
6. Galarraga, L., Hose, K., Schenkel, R.: Partout: A distributed engine for efficient rdf processing. In: WWW Companion, pp. 267–268 (2014)
7. Gallego, M.A., Fernández, J.D., Martínez-Prieto, M.A., de la Fuente, P.: An empirical study of real-world SPARQL queries. In: 1st International Workshop on Usage Analysis and the Web of Data at the 20th International World Wide Web Conference (2011)
8. Goasdoué, F., Kaoudi, Z., Manolescu, I., Quiané-Ruiz, J.A., Zampetakis, S.: Cliquesquare: flat plans for massively parallel RDF queries. In: ICDE, pp. 771–782. IEEE (2015)
9. Guo, Y., Pan, Z., Heflin, J.: LUBM: A benchmark for OWL knowledge base systems. Web Semant. Sci. Serv. Agents World Wide Web **3**(2), 158–182 (2005)
10. Harris, S., Lamb, N., Shadbolt, N.: 4store: The design and implementation of a clustered RDF store. In: SSWS (2009)
11. Hayes, P., McBride, B.: RDF semantics. In: W3C Rec. (2004)
12. Kaoudi, Z., Manolescu, I.: RDF in the clouds: a survey. VLDB J. **24**(1), 67–91 (2015)
13. Neumann, T., Weikum, G.: RDF-3X: a RISC-style engine for RDF. Proc. VLDB Endowment **1**(1), 647–659 (2008)
14. Odersky, M.: The scala language specification v 2.9 (2014)
15. Olston, C., Reed, B., Srivastava, U., Kumar, R., Tomkins, A.: Pig latin: a not-so-foreign language for data processing. In: SIGMOD, pp. 1099–1110. ACM (2008)
16. Punnoose, R., Crainiceanu, A., Rapp, D.: Rya: a scalable RDF triple store for the clouds. In: International Workshop on Cloud Intelligence, p. 4. ACM (2012)
17. Schätzle, A., Przyjaciel-Zablocki, M., Lausen, G.: PigSPARQL: Mapping SPARQL to pig latin. In: Proceedings of the International Workshop on Semantic Web Information Management, p. 4. ACM (2011)
18. Schätzle, A., Przyjaciel-Zablocki, M., Skilevic, S., Lausen, G.: S2RDF: RDF querying with SPARQL on spark. In: VLDB, pp. 804–815 (2016)
19. Shvachko, K., Kuang, H., Radia, S., Chansler, R.: The hadoop distributed file system. In: Mass Storage Systems and Technologies (MSST), pp. 1–10. IEEE (2010)
20. Thusoo, A., Sarma, J.S., Jain, N., Shao, Z., Chakka, P., Anthony, S., Liu, H., Wyckoff, P., Murthy, R.: Hive: a warehousing solution over a map-reduce framework. Proc. VLDB Endowment **2**(2), 1626–1629 (2009)
21. Zaharia, M., Chowdhury, M., Das, T., Dave, A., Ma, J., McCauley, M., Franklin, M.J., Shenker, S., Stoica, I.: Resilient distributed datasets: A fault-tolerant abstraction for in-memory cluster computing. In: NSDI, p. 2. USENIX Association (2012)

Querying Wikidata: Comparing SPARQL, Relational and Graph Databases

Daniel Hernández[1], Aidan Hogan[1(✉)], Cristian Riveros[2], Carlos Rojas[2], and Enzo Zerega[2]

[1] Center for Semantic Web Research, Department of Computer Science, Universidad de Chile, Santiago, Chile
ahogan@dcc.uchile.cl
[2] Center for Semantic Web Research, Department of Computer Science, Pontificia Universidad Católica de Chile, Santiago, Chile

Abstract. In this paper, we experimentally compare the efficiency of various database engines for the purposes of querying the Wikidata knowledge-base, which can be conceptualised as a directed edge-labelled graph where edges can be annotated with meta-information called qualifiers. We take two popular SPARQL databases (Virtuoso, Blazegraph), a popular relational database (PostgreSQL), and a popular graph database (Neo4J) for comparison and discuss various options as to how Wikidata can be represented in the models of each engine. We design a set of experiments to test the relative query performance of these representations in the context of their respective engines. We first execute a large set of atomic lookups to establish a baseline performance for each test setting, and subsequently perform experiments on instances of more complex graph patterns based on real-world examples. We conclude with a summary of the strengths and limitations of the engines observed.

Resource type: Benchmark and Empirical Study
Permanent URL: https://dx.doi.org/10.6084/m9.figshare.3219217

1 Introduction

Wikidata is a new knowledge-base overseen by the Wikimedia foundation and collaboratively edited by a community of thousands of users [21]. The goal of Wikidata is to provide a common interoperable source of factual information for Wikimedia projects, foremost of which is Wikipedia. Currently on Wikipedia, articles that list entities – such as top scoring football players – and the info-boxes that appear on the top-right-hand side of articles – such as to state the number of goals scored by a football player – must be manually maintained. As a result, for example, factual information in different locations will often be inconsistent. The aim of Wikidata is to instead keep such factual knowledge in one place: facts need be edited only once and can be drawn upon from multiple locations. Since the launch of Wikidata in October 2012, more than 80 thousand editors have contributed over 86 million statements about 17 million entities.

© Springer International Publishing AG 2016
P. Groth et al. (Eds.): ISWC 2016, Part II, LNCS 9982, pp. 88–103, 2016.
DOI: 10.1007/978-3-319-46547-0_10

To allow users issue bespoke queries over the knowledge-base, Wikimedia has begun hosting an official query service[1] which according to internal statistics[2] receives in the order of hundreds of thousands of queries per day. The query service runs over an RDF representation of Wikidata that is indexed in the Blazegraph SPARQL store (formerly known as BigData) [20].

In Wikidata, items are connected either to related items or datatype values by directed, named relations. There is thus a close correspondence between Wikidata and RDF. However, relations in Wikidata can be annotated with attribute–value pairs, such as qualifiers and references, to specify a further context for the relation (e.g., validity, cause, degree, etc.). This complicates the representation of Wikidata in RDF somewhat, requiring some form of *reification* [11] to capture the full knowledge-base in RDF. In previous work [12], we thus investigated various ways in which Wikidata can be represented in RDF and how that affects the query performance in various SPARQL engines; more specifically we tested n-ary relation, standard reification, singleton property and named graph representations of Wikidata against five SPARQL stores – 4store [7], Blazegraph (formerly BigData) [20], GraphDB (formerly (Big)OWLIM) [2], Jena TDB [22], and Virtuoso [4] – with respect to answering an initial selection of 14 real-world queries. Our results suggested that while engines struggled with the singleton property representation, no other representation was an outright winner. In terms of engines, we found that Virtuoso exhibited the best and most reliable performance, with GraphDB and Blazegraph following behind.

A follow-up question arising from our initial previous work was how the performance of these SPARQL engines would compare with that of other technologies. In this paper, we thus extend on our previous work by comparing a selection of SPARQL, relational and graph databases for the purposes of querying Wikidata. We select these families of databases since they correspond with the inherent structure of Wikidata *and* they offer support for comprehensively-featured declarative query languages as needed for the public query service (which rules out, for example, key–value stores, document stores and column-family stores). To conduct these comparisons, we design and apply a range of novel experiments.

However, there are still too many databases engines within these three families for all to be considered experimentally in the current scope; hence we must be selective in what we test. For SPARQL, we select Virtuoso [4], which performed best overall in our previous experiments, and Blazegraph [20], which is currently deployed in Wikimedia's official query service. For graph databases, we select Neo4J [19], which is based on *property graphs*: not only is it arguably the most popular graph database,[3] it is the only (non-SPARQL) graph database we know of that supports a mature query language (Cypher) in the declarative style of the example queries listed for the official query service. For relational databases, we select PostgreSQL [18] as a mature open-source solution supporting the SQL standard. Although we could consider other databases for testing – such as various relational

---

[1] https://query.wikidata.org/.

[2] https://grafana.wikimedia.org/dashboard/db/wikidata-query-service.

[3] We can refer (informally) to, e.g., http://db-engines.com/en/ranking/graph+dbms.

databases or other SPARQL engines such as Allegrograph [14] or Stardog [13] – our selection allows us to draw initial conclusions with respect to how well prominent databases from each family of technologies perform relative to each other, and indeed how the solution selected by Wikimedia (Blazegraph) compares with these other alternatives.

Towards testing these diverse engines, we first discuss some representations for encoding the Wikidata knowledge-base such that it can be loaded into each type of engine. We then introduce various novel experiments. Our first set of experiments that are based on the idea of performing sets of lookups for "atomic patterns" with exhaustive combinations of constants and variables; these results give an initial idea of the low-level performance of each configuration. Next we perform experiments on sets of instances of basic graph patterns that generalise the graph motifs we found to commonly occur in the use-case queries listed at the public query service. We also discuss how well the various engines support the query features commonly used for these use-case queries.

Before we continue, however, we first introduce Wikidata in more detail.

2 The Wikidata Model

In Fig. 1, we see an example Wikidata statement describing the U.S. presidency of Abraham Lincoln [12]. Internal identifiers are in grey: those beginning with Q refer to entities, and those referring to P refer to properties. These identifiers map to IRIs, where information about that entity or relationship can be found on Wikidata. Entities and relationships are also associated with labels, where the example shows labels in English. The statement contains a *primary relation* with Abraham Lincoln as *subject*, position held as *predicate*, and President of the United States of America as *object*; this relation is associated with pairs of *qualifier predicates* (e.g., start time, follows) and their *qualifier values* (e.g., "4 March 1861", James Buchanan); we call each such pair a *qualifier*. Statements are also often associated with one or more *references* that support the claims and with a *rank* that marks the most important statements for a given property, but herein we can treat references and ranks as special types of qualifiers. As such, we can define a Wikidata statement as a primary relation and a set of qualifiers.

At its core, Wikidata then consists of a set of such statements and a mapping from identifiers to labels in various languages. However, some statements may not contain any qualifiers (though in theory all statements should contain a reference, the knowledge-base is, by its nature, incomplete). Datatype values can themselves be associated with meta-data, where for example dates or times can be associated with a specific calendar. Likewise, objects can occasionally be existential (for example, when someone is known to have been murdered but their killer is unknown) or non-existential (to explicitly state that an entity has no such relation); however, these types of values are quite rare and hence for brevity we do not consider them directly.[4] Finally, it is important to note that Wikidata can contain multiple distinct statements with the same binary relation: for example, Clover Cleveland was

---

[4] See https://tools.wmflabs.org/wikidata-todo/stats.php.

the U.S. President for two non-consecutive terms (i.e., with different start and end times, different predecessors and successors). In Wikidata, this is represented as two separate statements whose primary relations are both identical, but where the qualifiers (start time, end time, follows, followed by) differ.

3   Wikidata Representations

In the context of querying Wikidata, our goal is to compare the performance of a selection of database *engines* from three different *families*: SPARQL, relational and graph. Each family is associated with a different *data model*: named (RDF) graphs, relations, and property graphs, respectively. Each data model may permit multiple possible *representations* of Wikidata, for example, different RDF reification schemes, different relational schema, etc. Although the representations we present (aside from corner cases [12]) encode the same data and do not change the computational complexity of query evaluation – since, loosely speaking, translating between the representations is cheaper than query evaluation – as we will see, different representations carry different performance costs, particularly for more low-level atomic lookups. We now discuss different representations of Wikidata suitable for each of the three data models we consider.

3.1   RDF/Named Graph Representations

With respect to representing Wikidata for querying in SPARQL, in our previous work, we considered four well-known reification schemes [12].[5] Each such scheme associates the primary relation of a statement with a statement identifier onto which qualifiers can be associated. In Fig. 2a, we give an example RDF graph where :X1 is the statement identifier and its four outgoing edges represent the four qualifiers from Fig. 1; we also show how complex datatypes can be represented. For each of the following representations, all we are left to do is associate the primary relation of Fig. 1 – namely (Q91, P39, Q11696) – with the statement identifier :X1 already associated with the qualifier information; in other words, we wish to encode a quad (s, p, o, i) where (s, p, o) is the primary relation of the statement and i is an identifier for the statement. We again recap these four schemes, where more details are available from our previous paper [12].

Abraham Lincoln [Q91]
    position held [P39]   President of the United States of America [Q11696]
        start time [P580]   "4 March 1861"
        end time [P582]   "15 April 1865"
        follows [P155]   James Buchanan [Q12325]
        followed by [P156]   Andrew Johnson [Q8612]

Fig. 1. A Wikidata statement about Abraham Lincoln (reused from [12])

---

[5] We refer here to "reification" in the general sense of describing triples, whereas we refer to the specific proposal in the RDF specifications as "standard reification" [11].

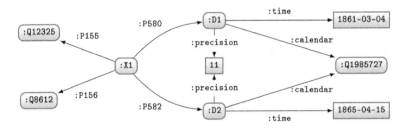

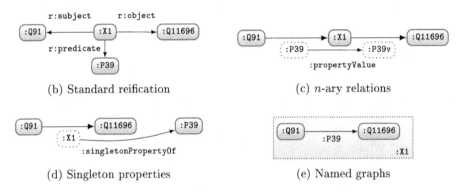

Fig. 2. RDF representations encoding data from Fig. 1 (adapted from [12])

Standard Reification (SR) [11]: an RDF resource is used to denote a triple. Using this scheme, we can use a reified triple to encode a Wikidata statement, as depicted in Fig. 2b. We can encode n quadruples with $3n$ triples.

n-ary Relation (NR) [5]: an RDF resource is used to identify a relation. Figure 2c shows such a scheme analogous to that proposed by Erxleben et al. [5] for representing Wikidata (modified slightly to reflect current data). The `:propertyValue` edge is important to explicitly link the original property with its twin. To represent n quads with m unique properties, we require $2n + m$ triples.

Singleton Properties (SP) [16]: a unique property represents each statement. The idea is captured in Fig. 2d. To encode n quads, we need $2n$ triples.

Named Graphs [8] (NG): this is a set of pairs of the form (G, i) where G is an RDF graph and i is an IRI (which can be omitted in the case of the *default graph*). We can "flatten" this representation by taking the union over $G \times \{i\}$ for each such pair, resulting directly in quads, as illustrated in Fig. 2e.

3.2 Relational Representations

Figure 3 exemplifies the relational representation we use for Wikidata, which involves three tables: **Statement** stores the primary relation and a statement id; **Qualifier** associates one or more qualifiers with a statement id; and **Label** associates each entity and property with one or more multilingual labels. We keep the **Label** table separate from **Statement** since we wish to keep a **lang** column for lookup/filtering, where labels are not qualified in Wikidata.

One complication with the relational model is that certain values – in particular the object (o) and qualifier value (v) – can have multiple types. Aside from entities, properties and logical values (e.g., exists, not exists), Wikidata contains four data-types – numeric, time, coordinates and strings – where as previously mentioned, each datatype can contain meta-information such as a calendar, precision, etc. One option is to store all values in one column as strings (e.g., JSON objects); however, this precludes the possibility of using such datatype values in filters, doing range queries, ordering, etc. Another possibility is to create separate columns, such as exemplified in Fig. 3 with o_{date} and v_{date}; however, we would need at least five such columns to support all Wikidata datatypes, leading to a lot of nulls in the table, and we still cannot store the meta-information. A third option would be to create separate tables for different datatype values, but this increases the complexity of joins. We currently use JSON strings to serialise datatype values, along with their meta-information, and store different types in different columns (i.e., as per Fig. 3 with, e.g., o_{date} and v_{date}).

Statement

id	s	p	o	o_{date}
X1	Q91	P39	Q11696	⊥

Qualifier

q	v	v_{date}	id
P155	Q12325	⊥	X1
P156	Q8612	⊥	X1
P580	⊥	1861-03-04	X1
P582	⊥	1865-04-15	X1

Label

e	label	lang
P39	position held	en
...	...	...
Q91	Abraham Lincoln	en
...	...	...

Fig. 3. Relational representation of data from Fig. 1

3.3 Property Graph Representations

Graph databases commonly model data as *property graphs* [9]: directed edge-labelled graphs where nodes and edges are associated with ids, where each node and edge id can be associated with a *type* and a set of *attributes* that provide additional meta-information.[6] In Fig. 4, we show how the Wikidata statement in Fig. 1 could be represented as a "direct" property graph. In this case, **Q91** and **Q11696** are *node ids*, **X1** is an *edge id*, **P39** is an *edge type*, pairs of the form **label_en=*** are *node attributes*, and pairs of the form **P*=*** are *edge attributes*. Though not used in this case, nodes can also have *node types*.

One may argue that property graphs offer a natural fit for Wikidata, where we first tried to represent Wikidata in Neo4J analogous to the "direct representation" given in Fig. 4, but we encountered a number of fundamental issues.

---

[6] Types are sometimes called "labels" and attributes are often called "properties" [9]. We avoid such terms, which clash with entity labels, RDF properties, etc.

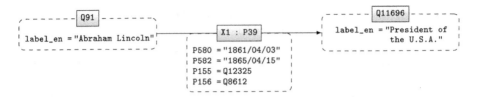

Fig. 4. A direct property graph encoding data from Fig. 1

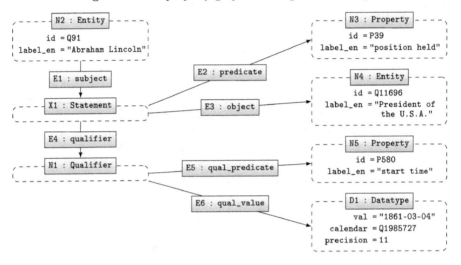

Fig. 5. A reified property graph encoding a subset of the data from Fig. 1

Firstly, Neo4J only allows one value per attribute property: this could be circumvented by using list values or special predicates such as label_en. Secondly, and more generally, Neo4J does not currently support queries involving joins or lookups on any information associated with edges, including edge ids, edge types or edge attributes. For example, one could not query for the president who followed (P155) James Buchanan (Q12325) without checking through all edges in the graph; this also means that we could not retrieve the labels of properties from such statements.[7] Hence, after initial experiments revealed these limitations for the direct representation, we sought another option.

Ultimately we chose to use a reified property graph representation to encode Wikidata in Neo4J, as depicted in Fig. 5. Conceptually the idea is similar to standard reification for RDF, where we use a node to represent the statement and use fixed edges to connect to the subject, predicate and object of the primary relation. We can likewise connect to a qualifier object, which in turn points to a qualifier predicate and value (which can be an item or a datatype). Most importantly, this model avoids putting any domain terms (e.g., P39, P580, Q8612, etc.)

---

[7] One option would be to store property meta-data such as labels locally using attributes, but this would require duplicating all such meta-data every time the property was used; we thus ruled this out as a possibility.

on edges, which avoids the aforementioned indexing limitations of Neo4J. Likewise, we had to store the domain identifiers (e.g., Q91, P39) as id attributes on nodes since node ids (and edge ids) in Neo4J are internally generated.

4 Experimental Setting

We now describe the setting for our experiments. Note that further details, including scripts, queries, raw data, configuration settings, instructions, etc., are provided at https://dx.doi.org/10.6084/m9.figshare.3219217. All original material and material derived from Wikidata is available under CC-BY (ⓒ①). The raw Wikidata dump is available under CC-0 (ⓒ⓪). The use of engines is governed by the respective vendor licences.

Data: We use the Wikidata dump in JSON format from 2016/01/04, which we converted to each representation using custom scripts. The dump contains 67 million claims describing 18 million entities, using 1.6 thousand properties, where 493 thousand claims are qualified with 1.5 million qualifiers (not including references or ranks), and 4 million claims have a datatype value.

Machine: All experiments were run on a single machine with 2× Intel Xeon Quad Core E5-2609 V3 CPUs, 32 GB of RAM, and 2 × 2 TB Seagate 7200 RPM 32 MB Cache SATA hard-disks in a RAID-1 configuration.

Engines: We used Blazegraph 2.1.0 (with Java SE "1.7.0_80"), and set 6 GB of RAM (per the vendor's recommendations to not use larger heaps, but rather to leave memory for OS-level caching).[8] Although Blazegraph supports a setting called RDF* to support reification [10], this model is non-standard and cannot *directly* encode multiple Wikidata statements with the same primary relation: in RDF*, each RDF triple must be unique. Though RDF* could be used to emulate named graphs, we instead use "Quad Mode" with standard SPARQL settings.

We used Virtuoso 7.2.3-dev.3215-pthreads, where we set NumberOfBuffers = 2720000 and MaxDirtyBuffers = 2000000 per vendor recommendations. We used the default indexes labelled PSOG, POGS, SP, OP and GS, where S, P and O correspond to the positions of a triple and G to the name of a graph. Each such index is sorted and allows for prefix lookups, where for example for the index PSOG, given a predicate p and a subject s as input, we can find all o and g in a single lookup such that (s, p, o) is the graph named g in the data.

We used PostgreSQL 9.1.20 set with maintenance_work_mem = 1920MB and shared_buffers = 7680MB. A secondary index (i.e. B-tree) was set for each attribute that stores either entities, properties or data values (e.g. dates) from Wikidata. Specifically, in the statement table (Fig. 3), each attribute s, p, o, and o_{date} has an index as well as the attributes q, v, and v_{date} in the qualifier table. In all other tables, only foreign keys attributes were indexed like, for example, attribute **e** in the **Label** table (Fig. 3).

---

[8] https://wiki.blazegraph.com/wiki/index.php/IOOptimization.

We used Neo4J-community-2.3.1 setting a 20GB heap. We used indexes to map from entity ids (e.g., Q42) and property ids (e.g., P1432) to their respective nodes. By default, Neo4J indexes nodes in the graph and their adjacent nodes in a linked-list style structure, such that it can navigate from a node to its neighbour(s) along a specific edge without having to refer back to the index.

Aside from this, we used default vendor-recommended settings. Given that the main use-case scenario we consider is to offer a public query service for Wikidata, we use REST APIs in the case of Blazegraph, Virtuoso and Neo4J. Unfortunately, however, we could not find a first-party REST API for PostgreSQL and thus resorted to using a direct command-line client.

We verified the completeness of (non-timeout) results by comparing result-sizes for the various representations. While we found minor differences (e.g., Blazegraph rejects dates like 2016-02-30), these were ≪1 % of all results.

5   Experimental Results

Atomic lookups: In our first experiments, we wish to test the performance of atomic lookups over Wikidata statements, where we focus in particular on qualified statements that require more complex representations to encode. Recall that qualified statements consist of five types of terms: subject (s), predicate (p), object (o), qualifier predicate (q), and qualifier value (v). Hence we can consider abstract query patterns based on quins of the form (s, p, o, q, v), where any term can be a variable or a constant.[9] For example, a pattern $(?, p, o, q, ?)$ may have an instance $(?u1, P39, Q4164871, P580, ?u5)$, asking for all US presidents and the date when their presidency started. We can consider $2^5 = 32$ abstract patterns of this form, where each position is either a constant or a unique variable.

To begin, we generated a set of 1.5 million (data) quins from Wikidata, shuffled them, and used them to randomly generate 300 unique instances for each of the 31 patterns (we exclude the open pattern $(?, ?, ?, ?, ?)$, which only has one instance). We select 300 since there are only 341 instances of $(?, p, ?, ?, ?)$ (i.e., there are 341 unique properties used on *qualified* statements); hence 300 allows us to have the same number of instances per pattern. In total, we have 9,300 instances, which we translate into concrete queries for our six experimental representations. Since instances of patterns such as $(?, p, ?, ?, ?)$ would generate millions of results, we set a limit of 10,000 results on all queries.

We then test these queries against our four engines, where we reset the engine, clear the cache, and start with the 300 instances of the first pattern, then moving to the second pattern, etc., in sequence. Given that we wish to run 9,300 queries for each representation, we set an internal server-side query timeout of 60 s to ensure a reasonable overall experiment time.

---

[9] Note that quins are not sufficient to represent Wikidata [12], where some form of statement identifier is needed to distinguish different statements with the same primary relation; however, such identifiers are not part of the domain but rather part of the representation, hence we do not consider them in our query patterns.

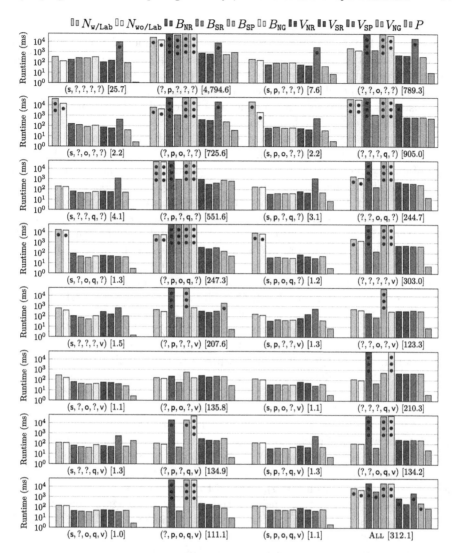

Fig. 6. Mean response times for the 31 patterns, and mean response times for ALL patterns, where N, B, V and P refer to Neo4J, Blazegraph, Virtuoso and Postgres, resp.; w/Lab and wo/Lab refer to Neo4J representations with and without label information, resp.; NR, SR, SP, and NG refer to n-ary relations, standard reification, singleton properties and named graphs, resp.; one, two and three (red) dots on a bar indicate that [0–33 %], [33–66 %] and [66–100 %] of queries timed out, resp.; timeouts are counted as 60 s; the x-axis labels indicate the pattern and the mean number of complete results for the instances (Color figure online)

Figure 6 gives the mean runtimes for each set of 300 instances per each pattern, where the y-axis is presented in log-scale (each horizontal line represents an order of magnitude), and where the presented order of patterns corresponds to the order of execution. We also include the mean considering all 9,300 instances. The x-axis labels indicate the pattern and the mean number of tuples that *should* be returned over the 300 instances (9,300 instances in the case of ALL). For the purposes of presentation, when computing the mean, we count a query timeout as 60 s (rather than exclude them, which would reduce the mean). Thus, for any set of instances that encountered a timeout – indicated by (red) dots in Fig. 6 – the mean values represent a lower-bound.

We see that PostgreSQL performs best for all but three patterns, and is often an order of magnitude faster than the next closest result, most often in the lower range of query times from 1–100 ms, where we believe that the choice of client may have an effect: the direct client that PostgreSQL offers does not incur the overhead of REST incurred by other engines, which may be significant in this range. In addition, PostgreSQL can answer instances of some patterns with a mean in the 1–10 ms range, suggesting that PostgreSQL is perhaps not touching the hard-disk in such cases (an SATA hard-disk takes in the order of 10 ms to seek), and is making more efficient use of RAM than other engines.

More conceptually, PostgreSQL also benefits from its explicit physical schema: every such pattern must join through the **id** columns of the **Statement** and **Qualifier** table, which are designated and indexed as primary/foreign keys. On the other hand, in other representations, no obvious keys exists. For example, in the named graphs representation, which is most similar to the relational version, a quin is encoded with a quad (s, p, o, i) (a triple in named graph i) and a triple (i, q, v) (in the default graph). There is no special key position, but Virtuoso, with its default indexes – PSOG, POGS, SP, OP, GS – is best suited to cases where predicates are bound along with a subject or object value. The pattern $(?, p, ?, ?, v)$ in named graphs then illustrates a nasty case for Virtuoso: if the engine starts with the much more selective v constant (starting with p would generate many more initial results), its only choice is to use the OP index (the only index where O is key) to get q from (i, q, v) using v, then POGS to get i from the same triple (i, q, v) using (q, v), then GS to get s from (s, p, o, i) using i, then SP to get p again from (s, p, o, i) using s, then PSOG to get o again from (s, p, o, i) using (s, p), and so we end up joining across all five indexes in Virtuoso instead of two in the case of PostgreSQL. Similar cases can be found for other patterns and representations, where the lack of explicit keys complicates performing joins.

On the other hand, we see from the mean of ALL queries that the gap between Virtuoso and PostgreSQL closes somewhat: this is because Virtuoso is often competitive with PostgreSQL for more expensive patterns, even outperforming it for the least selective pattern $(?, p, ?, ?, ?)$. These patterns with higher runtimes have a greater overall effect on the mean over all queries.

Conversely, we see that Neo4J and Blazegraph often fare the worst. In both cases, we found that one query would timeout and cause a subsequent series of queries to timeout, often early in the experiment. This occurred in quite a non-deterministic manner when dealing with low-selectivity instances: in repeated runs, different queries would instigate the issue. Given this non-determinism and our observations of memory usage and various settings, we believe this instability is caused by poor memory management, which leads to swapping and garbage collector issues. On the other hand, for patterns and representations where we did not encounter timeouts, we found that Blazegraph was often competitive and in some cases faster than Virtuoso, with Neo4J being the slowest.

Investigating Neo4J's performance, we speculated that the manner in which Neo4J operates – traversing nodes in a breadth-first manner – could be negatively affected by storing lots of embedded labels and aliases in different languages on the nodes, adding a large I/O overhead during such traversals. To test this, we built a version of Neo4J without labels. We found that this indeed improved performance, but typically by a constant factor: i.e., it was not the sole cause of slow query times. Investigating further, we found that the engine does not seem to make use of reliable selectivity estimates. One such example of this is to compare the results of $(s, ?, ?, ?, ?)$ versus $(s, ?, o, ?, ?)$ where the performance of the latter query is much worse than the former despite the fact that the latter could be run as the former, applying a simple filter thereafter. Investigating these curious results, we found that Neo4J often naively chooses to start with o rather than s, where objects often have low selectivity; for example, they can be countries of birth, or genders, etc. In general, we found the query planning features of Neo4J to be ill-suited to these sorts of queries.

Basic graph patterns: The atomic-lookup experiments show a clear distinction between two pairs of engines: Neo4J–Blazegraph and Virtuoso–PostgreSQL. Although PostgreSQL generally outperformed Virtuoso, these results may not hold for other types of queries. Thus having analysed atomic lookups involving qualifiers, next we wished to focus on queries involving joins. To guide the design of these queries, we studied the 94 example queries present on the official Wikidata query service[10] – queries submitted by the Wikidata community as exemplary use-cases of the service – to see what are the most common types of joins that appear. From this study, we identified that queries most commonly feature "s–s" joins – joins from subject to subject – and "s–o" joins – joins from subject to object, specifically on the primary relation. These observations correspond with other analyses of real-world SPARQL queries [6,17]. We also identified that for almost all queries, p is bound and that patterns form trees. About 80/94 queries followed this pattern: 11/94 involved qualifier information and 3/94 involved variables in the predicate position. We thus designed a set of experiments to generate a large number of instances of basic graph patterns (bgps) that follow the same abstract structure as most commonly observed in the use-case queries.

The first query pattern, which we refer to as DEPTH-1 "snowflake", starts with a subject variable and adds between 1 and 6 edges to it. The predicate of each edge is bound, and each object is made a constant or a variable with equal probability. The subject variable is projected; object variables are projected or not with equal probability. An example DEPTH-1 bgp is as follows, with projected variables underlined: $\{(\underline{?s}, P19, Q2280), (\underline{?s}, P106, \underline{?v1}), (\underline{?s}, P31, ?v2), (\underline{?s}, P27, \underline{?v3})\}$, which asks for

---

[10] https://query.wikidata.org/, see "Examples".

entities born in Minsk (P19,Q2280), their occupation(s) (P106), their citizenship(s) (P27), and checks that they are an instance (P31) of something. Each such instance thus tests s–s joins. We run an SQL query to generate 10,000 clusters of six unique primary relations from the data (joined by subject). We then randomly select 300 clusters to generate the DEPTH-1 bgps we use in the experiments. These are then converted into concrete queries for each representation. Each such query is guaranteed to return at least one result.

The second query pattern, which we call DEPTH-2 "snowflake", adds a second layer to the query. More specifically, we start with a DEPTH-1 query and then select an object variable at random to be used as the subject variable for a nested DEPTH-1 bgp (with out-degree of 0–4); e.g. {(?s, P140, Q9592), (?s, P27, Q30), (?s, P725, ?v1), (?s, P39, ?o), (?o, P511, Q1337757), (?o, P910, ?u1)} is a DEPTH-2 bgp that asks for entities with religion Catolicism (P140,Q9592), who are citizens of the U.S. (P27,Q30), who have a given name (P725), and who have a position (P39) with the title "His Eminence" (P511,Q1337757), additionally requesting the main category (P910) of that position. The result is a tree of depth two that tests s–s joins and s–o joins. We again use an SQL query to generate 10,000 clusters of data that can answer such patterns, randomly selecting 300 to generate instances that we convert into concrete queries in each representation.

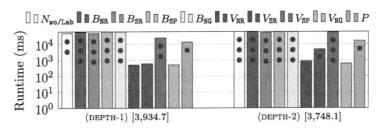

Fig. 7. Mean response times and result sizes for the snowflake queries of DEPTH-1 and DEPTH-2 (please see caption of Fig. 6 for explanation of legend, etc.)

For each setting, we run each batch of 300 queries sequentially and separately, clearing the cache and resetting the engine before each such batch. As before, each query is set to return a limit of 10,000 results and each engine is assigned an internal timeout of 60 s where, in Fig. 7, we see the results of these experiments, using similar conventions as before. The mean result sizes show that these queries tend to have low selectivity (i.e., lots of results).

We see that this time, Virtuoso performs best overall, where in particular the n-ary relations and named graph representations experience zero timeouts and outperform PostgreSQL (with a 15–19 % timeout rate) by an order of magnitude in both experiments. Regarding these two engines, the situation is almost the inverse of that for the previous experiments. In these experiments, under the standard reification and named graphs settings, the predicates are always bound, and the queries require a lot of self-joins (joins on the same table, for example for multiple distinct tuples about a given subject entity, which are particular common in SPARQL due to the structure of RDF). Virtuoso's indexing scheme – which in a relational sense considers predicates as a form of key for complete quads – is well-tuned for this morphology of query and excels for these settings. In more depth, Virtuoso can perform prefix lookups, meaning that a pattern (s, p, ?) or (?, p, o) can be performed in one lookup on its PSOG/POGS indexes respectively. On the other hand, PostgreSQL must join s/o with p (non-key attributes), through the statement ids (the key attribute), incurring a lot more work. In the case of singleton properties for Virtuoso, since the predicate encodes the statement identifier, these remain as a variable, which again means that Virtuoso struggles. We also observe that s–o joins cause some difficulty for Virtuoso in the n-ary relation setting for DEPTH-2 queries.

On the other hand, we see that most of the queries time-out for Neo4J and Blazegraph; more specifically, we encountered the same "domino effect" where one timeout causes a sequence of subsequent timeouts. Note that we only tested Neo4J without labels (the setting in which it is most efficient).

Query features: The aforementioned 94 use-case queries (taken from the public query service) use a mix of query features, where from SPARQL 1.0, 45/94 queries use ORDER BY, 38/94 use OPTIONAL, 21/94 use LIMIT, and 10/94 use UNION; and from SPARQL 1.1, 27/94 use GROUP BY, 18/94 use recursive property paths (i.e., * or +, mainly for traversing type hierarchies), 10/94 use BIND, and 5/94 use negation in the form of NOT EXISTS. Likewise, queries often use FILTER and occasionally sub-queries, though usually relating to handling labels. In any case, we see that a broad range of SPARQL (1.1) features are often used.

Testing all of these combinations of features in a systematic way would require extensive experiments outside the current scope but as an exercise, we translated a selection of eight such queries using a variety of features to our six representations. During that translation, we encountered a number of difficulties.

The first issue was with *recursive property paths*. With respect to SPARQL, these queries cannot be expressed for SR and SP due to how the RDF is structured, and are only possible in NG assuming the default graph contains the union (or merge) of all named graphs.[11] Although Neo4J specifically caters for these sorts of queries, the necessity to use a reified representation means that they are no longer possible. One solution in such cases would be to store the "direct" unqualified relations also. With PostgreSQL, it is in theory possible to use WITH_RECURSIVE to achieve paths, though being a more general operator, it is more difficult for the engine to optimise, thus incurring performance costs.

The second issue was with *datatype values*. In the case of PostgreSQL, a number of queries involve filters on dates, where currently we store datatype values as a JSON string, where SQL does not allow for parsing such objects. In order to better handle datatypes, one possibility is to create a different table for each datatype, with columns for the meta-information it can carry (calendar, precision, etc.). Another option is to use a non-standard feature for semi-structured data, where PostgreSQL has recently released support for accessing values in JSON objects stored in a table through its JSONB functionality.

---

[11] Paths cannot be traversed across named graphs in SPARQL unless loaded into the default graph by FROM clauses, which would require using specific statement ids.

Aside from these issues, the user queries can be expressed in any of the settings. We refer the interested reader to our previous work [12], where we present some example results for a selection of fourteen such queries over five different SPARQL engines and four representations.

6   Related Work

The goal of our work was to compare SPARQL, relational and graph databases for the purposes of querying Wikidata. Other authors have performed similar experimental comparisons across database families. Abreu et al. [1] compared a SPARQL engine (RDF-3X [15]) with a number of graph databases, showing that RDF-3X outperformed existing graph databases for the task of graph pattern matching. Bizer and Schultz [3] proposed the Berlin SPARQL Benchmark for comparing SPARQL engines with SPARQL-to-SQL engines and MySQL directly, concluding that MySQL easily outperformed the best performing SPARQL engine; however, these results were published in 2009 when SPARQL had only been standardised for one year, and likewise the dataset selected was inherently relational in nature. Our results complement such studies.

7   Conclusions

Our experiments revealed a number of strength and weaknesses for the tested engines and representations in the context of Wikidata. In terms of where systems could be improved, we recommend that Blazegraph and Neo4J should better isolate queries such that one poorly performing query does not cause a domino effect. Likewise, our results show that Neo4J would benefit from better query planning statistics and algorithms, as well as the provision of more customisable indices, particular on edge information. With respect to Virtuoso, it falls behind PostgreSQL in our first experiments due to the default indexing scheme chosen, but performs better in our second experiments based on real-world queries where predicates are bound and joins follow more "typical" patterns. On the other hand, PostgreSQL could benefit from better support for semi-structured information, where JSONB is indeed a step in the right direction. Regarding representations, we found that standard reification and named graphs performed best, with n-ary relations following in third, and singleton properties not being well-supported.

In this paper, we have not covered all pertinent issues for choosing an engine and representation – issues such as licensing, standardisation, support for inference, federation, etc. – but based on our results for query performance, the best configuration for a Wikidata query service would probably (currently) use Virtuoso and named graphs: a combination that performed well in experiments and that supports the various query features needed, including property paths.

Acknowledgments. This work was partially funded by the Millennium Nucleus Center for Semantic Web Research under Grant No. NC120004. The second author was supported by Fondecyt Grant No. 11140900 and the third author by Fondecyt Grant No. 11150653. We also thank Markus Krötszch for his contributions to the original workshop paper.

References

1. Abreu, D.D., Flores, A., Palma, G., Pestana, V., Piñero, J., Queipo, J., Sánchez, J., Vidal, M.: Choosing between graph databases and RDF engines for consuming and mining linked data. In: COLD (2013)
2. Bishop, B., Kiryakov, A., Ognyanoff, D., Peikov, I., Tashev, Z., Velkov, R.: OWLIM: a family of scalable semantic repositories. SWJ **2**(1), 33–42 (2011)
3. Bizer, C., Schultz, A.: The Berlin SPARQL benchmark. IJSWIS **5**(2), 1–24 (2009)
4. Erling, O.: Virtuoso, a hybrid RDBMS/graph column store. IEEE Data Eng. Bull. **35**(1), 3–8 (2012)
5. Erxleben, F., Günther, M., Krötzsch, M., Mendez, J., Vrandečić, D.: Introducing wikidata to the linked data web. In: Mika, P., et al. (eds.) ISWC 2014. LNCS, vol. 8796, pp. 50–65. Springer, Heidelberg (2014). doi:10.1007/978-3-319-11964-9_4
6. Gallego, M.A., Fernández, J.D., Martínez-Prieto, M.A., Fuente, P.D.L.: An empirical study of real-world SPARQL queries. In: USEWOD (2012)
7. Harris, S., Lamb, N., Shadbolt, N.: 4store: The design and implementation of a clustered RDF store. In: SSWS, pp. 94–109 (2009)
8. Harris, S., Seaborne, A., Prud'hommeaux, E. (eds.): SPARQL 1.1 query language. In: W3C Recommendation, 21 March 2013
9. Hartig, O.: Reconciliation of RDF* and property graphs. CoRR abs/1409.3288 (2014)
10. Hartig, O., Thompson, B.: Foundations of an alternative approach to reification in RDF. CoRR abs/1406.3399 (2014)
11. Hayes, P., Patel-Schneider, P.F. (eds.): RDF 1.1 semantics. In: W3C Recommendation, 25 February 2014
12. Hernández, D., Hogan, A., Krötzsch, M.: Reifying RDF: what works well with wikidata? In: SSWS, pp. 32–47 (2015)
13. Inc., C.: Stardog 4.0: The Manual (2016). http://docs.stardog.com/
14. Inc., F.: AllegroGraph RDFStore: Web 3.0's Database (2012). http://www.franz.com/agraph/allegrograph/
15. Neumann, T., Weikum, G.: x-RDF-3X: Fast querying, high update rates, and consistency for RDF databases. PVLDB **3**(1), 256–263 (2010)
16. Nguyen, V., Bodenreider, O., Sheth, A.: Don't like RDF reification? Making statements about statements using singleton property. In: WWW. ACM (2014)
17. Saleem, M., Ali, M.I., Hogan, A., Mehmood, Q., Ngomo, A.-C.N.: LSQ: the linked SPARQL queries dataset. In: Arenas, M., et al. (eds.) ISWC 2015. LNCS, vol. 9367, pp. 261–269. Springer, Heidelberg (2015). doi:10.1007/978-3-319-25010-6_15
18. Stonebraker, M., Kemnitz, G.: The Postgres next generation database management system. CACM **34**(10), 78–92 (1991)
19. The Neo4j Team: The Neo4j Manual v2.3.1 (2015). http://neo4j.com/docs/
20. Thompson, B.B., Personick, M., Cutcher, M.: The Bigdata RDF graph database. In: Linked Data Management, pp. 193–237 (2014)
21. Vrandečić, D., Krötzsch, M.: Wikidata: A free collaborative knowledgebase. Comm. ACM **57**, 78–85 (2014)
22. Wilkinson, K., Sayers, C., Kuno, H.A., Reynolds, D., Ding, L.: Supporting scalable, persistent Semantic Web applications. IEEE Data Eng. Bull. **26**(4), 33–39 (2003)

Clinga: Bringing Chinese Physical and Human Geography in Linked Open Data

Wei Hu[(✉)], Haoxuan Li, Zequn Sun, Xinqi Qian, Lingkun Xue,
Ermei Cao, and Yuzhong Qu

State Key Laboratory for Novel Software Technology,
Nanjing University, Nanjing, China
{whu,yzqu}@nju.edu.cn, hxli.nju@gmail.com, zqsun.nju@gmail.com,
xqqian.nju@gmail.com, lkxue.nju@gmail.com, emcao.nju@gmail.com

Abstract. While the geographical domain has long been involved as an important part of the Linked Data, the small amount of Chinese linked geographical data impedes the integration and sharing of both Chinese and cross-lingual knowledge. In this paper, we contribute to the development of a new Chinese linked geographical dataset named Clinga, by obtaining data from the largest Chinese wiki encyclopedia. We manually design a new geography ontology to categorize a wide range of physical and human geographical entities, and carry out an automatic discovery of links to existing knowledge bases. The resulted dataset contains over half million Chinese geographical entities and is open access.

Resource Type: Dataset and ontology
Permanent URL: http://w3id.org/clinga

1 Introduction

To embrace the vision of the Semantic Web, significant efforts have been devoted towards creating linked data for the geographical domain, such as GeoNames[1], GeoLink[2], GeoWordNet [3] and LinkedGeoData [7], in addition to a few multilingual, open-domain knowledge bases like DBpedia [5], Freebase [2], YAGO [4] and Wikidata [8] involving geographical data as well. Although they are being widely used by many semantic applications, the amount of Chinese geographical data is relatively limited and mainly exist in their labels. For example, according to our analysis of GeoNames 2015, only 4.6 % of its geographical features involve Chinese names, even for the features in China, this proportion is still 63 %. Despite two early works, Zhishi.me [6] and XLore [9], which extracted knowledge from Chinese wiki encyclopedias, to the best of our knowledge, there is no prior work on building Chinese linked geographical data.

---

[1] http://www.geonames.org/ontology.
[2] http://www.geolink.org.

© Springer International Publishing AG 2016
P. Groth et al. (Eds.): ISWC 2016, Part II, LNCS 9982, pp. 104–112, 2016.
DOI: 10.1007/978-3-319-46547-0_11

In this paper, we present a new **Chinese linked** geographical dataset, Clinga[3], which contains generally but not exclusively a large number of geographical entities in China (e.g. cities) and their relations (e.g. has-capital). Compared with existing geographical datasets, Clinga has three distinguished features:

- We obtain our data from Baidu Baike[4], the largest collaboratively-built Chinese wiki encyclopedia. Both structural data and textual description (e.g. in main text) of an article are extracted and translated to RDF using our structure ontology, to achieve the completeness of the article at our best effort.
- We manually design a physical and human geography ontology to categorize various geographical entities. Following the Chinese naming conventions, we combine the type-based heuristic rules and an SVM classifier to obtain good categorization accuracy.
- We link Clinga to existing knowledge bases like DBpedia and GeoNames. An automatic discovery of entity links is conducted based on bilingual (Chinese and English) labels and manually-defined ontology mappings.

Clinga is expected to be useful not only as a complementary source for location-based semantic applications such as DBpedia Mobile [1], but also as our primary knowledge base for answering geographical questions in the national matriculation examination of China (called GaoKao), under the support of the National High-tech R&D Program of China.

2 Development Methods

Figure 1 shows the general steps to develop the Clinga dataset. To achieve a satisfactory qualitative and quantitative result, the methodology that we follow is largely automatic, with a little amount of human intervention in critical parts.

We choose Baidu Baike rather than the Chinese Wikipedia[5] as the main data source due to its larger scale and richer content, especially about geography and contemporary people in China. At the time of writing, Baidu Baike has 13 million articles and is 15 times more than the Chinese Wikipedia. For another Chinese wiki encyclopedia, Hudong Baike[6], it significantly overlaps with Baidu Baike but has less influence in the Chinese language community[7]. It worth noting that the proposed methods can be smoothly applied to these wiki encyclopedias.

Technically, our data extraction process is similar to DBpedia, but having geographical data in Baidu Baike is not a trivial task. First, the category structure of Baidu Baike is often incorrect and inconsistent. For example, "Administrative division" and "Administrative region" categories co-exist

---

[3] Clinga is publicly available at http://w3id.org/clinga, under the Creative Commons BY-NC 4.0 license. It is also registered on https://datahub.io. Documentation and online services including SPARQL endpoint and keyword search are accessible at http://ws.nju.edu.cn/clinga.

[4] http://baike.baidu.com.

[5] http://zh.wikipedia.org/.

[6] http://www.baike.com/.

[7] https://strategy.wikimedia.org/wiki/Case_studies/Baidu_and_Hudong.

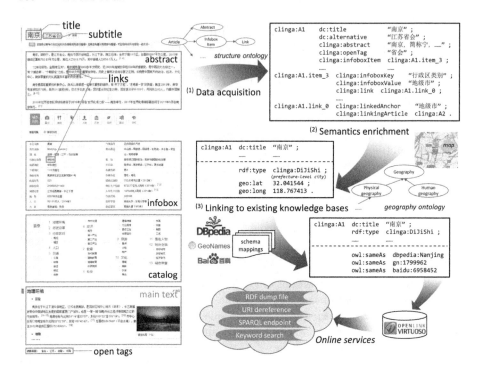

Fig. 1. Left part is a typical article in Baidu Baike, and right part is the methodological steps to develop Clinga, including: (1) data acquisition, (2) semantics enrichment, and (3) linking to existing knowledge bases.

with a very similar meaning; "Ancient place name" appears redundantly under "Place name". Second, mismatches frequently happen between categories and articles. For instance, several articles about people are assigned to "Administrative division"; hundreds of mountains are not categorized under "Mountain". These problems prevent us from directly using the category structure of Baidu Baike to obtain its geographical data. Besides, the main text is often causal and complex.

Therefore, we exhaustively crawl all the articles in Baidu Baike, and design a geography ontology to identify physical and human geographical entities. We also conduct an automatic discovery of links to knowledge bases such as DBpedia and GeoNames. The details of our methods are described below.

2.1   Data Acquisition

For an article in Baidu Baike (see the left part of Fig. 1), a unique and sequential ID is generated for entity URI (e.g. http://ws.nju.edu.cn/clinga/A1). Then, the following items are extracted and translated using our structure ontology.

Title and subtitle. An article has exactly one title and optionally one subtitle, which are converted as the values of dc:title and dc:alternative, respectively.

Infobox. An infobox presents the structural facets of an article using key-value pairs, and keys are later converted to properties in our geography ontology. A compound structure is used to model a key, a value, and links in the value.

Abstract. The first few paragraphs before infobox form the abstract of an article, which is described as the value of clinga:abstract.

Catalog and main text. An article organizes its main text by separated sections, and section titles form the catalog of the article. The correspondence between the title and paragraphs of a section is reserved, which is especially useful for geographical question answering, e.g. searching a related section and performing natural language understanding. However, the text in each paragraph is not RDFized yet.

Open tag. An article may have a few plain-text tags indicating its topics, which are converted as the values of clinga:openTag. Note that these tags are much more casual than the categories in the Chinese Wikipedia.

Link. Links in an article refer to other articles within Baidu Baike or webpages outside, which are described using clinga:link.

Furthermore, two other kinds of information are particularly considered.

Redirect. Baidu Baike uses redirects to handle the synonym problem. Two articles with a redirect relation are presented using clinga:redirects.

Disambiguation. Baidu Baike does not offer disambiguation pages, but rather puts all meanings together in a polysemy list. This relation is presented using clinga:disambiguates.

2.2 Semantics Enrichment

Before developing our own ontology, we investigate some existing geographical ontologies. The ontology in GeoNames has no class hierarchy, but uses feature codes for categorization. DBpedia and XLore automatically extracted their ontologies from categories, but Baike categories are incomplete and inconsistent to be reused directly. By discussing with the professors at the School of Geographic and Oceanographic Sciences in our university and referencing the DBpedia ontology and textbooks, we design a new geography ontology involving two main branches, physical geography and human geography, to categorize the geographical entities in Clinga. See Fig. 2 for an excerpt of our geography ontology.

We introduce a two-step method for entity categorization. In the first step, we define 213 heuristic rules to extract the candidate entity set for each most-specific type (i.e. the "leaf" class at the class hierarchy). The rationale is based on the Chinese naming conventions and distinguished infobox keys. For example, the names of mountains in Chinese usually end with the same Chinese character "山". With these rules, we generate candidate entities for each type.

However, the heuristics-based categorization may also contain wrongly-typed entities. For instance, the last character of many Chinese people's names is also "山". So in the second step, we adopt a machine learning method to filter these errors.

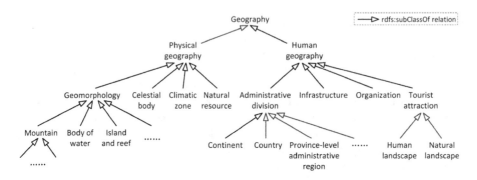

Fig. 2. Excerpt of the geography ontology. The actual version contains 130 classes with preferred labels in Chinese and alternative labels in both Chinese and English, where 35 leaf classes are newly involved as compared with GeoNames.

Specifically, an SVM classifier is employed with the RBF (Radial Basis Function) kernel[8], which performs better than the linear kernel based on human observation on some sample data. Article titles, subtitles, keys in infoboxes, first sentences of abstracts and open tags are pre-processed by word segmentation, stop word removal and TF-IDF scoring, and form the feature vectors for SVM. By manually creating a training set for each type, we find that the SVM-based filtering performs very well. More details are provided in Sect. 3.

After identifying all geographical entities for Clinga, we extract all the properties from their infobox keys. We follow the idea of DBpedia to distinguish raw properties and mapping-based properties. In total, there are 41,827 raw properties in all the infoboxes, in which 124 properties are mandated by Baidu Baike when creating articles, while the rest of them are user-customized and not validated. By considering the infobox values and manual refinement, 54 properties are identified as object properties and 70 are datatype properties.

To ensure the coverage of the Clinga dataset, we also use a gazetteer about administrative divisions and landforms in China. Furthermore, coordinates are complemented (if not exists in infoboxes) by querying Baidu Map.

2.3 Linking to Existing Knowledge Bases

Following the Linked Data principles, we connect Clinga to DBpedia and GeoNames to promote the integration and sharing of both Chinese and cross-lingual geographical knowledge. DBpedia has become the hub of the Linked Data, and GeoNames is one of the most famous geo-spatial database, but their schemas are completely different. To overcome the heterogeneity and achieve good accuracy, we first manually exploit 118 mappings among their classes with our best effort. Two class mappings between Clinga, DBpedia and GeoNames are below:

---

[8] We use scikit-learn 0.17.1 (via LIBSVM) with default parameters.

clinga:JiChang $=_M$ dbpedia:Airport $=_M$ (gn:featureCode value 'S.AIRP'),

where $=_M$ denotes the exact match relation between two classes.

Then, we use these class mappings to guide the entity linking process. Let **C** be the Clinga dataset and **D** be DBpedia or GeoNames. The links between them, denoted by $link(\mathbf{C}, \mathbf{D})$, are defined as the set of entity pairs having compatible types and similarities larger than $\eta \in [0, 1)$:

$$link(\mathbf{C}, \mathbf{D}) = \{(c, d) \in \mathbf{C} \times \mathbf{D} \mid type(c) \simeq_M type(d), sim(c, d) > \eta\}, \quad (1)$$

where $\simeq_M$ denotes the compatible relation derived from the class mappings and subclass relations. $sim()$ calculates similarities based on name comparison:

$$sim(c, d) = \begin{cases} 1 & \exists a \in alias(c) \, \exists b \in alias(d) \, (a = b) \\ 0 & \text{otherwise} \end{cases}, \quad (2)$$

where $alias()$ involves various aliases, such as official names and English names. Besides, Traditional Chinese names are translated to Simplified Chinese.

For the entities holding the disambiguation relation, an *ad hoc* post filtering method is used by considering subtitles and coordinates, which keeps only one link with highest similarity for each entity.

3 Dataset Statistics

In this section, we report the statistical data of Clinga and a preliminary evaluation on its accuracy. Overall, we obtained 13,068,395 articles from Baidu Baike, and the uncompressed raw data is about 180 GB. From the articles, we identified more than 624 thousand geographical entities and generated around 73 million RDF triples (see Table 1). As compared with the Chinese portion of GeoNames 2015, Clinga contains a larger number of Chinese geographical entities and much more triples. Moreover, XLore reported in 2012 [9] that it obtained 185 thousand Chinese geographical entities using the original "Geography" category of Baidu Baike. We argue that this smaller quantity is caused by the increase of articles in recently years and the mismatch between categories and articles.

Table 2 lists the numbers of entities and links w.r.t. eight upper-level classes. We can see that the numbers vary between classes. The three classes with most entities are "Infrastructure", "Administrative division" and "Organization". Except "Administrative division", the other two have little overlap with DBpedia and GeoNames, indicating the complementarity of the three datasets.

To evaluate the accuracy of entity categorization, for each class we manually set up a training set containing 500 examples (the ratio of positive and negative examples is 5:1), and conducted 5-fold cross-validation. Classes with less than 500 entities were not counted. The F1-score in average is 0.94 and the standard deviation is 0.01. This good result was achieved because of the significant difference between entities in geography and other classes (e.g. people and movies).

Table 1. General statistics and comparison with GeoNames and XLore

	No. of entities	No. of RDF triples
Clinga	624,391	73,326,425
GeoNames (gn:alternateName lang 'zh')	467,679	546,824
XLore ("Geography" category of Baidu Baike)	185,204	665,287

Last accessed date of GeoNames RDF dump is April 21, 2015.

Table 2. Class-based statistics and entity links to DBpedia and GeoNames

Classes	No. of entities	No. of links to	
		DBpedia	GeoNames
Geomorphology	41,999	26,914	33,076
Celestial body	365	272	0
Climatic zone	28	17	0
Natural resource	1,324	128	56
Administrative division	128,697	50,694	63,820
Infrastructure	356,937	38,968	12,311
Organization	88,848	2,789	250
Tourist attraction	8,801	1,497	1,567

Note that an entity can have multiple types, e.g. Mount Huang can be both a mountain and a tourist attraction.

However, for entities not included using our heuristic rules, they would be missed by the entity categorization algorithm, which is a limitation of our method.

For entity linking, we randomly chose 100 entities for each class and manually judged the correctness of 17,381 entity links in total. The precision of entity links is 0.81. We observed that the precision on "Street" and "Administrative village" is not good, because there exist a number of articles with exactly the same title but no more information for disambiguation.

4   Related Work

The vast increase of data sources containing geographical information is bound up with the diversity of geo-spatial applications such as location-based services, among which GeoNames (see Footnote 1) is perhaps the most famous geo-spatial database collecting data like place names in many languages from many sources. GeoWordNet [3] is a semantic and linguistic resource developed by the full integration of GeoNames with WordNet and MultiWordNet Italian portion. Linked-GeoData [7] transformed and represented OpenStreetMap using RDF, and linked itself with DBpedia, GeoNames and others. GeoLink (see Footnote 2) aims to improve data retrieval, reuse, and integration of seven geoscience data repositories with ontologies. Moreover, GeoLinkedData.es[9] and Ordnance Survey[10]

---

[9] http://geo.linkeddata.es.

[10] http://data.ordnancesurvey.co.uk/datasets/os-linked-data.

created linked geographical data for Spain and Great Britain, respectively. Currently, Chinese data in these datasets are still sparse and often limited to names.

Zhishi.me [6] and XLore [9] are two early works on extracting open-domain knowledge from the Chinese Wikipedia, Hudong Baike and Baidu Baike. Their difference is that Zhishi.me did not provide an ontology to describe the crawled data, while XLore automatically built an ontology using categories and infobox keys. By focusing on the geographical domain, we manually constructed a geography ontology having a more consistent category structure and covering more physical and human geographical topics. Also, the amount of entities and RDF triples in the Clinga dataset is larger, due to our exhaustive ways of entity extraction and content RDFization.

5   Conclusion and Future Work

Clinga is our ongoing effort to make Chinese geographical data easily findable, accessible, interoperable and reusable. We obtained the data from Baidu Baike, and built a geography ontology to categorize physical and human geographical entities, which we further linked to DBpedia and GeoNames. At present, Clinga is freely accessible through RDF dump, URI dereference, SPARQL endpoint and keyword search. The challenges for future work include a continual improvement of the Clinga's quality, and the study of new knowledge extraction and integration methods, to better support semantic geographical applications.

Acknowledgments. This work is supported by the National High-tech R&D Program of China (No. 2015AA015406) and the National Natural Science Foundation of China (Nos. 61370019, 61223003 and 61321491).

References

1. Becker, C., Bizer, C.: Exploring the geospatial semantic web with DBpedia mobile. J. Web Seman. **7**(4), 278–286 (2009)
2. Bollacker, K., Evans, C., Paritosh, P., Sturge, T., Taylor, J.: Freebase: a collaboratively created graph database for structuring human knowledge. In: SIGMOD 2008, pp. 1247–1250. ACM (2008)
3. Giunchiglia, F., Maltese, V., Farazi, F., Dutta, B.: GeoWordNet: a resource for geo-spatial applications. In: Aroyo, L., Antoniou, G., Hyvönen, E., Teije, A., Stuckenschmidt, H., Cabral, L., Tudorache, T. (eds.) ESWC 2010. LNCS, vol. 6088, pp. 121–136. Springer, Heidelberg (2010). doi:10.1007/978-3-642-13486-9_9
4. Hoffart, J., Suchanek, F., Berberich, K., Weikum, K.: YAGO2: a spatially and temporally enhanced knowledge base from wikipedia. Artif. Intell. **194**, 28–61 (2013)
5. Lehmann, J., Isele, R., Jakob, M., Jentzsch, A., Kontokostas, D., Mendes, P., Hellmann, S., Morsey, M., van Kleef, P., Auer, S., Bizer, C.: DBpedia - a large-scale, multilingual knowledge base extracted from wikipedia. Seman. Web J. **6**(2), 167–195 (2015)
6. Niu, X., Sun, X., Wang, H., Rong, S., Qi, G., Yu, Y.: Zhishi.me - weaving chinese linking open data. In: Aroyo, L., Welty, C., Alani, H., Taylor, J., Bernstein, A., Kagal, L., Noy, N., Blomqvist, E. (eds.) ISWC 2011. LNCS, vol. 7032, pp. 205–220. Springer, Heidelberg (2011). doi:10.1007/978-3-642-25093-4_14

7. Stadler, C., Lehmann, J., Höffner, K., Auer, S.: LinkedGeoData: a core for a web of spatial open data. Seman. Web J. **3**(4), 333–354 (2012)
8. Vrandečić, D., Krötzsch, M.: Wikidata: a free collaborative knowledgebase. Commun. ACM **57**(10), 78–85 (2014)
9. Wang, Z., Wang, Z., Li, J., Pan, J.Z.: Knowledge extraction from chinese wiki encyclopedias. J. Zhejiang Univ. Sci. C **13**(4), 268–280 (2012)

LinkGen: Multipurpose Linked Data Generator

Amit Krishna Joshi[✉], Pascal Hitzler, and Guozhu Dong

Data Semantics Lab, Wright State University, Dayton, OH, USA
{joshi.35,pascal.hitzler,guozhu.dong}@wright.edu

Abstract. The paper presents LinkGen, a synthetic linked data generator that can generate a large amount of RDF data based on certain statistical distribution. Data generation is platform independent, supports streaming mode and produces output in N-Triples and N-Quad format. Different sets of output can be generated using various configuration parameters and the outputs are reproducible. Unlike existing generators, our generator accepts any vocabulary and can supplement the output with noisy and inconsistent data. The generator has an option to inter-link instances with real ones provided that the user supplies entities from real datasets.

1 Introduction

In recent years, we have seen a rapid adoption of semantic technologies by a number of large organizations such as BBC, Thomson Reuters, New York Times and Library of Congress [3]. Linked Open Data (LOD) cloud consists of more than 30+ billion triples and hundreds of datasets[1]. These datasets use a number of vocabularies to describe the group of related resources and relationships between them. According to [16], Linked Open Vocabularies (LOV) dataset now consists of more than 500 vocabularies, 20,000 classes and almost 30,000 properties. The vocabularies are modeled using either RDF Schema (RDFS) or richer ontology languages such as OWL [5].

Linking enterprise data is also gaining popularity and industries are perceiving semantic technologies as a key contributor for effective information and knowledge management [14,18]. One of the major obstacles for building a linked data application is generating a synthetic dataset to test against specific vocabularies. In this paper, we present LinkGen, a synthetic linked data generator that generates arbitrarily large datasets for a given vocabulary. Generating synthetic data is not a new concept. It has been widely used in database field for testing database design and software applications as well as database benchmarking and data masking [2]. In the semantic web field, it has been primarily used for benchmarking Triplestores. Existing generators [4,7,9,13] are designed for specific use cases and work well with certain vocabularies but cannot be re-purposed for other vocabularies. LinkGen, on the other hand, can work with widely available vocabularies and can be used in multiple scenarios including:

---

[1] http://lod-cloud.net/.

P. Groth et al. (Eds.): ISWC 2016, Part II, LNCS 9982, pp. 113–121, 2016.
DOI: 10.1007/978-3-319-46547-0_12

(1) Testing new vocabulary (2) Querying datasets (3) Diagnosing data inconsistencies (4) Evaluating performance of datasets (5) Testing Linked Data aggregators (6) Evaluating various compression methods.

Creating synthetic datasets that closely resemble real world datasets is very important. Numerous studies including [6,15] found that URIs in real world linked datasets exhibit a power-law distribution. In order to automatically generate synthetic data that exhibit such power-law distribution, LinkGen employs random data generation based on various statistical distributions including Zipf's Law[2].

Real world linked datasets are by no means free of noise and redundancy. Linked Data quality and noise in Linked Data has been studied extensively in [10,11,17,19]. The noise can be in the form of invalid data, syntactic errors, inconsistent data and wrong statements. LinkGen provides some of these options to add noise in the synthetic dataset. LinkGen also has the option to specify the number of triples to generate. It aids in testing existing linked data compression methods such as [6,8] against varying database size and scenarios.

Specifically, the contribution of this work is a tool that can automatically generate synthetic datasets with the following properties:

- Dataset can be generated based on power-law distribution to resemble real world datasets
- Noise can be added to the synthetic dataset
- Dataset can be generated in both streaming and on-disk mode
- Synthetic instances can be linked to real-world entities if dictionary of real world entities is available.

The rest of this paper is organized as follows. Section 2 describes related work and existing generators. Section 3 describes the LinkGen generator with details on various parameters including data distribution and noisy data. Section 4 reports on experimental results and finally, Sect. 5 concludes the paper and identifies topics for further research. The tool is open source and available at GitHub[3] under GNU License[4].

2   Related Work

To the best of our knowledge, this is the first work that generates synthetic linked dataset for any vocabulary that can mimic real world datasets with features such as statistical distribution and noisy data. Quite a few synthetic generators exist that have been developed for benchmarking RDF stores using specific vocabularies. The Lehigh University Benchmark (LUBM) [7] consists of a data generator that produces repeatable and customizable synthetic dataset using Univ-Bench Ontology in the unit of a university. Different set of data can be generated by

---

[2] To review Zipf's and Pareto's Law, see [1].

[3] http://www.w3id.org/linkgen.

[4] https://opensource.org/licenses/GPL-3.0.

specifying the seed for random number generation, number of universities and the starting index of the universities.

Berlin SPARQL Benchmark (BSBM) [4] is built around an e-commerce use-case in which a set of products is offered by different vendors and consumers have posted reviews about products. BSBM constitutes a data generator that supports the creation of large datasets using number of products as the scale factor and can output in an RDF representation as well as relational representation.

SP^{2}Bench [13] has a data generator for creating DBLP[5]-like RDF triples and mimics correlations between entities using power law distributions and growth curves. The Social Intelligence Benchmark (SIB) [12] contains an S3G2 (Scalable Structure-correlated Social Graph Generator) that creates a synthetic social graph with correlations. Tontogen[6] is a protege-plugin that can create synthetic dataset using a uniform distribution of instances for relationships. WatDiv[7] and Sygenia[8] are two other tools that can generate data based on user supplied queries.

As noted above, none of the existing generators are suitable for creating synthetic data for different vocabularies. They have a little or no option to configure the output in regards to data distribution, noise and alignments.

3   Data Generator

In this section, we describe different concepts related to the data generator and provide details on how it works. At the core of data generation is a random data generator used for generating unique identifiers for each entity. In order to create different sets of output, LinkGen creates random data based on the seed value supplied by the user.

3.1   Entity Distribution

There are different statistical methods to generate and distribute entities in a dataset. LinkGen provides two statistical distribution techniques namely Gaussian distribution and Zipf's power-law distribution. Example of Gaussian distribution includes those in real life phenomena such as heights of people, errors in measurement and marks on a test. Examples of Power-law distributions include the frequencies of words and frequencies of family names. [6,15] found that subject URIs in real world linked datasets exhibit a power-law distribution. LinkGen use zipf's law as a default option for entity distribution. Figure 1 taken from [6] shows the power-law distribution of subjects in a Wikipedia dataset.

---

[5] http://dblp.uni-trier.de/db/.
[6] http://lsdis.cs.uga.edu/projects/semdis/tontogen/.
[7] http://dsg.uwaterloo.ca/watdiv/#download.
[8] https://sourceforge.net/projects/sygenia/.

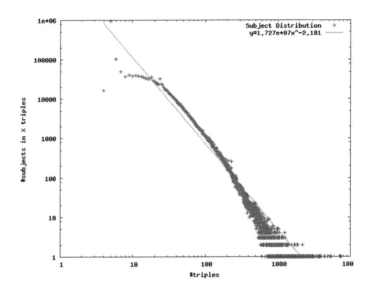

Fig. 1. Power-law distribution of subjects in Wikipedia

3.2   Noisy Data

Noisy data plays a critical role in applications that aggregate data from multiple sources and those that deal with semi-structured and unstructured data [10]. LinkGen creates noisy data by:

- Adding inconsistent data, for instance writing two conflicting values for a given dataType property
- Adding triples with syntactic errors, ex: typos in subjectURI or rdfs:Label
- Adding wrong statement by assigning invalid domain and range, ex: ns:PlaceInstance rdf:type ns:Person
- Creating instances with no type information

Users can specify a combination of parameters for generating noisy data. All parameters related to noise are prefixed with *noise.data* text in the configuration file ex: *noise.data.total* and *noise.data.num.notype*. If the output is in N-Quads format, the noisy data are added to a separate named graph.

3.3   Inter-linking Real World Entities

LinkGen allows mapping real world entities with automatically generated entities. For this, the user has to supply a set of real world entities expressed in RDF format: <ns:entityuri> rdf:type <ns:class>. LinkGen will then inter-link by using owl:sameAs triple, such as: <ns:entityuri> owl:sameAs <ns:classInstance>. This enables users to create a mixed dataset by combining synthetic dataset with the real dataset. This is important in scenarios where you would need to study the effect of adding new triples in current live dataset. Existing SPARQL queries can be slightly modified to fetch additional results from test dataset by adding owl:sameAs statement in the query.

3.4   Output Data and Streaming Mode

LinkGen creates a VoID[9] dump once the synthetic data is generated. VoID, the Vocabulary of Interlinked Datasets, is used to express metadata about RDF dataset and provides a number of properties for expressing numeric statistical characteristics about a dataset, such as the number of RDF triples, classes, properties or, the number of entities it describes.

LinkGen supports N-Triples and N-Quads format for output data. By default, the tool will save output to a file but it can be run in streaming mode, enabling users to pipe the output of RDF streams to other custom applications.

3.5   Config Parameters

There's an array of configuration parameters available to create unique synthetic datasets. The output is reproducible so running LinkGen multiple times with same set of input parameters will yield same output. Most useful configuration parameters include: (a) distribution type which can be gaussian or zipf and (b) seed values for creating different datasets.

3.6   Data Generation Steps

The first step in data generation involves loading ontology and gathering statistics about all ontology components such as number of classes, datatype properties, object properties and properties for which domain and range are not defined. We also store the connectivity of each class and order the classes based on the frequency. Most connected class will lead to generation of larger number of corresponding entities.

The second step involves using statistical distribution to generate large number of entities and associating the weights for each one of them. Parameters for Zipf and Gaussian distribution are configurable and can be used to create different sets of output. For Zipf's distribution, sample size is equal to the size of maximum number of triples to be generated. For Gaussian distribution, two parameters viz. mean and standard deviation are required.

Next step involves going through each class and generating synthetic triples for associated properties using weighted entities. For each entity, at least two triples are added to denote its type. They are: *instance* rdf:type *Classs* and, *instance* rdf:type *owl:Thing*. It should be noted that not all properties have well defined domain and range. For instance, in DBpedia, more than 600 properties including the ones in Table 1 have either missing domain or range information

---

[9] https://www.w3.org/TR/void/.

Table 1. Properties with no domain or range info in DBpedia ontology

DataTypeProperty with no domain	ObjectProperty with no range
http://dbpedia.org/ontology/number	http://dbpedia.org/ontology/teachingStaff
http://dbpedia.org/ontology/width	http://dbpedia.org/ontology/daylightSavingTimeZone
http://dbpedia.org/ontology/distance	http://dbpedia.org/ontology/simcCode
http://dbpedia.org/ontology/fileSize	http://dbpedia.org/ontology/uRN

in the vocabulary. In such cases, RDF Semantics[10] permits using any resources as a domain of the property. Similarly, the range can be any Literal or resource depending on whether the property is datatypeProperty or objectProperty.

For datatypeProperties which have range of XSD datatypes, we used a simple random generator to create literal values.

4   Evaluation

To evaluate our work, we generated varying number of synthetic datasets for two general purpose vocabularies: DBpedia[11] and schema.org[12]. For schema.org, we used an owl version available from TopBraid[13]. We built LinkGen using Apacha Jena[14], a widely used free and open source Java framework for building Semantic Web and Linked Data applications. At the current state, LinkGen supports only RDFS vocabularies. Although it can generate synthetic dataset for any vocabulary expressed in RDFS or OWL, it does not implement all class descriptions and property restrictions specified in the OWL ontology. Also, the support for blank nodes is not provided.

Table 2 shows the general characteristics of the dataset used for the experiment. For both DBpedia and Schema.org, the most connected classes were Person, Place and owl:Thing.

Table 2. Characteristics of the datasets used for evaluation

	DBpedia	Schema.org
Number of distinct classes	147	158
Number of distinct properties	2891	1002
Number of distinct object properties	1734	463
Number of distinct data properties	1100	490
distinct properties without domain and/or range specification	685	11

---

[10] https://www.w3.org/TR/2000/CR-rdf-schema-20000327/.
[11] http://www.dbpedia.org.
[12] http://www.schema.org.
[13] http://topbraid.org/schema/.
[14] http://jena.apache.org/.

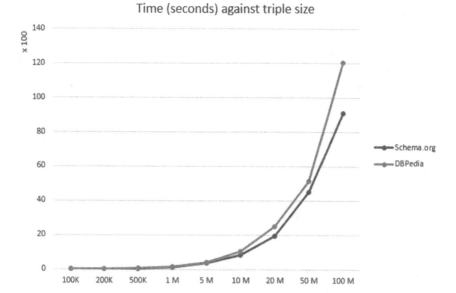

Fig. 2. Time taken for generating datasets of various sizes

Figure 2 is the performance chart depicting the total time taken to create synthetic datasets of varying size for both vocabularies. There's a slight increase in time for DBpedia which may be due to the relatively high number of properties.

5   Conclusion

In this paper, we have introduced a multipurpose synthetic linked data generator. The system can be configured to generate various sets of output to test semantic web applications under different scenarios. This includes defining a statistical distribution type for instances, adding inconsistent and noisy data, and integrating real world entities. The system supports streaming mode which can be used for evaluating applications that deal with streaming data. By generating a large amount of RDF data, it can aid in testing the performance of various applications that deal with querying, storage, visualization, compression and reporting. Experimental results show that out generator is highly performant and scalable. In the future, we will explore supporting OWL constraints as well as using parallel and distributed algorithms to generate massive dataset in short duration.

References

1. Adamic, L.A.: Zipf, Power-Laws, and Pareto - A Ranking Tutorial. Xerox Palo Alto Research Center, Palo Alto (2000). http://ginger.hpl.hp.com/shl/papers/ranking/ranking.html

2. Arasu, A., Kaushik, R., Li, J.: Data generation using declarative constraints. In: Proceedings of the 2011 ACM SIGMOD International Conference on Management of data, pp. 685–696. ACM (2011)

3. Bizer, C., Heath, T., Berners-Lee, T.: Linked data-the story so far. In: Semantic Services, Interoperability and Web Applications: Emerging Concepts, pp. 205–227 (2009)

4. Bizer, C., Schultz, A.: Benchmarking the performance of storage systems that expose SPARQL endpoints. In: World Wide Web Internet and Web Information Systems (2008)

5. Brickley, D., Guha, R.V.: RDF vocabulary description language 1.0: RDF schema (2004)

6. Fernández, J.D., Martínez-Prieto, M.A., Gutierrez, C.: Compact representation of large RDF data sets for publishing and exchange. In: Patel-Schneider, P.F., Pan, Y., Hitzler, P., Mika, P., Zhang, L., Pan, J.Z., Horrocks, I., Glimm, B. (eds.) ISWC 2010. LNCS, vol. 6496, pp. 193–208. Springer, Heidelberg (2010). doi:10.1007/978-3-642-17746-0_13

7. Guo, Y., Pan, Z., Heflin, J.: LUBM: a benchmark for OWL knowledge base systems. Web Seman. Sci. Serv. Agents World Wide Web 3(2), 158–182 (2005)

8. Joshi, A.K., Hitzler, P., Dong, G.: Logical linked data compression. In: Cimiano, P., Corcho, O., Presutti, V., Hollink, L., Rudolph, S. (eds.) ESWC 2013. LNCS, vol. 7882, pp. 170–184. Springer, Heidelberg (2013). doi:10.1007/978-3-642-38288-8_12

9. Morsey, M., Lehmann, J., Auer, S., Ngonga Ngomo, A.C.: DBpedia SPARQL benchmark-performance assessment with real queries on real data. In: The Semantic Web-ISWC 2011, pp. 454–469 (2011)

10. Paulheim, H., Bizer, C.: Improving the quality of linked data using statistical distributions. Int. J. Seman. Web Inf. Syst. (IJSWIS) 10(2), 63–86 (2014)

11. Péron, Y., Raimbault, F., Ménier, G., Marteau, P.F.: On the detection of inconsistencies in RDF data sets and their correction at ontological level (2011)

12. Pham, M.-D., Boncz, P., Erling, O.: S3G2: a scalable structure-correlated social graph generator. In: Nambiar, R., Poess, M. (eds.) TPCTC 2012. LNCS, vol. 7755, pp. 156–172. Springer, Heidelberg (2013). doi:10.1007/978-3-642-36727-4_11

13. Schmidt, M., Hornung, T., Lausen, G., Pinkel, C.: SP²Bench: a SPARQL performance benchmark. In: IEEE 25th International Conference on Data Engineering, ICDE 2009, pp. 222–233. IEEE (2009)

14. Semeraro, G., Basile, P., Basili, R., De Gemmis, M., Ghidini, C., Lenzerini, M., Lops, P., Moschitti, A., Musto, C., Narducci, F., et al.: Semantic technologies for industry: from knowledge modeling and integration to intelligent applications. Intelligenza Artificiale 7(2), 125–137 (2013)

15. Tummarello, G., Morbidoni, C., Bachmann-Gmür, R., Erling, O.: RDFSync: efficient remote synchronization of RDF models. In: Aberer, K., et al. (eds.) ASWC/ISWC -2007. LNCS, vol. 4825, pp. 537–551. Springer, Heidelberg (2007). doi:10.1007/978-3-540-76298-0_39

16. Vandenbussche, P.Y., Atemezing, G.A., Poveda-Villalón, M., Vatant, B.: Linked Open Vocabularies (LOV): a gateway to reusable semantic vocabularies on the web. Seman. Web, 1–16 (2015, preprint)

17. Wienand, D., Paulheim, H.: Detecting incorrect numerical data in DBpedia. In: Presutti, V., d'Amato, C., Gandon, F., d'Aquin, M., Staab, S., Tordai, A. (eds.) ESWC 2014. LNCS, vol. 8465, pp. 504–518. Springer, Heidelberg (2014). doi:10.1007/978-3-319-07443-6_34

18. Wood, D.: Linking Enterprise Data. Springer Science & Business Media, New York (2010)

19. Zaveri, A., Rula, A., Maurino, A., Pietrobon, R., Lehmann, J., Auer, S.: Quality assessment for linked data: a survey. Seman. Web **7**(1), 63–93 (2015)

OntoBench: Generating Custom OWL 2 Benchmark Ontologies

Vincent Link[1], Steffen Lohmann[2(✉)], and Florian Haag[1]

[1] Institute for Visualization and Interactive Systems, University of Stuttgart,
Universitätsstraße 38, 70569 Stuttgart, Germany
[2] Fraunhofer Institute for Intelligent Analysis and Information Systems (IAIS),
Schloss Birlinghoven, 53757 Sankt Augustin, Germany
`steffen.lohmann@iais.fraunhofer.de`

Abstract. A variety of tools for visualizing, editing, validating, and documenting OWL ontologies have been developed in the last couple of years. The OWL coverage and conformance of these tools usually needs to be tested during development for evaluation and comparison purposes. However, in particular for the testing of special OWL concepts and concept combinations, it can be tedious to find suitable ontologies and test cases. We have developed OntoBench, a generator for OWL benchmark ontologies that can be used to test and compare ontology tools. In contrast to existing OWL benchmarks, OntoBench does not focus on scalability and performance but OWL coverage and concept combinations. Consistent benchmark ontologies are dynamically generated based on any combination of OWL 2 language constructs selected in a graphical user interface. OntoBench is available on GitHub and as a public service, making it easy to use the tool to generate custom benchmark ontologies and ontology fragments.

Keywords: Ontology · Benchmark · Generator · OWL 2 · Coverage · Conformance

1 Introduction

A large number of tools that support the visualization, editing, validation, and documentation of OWL ontologies have been developed in the last couple of years. During the development of such tools, it is important to test them with ontologies representing different test cases (henceforth called *benchmark ontologies*) in order to ensure that the language constructs of OWL are adequately represented. Benchmark ontologies are also useful to support the comparison and evaluation of existing tools in order to assess the features of the tools and to check whether they provide adequate support for a certain use case.

In our previous work [7,8], we developed a static benchmark ontology for the purpose of testing feature completeness of ontology visualization tools. With OntoBench, we took this idea one step further and made it generally applicable in different use cases. Many ontology tools do not aim for a complete coverage

© Springer International Publishing AG 2016
P. Groth et al. (Eds.): ISWC 2016, Part II, LNCS 9982, pp. 122–130, 2016.
DOI: 10.1007/978-3-319-46547-0_13

of OWL but focus on specific aspects, or are designed to cover only some of the OWL profiles. To overcome the inflexibility and overhead caused by a static benchmark ontology, we developed a systematic approach to dynamically generate benchmark ontologies tailored to the OWL coverage and feature set that a tool intends to support.

As opposed to most other ontology benchmarks, OntoBench is not meant for testing the scalability or performance in terms of the number of elements contained in an ontology, but it rather aims to test the scope of ontology tools in terms of supported features and OWL constructs. Accordingly, it focuses on the representation of the TBox of ontologies (i.e., the classes, properties, datatypes, and a few key individuals), while it does not support the testing of ABox information (i.e., larger collections of individuals and data values), which is the focus of most of the related work.

2   Related Work

One well-known ontology benchmark is the Lehigh University Benchmark (LUBM) [6], a test suite for ontology-based systems. LUBM extends benchmarks for databases with a focus on the Semantic Web. It contains an ontology describing the university domain, a tool that generates instance data for the university ontology, and test queries for the data whose performance can be evaluated by using a couple of metrics provided by LUBM. Wang et al. extended the LUBM benchmark by implementing a domain-agnostic generator for instance data [18]. It uses a probabilistic model to generate a user-given number of instances based on representative data from the domain in focus. This enables testing a broader range of possible topics and, consequently, different kinds of ontology structures. As an example, they created the Lehigh BibTeX Benchmark (LBBM) on the basis of a BibTeX ontology.

Another extension of LUBM has been proposed by Ma et al. with the University Ontology Benchmark (UOBM) [12]. UOBM aims to contain the complete set of OWL 1 language constructs and defines two ontologies, one being compliant with OWL 1 Lite and the other with OWL 1 DL. In addition, several links were added between the generated instances in order to create more realistic data.

The W3C Ontology Working Group also published a number of small ontologies and ontology fragments together with the specifications of OWL 1 and 2, providing normative test cases [5,16]. These test cases are intended for validating applications in terms of conformance to the respective OWL version and to demonstrate the correct usage of OWL. The majority of these test cases aim at testing different syntaxes, specific combinations of OWL constructs, or reasoners.

Furthermore, there are benchmarks addressing certain aspects of ontology engineering. For instance, a number of datasets and test cases has been created in the context of the Ontology Alignment Evaluation Initiative (OAEI) [1]. They are intended to evaluate and compare the quality and performance of ontology matching methods in particular. Finally, benchmarks for comparing the performance of SPARQL endpoints are available that feature RDF data generators and provide sets of benchmark queries [4,15].

To conclude, there is currently no benchmark—except for the static OntoViBe ontology we developed in our previous work [7,8]—that focuses on testing ontology tools for feature-completeness with regard to OWL coverage, and which supports a major part of the concepts specified in OWL 2. For generating custom ontologies, one could use ontology editors like Protégé to manually create a benchmark ontology tailored to one's needs. However, despite the modeling support provided by ontology editors, reliably covering all relevant test cases can still be very error-prone and tedious for users.

3   Requirements and Design Considerations

To fill this gap, we developed OntoBench, a generator for benchmark ontologies with a focus on testing the OWL 2 coverage of ontology tools. The OWL language constructs contained in the generated ontologies are selected by the user. For this purpose, we have defined abstract features that encompass one or more OWL language constructs and thereby form a test case. The features can be individually enabled or disabled by the user when generating the benchmark ontology.

3.1   Requirements

The features defined for OntoBench were drawn from two main considerations:

1. A complete (as far as possible) coverage of OWL 2 language constructs was to be achieved, hence features were built around the list of language constructs.
2. OWL includes some concepts that cannot be represented by single language constructs but require combinations of constructs. Such combinations were also included as test cases.

Each test case can be seen as a fragment of the ontology to be generated. In order to optimally embed the fragments in the resulting ontology, we specified that the test cases should satisfy a couple of requirements:

Compactness. All test cases have to be designed compactly with regards to the amount of OWL constructs they require. On the one hand, this reduces side effects due to language constructs that are not in the focus of the test case. On the other hand, it improves the readability of the test cases in the generated ontology.

Independence. All test cases shall be defined as independently as possible in order to avoid that they interfere with each other. However, this goes along with a larger number of *helper* constructs in the ontology. For instance, properties can only reasonably be tested if classes are added that the properties are linked with. Since adding a pair of classes for each property would result in a large number of additional ontology elements, we used the same class as a domain for all properties in OntoViBe [8]. OntoBench builds upon this approach by reusing a domain class several times, but creating a new one once a given number of properties has been linked to that class.

Self-Descriptiveness. The elements for all used OWL language constructs have to be named in a way that eases the verification process. Like OntoViBe, OntoBench names elements according to their role in the benchmark ontology. However, where OntoViBe was static and could thus rely on the uniqueness of self-explanatory, yet non-systematically assigned names, OntoBench introduces a uniform naming scheme that ensures uniqueness of names despite the variation in generated ontologies. This is accomplished by prefixing all elements with the name of their corresponding test case. The suffix of the name indicates the role in the test case. For example, the class that serves as the range for the OWL construct `owl:ReflexiveProperty` is named *OwlReflexiveProperty_Range*.

3.2   OWL Profiles and Specific Test Cases

OWL 1 and 2 define multiple profiles of different expressiveness. These profiles restrict the set of eligible language constructs and the way the constructs can be combined. Some ontology tools do not support all elements defined by OWL but are limited by design to one of the less powerful profiles. Accordingly, OntoBench is able to generate ontologies that are conformant with the selected OWL profiles. It provides a preselection of test cases for OWL Lite and DL as defined in the OWL reference [3] as well as OWL 2 EL, RL and QL from the OWL 2 profiles [17]. There is no separate profile for OWL Full, since it does not contain new OWL constructs in comparison to OWL DL.

Since ontologies can make use of multilingualism, for instance in `rdfs:label` annotations, there are also test cases for this aspect.

4   Implementation as a Web Application

OntoBench is implemented as a web application to ease access and reuse [10].[1] The frontend implementation is based on HTML, CSS, and JavaScript in combination with SemanticUI and jQuery. A REST interface is used for communication with the backend, which is implemented as a Java server using the Spring Framework. This server contains the business logic for generating the customized ontologies by means of the OWL API [9]. Additionally, it manages a database of previously generated benchmark ontologies that can be restored via short URIs. These short URIs are provided for easy reference of the generated ontologies, whereas their long URIs are more persistent and transparent, as they include a list of the features contained in the ontology.

---

[1] OntoBench is released under the MIT license and available on GitHub at https://github.com/VisualDataWeb/OntoBench A public OntoBench service is available at http://ontobench.visualdataweb.org.

As an example, the main part of the ontology generated for the test case of the OWL construct `owl:AllDisjointClasses` is depicted in Listing 1.1. The order and indentation of all statements in the ontology is determined by the Turtle syntax formatter of the OWL API.

Listing 1.1. Main part of the ontology generated for the test case of the OWL construct `owl:AllDisjointClasses` formatted in Turtle syntax.

```
:AllDisjointClasses_Class1 rdf:type owl:Class .
:AllDisjointClasses_Class2 rdf:type owl:Class .
:AllDisjointClasses_Class3 rdf:type owl:Class .

[ rdf:type owl:AllDisjointClasses ;
owl:members ( :AllDisjointClasses_Class1
:AllDisjointClasses_Class2
:AllDisjointClasses_Class3
)
] .
```

4.1 Graphical User Interface

The graphical user interface (GUI) of OntoBench consists of two panels organized in tabs, one allows to configure the benchmark ontology and the other displays the generated output. Figure 1 shows screenshots of parts of the two panels. The configuration panel lists all eligible features grouped into categories, inspired by the grouping of the OWL 2 quick reference guide [2]. The categories are organized into frames in the GUI, while predefined buttons allow to immediately select certain presets, such as all elements of a category or all elements matching a particular OWL profile.

When the user hits the *generate* button or switches to the *generator* tab, the second panel is opened which displays the generated ontology. The ontology is shown on screen and can be downloaded as a file (cp. Fig. 1). It is provided in Turtle syntax by default, but the user can also chose other OWL serializations from a drop-down menu. OntoBench provides all OWL serializations supported by the OWL API (which are Turtle, Manchester, Functional, OWL/XML, and RDF/XML at the moment). The generated output as well as the endings of the ontology URIs change accordingly, so that a particular serialization can be directly accessed from remote via its URI.

OntoBench has been designed for a target group that is at least somehow familiar with OWL and/or wants to use or learn OWL. In some informal user tests, the user interface was praised for its ease of use. The test users liked that they could create OWL ontologies with only a few clicks and found the user interface very self-explanatory.

Fig. 1. Screenshots of the user interface of OntoBench showing parts of the two main panels

4.2 Extensibility

In the case that OntoBench does not provide a test case required in a certain situation, users can manually edit and extend the generated ontologies according to their needs. Alternatively, they can edit the source code of OntoBench and add the required test cases to the generator. The source code has an object-oriented design: Each feature is described by a class which is derived from a superclass containing helpers and providing access to the ontology. The feature can either be modeled by directly accessing the OWL API or by using the provided helper classes. Each feature has additionally a name and a token (for the URI) and is assigned to a category for grouping in the user interface. The user interface is automatically generated from the modeled features.

4.3 Limitations

Limitations in using and extending OntoBench result mainly from the OWL API that OntoBench is using to create the OWL ontologies. For instance, the OWL API makes use of the *OWL functional syntax* internally, which represents the OWL constructs `owl:AllDifferent` and `owl:differentFrom` both by the functional concept `DifferentIndividuals`. When having the OWL API output an ontology in Turtle syntax, it will always use `owl:AllDifferent` and never `owl:differentFrom`, both of which imply the same assertion with two individuals.

5   Validation of the Generated Ontologies

It is not feasible to validate the correctness of the generated ontologies in all possible combinations, but we systematically checked a representative subset using test classes and manual inspection. However, to some extent, we have to trust the OWL API that is used by OntoBench for generating the ontologies. Since it has "widespread usage in a variety of tools" [9] and intends to be a "reference implementation for creating, manipulating and serializing OWL Ontologies"[2], it can be assumed that the generated ontologies are mostly correct in terms of syntax and general structure.

Nevertheless, we applied syntax validators, such as the W3C RDF Validation Service [14], to the representative subset of generated ontologies. The tests showed that all ontologies were valid RDF documents. To validate whether the contents of the generated ontologies are correct, we tested the representative subset by loading the ontologies into different tools, including ontology editors like Protégé and reasoners like Pellet. These checks all showed that the generated ontologies are correct and contain the test cases that were selected in the user interface.

The presets matching OWL profiles were evaluated with the validators built into the OWL API and further refined by manual inspection and comparison with the OWL profiles specifications [3,17]. Some issues with the OWL API were found during these validations and reported to the issue tracker of that project on GitHub.[3] They could be quickly fixed by the developers of the OWL API so that we could finally include the corrected version of the API in OntoBench.

The scalability of OntoBench was tested by selecting different subsets of test cases in the user interface and run the generator. The ontologies are not cached but generated at runtime, which takes less than two seconds on the public OntoBench instance we provide, even if all elements or a large subset of them are selected. The resulting ontology can consist of more than 2000 lines in Turtle syntax in those cases.

6   Application in a Visualization Use Case

During the development of the latest version of the ontology visualization tool WebVOWL [11], we regularly used OntoBench to check whether WebVOWL displays all OWL language constructs according to the VOWL 2 specification [13]. Testing the generated ontologies with WebVOWL was very convenient, as we only had to append their URIs to the URL of WebVOWL. Since we noticed that a visualization like VOWL can help to better understand the generated ontologies, we integrated it into OntoBench by adding a button to the generator panel that directly opens the WebVOWL visualization of each ontology (cf. Fig. 1).

---

[2] http://owlapi.sourceforge.net.
[3] https://github.com/owlcs/owlapi/issues/435.

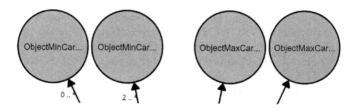

Fig. 2. Example issue found in the use case: while `owl:minCardinality` was shown in the beta version of WebVOWL as expected, `owl:maxCardinality` was not shown at all

In one of the test cases for WebVOWL, we generated an ontology with Onto-Bench containing all OWL language constructs that are supported according to the VOWL 2 specification [13]. We could uncover two issues this way: (1) The VOWL 2 specification states that `owl:Nothing` should either be visualized the same way as `owl:Thing` or should not be visualized at all, the latter being recommended. However, when we visualized the generated ontology with a beta version of WebVOWL, we discovered that `owl:Nothing` was incorrectly displayed as an external class. (2) While inspecting the indicated cardinalities in the visualization, we realized that in contrast to `owl:minCardinality`, `owl:maxCardinality` was not displayed at all in the beta version of WebVOWL (cf. Fig. 2). As can be seen in the figure, the generated names help to spot the concepts in the visualization and to interpret them correctly.

OntoBench can be extended to also include cases for ABox testing, without affecting the general approach. For instance, we added some specific test cases for the WebVOWL tool, among others a test case generating 1000 instances for a class. However, there is a high variety of such ABox test cases and its systematic investigation constitutes a considerable research effort that would warrant a separate project.

7   Conclusion

The application example illustrates how easy it is to test ontology tools like WebVOWL with OntoBench. Instead of creating suitable test cases from scratch with quite some effort, or searching for existing ontologies that contain those or similar cases, OntoBench allows to quickly generate tailored and consistent ontologies with only a few clicks.

References

1. Ontology alignment evaluation initiative. http://oaei.ontologymatching.org
2. Bao, J., Kendall, E.F., McGuinness, D.L., Patel-Schneider, P.F.: OWL 2 web ontology language quick reference guide (2nd edn.) (2012). https://www.w3.org/TR/owl2-quick-reference/

3. Bechhofer, S., van Harmelen, F., Hendler, J., Horrocks, I., McGuinness, D.L., Patel-Schneider, P.F., Stein, L.A.: OWL web ontology language reference. In: W3C Recommendation (2004). http://www.w3.org/TR/2004/REC-owl-ref-20040210/

4. Bizer, C., Schultz, A.: The Berlin SPARQL benchmark. Int. J. Semant. Web Inf. Syst. **5**(2), 1–24 (2009)

5. Carroll, J.J., Roo, J.D.: OWL web ontology language test cases (2004). http://www.w3.org/TR/owl-test/

6. Guo, Y., Pan, Z., Heflin, J.: LUBM: a benchmark for OWL knowledge base systems. Web Semant. **3**(2–3), 158–182 (2005)

7. Haag, F., Lohmann, S., Negru, S., Ertl, T.: OntoViBe: an ontology visualization benchmark. In: International Workshop on Visualizations and User Interfaces for Knowledge Engineering and Linked Data Analytics (VISUAL 2014), CEUR-WS, vol. 1299, pp. 14–27 (2014)

8. Haag, F., Lohmann, S., Negru, S., Ertl, T.: OntoViBe 2: advancing the ontology visualization benchmark. In: Lambrix, P., et al. (eds.) EKAW 2014. Lecture Notes in Artificial Intelligence (LNAI), vol. 8982, pp. 83–98. Springer, Heidelberg (2015). doi:10.1007/978-3-319-17966-7_9

9. Horridge, M., Bechhofer, S.: The OWL API: a java API for OWL ontologies. Semant. Web **2**(1), 11–21 (2011)

10. Link, V., Lohmann, S., Haag, F.: OntoBench: ontology benchmark generator (2016). http://ontobench.visualdataweb.org

11. Lohmann, S., Link, V., Marbach, E., Negru, S.: WebVOWL: web-based visualization of ontologies. In: Lambrix, P., et al. (eds.) EKAW 2014. Lecture Notes in Artificial Intelligence (LNAI), vol. 8982, pp. 154–158. Springer, Heidelberg (2015). doi:10.1007/978-3-319-17966-7_21

12. Ma, L., Yang, Y., Qiu, Z., Xie, G., Pan, Y., Liu, S.: Towards a complete OWL ontology benchmark. In: Sure, Y., Domingue, J. (eds.) ESWC 2006. LNCS, vol. 4011, pp. 125–139. Springer, Heidelberg (2006). doi:10.1007/11762256_12

13. Negru, S., Lohmann, S., Haag, F.: VOWL: visual notation for OWL ontologies (2014). http://purl.org/vowl/spec/v2/

14. Prud'hommeaux, E., Lee, R.: W3C RDF validation service (2004). http://www.w3.org/RDF/Validator

15. Schmidt, M., Hornung, T., Meier, M., Pinkel, C., Lausen, G.: Sp^{2}bench: a SPARQL performance benchmark. In: Virgilio, R.D., Giunchiglia, F., Tanca, L. (eds.) Semantic Web Information Management, pp. 371–393. Springer, Heidelberg (2009)

16. Smith, M., Horrocks, I., Krötzsch, M., Glimm, B.: OWL 2 web ontology language conformance (2nd edn.) (2012). http://www.w3.org/TR/owl2-conformance/

17. W3C OWL Working Group: OWL 2 web ontology language profiles (2nd edn.) (2012). https://www.w3.org/TR/owl2-profiles/

18. Wang, S.-Y., Guo, Y., Qasem, A., Heflin, J.: Rapid benchmarking for semantic web knowledge base systems. In: Gil, Y., Motta, E., Benjamins, V.R., Musen, M.A. (eds.) ISWC 2005. LNCS, vol. 3729, pp. 758–772. Springer, Heidelberg (2005). doi:10.1007/11574620_54

Linked Data (in Low-Resource) Platforms: A Mapping for Constrained Application Protocol

Giuseppe Loseto, Saverio Ieva, Filippo Gramegna, Michele Ruta$^{(\boxtimes)}$, Floriano Scioscia, and Eugenio Di Sciascio

Politecnico di Bari, via E. Orabona 4, 70125 Bari, Italy
{giuseppe.loseto,saverio.ieva,filippo.gramegna,
michele.ruta,floriano.scioscia,eugenio.disciascio}@poliba.it

Abstract. This paper proposes a mapping of the Linked Data Platform (LDP) specification for Constrained Application Protocol (CoAP). Main motivation stems from the fact that LDP W3C Recommendation presents resource management primitives for HTTP only. A general translation of LDP-HTTP requests and responses is provided, as well as a framework for HTTP-to-CoAP proxying. Experiments have been carried out using the LDP W3C Test Suite.

Keywords: Linked Data Platform · CoAP · Semantic web of things

Resource type: Software
Permanent URL: http://dx.doi.org/10.5281/zenodo.50701

1 Introduction and Motivation

The World Wide Web Consortium (W3C) has standardized the Linked Data (LD) management on the Web with the Linked Data Platform (LDP) specification [8]. Unfortunately, this effort leaves out the so-called Web of Things (WoT) where HTTP is replaced by simpler protocols, *e.g.*, CoAP (Constrained Application Protocol) [12], suitable for resource-constrained scenarios. CoAP adopts a loosely coupled client/server model, based on stateless operations on *resources* [2] identified by URIs (Uniform Resource Identifiers). Clients access them via asynchronous request/response interactions through HTTP-derived methods mapping the Read, Create, Update and Delete operations of data management. Section 3.12 of Linked Data Platform Use Cases and Requirements [1] reports on a possible one-to-one translation of HTTP primitives toward CoAP, nevertheless the proposed solution appears quite limited. The given mapping [3] only considers basic HTTP interactions: `options`, `head` and `patch` methods are not allowed and various MIME content-format types are missing.

Main motivation of this resource is to enable the extension of the Linked Data Platform standard to Web of Things contexts. A specific variant of the HTTP-CoAP mapping is proposed, preserving LDP features and capabilities:

P. Groth et al. (Eds.): ISWC 2016, Part II, LNCS 9982, pp. 131–139, 2016.
DOI: 10.1007/978-3-319-46547-0_14

Table 1. Current LDP implementations

Name	Status	Last Version	License	Language	Supported LDP Resources
RWW.IO	Pending	1.2 (Nov 2014)	MIT	PHP	RS, BC
Apache Marmotta	Full release	3.3.0 (Dec 2014)	APL 2.0	Java	RS, NR, BC
Bygle	In progress	Feb 2015	APL 2.0	Java	RS, BC
Eclipse Lyo	Completed	2.1.0 (Mar 2015)	EPL 1.0	Java	RS, NR, BC, DC
LDP.js	Completed	Apr 2015	APL 2.0	JavaScript	RS, BC, DC
Glutton	In progress	Apr 2015	GPLv3	Python	RS, BC
Carbon LDP	In progress	0.5.7 (Oct 2015)	BSD	JavaScript	RS, NR, BC, DC, IC
LDP4j	In progress	0.2.0 (Dec 2015)	APL 2.0	Java	RS, BC, DC, IC
RWW Play	In progress	2.3.6 (Dec 2015)	APL 2.0	Scala	RS, NR, BC
Fedora	Full release	4.5.0 (Jan 2016)	APL 2.0	Java	RS, NR, BC, DC, IC
Callimachus	Full release	1.5.0 (Mar 2016)	APL 2.0	Java	RS, NR, IC
Gold	In progress	1.0.1 (Apr 2016)	MIT	Go	RS, BC
OpenLink Virtuoso	Full release	7.2.5 (Apr 2016)	GPLv2	C/C++	RS, BC
ldnode	In progress	0.2.31 (Apr 2016)	MIT	JavaScript	RS, BC

the envisioned HTTP-CoAP proxy makes objects networks first-class Linked Data providers on the Web. Novel features are also giving added value to the strongest peculiarities of CoAP with respect to HTTP, *e.g.*, resource discovery via CoRE Link Format. The proposed solution is released as open source. Performance tests evidence LDP-CoAP supports all types of LDP resources keeping computational performances comparable with other frameworks. Results of the W3C LDP conformance test suite show the proposal does not completely cover LDP specification yet.

2   Coping with Lightweight Linked Data Platform

The LDP W3C Recommendation provides standard rules for accessing and managing Linked Data on the Web *LDP servers*. Basically, it defines seven types of LDP *Resources* as well as patterns of HTTP methods and headers for CRUD (Create, Read, Update, Delete) operations[1]. W3C LDP implementations web page (http://www.w3.org/wiki/LDP_Implementations) lists several software tools: Table 1 reports the most relevant ones along with main properties and supported resource types, in order of release date. All solutions are based on the HTTP protocol, with no current support for WoT standards such as CoAP.

The W3C suggests explicit use cases [1] aiming to integrate LDP in resource-constrained devices and networks with specific reference to CoAP [12], a compact protocol conceived for machine-to-machine (M2M) communication. Some CoAP options are derived from HTTP header fields (*e.g.*, content type, headers and proxy support), while some other ones have no counterpart in HTTP. So an HTTP-CoAP mapping is needed to exploit all LDP features with CoAP. An early mapping proposal was defined in [3], but it only worked with basic HTTP interactions. The HTTP-CoAP mapping for LDP envisioned in [7] and outlined here,

---

[1] Due to space constraints, details of LDP specification are not recalled here; basic knowledge of LDP is assumed, while the reader is referred to [8] for details.

Table 2. HTTP-CoAP mapping of preference headers

HTTP Header	LDP-CoAP
Prefer: return=representation; include="*pref*"	ldp-incl=*pref* Core Link Format attribute
Prefer:return=representation; omit="*pref*"	ldp-omit=*pref* Core Link Format attribute
Preference-Applied: return=representation	*pref* returned using location-query CoAP option

enables a direct CoAP-to-CoAP interaction. *HTTP methods* mapping is applied
for each CoAP method (if present). HEAD and OPTIONS, undefined in CoAP, are
mapped to existing GET and PUT methods, by adding the new Core Link Format
attribute ldp. There is full backward compatibility with the standard protocol, while extending the basic CoAP functionalities. W.r.t. the early proposal
[7], additional features have been also defined to support: (i) PATCH method;
(ii) *RDF Patch* format [10] along with application/rdf-patch content-format
media type; (iii) LDP *Prefer headers* of request/reply messages (Table 2).

LDP-CoAP mapping was implemented in a Java-based framework providing
the basic components required to publish Linked Data on the WoT according to
LDP-CoAP specification. It consists of several modules, as shown in Fig. 1a.

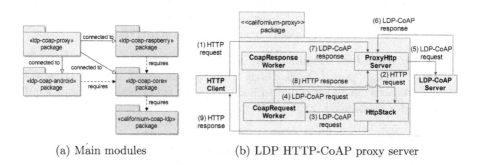

(a) Main modules   (b) LDP HTTP-CoAP proxy server

Fig. 1. LDP-CoAP framework architecture

ldp-coap-core: includes the implementation of all LDP-CoAP resources and a
basic LDP-CoAP server handling CoAP-based communication and RDF data
management. The main Java package coap.ldp was partitioned in the following
sub-packages each providing a specific functionality.
– coap.ldp.server: the reference CoAPLDPServer implementation. It extends
the CoAPServer provided by *californium-core-ldp* module (described below) and
exposes methods to create and manage LDP resources. The package also includes
the CoAPLDPTestSuiteServer, used for experiments described in Sect. 3.

– `coap.ldp.resources`: according to the LDP resource hierarchy [8], several Java classes were developed extending the `CoAPLDPResource` base class providing common methods and attributes. For each resource class, a specific data handler can be implemented to retrieve whatever kind of data (*e.g.*, observation from a sensor) and update the RDF repository with user-defined periodicity. Handlers can be defined starting from the `LDPDataHandler` abstract class. In this way, developers can build specific applications implementing the whole business logic and data management procedures within the `handleData` method of the handler, without any other modification of the source code. `CoAPLDPResourceManager` implements read-write operations on the RDF data storage exploiting an *Open-RDF Sesame* (http://rdf4j.org) in-memory RDF repository.

– `coap.ldp.handler`: two simple handlers were defined as usage examples to expose real-time system CPU load and RAM usage ratio as LDPRDFResource. Data are collected through the operating system interfaces of Java 7 (or later).

– `coap.ldp.exception`: a `CoAPLDPException` class was defined to catch errors due to incorrect usage of LDP methods, headers or attributes. Its subclasses represent typical problems (*e.g.*, *content format* or *precondition failed*).

– `rdf.vocabulary`: contains RDF ontologies mapped as Java classes to simplify creation and querying of RDF triple. As an example, *SSN-XG* ontology [4] was mapped through the *Sesame Vocabulary Builder* (http://github.com/tkurz/sesame-vocab-builder) tool and included here.

The following libraries are required to correctly compile *ldp-coap-core*: *JSON-java* (http://github.com/stleary/JSON-java) to format data in JSON; *jsonld-java* (http://github.com/jsonld-java) to support the `json-ld` specification [6]; *Apache Marmotta RDF Patch Util* (http://marmotta.apache.org/sesame.html) to update RDF statements of a Sesame repository according to the `rdf-patch` [10] format.

californium-core-ldp: a modified version of the *Californium* CoAP framework [5], extended to support LDP features. Main modifications include: (i) novel content-format media types added to `MediaTypeRegistry` class; (ii) additional response codes introduced within `CoAP` main class.

ldp-coap-proxy: a modified version of *californium-proxy* implementing the mapping rules defined before and translating LDP-HTTP request to the corresponding LDP-CoAP ones. As shown in Fig. 1b, LDP-CoAP mapping procedures take advantage of the classes in this module. In particular, *ProxyHttpServer* is responsible for processing a request –coming from a generic HTTP client– through its *HttpStack* member class where the mapping occurs. *HttpStack* transforms an HTTP request into a compatible LDP-CoAP one and for each CoAP request it starts two threads, *CoapRequestWorker* and *CoapResponseWorker*, synchronized according to the producer-consumer pattern. The *CoapRequestWorker* thread produces the LDP-CoAP translated request for the *ProxyHttpServer* class instance which forwards that request to the proper LDP-CoAP server. The *CoapResponseWorker* is responsible for consuming and translating the LDP-CoAP response coming from the *ProxyHttpServer* into the HTTP response which is returned to the client.

In addition to the basic framework, the following two packages were developed to build LDP-CoAP applications on embedded and resource-constrained devices.

ldp-coap-raspberry: *ldp-coap-core* was tested on a Raspberry Pi (http://www.raspberrypi.org) board. W.r.t. other LDP implementations, LDP-CoAP is very lightweight and simple to run on low-resource environments like Raspberry Pi, having a minimum number of dependencies and low system requirements in terms of memory and processing capabilites. As a reference example, two handlers were implemented to publish CPU temperature and free RAM as LDP resources. Data are retrieved using the *Pi4J* (http://pi4j.com) library.

ldp-coap-android: a simple project exploiting *ldp-coap-core* on Android devices. It runs unmodified on all platforms supporting modules compiled with Java SE runtime environment, version 7 or later, so it can be directly used as a library also by Android applications. Android OS provides a uniform interface (the Android sensor framework, http://developer.android.com/guide/topics/sensors/sensors_overview.html) to access sensor data. Therefore, a single handler (named `GenericSensorHandler`) was implemented to manage both hardware and software-based device sensors. The project includes a basic activity starting a LDP-CoAP server exposing data from interface sensors modeled as LDP resources. Source code is available on [9], including *Javadoc* documentation; quick usage examples are on the project website http://sisinflab.poliba.it/swottools/ldp-coap. All modules were developed as Eclipse (http://eclipse.org) projects using Apache Maven (http://maven.apache.org) to manage dependencies. Only *ldp-coap-android* is a project for Android Studio (http://developer.android.com/tools/studio/index.html), the Google official IDE for app development. In this case, all dependencies can be defined through a Gradle (http://gradle.org) configuration file.

A few validation examples are reported here, in order to clarify the proposal. Full examples are on the LDP-CoAP project website.

Ex. 1 – Basic HTTP GET request on an LDP resource. HTTP-CoAP mapping is shown in Fig. 2. As described in [7], a single CoAP `GET` request cannot produce all the needed headers. So the original HTTP request (Fig. 2a) is translated to three LDP-CoAP packets: a `GET` message (Fig. 2d), a CoAP discovery message (Fig. 2c), and an `OPTIONS` message (Fig. 2b). In particular, since `Allow`, `Accept-Post` and `Accept-Patch` response headers are not defined in CoAP, their values are set in the LDP-CoAP `OPTIONS` response body in JSON syntax and then mapped to the corresponding HTTP headers. As per the CoRE Link Format specification [11], the CoAP discovery request maps the HTTP `Link` header with the resource type (`rt`) retrieved via the /.well-known/core reserved resource path.

Ex. 2 – Create a new LDP resource through an HTTP POST request. In this case, the HTTP request (Fig. 3a) is translated to a single CoAP POST message, as in Fig. 3b (see [7] for details).

```
GET /alice/ HTTP/1.1
Host: example.org Accept: text/turtle

HTTP/1.1 200 OK
Content-Type: text/turtle; charset=UTF-8
Link: <http://www.w3.org/ns/ldp#BasicContainer> rel="type",
<http://www.w3.org/ns/ldp#Resource> rel="type"
Allow: OPTIONS,HEAD,GET,POST,PUT,PATCH
Accept-Post: text/turtle, application/ld+json
Accept-Patch: application/rdf-patch
Content-Length: 250
ETag: W/'123456789'
...RDF payload...
```

(a) HTTP GET

```
GET coap://example.org/alice?ldp=options

2.05 Content
Content-Format (ct): application/json
{
"Allow": ["OPTIONS", "HEAD", "GET",
"POST", "PUT", "PATCH"],
"Accept-Post": [ "text/turtle",
"application/ld+json"],
"Accept-Patch": "application/rdf-patch"
}
```

(b) CoAP OPTIONS

```
GET coap://example.org/.well-known/core?title=alice

2.05 Content
Content-Format (ct): application/link-format
</alice>
rt="http://www.w3.org/ns/ldp#BasicContainer
http://www.w3.org/ns/ldp#Resource";
ct=4; title="alice"
```

(c) CoAP Discovery

```
GET coap://example.org/alice/
Accept: text/turtle

2.05 Content
Content-Format (ct): text/turtle
ETag: W/'123456789'
...RDF payload...
```

(d) CoAP GET

Fig. 2. HTTP-CoAP mapping for an LDP GET request/response

```
POST /alice/ HTTP/1.1
Host: example.org Slug: foaf
Content-Type: text/turtle
...RDF payload...

HTTP/1.1 201 Created
Location: http://example.org/alice/foaf
Link: <http://www.w3.org/ns/ldp#Resource> rel='type'
Content-Length: 0
```

(a) HTTP POST

```
POST coap://example.org/alice?title=foaf
Content-Format (ct): text/turtle
...RDF payload...

2.01 Created
Location-Path:
coap://example.org/alice/foaf
```

(b) CoAP POST

Fig. 3. HTTP-CoAP mapping for an LDP POST request/response

3   Experiments

The W3C LDP Test Suite (http://w3c.github.io/ldp-testsuite/) is used to evaluate the functionality of the proposed framework and to compare it with existing solutions. The suite consists of 236 tests which query an LDP server by means of HTTP messages; only for LDP-CoAP requests were sent to the server through an LDP-CoAP proxy as in Fig. 1a. Obtained results are grouped by supported LDP resources (RDF Sources, Non-RDF Sources and Basic, Direct, Indirect Containers – see [8] for definitions) and compliance levels (MUST, SHOULD, MAY). For each resource/level pair, Table 3 compares the score of LDP-CoAP with the highest value obtained by other LDP tools. Full LDP-CoAP results are on the project website. Overall, LDP-CoAP presents good scores, when considering 17 manual tests were skipped in this first experimental campaign and only automated ones were executed.

In addition to LDP-CoAP 7 tools were evaluated: *Virtuoso, LDP.js, Apache Marmotta, LDP4j, RWW.IO, Fedora4* and *Eclipse Lyo*. They were selected according to the features listed in Table 1: current status, completeness, open license, last update and supported resources (in particular RDF Source and Basic

Table 3. Comparison of implementation conformance tests

Feature	MUST		SHOULD		MAY	
	LDP-CoAP	Highest Val.	LDP-CoAP	Highest Val.	LDP-CoAP	Highest Val.
LDP-RS	91.7 % (22/24)	100 % [a,b,c,d,e]	71.4 % (5/7)	100 % [a,b,c,d]	100 % (1/1)	100 % [all]
LDP-BC	86.5 % (32/37)	100 % [b,c,d,e]	88.2 % (15/17)	100 % [b,c]	100 % (4/4)	100 % [b,c,e,f]
LDP-DC	88.1 % (37/42)	100 % [b,d,e]	89.5 % (17/19)	100 % [b]	100 % (4/4)	100 % [b,d,f]
LDP-IC	84.6 % (33/39)	97.4 % [a]	88.2 % (15/17)	88.2 % [d]	100 % (4/4)	100 % [f]
LDP-NR	80.0 % (12/15)	100 % [a,b,c]	100 % (1/1)	100 % [a,b,c,f,i]	66.7 % (4/6)	100 % [b,c,f]

(a) Callimachus, (b) Eclipse Lyo, (c) Apache Marmotta, (d) LDP4j, (e) LDP.js, (f) Fedora4, (g) ldphp, (h) Virtuoso, (i) rww-play

Container). *Gold* was tested and discarded due to the limited compatibility with LDP specification. Only supported resources were taken into account to retrieve processing time. Each test was repeated three times on the same PC and (only for tests passed by all tools) the average value was reported in Fig. 4. Fedora4 and LDP-CoAP support all LDP resources. Eclipse Lyo and LDP4j manage four resources groups, whereas remaining frameworks only operate on RDF Sources and Basic Containers. LDP-CoAP has good processing times, as results are comparable with the other implementations even while involving the HTTP-CoAP proxy. Only for non-RDF Source tests performance is slightly worse.

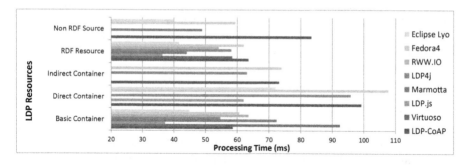

Fig. 4. Comparison of processing time for tested LDP implementations

To evaluate the feasibility of exploiting LDP in mobile and pervasive computing scenarios, LDP-CoAP performance was tested on three different Java-compatible platforms: a PC[2], an Android smartphone (LG Google E960 *Nexus 4*, specifications at http://www.lg.com/us/cell-phones/lg-LGE960-nexus-4) and a Raspberry Pi 1 Model B+ board (http://www.raspberrypi.org/products/model-b-plus/) All requests were originated from a PC client running both the LDP Test Suite and the LDP HTTP-CoAP proxy, connected through a local IEEE 802.11 network to one of the three LDP-CoAP servers for each test. The overall processing time, shown in Fig. 5, is defined as the time elapsed from

---

[2] With Intel Core i7 CPU 3770K at 3.50 GHz (4 cores/8 threads), 12 GB DDR3-SDRAM (1333 MHz), 2 TB SATA (7200 RPM) HD, 64-bit Microsoft Windows 7 Professional and 64-bit Java 8 SE Runtime Environment (build 1.8.0_65-b17).

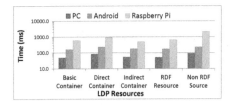

Fig. 5. Device time comparison

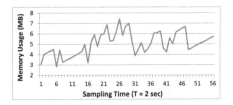

Fig. 6. Memory use on Raspberry Pi

sending the request until receiving a response by the client, including communication and HTTP-CoAP message translation times. Values on Android are roughly 3 times higher than on PC, whereas performance on Raspberry are an order of magnitude higher with respect to PC. However, average response times are under 1 second both on Android and Raspberry (except for LDP-NR responses on Raspberry). Memory usage was also measured every 2 s during the execution of the test suite for the three platforms. Memory allocation peak of the LDP-CoAP server was about 44.7 MB on PC, 18.3 MB on Android and 7.4 MB on Raspberry. Stricter memory constraints on smartphones and embedded devices imposes to have as much free memory as possible at any time. Consequently, on these platforms Java virtual machines perform more frequent and aggressive garbage collection (see Fig. 6). The garbage collector was invoked many times, corresponding to the falling edges in the chart. This behavior reduces memory usage, but on the other hand it causes the processing time gap found on the different platforms.

4   Future Directions

This paper presented an LDP-CoAP mapping and framework for managing Linked Data in the Web of Things. Performance tests evidence LDP-CoAP supports all types of LDP resources and its computational performances are comparable with those of other frameworks. Future revisions will extend compliance as much as possible; progress will be measured through test suite adopted here. Planned developments also include: evolving the forks of Californium core and proxy modules to merge them with the original codebase eventually; adding the capability to manage RDF resources on persistent storage in addition to in-memory ones; porting LDP-CoAP server to more languages (*e.g.*, C/C++, Python, Go) and computing platforms (*e.g.*, Arduino).

References

1. Battle, S., Speicher, S.: Linked data platform use cases and requirements. W3C Working Group Note, W3C, March 2014. http://www.w3.org/TR/ldp-ucr/
2. Bormann, C., Castellani, A., Shelby, Z.: CoAP: an application protocol for billions of tiny internet nodes. IEEE Internet Comput. **16**(2), 62–67 (2012)

3. Castellani, A., Loreto, S., Rahman, A., Fossati, T., Dijk, E.: Guidelines for HTTP-CoAP Mapping Implementations. Internet-Draft 07, IETF Secretariat, July 2015
4. Compton, M., Barnaghi, P., Bermudez, L., Garcia-Castro, R., Corcho, O., Cox, S., Graybeal, J., Hauswirth, M., Henson, C., Herzog, A., et al.: The SSN Ontology of the W3C Semantic Sensor Network Incubator Group. Web Semantics: Science, Services and Agents on the World Wide Web **17**, (2012)
5. Kovatsch, M., Lanter, M., Shelby, Z.: Californium: Scalable cloud services for the Internet of Things with CoAP. In: International Conference on the Internet of Things, 2014, pp. 1–6. IEEE (2014)
6. Lanthaler, M., Sporny, M., Kellogg, G.: JSON-LD 1.0. W3C Recommendation, W3C, January 2014. http://www.w3.org/TR/json-ld/
7. Loseto, G., Ieva, S., Gramegna, F., Ruta, M., Scioscia, F., Di Sciascio, E.: Linking the web of things: LDP-CoAP mapping. In: Shakshuki, E., (ed.) 7th International Conference on Ambient Systems, Networks and Technologies (ANT 2016)/Affiliated Workshops. Procedia Computer Science, vol. 83, pp. 1182–1187. Elsevier, May 2016
8. Malhotra, A., Arwe, J., Speicher, S.: Linked Data Platform 1.0. W3C Recommendation, W3C, February 2015. http://www.w3.org/TR/ldp/
9. Ruta, M., Scioscia, F., Loseto, G., Ieva, S., Gramegna, F., Sciascio, E.D.: LDP-CoAP: Linked Data Platform for the Constrained Application Protocol (v1.0) (2016). http://dx.doi.org/10.5281/zenodo.50701
10. Seaborne, A., Vesse, R.: RDF Patch Describing Changes to an RDF Dataset. Unofficial Draft, August 2014. https://afs.github.io/rdf-patch/
11. Shelby, Z.: Constrained RESTful Environments (CoRE) Link Format. RFC 6690, August 2012. http://www.ietf.org/rfc/rfc6690.txt
12. Shelby, Z., Hartke, K., Bormann, C.: The Constrained Application Protocol (CoAP). RFC 7252, June 2014. http://www.ietf.org/rfc/rfc7252.txt

TripleWave: Spreading RDF Streams on the Web

Andrea Mauri[1], Jean-Paul Calbimonte[2,3], Daniele Dell'Aglio[1(✉)],
Marco Balduini[1], Marco Brambilla[1], Emanuele Della Valle[1], and Karl Aberer[2]

[1] DEIB, Politecnico di Milano, Milan, Italy
{andrea.mauri,daniele.dellaglio,marco.balduini,marco.brambilla,
emanueledella.valle}@polimi.it
[2] EPFL, Lausanne, Switzerland
{jean-paul.calbimonte,karl.aberer}@epfl.ch
[3] HES-SO Valais, Sierre, Switzerland
jean-paul.calbimonte@hevs.ch

Abstract. Processing data streams is increasingly gaining momentum, given the need to process these flows of information in real-time and at Web scale. In this context, RDF Stream Processing (RSP) and Stream Reasoning (SR) have emerged as solutions to combine semantic technologies with stream and event processing techniques. Research in these areas has proposed an ecosystem of solutions to query, reason and perform real-time processing over heterogeneous and distributed data streams on the Web. However, so far one basic building block has been missing: a mechanism to disseminate and exchange RDF streams on the Web. In this work we close this gap, proposing TripleWave, a reusable and generic tool that enables the publication of RDF streams on the Web. The features of TripleWave were selected based on requirements of real use-cases, and support a diverse set of scenarios, independent of any specific RSP implementation. TripleWave can be fed with existing Web streams (e.g. Twitter and Wikipedia streams) or time-annotated RDF datasets (e.g. the Linked Sensor Data dataset). It can be invoked through both pull- and push-based mechanisms, thus enabling RSP engines to automatically register and receive data from TripleWave.

1   Introduction

Semantic streams represent flows of knowledge over time, in diverse domains including social networks, health monitoring, financial markets or environmental monitoring, to name only a few. The Semantic Web community has studied the problems associated with the processing of and reasoning over these complex streams of data, leading to the emergence of *RDF Stream Processing* and *Stream Reasoning* techniques.

The Web is a natural context for semantic streams, due to the quantity of dynamic data it contains, generated for example by social networks or the Web of Things [11]. RDF streams emerged as a model to realize semantic streams on the Web: they are (potentially infinite) sequences of time-annotated RDF items ordered chronologically.

© Springer International Publishing AG 2016
P. Groth et al. (Eds.): ISWC 2016, Part II, LNCS 9982, pp. 140–149, 2016.
DOI: 10.1007/978-3-319-46547-0_15

While several definitions of RDF streams have been proposed in the past, thanks to the efforts of the W3C RSP Community Group[1] these are converging towards a general formalization based on time-annotated RDF graphs. However, the model is not the only aspect about RDF streams that needs agreement in this community. Standard protocols and mechanisms for RDF stream exchange are currently missing, therefore limiting the adoption and spread of RSP technologies on the Web.

Existing systems such as C-SPARQL [3], CQELS [9] and EP-SPARQL [1] do not tackle directly this problem, but delegate the task of managing the stream publication and ingestion to the developer. Other approaches have proposed to create RDF datasets fed from unstructured streams [8,16], to lift streaming data as Linked Data [2,10], or to provide virtual RDF Streams [6]. To improve scalability, systems like Ztreamy [7] are designed for efficient transmission of compressed data streams, although they do not address the heterogeneity of data sources, declarative transformation and consumption modes. Nevertheless, so far there is still a need for a generic and flexible solution for making RDF streams available on the Web. Such a solution needs to follow Semantic Web standards and best practices, and to allow different data source configurations and data access modes.

In this work, we propose TripleWave [2], an open-source framework for *creating RDF streams and publishing them over the Web*. Triplewave facilitates the dissemination and consumption of RDF streams, in a similar manner as is already common for static RDF datasets with RDF graphs and datasets. In order to do so, we first elicit a set of requirements (Sect. 2) identified from real scenarios and use cases reported in the literature and the W3C RSP group. Then, we extend the prototype in [13] to address the new requirements. These include a flexible configuration of different types of data sources, and different stream generation modes, including transformation (via mappings) of Web streams and replay of existing RDF datasets and RDF sub-streams. Moreover, TripleWave offers a hybrid consumption mechanism that allows both pull-based consumption of RDF streams [2], and push communication through WebSockets.

2   Requirements

To elicit the requirements for TripleWave, we have taken into account a set of scenarios based on real-world use cases[3]. In the following, we highlight the most important of these requirements, organized along four main axes: data sources, data models, data provisioning, and management of contextual data (schema and metadata).

Data Sources. The first set of requirements focuses on the data sources that TripleWave should support to ensure wide adoption and reuse of the tool.

---

[1] Cf. https://www.w3.org/community/rsp/.

[2] TripleWave: http://streamreasoning.github.io/TripleWave/.

[3] Cf. https://www.w3.org/community/rsp/wiki/Use_cases.

[**R1**] *TripleWave may use streams available on the Web as input.* Examples of this kind of data may be found in Twitter, Wikipedia, etc. Twitter supplies data through its streaming API[4]; similarly, Wikipedia publishes a change stream through an IRC-based or Websocket API[5]. Online streaming data is not the only kind of data that TripleWave may support. In the context of testing and benchmarking, system developers & designers have an intrinsic need to feed the engines in a reproducible and repeatable way. That means, they aim at streaming previously generated data, one or multiple times, to assess the behavior of the system.

[**R2**] *TripleWave shall be able to process existing time-aware (RDF) datasets,* which could or could not be formatted as streams.

[**R3**] *TripleWave shall provide format conversion mechanisms towards RDF streams,* in case the input is not formatted as a stream. A typical usage scenario where these features are needed is testing. In this context, sharing the test data may not be enough, as the time dimension plays a key role and streaming the same data in different ways may influence the behavior of the engines. An open, reusable tool to stream data is therefore needed to enable a fair and reproducible execution of the tests.

Data Models. Although recommendations on data models and serialization formats for RDF streams are still under specification[6], it is important to identify and reuse formats that adhere as much as possible to existing recommendations and standards.

[**R4**] *TripleWave should adopt a data format compatible with RDF,* since RDF streams are heavily based on the RDF building blocks. In this way, it would be possible to increase the potential data reuse and tool usage itself in a wider set of scenarios.

Data Provisioning. This category of requirements describes the different ways of consuming RDF streams by RSP client applications.

[**R5**] *TripleWave shall be capable of pro-actively supplying streaming data to processing engines.* Indeed, stream processing applications are usually designed to be fed with streaming data. For instance, the SLD framework [2] receives and analyzes real-time data from social networks or sensor networks, while Star-City [12] is fed with Dublin public transportation data to compute urban analyses.

[**R6**] *TripleWave shall offer the data accordingly to existing W3C recommendations.* In particular, RDF streams should be accessible not only for stream processing and reasoning engines, but also for other applications based on Semantic Web technologies (e.g. SPARQL and Linked Data).

---

[4] Cf. https://dev.twitter.com/streaming/overview.
[5] Cf. https://www.mediawiki.org/wiki/API:Recent_changes_stream.
[6] W3C RSP Design Principles draft http://streamreasoning.github.io/RSP-QL/RSP_Requirements_Design_Document.

Requirements [R4] and [R6] ensure the compatibility with tools and frameworks already developed and available to process RDF data.

Contextual Data Management. In the stream processing context, continuous execution models are often adopted.

[**R7**] *TripleWave shall be able to publish the schema and metadata about the stream independently from the actual transmission of the stream itself.* Indeed, in the case of continuous query evaluation, the steps of query registration, schema provisioning, and metadata provisioning can be performed separately from the streaming itself.

3   The TripleWave Approach

In this section we describe how TripleWave enables the publication and consumption of RDF streams on the Web, following the requirements listed in Sect. 2.

Figure 1 represents a high-level architectural view of our solution. As we saw previously, RDF stream consumers may have different requirements on how to ingest the incoming data. According to [R1] and [R2], in TripleWave we consider two main types of data sources: (i) Non-RDF live streams on the Web, and (ii) RDF datasets with time-annotations. While the former mainly requires a conversion of existing streams to RDF, the latter is focused on streaming RDF data, provided that it has timestamped data elements. When performing the transformation to RDF streams, TripleWave makes use of R2RML mappings in order to allow customizing the shape of the resulting stream. In the case of streaming time-annotated RDF datasets, TripleWave also re-arranges the data if necessary, so that it is structured as a sequence of timestamped RDF graphs, following the W3C RSP Group design principles.

As output, TripleWave produces a JSON stream in the JSON-LD format: each stream element is described by an RDF graph and the time annotation is modeled as an annotation over the graph[7]. Using this format compliant with existing standards, TripleWave enables processing RDF streams not only through specialized RSP engines, but also with existing frameworks and techniques for standard RDF processing.

3.1   Running Modes

In order to address requirements [R1, R2, R3], TripleWave supports a flexible set of data sources, namely non-RDF streams from the Web, as well as timestamped RDF datasets. TripleWave also provides different use modes for RDF stream generation, detailed below.

Converting Web streams. Existing streams can be consumed through the TripleWave JSON and CSV connectors. Extensions of these can be easily incorporated in order to support additional formats. These feeds or streams (e.g.

---

[7] The time annotation is stored in the default graph, as in http://www.w3.org/TR/json-ld#named-graphs, Example 49.

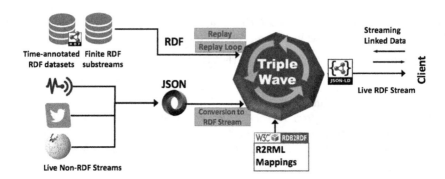

Fig. 1. The architecture of TripleWave: generating RDF streams from non-RDF data sources and time-annotated datasets. R2RML mappings allow customizing the transformation from non-RDF streams. The RDF stream output can be pushed or pulled toward the client as a JSON-LD dataset.

Twitter, earthquakes, live weather, Wikipedia updates, etc.) can be directly plugged to the TripleWave pipeline, which then uses R2RML mappings in order to construct RDF triples that will be output as part of an RDF stream. The mappings can be customized to produce RDF triples of arbitrary structure, and using any ontology.

```
{"type":"Feature",
"properties":{
"time":1388620046000,
"url":"http://earthquake.usgs.gov/earthquakes/eventpage/ak10992887",
"mag":1.1, "magType":"ml",
"type":"earthquake",
"title":"M 1.1 - 117km NW of Talkeetna, Alaska"},
"geometry":{"type":"Point", "coordinates":[-151.6458,63.102,14.1]},
"id":"ak10992887" }
```

Listing 1.1. Example of GeoJSON feed retrieved from USGS API

As an example of input, consider the following GeoJSON feed item from the USGS earthquake API[8]. It contains information about the last reported earthquakes around the world, including the magnitude, location, type and other observed annotations.

Replaying RDF Datasets. RDF data is commonly available as archives and Linked Data endpoints, which may contain timestamp annotations and that can be replayed as a stream. Examples of these include sensor data archives, event datasets, transportation logs, update feeds, etc. These datasets typically contain a time-annotation within the data triples, and one or more other triples are connected to this timestamp. Replaying such datasets means converting an otherwise static dataset into a continuous flow of RDF data, which can then be used by an RDF Stream Processing engine. Common use cases include evaluation, testing, and benchmarking applications, as well as simulation systems. As

---

[8] Cf. http://earthquake.usgs.gov/earthquakes/feed/v1.0.

an example consider the example air temperature observation extracted from
the Linked Sensor Data [14] dataset. Each observation is associated to a particular instant, represented as an XSD `dateTime` literal. Using TripleWave we can
replay the contents of this dataset as a stream, and the original timestamps can
be tuned so that they can meet any test or benchmarking requirements.

```
ssw:Observation_AirTemperature_JEMC1_2003-04-02T06:00:00.0 sobs:samplingTime
ssw:Instant_2003-04-02T06:00:00.0 .
ssw:Instant_2003-04-02T06:00:00.0 time:inXSDDateTime "2003-04-02 06:00:00.0"^^xsd:dateTime
    .
ssw:Observation_AirTemperature_JEMC1_2003-04-02T06:00:00.0 sobs:result ssw:
    MeasureData_AirTemperature_JEMC1_2003-04-02T06:00:00.0
ssw:MeasureData_AirTemperature_JEMC1_2003-04-02T06:00:00.0 sobs:uom weather:fahrenheit
ssw:MeasureData_AirTemperature_JEMC1_2003-04-02T06:00:00.0 sobs:floatValue 7.0
ssw:MeasureData_AirTemperature_JEMC1_2003-04-02T06:00:00.0 rdf:type sobs:MeasureData
```

Listing 1.2. Example of an observation contained in the Linked Sensor Data dataset

Replay Loop. In certain cases, the replay of RDF datasets as streams can be
set up in a way that the data is re-fed to the system after it has been entirely consumed. This is common in testing and benchmarking scenarios where data needs
to be endlessly available until a break point, or in simulation use-cases where an
infinite data stream is required [15]. Similar to the previous scenario, the original RDF dataset is pre-processed in order to structure the stream as a sequence
of annotated graphs, and then it is continuously streamed through TripleWave
as a JSON-LD RDF stream. The main difference is that the timestamps are
cyclically incremented, when the dataset is replayed, so that they provide the
impression of an endless stream.

3.2   R2RML to Generate RDF Streams

Streams on the Web are available in a large variety of formats, so in order to
adapt and transform them into RDF streams we use a generic transformation
process that is specified as R2RML[9] mappings. Although these mappings were
originally conceived for relational database inputs, we use light extensions that
support other formats such as CSV or JSON (as in RML extensions[10]).

The example in Listing 1.3 specifies how earthquake stream data items
can be mapped to a graph of an RDF stream[11]. This mapping defines first
a triple that indicates that the generated subject is of type `ex:Earthquake`. The
`predicateObjectMap` clauses add two more triples, one specifying the URL of the
earthquake (e.g. the reference USGS page) and its description.

```
:earthquakeMap a rr:TriplesMap; rr:logicalTable :quakestream;
rr:subjectMap [rr:template "http://streamreasoning.org/TripleWave/{id}"; rr:class ex:
    Earthquake;];
rr:predicateObjectMap [rr:predicate schema:url; rr:objectMap [ rr:column "url" ]];
rr:predicateObjectMap [rr:predicate schema:description; rr:objectMap [ rr:column "title"]];
rr:predicateObjectMap [rr:predicate schema:location; rr:objectMap [ rr:parentTriplesMap :
    locMap]].
```

Listing 1.3. Example of R2RML mapping

---

[9] R2RML W3C Recommendation: http://www.w3.org/TR/r2rml/.
[10] http://rml.io.
[11] We use schema.org as the vocabulary in the example.

A snippet of the resulting RDF Stream graph, serialized in JSON-LD, is shown in in Listing 1.4. As can be observed, a stream element is contained in a timestamped graph, using the `generatedAtTime` property of the PROV ontology[12].

```
{"http://www.w3.org/ns/prov#generatedAtTime": "2015-06-30T16:44:59.587Z",
 "@id": "http://streamreasoning.org/TripleWave/ak10992887",
 "@graph": [
 { "@id": "http://streamreasoning.org/TripleWavee/ak10992887",
 "@type": "http://example.org/onto/earth#Earthquake",
 "url": "http://earthquake.usgs.gov/earthquakes/eventpage/ak10992887",
 "location": {"@id": "http://streamreasoning.org/TripleWave/ak10992887Location"},
 "description": "M 1.1 - 117km NW of Talkeetna, Alaska" },
 { "@id": "http://streamreasoning.org/TripleWave/ak10992887Location",
 "@type": "https://schema.org/Place",
 "longitude": "-151.6458",
 "latitude":  "63.102" } ],
 "@context": "https://schema.org/"  }
```

Listing 1.4. Portion of the timestamped element in the RDF stream.

3.3   Consuming TripleWave RDF Streams

TripleWave is implemented in Node.js and produces the output RDF stream using HTTP with chunked transfer encoding by default, or alternatively through WebSockets. Consumers can register to a TripleWave endpoint and receive the data following a push paradigm. In cases where consumers may want to pull the data, TripleWave allows publishing the data according to the Linked Data principles [5]. Given that the stream supplies data that changes very frequently, data is only temporarily available for consumption, assuming that recent stream elements are more relevant. We describe both cases below.

Publishing Stream Elements as Linked Data. TripleWave allows consuming RDF Streams following the Linked Data principles, extending the framework proposed in [4]. According to this scheme, for each RDF Stream TripleWave distinguishes between two kinds of Named Graphs: the Stream Graph (*sGraph*) and Instantaneous Graphs (*iGraphs*). Intuitively, an iGraph represents one stream element, while the sGraph contains the descriptions of the iGraphs, e.g. their timestamps.

As an example, the sGraph in Listing 1.5 describes the current content that can be retrieved from a TripleWave RDF stream. The ordered list of iGraphs is modeled as an `rdf:list` with the most recent iGraph as the first element, and with each iGraph having its relative timestamp annotation. By accessing the sGraph, consumers discover which are the stream elements (identified by iGraphs) available at the current time instants. Next, the consumer can access the iGraphs dereferencing the iGraph URL address. The annotations on the sGraph use a dedicated vocabulary[13].

---

[12] Cf. https://www.w3.org/TR/prov-o/.

[13] Cf. http://streamreasoning.org/ontologies/SLD4TripleWave#.

```
{"@context": {
"sld": "http://streamreasoning.org/ontologies/SLD4TripleWave#",
"generatedAt": { "@id": "http://www.w3.org/ns/prov#generatedAtTime",
"@type": "http://www.w3.org/2001/XMLSchema#dateTime" }},
"@type": "sld:sGraph",
"sld:streamLocation": "ws://localhost:8101/TripleWave/replay",
"sld:tBoxLocation": {"@id":"http://purl.oclc.org/NET/ssnx/ssn"},
"sld:contains": {"@list": [
{ "generatedAt": "2016-04-21T13:01:18.663Z", "@id": "tr:1461243678663" },
{ "generatedAt": "2016-04-21T13:01:19.784Z", "@id": "tr:1461243679784" } ]},
"sld:lastUpdated": "2016-04-21T13:02:06.575Z" }
```

Listing 1.5. The sGraph pointing to the iGraph described in Listing 1.4.

RDF Stream Push. An RSP engine can consume an RDF stream from Triple-Wave, extending the rsp-services framework[14] as follows (with C-SPARQL as a sample RSP): (1) the client identifies the stream by its IRI (which is the URL of the sGraph). (2) rsp-services registers the new stream in the C-SPARQL engine. (3) rsp-services looks at the sGraph URL, parses it and gets the information regarding the TBox and WebSocket. (4) The TBox is associated to the stream. (5) A WebSocket connection is established and the data flows into C-SPARQL. (6) The user registers a new query for the registered stream. (7) The TBox is loaded into the reasoner (if available) associated to the query. (8) The query is performed on the flowing data.

4   Conclusion

In this work we have described TripleWave, an open-source framework for publishing and sharing RDF streams on the Web. This work fills an important gap in RDF stream processing as it provides flexible mechanisms for plugging in diverse Web data sources, and for consuming streams in both push and pull mode. TripleWave covers a set of crucial requirements for the stream reasoning community and the semantic Web community at large, including: reusing available streams on the Web [R1], as well as time annotated RDF datasets [R2], which are transformed to follow a homogenized RDF stream structure [R3]. TripleWave adopts a stream format compatible with Semantic Web standards, including RDF for data modeling, and Linked Data principles for publishing [R4] [R6]. The proposed tool also provides pull and push data access to client applications and RSP engines [R5], as well as context information about the stream [R7].

The inherent flexibility of TripleWave makes it suitable for reuse in a wide range of streaming data applications, and it has the potential of enabling the integration of RSP query engines, stream reasoners, RDF stream filters, semantic complex event processors, benchmark platforms, and stored RDF data sources. This versatility, combined with a standards-driven design, and aligned with the

---

[14] Cf. https://github.com/streamreasoning/rsp-services.

requirements and design principles discussed in the W3C RSP Group, can help spreading the adoption of RDF for streaming data scenarios and applications.

Show Cases. We developed two show cases in order to illustrate the capabilities of TripleWave. In the first case we set up TripleWave for converting Web streams and we configured it to transform the stream generated by the changes in Wikipedia. We developed the component to listen to the Wikipedia endpoint and the R2RML mapping.

In the second case we started another instance of TripleWave and we configured it to endlessly replay as a stream the Linked Sensor Data [14] dataset as a stream. Furthermore for this scenario we also set up a instance of the C-SPARQL engine to consume the data produced by TripleWave. Links to both the show cases are available on the project website[15].

Availability. TripleWave is available under the Apache 2.0 license[16], its code is accessible on Github[17], and accompanied by user and developer guides. It is maintained and supported by the Stream Reasoning initiative[18].

References

1. Anicic, D., Fodor, P., Rudolph, S., Stojanovic, N.: EP-SPARQL: a unified language for event processing and stream reasoning. In: WWW, pp. 635–644. ACM (2011)
2. Balduini, M., Della Valle, E., Dell'Aglio, D., Tsytsarau, M., Palpanas, T., Confalonieri, C.: Social listening of city scale events using the streaming linked data framework. In: Alani, H., et al. (eds.) The Semantic Web – ISWC 2013. LNCS, vol. 8219, pp. 1–16. Springer, Heidelberg (2013)
3. Barbieri, D.F., Braga, D., Ceri, S., Della Valle, E., Grossniklaus, M.: C-sparql: a continuous query language for rdf data streams. Intl. J. Semant. Comput. **4**(01), 3–25 (2010)
4. Barbieri, D.F., Della Valle, E.: A proposal for publishing data streams as linked data - A position paper. In: LDOW (2010)
5. Berners-Lee, T., Bizer, C., Heath, T.: Linked data-the story so far. IJSWIS **5**(3), 1–22 (2009)
6. Calbimonte, J.-P., Jeung, H., Corcho, O., Aberer, K.: Enabling query technologies for the semantic sensor web. Int. J. Semant. Web Inf. Syst. **8**, 43–63 (2012)
7. Fisteus, J.A., Garcia, N.F., Fernandez, L.S., Fuentes-Lorenzo, D.: Ztreamy: A middleware for publishing semantic streams on the web. J. Web Semant. **25**, 16–23 (2014)
8. Gerber, D., Hellmann, S., Bühmann, L., Soru, T., Usbeck, R., Ngonga Ngomo, A.-C.: Real-Time RDF extraction from unstructured data streams. In: Alani, H., et al. (eds.) ISWC 2013. LNCS, vol. 8218, pp. 135–150. Springer, Heidelberg (2013). doi:10.1007/978-3-642-41335-3_9

---

[15] Cf. http://streamreasoning.github.io/TripleWave/.
[16] Cf. https://www.apache.org/licenses/LICENSE-2.0.html.
[17] Cf. https://github.com/streamreasoning/TripleWave.
[18] Cf. http://streamreasoning.org.

9. Le-Phuoc, D., Dao-Tran, M., Xavier Parreira, J., Hauswirth, M.: A native and adaptive approach for unified processing of linked streams and linked data. In: Aroyo, L., Welty, C., Alani, H., Taylor, J., Bernstein, A., Kagal, L., Noy, N., Blomqvist, E. (eds.) ISWC 2011. LNCS, vol. 7031, pp. 370–388. Springer, Heidelberg (2011). doi:10.1007/978-3-642-25073-6_24

10. Le-Phuoc, D., Nguyen-Mau, H.Q., Parreira, J.X., Hauswirth, M.: A middleware framework for scalable management of linked streams. J. Web Semant. **16**, 42–51 (2012)

11. Le-Phuoc, D., Quoc, H.N.M., Quoc, H.N., Nhat, T.T., Hauswirth, M.: The graph of things: A step towards the live knowledge graph of connected things. J. Web Semant. **37**, 25–35 (2016)

12. Lécué, F., Tallevi-Diotallevi, S., Hayes, J., Tucker, R., Bicer, V., Sbodio, M.L., Tommasi, P.: Star-city: semantic traffic analytics and reasoning for city. In: ACM IUI, pp. 179–188 (2014)

13. Mauri, A., Calbimonte, J.-P., Dell'Aglio, D., Balduini, M., Della Valle, E., Aberer, K.: Where are the rdf streams?: Deploying rdf streams on the web of data with triplewave. In: Poster Proceedings of ISWC (2015)

14. Patni, H., Henson, C., Sheth, A.: Linked sensor data. In: IEEE CTS, pp. 362–370 (2010)

15. Scharrenbach, T., Urbani, J., Margara, A., Valle, E., Bernstein, A.: Seven commandments for benchmarking semantic flow processing systems. In: Cimiano, P., Corcho, O., Presutti, V., Hollink, L., Rudolph, S. (eds.) ESWC 2013. LNCS, vol. 7882, pp. 305–319. Springer, Heidelberg (2013). doi:10.1007/978-3-642-38288-8_21

16. Trinh, T.-D., Wetz, P., Do, B.-L., Anjomshoaa, A., Kiesling, E., Tjoa, A.M.: A web-based platform for dynamic integration of heterogeneous data. In: IIWAS, pp. 253–261 (2014)

Conference Linked Data: The ScholarlyData Project

Andrea Giovanni Nuzzolese[1]([⊠]), Anna Lisa Gentile[2], Valentina Presutti[1],
and Aldo Gangemi[1,3]

[1] Semantic Technology Lab, ISTC-CNR, Padova, Italy
andrea.nuzzolese@istc.cnr.it, {valentina.presutti,aldo.gangemi}@cnr.it
[2] Data and Web Science Group, University of Mannheim, Mannheim, Germany
annalisa@informatik.uni-mannheim.de
[3] LIPN, Université Paris 13, Sorbone Cité, UMR CNRS, Paris, France

Abstract. The Semantic Web Dog Food (SWDF) is the reference linked dataset of the Semantic Web community about papers, people, organisations, and events related to its academic conferences. In this paper we analyse the existing problems of generating, representing and maintaining Linked Data for the SWDF. With this work (i) we provide a refactored and cleaned SWDF dataset; (ii) we use a novel data model which improves the Semantic Web Conference Ontology, adopting best ontology design practices and (iii) we provide an open source workflow to support a healthy growth of the dataset beyond the Semantic Web conferences.

Permanent URL: https://w3id.org/scholarlydata
Resource type: Ontology and dataset.

1   Introduction

A good practise in the Semantic Web community is to encourage the publication of Linked Data about scientific conferences in the field, as a way of "eating our own dog food" [8]. The main example is the *Semantic Web Dog Food*[1] (SWDF), a corpus that collects Linked Data about papers, people, organisations, and events related to academic conferences. Currently, all main Semantic Web conferences and related events publish their data as Linked Data on SWDF, but for many other conferences, events and publication venues information is still not available in a structured and linked form. On the other hand the growth of available content with respect to the early times of SWDF poses data management issues and reveals design problems which where not foreseen when the dataset was at its initial stage. There are several challenges to pursue the maintenance of a healthy and sustainable SWDF for the future: (i) the availability of appropriate vocabularies to express the current state of the data; (ii) the shared knowledge

---

[1] SWDF: http://data.semanticweb.org.

© Springer International Publishing AG 2016
P. Groth et al. (Eds.): ISWC 2016, Part II, LNCS 9982, pp. 150–158, 2016.
DOI: 10.1007/978-3-319-46547-0_16

of such vocabularies; (iii) the availability of tools to ease the task of data acquisition, conversion, integration, augmentation, verification and finally publication; (iv) the ongoing maintenance of the dataset.

In this work we address these issues and we propose a refactoring of the Semantic Web Conference (SWC) Ontology[2]. The new ontology, named *conference-ontology* [12], adopts best ontology design practices (e.g. Ontology Design Patterns, ontology reuse and interlinking) and guarantees interoperability with SWC ontology and all other pertinent vocabularies. We use cLODg[3] (conference Linked Open Data generator) [6] to regenerate the SWDF dataset according to *conference-ontology* and provide a sustainable solution for the growth of the dataset in the future.

The main advantage of the proposed approach is the availability of a shared procedure and open source tools for conference data generation, with the primary goal to ensure the sustainability and usability of our own Semantic Web Dog Food and the ease of data contribution from beyond our community. We make the new resource available at https://w3id.org/scholarlydata as data dump, SPARQL endpoint and we offer the facilities to generate data about new conferences using cLODg and submit it for addition to *scholarlydata*. The newly submitted data is manually checked before inclusion to avoid corruption of the dataset and general spam.

2   State of the Art

The first considerable effort to offer comprehensive semantic descriptions of conference events is represented by the metadata projects at ESWC 2006 and ISWC 2006 conferences [11], with the Semantic Web Conference Ontology[4] being the vocabulary of choice to represent such data. Increasing number of initiatives are pursuing the publication about conferences data as Linked Data, mainly promoted by publishers such as Springer[5] or Elsevier[6] amongst many others. For example, the knowledge management of scholarly products is an emerging research area in the Semantic Web field known as Semantic Publishing [14]. Semantic Publishing aims at providing access to semantic enhanced scholarly products with the aim of enabling a variety of semantically oriented tasks, such as knowledge discovery, knowledge exploration and data integration. The Semantic Publishing challenge [9] is a breakthrough in this direction. Its objective is assessing the quality of systems that extract meaningful metadata from scholarly articles and represent them as RDF. Similarly, the *Jailbreaking the PDF* initiative [5] is aimed at creating a formal flexible infrastructure to extract semantic

---

[2] http://data.semanticweb.org/ns/swc/swc_2009-05-09.html.

[3] cLODG is an Open Source tool that provides a formalised process for the conference metadata publication workflow https://github.com/anuzzolese/cLODg2.

[4] Semantic Web Conference Ontology http://data.semanticweb.org/ns/swc/swc_2009-05-09.html.

[5] http://lod.springer.com/wiki/bin/view/Linked+Open+Data/About.

[6] http://data.elsevier.com/documentation/index.html.

information from PDF documents as domain-specific annotations. Despite these continuous efforts, it has been argued that lots of information about academic conferences is still missing or spread across several sources in a largely chaotic and non-structured way [1]. Besides the problem of missing content, one of the other major challenges with scholarly data is to ensure data quality, which means dealing with data-entry errors, disparate citation formats, lack of (enforcement of) standards, imperfect citation-gathering software, ambiguous author names and abbreviations of publication venue titles [10]. Currently the generation of data for the SWDF corpus still relies on little or no strategies to deal with duplicates, inconsistencies, misspelling and name variations. In this work we aim to close these gaps by making available a solid data model and a shared and open workflow (available Open Source) as a long term solution for the population and maintenance of an enhanced version of the SDWF dataset.

3   The SWDF and Its Current Issues

The SWDF uses the Semantic Web Conference (SWC) ontology as the reference ontology for modelling data about academic conferences. The SWC ontology combines existing widely accepted vocabularies (i.e. FOAF[7], SIOC[8] and Dublin Core[9]) and relies on the SWRC[10] (Semantic Web for Research Communities) ontology for modelling entities such as accepted papers, authors, their affiliations, talks and other events, the organising committee and all other roles involved. The core types of SWC ontology are `foaf:Person` for describing people, `foaf:Organization` for organisations (e.g. universities, research institutions, etc.), `swc:Artefact` for documents (e.g. papers, proceedings, etc.), `swc:OrganisedEvent` for events and `swc:Role` for the people roles at the conference. Unfortunately, the lack of clear guidelines for data generation and maintenance and some modelling choices of the SWC ontology affect the current quality of SWDF. The data generation is based on a collaborative model that delegates the metadata chairs of each conference to independently deal with the process of generating conference Linked Data. Linked Data are generated from a variety of formats typically provided by a conference management system (e.g. EasyChair). While the collaborative process is beneficial to the growth of the dataset and its adoption in the community, the lack of clear guidelines and of standard tools supporting the generation process affects the quality of generated data. Examples are: (i) a portion of the included conference/workshop data use vocabularies or ontologies which are not aligned to the SWC ontology and, in some cases, no longer maintained or existing (e.g. `swrc-ext`[11] or `xmllondon`[12]); (ii) the usage of classes and properties not defined

---

[7] http://xmlns.com/foaf/spec/.

[8] http://rdfs.org/sioc/spec/.

[9] http://dublincore.org/documents/dcmi-terms/.

[10] http://ontoware.org/swrc/swrc/SWRCOWL/swrc_updated_v0.7.1.owl.

[11] http://www.cs.vu.nl/~mcaklein/onto/swrc_ext/2005/0.

[12] http://xmllondon.com/ns/swc/ontology.

in the SWC ontology and introduced without providing an extension of the ontology (e.g. `swc:room`, `swc:editorList`, `swc:completeGraph`, `swc:IW3C2Liaison`, `swc:SemanticWebTechnologiesCo-ordinator`, etc.); (iii) the misuse of properties (either defined in the SWC ontology or in other vocabularies/ontologies) with respect to their domain and range; (iv) typos (e.g. the materialisation of triples having the predicate `swc:partOf` instead of `swc:isPartOf`). In addition, we argue that the SWC ontology itself has intentional issues, mainly concerning the modelling of *affiliations*, *roles* and *lists*. *Affiliations* (of people to organisation) are represented via the object property `swrc:affiliation` from the SWRC ontology while the membership relation (organisation to people) via the property `foaf:member`. Although intuitive, this representation ignores the temporal dimension (i.e. the time when a given affiliation is held by an actor) that is relevant to interpret affiliations correctly. For example, with this model it is not possible to provide a correct answer to a simple competency question, such as "What was the affiliation of a person when participating to a certain conference?". *Roles* such as program chair, track chair, etc. are currently modelled using an ontology pattern based on the reification of a *n*-ary relation. The *n*-ary relation is identified by individuals of the class `swc:Role` which are used to associate people to events. The SWC ontology contains a very basic set of role classes (i.e. `swc:Chair`, `swc:Delegate`, `swc:Presenter` and `swc:ProgrammeCommitteeMember`) represented as sub-classes of `swc:Role`. This choice allows to instantiate the small set of different Role classes and cover the roles at specific events. For example, instead of sub-classing the `swc:Chair` class with `MainChair`, `WorkshopChair`, `TutorialChair`, etc., the different types of chairs should simply be instances of the generic `swc:Chair` and labelled appropriately (e.g. `iswc2015:general-chair`[13]). The problem with this solution is that the (individuals representing) roles are defined locally to each conference, e.g. there is a different individual for representing the role "general chair" for each conference in the dataset. This causes the presence of 1,717 distinct individuals in the current dataset that truly represent a set of only 34 unique roles (cf. Sect. 4). Hence, it is difficult to answer simple queries like "Who was the general chair at each edition of ISWC?" without using regular expressions on roles' labels (such labels are heterogeneous and not always provided). Finally, *lists of authors* are represented via the property `bibo:authorList`, which accepts `rdf:List` or `rdf:Seq` as range. Therefore lists of authors in the SWDF are expressed via the properties `rdf:_1`, `rdf:_2`, `rdf:_3`, etc., based on `rdfs:ContainerMembershipProperty`. This solution makes querying and reasoning on ordered list of authors very hard [2].

4   A Sustainable SWDF

Our solution for enhancing the SWDF and solving the issues described in Sect. 3 is based on (i) the refactoring of the SWC ontology, (ii) the refactoring of the

---

[13] The prefix `iswc2015:` stands for the namespace http://data.semanticweb.org/conference/iswc/2015/.

current SWDF dataset and (iii) a fully implemented open source workflow to generate, verify and add data to SWDF. The proposed refactoring of the SWC ontology, *conference-ontology*[14], is a new self-contained ontology, which exploits Ontology Design Patterns (ODP) [3]. We model *affiliations* reusing the time indexed situation ODP[15] and the *roles* held by people at a conference with the time indexed person role ODP[16]. Both patterns provide commonly accepted solutions to model complex situations as n-ary relations, amongst many other available ones [4].

The classes `conf:AffiliationDuringEvent` and `conf:Affiliation AtTimeOfSubmission` model situations where a person (an individual of the class `conf:Person`) is affiliated to an organisation (an individual of the class `conf:Organisation`) at a specific time, which can be either an interval (coinciding with the conference dates) or the instant when the paper was submitted. This allows the representation of cases where a person changes affiliation in the time interval between paper submission and conference event. Similarly, the class `conf:RoleDuringEvent` associates a person with a role (an individual of the class `conf:Role`) at a conference. Additionally, `conf:AffiliationDuringEvent` and `conf:AffiliationAtTimeOfSubmission` can be associated with `conf:AffiliationRole`, a subclass of `conf:Role`, to represent the role held by a person within an organisation. We reused the Sequence ODP[17] to represent ordered *lists of authors*. We represent a list with `conf:List`, whose items are individuals of the class `conf:ListItem`. The association between `conf:ListItem` and `conf:List` is done via the property `conf:isItemOf`. A `conf:List` has pointers to the first (`conf:hasFirstItem`) and the last item (`conf:hasLastItem`). Each `conf:ListItem` is linked to its predecessor (`conf:hasPreviousItem`) and successor (`conf:hasNextItem`). This new modelling overcomes the limitation in the current SWDF offering a new range of services for scholarly monitoring, such as statistics on career development, change of affiliations over time, covered roles at conferences in order to monitor their involvement and impact at different granularity levels, ranging from a broader scientific area to specific communities or conferences. An example of query to obtain all roles covered overtime by a specific researcher is the following:

```
PREFIX person: <http://w3id.org/scholarlydata/person/>
PREFIX conf: <http://w3id.org/scholarlydata/ontology/conference-ontology.owl#>
SELECT ?role ?during
WHERE{ person:valentina-presutti conf:holdsRole ?roleAt .
?roleAt conf:withRole ?role .
?roleAt conf:during ?during}
```

To guarantee interoperability with SWC and all other already used vocabularies in the SWDF dataset, we produced extensive alignments[18], which allow the materialisation of triples via reasoning. We include alignments to:

---

[14] The ontology diagram, description and specification are available at http://w3id. org/scholarlydata/ontology.

[15] http://ontologydesignpatterns.org/cp/owl/timeindexedsituation.owl.

[16] http://ontologydesignpatterns.org/cp/owl/timeindexedpersonrole.owl.

[17] http://ontologydesignpatterns.org/cp/owl/sequence.owl.

[18] http://w3id.org/scholarlydata/ontology/conference-ontology-alignments.owl.

- the SWC ontology itself to guarantee backward interoperability with SWDF;
- the top level classes of Dolce D0[19] for interoperability with a series of linked datasets aligned to it (e.g. DBpedia);
- all relevant SPAR ontologies[20] such as: FaBIO [13] for compliance with FRBR; DoCO for modelling the part relations (`conf:hasPart` and its inverse `conf:isPartOf` existing between abstracts `conf:Abstract`, articles `conf:-InProceedings` and the books of proceedings `conf:Proceedings`); PRO and SCORO for modelling roles as defined in SPAR;
- the Organization Ontology[21] for modelling organisations, roles and affiliations;
- FOAF for modelling people;
- SKOS[22] for modelling broader/narrower relations;
- ICATZD[23] for events;
- the Collection Ontology [2] for modelling the sequences represented by the lists of authors.

5   Scholarlydata.org

Using cLODg and our new *conference-ontology* we performed a batch cleaning of the whole SWDF dataset, consisting of 48 conferences and 235 workshops. The new dataset contains 93,519 individuals. The distribution of classes is reported in Table 1.

Table 1. Number of unique individuals for each class of *conference-ontology* generated with cLODg.

Type	Individuals	Type	Individuals
conf:TimeIndexedSituation	20,998	conf:RoleDuringEvent	6,510
conf:ListItem	14,805	conf:List	4,463
conf:AffiliationDuringEvent	14,488	conf:InProceedings	4,393
conf:Agent	12,490	conf:OrganisedEvent	2,882
conf:Person	9,682	conf:Organisation	2,808

For the role definitions we corrected the current 1,717 roles in SWDF, defined at conference level, by generating 34 roles at global level and reusing them at conference level. E.g. the role `role:general-chair`[24] is one individual which can be reused in all conferences with the relation `conf:withRole`. These 34 roles

---

[19] http://www.ontologydesignpatterns.org/ont/dul/d0.owl.
[20] http://www.sparontologies.net/ontologies.
[21] https://www.w3.org/TR/vocab-org/.
[22] https://www.w3.org/TR/2005/WD-swbp-skos-core-spec-20051102/.
[23] http://www.w3.org/2002/12/cal/icaltzd.
[24] The prefix *role:* stands for the namespace http://w3id.org/scholarlydata/role/.

are organised in a hierarchy by using SKOS to express broader and narrower relations between them, e.g. the role `role:chair` is defined as `skos:broader` `role:general-chair`. The current list of roles can be obtained using the query:

```
PREFIX conf: <http://w3id.org/scholarlydata/ontology/conference-ontology.owl#>
SELECT distinct ?role
WHERE{ ?person conf:holdsRole ?roleAt .
?roleAt conf:withRole ?role }
```

Using cLODg to produce metadata about a new conference guarantees that pertinent roles are reused if already existing the dataset.

We produced instance level alignments of (i) individuals of `conf:Person` to ORCID[25] (Open Researcher and Contributor ID) and (ii) individuals of `conf:InProceedings` to DOI[26] (Digital Object Identifier), whenever possible. ORCID provides persistent digital identifiers for scientific researchers and academic authors. A DOI is a serial code used to uniquely identify digital objects, particularly used for electronic documents. The alignments to ORCID were produced by using the public API provided by ORCID[27]. The references to DOI were produced by using the API provided by Crossref[28], performing a search on each article title.

All data is uploaded on https://w3id.org/scholarlydata where can be accessed in different formats (i.e. HTML+RDFa, RDF/XML, Turtle, N-TRIPLES, and JSON-LD) via URI dereferencing, queried via SPARQL or downloaded as single RDF dumps for each conference and workshop. Each dump is provided in two versions: a simple one, where data is represented by the *conference-ontology* only and one containing all the alignments (and therefore also complaint to SWDF), which have been materialised using a reasoner. These dumps are released with the "creative commons by 3.0" license[29] and are described by using the VOID vocabulary[30]. Additionally, we explicitly state the primary source of our data is the SWDF by using the property `prov:hadPrimarySource` of PROV-O[31]. Dump data is also publicly available on datahub[32]. It is worth remarking that cLODg is released as an open source software with the MIT License[33] and can be used by metadata curator to add data about a new conference. In fact, cLODg provides a nearly one-click process to produce conference Linked Data and includes all the components for data transformation, deduplication, URIs reuse, alignment of individuals to external resources, etc. and assures that data is produced according to the *conference-ontology* and compliant with the SWDF. An early description of cLODg can be found in [7] and in the github repository for its

---

[25] http://orcid.org/.

[26] http://www.doi.org/index.html.

[27] http://members.orcid.org/api/introduction-orcid-public-api.

[28] http://www.crossref.org/guestquery/.

[29] http://creativecommons.org/licenses/by/3.0.

[30] http://vocab.deri.ie/void.

[31] https://www.w3.org/TR/prov-o/.

[32] https://datahub.io/dataset/scholarlydata.

[33] https://opensource.org/licenses/MIT.

newer version[34]. By providing a user friendly data generation tool we aim at encouraging the growth of the dataset beyond the Semantic Web community.

6   Conclusions and Future Work

This paper analyses the Semantic Web Dog Food dataset and discusses its quality and sustainability issues. As the main scholarly dataset for the Semantic Web community, we believe it is important that the dataset is maintained in good health. We therefore perform a refactoring on the dataset addressing its current issues and we make the cLODg workflow publicly available as potential solution for future maintenance. The new resource https://w3id.org/scholarlydata is publicly available both as dump download and SPARQL endpoint, with facilities to upload new data. With the availability of cLODg as standard Linked Data publication workflow, we believe that *scholarlydata* has the potential to grow way beyond the Semantic Web conferences. As future work we plan a systematic evaluation of the resource and the introduction of more sophisticated components to deal with instance matching in the cLODg workflow. Moreover we will work on fostering collaboration with Conference Management System providers, to provide cLODg as a build-in facility in the systems.

References

1. Bryl, V., Birukou, A., Eckert, K., Kessler, M.: What is in the proceedings? combining publishers and researchers perspectives. In: Proceedings of SePublica 2014, Anissaras, Greece, May 25th 2014
2. Ciccarese, P., Peroni, S.: The collections ontology: creating and handling collections in OWL 2 DL frameworks. Semant. Web 5(6), 515–529 (2014)
3. Gangemi, A., Presutti, V.: Ontology design patterns. In: Staab, S., Studer, R. (eds.) Handbook on Ontologies, 2nd edn. Springer, Heidelberg (2009)
4. Gangemi, A., Presutti, V.: A multi-dimensional comparison of ontology design patterns for representing n-ary relations. In: Emde Boas, P., Groen, F.C.A., Italiano, G.F., Nawrocki, J., Sack, H. (eds.) SOFSEM 2013. LNCS, vol. 7741, pp. 86–105. Springer, Heidelberg (2013). doi:10.1007/978-3-642-35843-2_8
5. Garcia, A., Murray-Rust, P., Burns, et al.: Pdfjailbreak-a communal architecture for making biomedical pdfs semantic. In: Proceedings of BioLINK SIG 2013, p. 13 (2013)
6. Gentile, A.L., Acosta, M., Costabello, L., Nuzzolese, A.G., Presutti, V., Recupero, D.R., Live, C.: Accessible and sociable conference semantic data. In: Proceedings of WWW 2015 (Companion Volume), pp. 1007–1012. ACM (2015)
7. Gentile, A.L., Nuzzolese, A.G.: cLODg - conference linked open data generator. In: Proceedings of the ISWC 2015 Posters and Demonstrations Track. CEUR-WS.org (2015)
8. Harrison, W.: Eating your own dog food. In: Industrial, Organizational Psychology, 5–7 June 2011

---

[34] https://github.com/anuzzolese/cLODg2.

9. Lange, C., Iorio, A.: Semantic publishing challenge – assessing the quality of scientific output. In: Presutti, V., et al. (eds.) SemWebEval 2014. CCIS, vol. 475, pp. 61–76. Springer, Heidelberg (2014). doi:10.1007/978-3-319-12024-9_8

10. Lee, D., Kang, J., Mitra, P., Giles, C.L., On, B.-W.: Are your citations clean? Commun. ACM **50**(12), 33–38 (2007)

11. Möller, K., Heath, T., Handschuh, S., Domingue, J.: Recipes for semantic web dog food: The eswc and iswc metadata projects. In: Proceedings of ISWC'07/ASWC 2007, pp. 802–815. Springer-Verlag, Berlin, Heidelberg (2007)

12. Nuzzolese, A.G., Gentile, A.L., Presutti, V., Gangemi, A.: Semantic web conference ontology - a refactoring solution. In: The Semantic Web: ESWC 2016 Satellite Events, Springer, Heidelberg (2016). page to appear

13. Peroni, S., Shotton, D.: FaBiO and CiTO: Ontologies for describing bibliographic resources and citations. J. Web Semant. **17**, 33–43 (2012)

14. Shotton, D.: Semantic publishing: the coming revolution in scientific journal publishing. Learned Publishing **22**(2), 85–94 (2009)

The OWL Reasoner Evaluation (ORE) 2015 Resources

Bijan Parsia[1], Nicolas Matentzoglu[1], Rafael S. Gonçalves[2(✉)], Birte Glimm[3], and Andreas Steigmiller[3]

[1] Information Management Group, University of Manchester, Manchester, UK
rafael.goncalves@stanford.edu
[2] Stanford Center for Biomedical Informatics Research,
Stanford University, Stanford, USA
[3] Institute of Artificial Intelligence, University of Ulm, Ulm, Germany

Abstract. The OWL Reasoner Evaluation (ORE) Competition is an annual competition (with an associated workshop) which pits OWL 2 compliant reasoners against each other on various standard reasoning tasks over naturally occurring problems. The 2015 competition was the third of its sort and had 14 reasoners competing in six tracks comprising three tasks (consistency, classification, and realisation) over two profiles (OWL 2 DL and EL). In this paper, we outline the design of the competition and present the infrastructure used for its execution: the corpora of ontologies, the competition framework, and the submitted systems. All resources are publicly available on the Web, allowing users to easily re-run the 2015 competition, or reuse any of the ORE infrastructure for reasoner experiments or ontology analysis.

Keywords: OWL · Ontologies · Reasoning

1 Introduction

The Web Ontology Language (OWL) is in its second iteration (OWL 2) and has seen significant adoption especially in Health Care and Life Sciences. OWL 2 DL can be seen as a variant of the description logic $\mathcal{SROIQ}$, with the various other profiles being either subsets (e.g., OWL 2 EL) or extensions (e.g., OWL 2 Full). Description logics generally are designed to be *computationally practical* so that, even if they do not have tractable worst-case complexity for key services, they nevertheless admit implementations which seem to work well in practice [2]. Unlike in the early days of description logics or even of the direct precursors of OWL (DAML+OIL), the reasoner landscape for OWL is rich, diverse, and highly compliant with a common, detailed specification. Thus, we have a large number of high performance, production-quality reasoners with similar core capacities (with respect to language features and standard inference tasks).

© Springer International Publishing AG 2016
P. Groth et al. (Eds.): ISWC 2016, Part II, LNCS 9982, pp. 159–167, 2016.
DOI: 10.1007/978-3-319-46547-0_17

Research on optimising OWL reasoning continues apace, though empirical work still lags both theoretical and engineering work in breadth, depth, and sophistication. There is, in general, a lack of shared understanding of test cases, test scenarios, infrastructure, and experiment design. A common strategy in research communities to help address these issues is to hold competitions, that is, experiments designed and hosted by third parties on an independent (often constrained, but sometimes expanded) infrastructure. Such competitions (in contrast to published benchmarks) typically do not directly provide strong empirical evidence about the competing tools. Instead, they serve two key functions: (1) they provide a clear, motivating event that helps drive tool development (e.g., for correctness or performance) and (2) *components* of the competition are useful for subsequent research. Finally, competitions can be great fun and help foster a strong community. They can be especially useful for newcomers by providing a simple way to gain some prima facie validation of their tools without the burden of designing and executing complex experiments themselves.

Toward these ends, we have been running a competition for OWL reasoners (with an associated workshop): The OWL Reasoner Evaluation (ORE) competition. ORE has been running, in substantively its current form, for three years, and this year it was held in conjunction with the 28th International Description Logic Workshop (DL 2015)[1] in June 2015. A report on the ORE 2015 competition results and analysis is under submsission [15]. In this paper we focus on the elements of the ORE 2015 that are reusable by the general public. To that end, we describe the design of the 2015 competition, which provides a reasonable default structure for reasoner comparison. We also describe the competition infrastructure: the corpora of ontologies, the competition framework, and the submitted systems. All resources are publicly available on the Web, allowing users to easily re-run the 2015 competition, or reuse any of the ORE infrastructure for reasoner experiments, benchmarks, debugging, or improvement, or for ontology analysis.

2   Competition Design

The ORE competition is inspired by and modeled on the CADE ATP System Competition (CASC) [16, 23] which has been running for 25 years and has been heavily influential in the automated theorem proving community[2] (especially for first order logic). The key common elements between ORE and CASC are:

1. A number of distinct tracks/divisions/disciplines characterised by problem type (e.g., "effectively propositional" or "OWL 2 EL ontology").
2. The test problems are derived from a large, neutral, updated yearly set of problems (e.g., for CASC, the TPTP library [22]).

---

[1] The websites for DL2015 and ORE2015 are archived at http://dl.kr.org/dl2015/ and https://www.w3.org/community/owled/ore-2015-workshop/ respectively.

[2] See the CASC website for details on past competitions: http://www.cs.miami.edu/~tptp/CASC/. Also of interest, though not directly inspirational for ORE, is the SAT competition http://www.satcompetition.org/.

3. Reasoners compete (primarily) on number of problems solved with a tight
   per problem timeout.

The last point is worth some comment. Most evaluations of reasoner performance in the literature use some form of time in their evaluations (e.g., CPU or
wall clock time). This has several advantages, including capturing the primary
quantity of interest for most users in most situations. However, it is vulnerable
to a number of problems esp. as one starts testing on large numbers of diverse
ontologies. For example, one reasoner might perform very well on a large number of small ontologies and comparatively poorly on a few larger ones, whereas
another might lose on all the small ones (due to start up overhead) and do much
better on the larger ones. Yet, their averages and even medians might be similar.

More critically, timeouts severely distort aggregate statistics about time. If
we include timeouts in the statistics, they *crop* times. That is, given a timeout of
two hours, we cannot distinguish between a reasoner that would take two hours
and one minute from one that would take days. Reasoner errors can cause similar
issues. If we drop those times, buggy reasoners seem to do better. Even if we
include them, less buggy reasoners that time-out, or just take longer to correctly
finish than the buggy reasoners take to hit an error, will be penalised. Measuring
problems solved does not fully ameliorate these problems and introduces some
new ones, but it seems more robust for simple comparisons. So it serves as a
better default experiment design.

As description logics have a varied set of core inference services supported
by essentially all reasoners, ORE also has track distinctions based on task (e.g.,
classification or realisation). ORE 2015 had both a live as well as an offline competition. The offline competition is executed with more relaxed time constraints
against user-submitted ontologies, while the live competition is executed with a
tight timeout against a corpus of ontologies we constructed.

3   Ontology Corpora

In the following sections we present the publicly available corpora of ontologies
constructed for the live competition and the user-submitted ontologies. Ontology
pre-processing was done using the OWL API (v3.5.1) [4].

3.1   Live Competition Corpus

The full live competition corpus contains 1,920 ontologies, sampled from three
source corpora: A January 2015 snapshot of Bioportal [12] containing 330 biomedical ontologies, the Oxford Ontology Library[3] with 793 ontologies that
were collected for the purpose of ontology-related tool evaluation, and MOWL-
Corp [6], a corpus based on a 2014 snapshot of a Web crawl containing around
21,000 unique ontologies. As a first step, the ontologies of all three source corpora
were collected and serialised into OWL/XML with their imports closure merged

---

[3] http://www.cs.ox.ac.uk/isg/ontologies/.

into a single ontology. The merging is, from a competition perspective, necessary to mitigate the bottleneck of loading potentially large imports repeatedly over the network, and because the hosts of frequently imported ontologies sometimes impose restrictions on the number of simultaneous accesses.[4] After the collection, the entire pool of ontologies is divided into three groups: (1) Ontologies with less than 50 axioms, (2) OWL 2 DL ontologies, and (3) OWL 2 Full ontologies. The first group was removed from the pool.

As reasoner developers could tune their reasoners towards the ontologies in the three publicly available source corpora, we included a number of approximations into our pool. The entire set of OWL 2 Full ontologies was approximated into OWL 2 DL, i.e., we used a (slightly modified) version of the OWL API profile checker to drop DL profile-violating axioms so that the remainder is in OWL 2 DL [8]. Because of some imperfections in the "DLification" process, this process had to be performed twice. For example, in the first round, the DL expressivity checker may have noted a missing declaration *and* an illegal punning. Fixing this would result in dropping the axiom(s) causing the illegal punning *as well as* injecting the declaration—which could result again in an illegal punning.

The OWL 2 DL group was then approximated into OWL 2 EL using the approximation method employed by TrOWL [17]. As some ontologies are included in more than one of the source corpora, we excluded at this point (as a last pre-processing step) all duplicates[5] from the entire pool of ontologies, and removed ontologies with TBoxes containing less than 50 axioms. This left us with the full competition dataset of 1,920 unique OWL 2 DL ontologies. The full competition corpus can be obtained from Zenodo [9].

3.2   User Submitted Ontologies

We had four user submissions to ORE 2015, consisting of a total of 7 ontologies. The user submissions underwent the same pre-processing procedures as the corpus (Sect. 3.1). This had occasionally large consequences on the ontologies, most importantly with respect to rules (they were stripped out) and any axiom beyond OWL 2 DL (for example, axioms redefining built-in vocabulary or violating the global constraints on role hierarchies, see [8]).

We make all user submitted ontologies for which we have permission to redistribute, redistributable. Occasionally, we have some user submitted ontologies which are proprietary and so cannot be redistributed. On the one hand, we prefer all ontologies be fully shareable. On the other, we want the widest reach possible. Currently, the number of "restricted" ontologies that have been submitted are very few, so it seems worth the outreach. We do work with submitters to make those ontologies as accessible as possible. Some basic metrics for the ontologies can be found in Table 1. The ontology archive is published on Zenodo, and linked to from http://owl.cs.manchester.ac.uk/publications/supporting-material/ore-2015-report.

---

[4] Which may be exceeded considering that all reasoners in the competition run in parallel.

[5] Duplicates are those that are *byte identical* after being "DLified" and serialised into Functional Syntax.

Table 1. Breakdown of user-submitted ontologies in the ORE 2015 corpus

Ontology name	#Axioms	OWL	Expressivity
Drug-Drug Interactions Ontology (DINTO)[6]	123,930	Pure DL	$\mathcal{ALCRIQ(D)}$
Cell Ontology (CO)[7]	7,527	Pure DL	$\mathcal{SRI}$
Drosophila Phenotype Ontology (DPO)[8] [13]	917	Pure DL	$\mathcal{SRIF(D)}$
Gene Ontology Plus (GO+)[9]	150,955	Pure DL	$\mathcal{SRI}$
Virtual Fly Brain Ontologies (VFB-KB)[10]	168,183	Pure DL	$\mathcal{SRI}$
Virtual Fly Brain Ontologies Ext. 1 (VFB-EPNT)	96,907	Pure DL	$\mathcal{SRI}$
Virtual Fly Brain Ontologies Ext. 2 (VFB-NCT)	96,907	Pure DL	$\mathcal{SRI}$

4 Competing Systems

There were 14 reasoners submitted with 11 purporting to cover OWL 2 DL, and 3 being OWL 2 EL specific. The set of competing systems (as submitted)

Table 2. Reasoners submitted to the ORE 2015 competition.

Reasoner	Version	Language	Maintained by
Chainsaw [26]	1.0	OWL 2 DL	University of Manchester, UK
ELepHant [19]	0.5.7	OWL 2 EL	Not given
ELK [5]	0.5.0	OWL 2 EL	University of Ulm, Germany
FaCT++ [25]	1.6.4	OWL 2 DL	University of Manchester, UK
HermiT [1]	1.3.8.5	OWL 2 DL	University of Oxford, UK
jcel [10]	0.21.0	OWL 2 EL	Technische Universität Dresden, Germany
Jfact [14]	4.0.1	OWL 2 DL	University of Manchester, UK
Konclude [21]	0.6.1	OWL 2 DL	University of Ulm, derivo GmbH, Germany
MORe [18]	0.1.6	OWL 2 DL	University of Oxford, UK
PAGOdA [27]	-	OWL 2 DL	University of Oxford, UK
Pellet (OA4) [20]	2.4.0	OWL 2 DL	Complexible (Original version)
Racer [3]	2.0	OWL 2 DL	Concordia University, Montreal, Canada
TrOWL [24]	1.5	OWL 2 DL	University of Aberdeen, UK

---

[6] https://code.google.com/archive/p/dinto/. Submitted by María Herrero, Computer Science Department, Univesidad Carlos III de Madrid. Leganés, Spain.

[7] https://github.com/obophenotype/cell-ontology. Submitted by Dr. David Osumi-Sutherland, GO Editorial Office, European Bioinformatics Institute, European Molecular Biology Laboratory, Cambridge, UK.

[8] https://github.com/FlyBase/flybase-controlled-vocabulary.

[9] http://bioportal.bioontology.org/ontologies/GO-PLUS.

[10] https://github.com/VirtualFlyBrain.

is available on the Web.[11] In Table 2, we briefly summarize the participating reasoning systems. More detailed information about each reasoner can be found online[12] as well as in our recently conducted OWL reasoner survey [7]. The version information reflect the state of the system as it was submitted to ORE.

5   Test Framework

The test framework used in ORE 2015 is a slightly modified version of the one used for ORE 2014, which is implemented in Java, open sourced under the LGPL license, and versioned and distributed on Github.[13]

The framework takes a "script wrapper" approach to execute reasoners, instead of, for example, requiring all reasoners to use (a specific version of) the OWL API. While this puts some extra burden on established reasoners with good OWL API bindings, this, combined with the requirement only to handle *some* OWL 2 standard syntax (with the very easy to parse and serialise functional syntax [11] as a common choice), makes it very easy for new reasoners to participate even if they are written in hard-to-integrate with the JVM languages. There is a standard script for OWL API based reasoners so it is fairly trivial to prepare an OWL API wrapped reasoner for competition. On the other hand, this is not necessarily a desirable outcome as encouraging reasoners to provide good OWL API support (thus supporting access to those reasoners by the plethora of tools which use the OWL API) is an outcome we want to encourage.

Reasoners report times, results, and any errors through the invocation script. Times are in wall clock time (CPU time is inappropriate because it will penalise parallel reasoners) and exclude "standard" parsing and loading of problems (i.e., without significant processing of the ontology). The framework enforces (configurable) timeouts for each reasoning problem. Results are validated by comparison between competitors with a majority vote/random tie-breaking fallback strategy.

The framework supports both serial and parallel execution of a competition. Parallel distributed mode is used for the live competition, but serial mode is sufficient for testing or offline experiments. The framework also logs sufficient information to allow "replaying" the competition, and includes scripts for a complete replay as well as jumping to the final results.

5.1   Usage

The framework project can be cloned with git,[14] after which users can use the *build-evaluator* script to build the project with Maven.[15] Subsequently, the configuration work will primarily focus on the *data* folder (the paragraphs below

---

[11] See links in supplementary materials website: http://owl.cs.manchester.ac.uk/publications/supporting-material/ore-2015-report.

[12] http://owl.cs.manchester.ac.uk/tools/list-of-reasoners/.

[13] https://github.com/andreas-steigmiller/ore-competition-framework/.

[14] https://git-scm.com/.

[15] http://maven.com.

discuss folders within *data*), and minor changes to the scripts in the *scripts* folder may be necessary to modify the amount of memory allocated to the JVM.

Global settings. The global settings of the framework are defined in a file in the *configs* folder. Possible settings include the timeout (in milliseconds) for the execution each reasoner, the processing timeout (used to cut the reported processing time of the reasoners for the evaluation), the memory limit, output options, and execution options such as forcing the competition to only take place if there exists one client-machine available for each reasoner.

Competition settings. The competition settings are defined in files in the *competitions* folder. The key competition settings are its name, output folder, list of participating reasoners, query folder, and execution and processing timeouts.

Reasoner settings. Each reasoner under test needs to be accompanied with two elements: a starter script that the framework executes to start the reasoner, and a configuration file that defines the reasoner name, response output folder, starter script location, accepted ontology format, supported OWL 2 profiles, and whether the reasoner supports datatypes and rules. Multiple versions of the same reasoner can be benchmarked, so long as their respective configuration files differ in name and input-output information.

Inputs. The inputs for evaluating a competition are: a corpus of ontologies (which should be placed in the *ontologies* folder), and a collection of reasoners configured as above (each of which should be a folder within the *reasoners* folder). Next, the framework needs the queries that are meant to be evaluated in the benchmark; these files specify the reasoning task, and the ontology location, its profile(s), and whether it contains rules or datatypes. Query files are generated by the framework using the appropriate *create-queries* scripts for the task.

There is further documentation on the framework's GitHub repository.

6   Conclusion

The ORE 2015 Reasoner Competition continues the success of its predecessors. And with it, the general public can benefit from the resources we presented here for their own experimentation. The ORE 2015 corpus, whether used with the ORE framework or in a custom test harness, is a significant and distinct corpus for reasoner experimentation. Developers can easily rerun this year's competition with new or updated reasoners to get a sense of their relative progress and we believe that solving all the problems in that corpus in similar or somewhat relaxed time constraints is a reliable indicator of a very high quality implementation.

Ideally, the ORE toolkit and corpora will serve as a nucleus for an infrastructure for common experimentation. To that end, every relevant resource (from corpora to test framework) has been appropriately published, and where appropriate versioned, on the Web. The test harness seems perfectly well suited for black box head-to-head comparisons, and we recommend experimenters consider it before writing a home grown one. This will improve the reliability of the test harness as well as reproducibility of experiments. Even for cases where more elaborate internal measurements are required, the ORE harness can serve as the command and control mechanism. For example, separating actual calculus activity from other behavior (e.g., parsing) requires a deep delve into the reasoner internals. However, given a set of reasoners that could separate out those timings, it would be a simple extension to the harness to accommodate them.

References

1. Glimm, B., Horrocks, I., Motik, B., Stoilos, G., Wang, Z.: HermiT: an OWL 2 reasoner. J. Autom. Reasoning **53**(3), 245–269 (2014)
2. Gonçalves, R.S., Matentzoglu, N., Parsia, B., Sattler, U.: The empirical robustness of description logic classification. In: Proceedings of ISWC (2013)
3. Haarslev, V., Hidde, K., Möller, R., Wessel, M.: The RacerPro knowledge representation and reasoning system. Semant. Web J. **3**(3), 267–277 (2012)
4. Horridge, M., Bechhofer, S.: The OWL API: a java API for OWL ontologies. Semant. Web J. **2**(1), 11–21 (2011)
5. Kazakov, Y., Krötzsch, M., Simancik, F.: The incredible ELK - From polynomial procedures to efficient reasoning with EL ontologies. J. Autom. Reasoning **53**(1), 1–61 (2014)
6. Matentzoglu, N., Bail, S., Parsia, B.: A snapshot of the OWL web. In: Alani, H., et al. (eds.) ISWC 2013. LNCS, vol. 8218, pp. 331–346. Springer, Heidelberg (2013). doi:10.1007/978-3-642-41335-3_21
7. Matentzoglu, N., Leo, J., Hudhra, V., Sattler, U., Parsia, B.: A survey of current, stand-alone OWL reasoners. In: Proceedings of ORE (2015)
8. Matentzoglu, N., Parsia, B.: The OWL Full/DL gap in the field. In: Proceedings of OWLED (2014)
9. Matentzoglu, N., Parsia, B.: ORE 2015 reasoner competition dataset (2015). http://dx.doi.org/10.5281/zenodo.18578
10. Mendez, J.: jcel: A modular rule-based reasoner. In: Proceedings of ORE (2012)
11. Motik, B., Patel-Schneider, P.F., Parsia, B.: OWL 2 Web Ontology Language: Structural specification and functional-style syntax. In: W3C Recommendation (2009)
12. Noy, N.F., Shah, N.H., Whetzel, P.L., Dai, B., Dorf, M., Griffith, N., Jonquet, C., Rubin, D.L., Storey, M.A., Chute, C.G., Musen, M.A.: BioPortal: ontologies and integrated data resources at the click of a mouse. Nucleic Acids Res. **37**, 170–173 (2009)
13. Osumi-Sutherland, D., Marygold, S.J., Millburn, G.H., McQuilton, P., Ponting, L., Stefancsik, R., Falls, K., Brown, N.H., Gkoutos, G.V.: The drosophila phenotype ontology. J. Biomed. Semant. **4**(1), 1–10 (2013)
14. Palmisano, I.: JFact repository (2015). https://github.com/owlcs/jfact
15. Parsia, B., Matentzoglu, N., Gonçalves, R.S., Glimm, B., Steigmiller, A.: The OWL reasoner evaluation (ORE) 2015 competition report. J. Autom. Reasoning **53**(3), 245–269 (2016). in submission
16. Pelletier, F., Sutcliffe, G., Suttner, C.: The development of CASC. AIC **15**(2–3), 79–90 (2002)
17. Ren, Y., Pan, J.Z., Zhao, Y.: Soundness preserving approximation for TBox reasoning. In: Proceedings of AAAI (2010)

18. Armas Romero, A., Cuenca Grau, B., Horrocks, I.: MORe: modular combination of OWL reasoners for ontology classification. In: Cudré-Mauroux, P., et al. (eds.) ISWC 2012. LNCS, vol. 7649, pp. 1–16. Springer, Heidelberg (2012). doi:10.1007/978-3-642-35176-1_1

19. Sertkaya, B.: The ELepHant reasoner system description. In: Proceedings of ORE (2013)

20. Sirin, E., Parsia, B., Cuenca Grau, B., Kalyanpur, A., Katz, Y.: Pellet: a practical OWL-DL reasoner. J. Web Semant. 5(2), 51–53 (2007)

21. Steigmiller, A., Liebig, T., Glimm, B.: Konclude: system description. J. Web Semant. 27, 78–85 (2014)

22. Sutcliffe, G.: The TPTP problem library and associated infrastructure: the FOF and CNF parts, v3.5.0. J. Autom. Reasoning 43(4), 337–362 (2009)

23. Sutcliffe, G., Suttner, C.: The state of CASC. AIC 19(1), 35–48 (2006)

24. Thomas, E., Pan, J.Z., Ren, Y.: TrOWL: tractable OWL 2 reasoning infrastructure. In: Aroyo, L., Antoniou, G., Hyvönen, E., Teije, A., Stuckenschmidt, H., Cabral, L., Tudorache, T. (eds.) ESWC 2010. LNCS, vol. 6089, pp. 431–435. Springer, Heidelberg (2010). doi:10.1007/978-3-642-13489-0_38

25. Tsarkov, D., Horrocks, I.: FaCT++ description logic reasoner: system description. In: Furbach, U., Shankar, N. (eds.) IJCAR 2006. LNCS (LNAI), vol. 4130, pp. 292–297. Springer, Heidelberg (2006). doi:10.1007/11814771_26

26. Tsarkov, D., Palmisano, I.: Chainsaw: a metareasoner for large ontologies. In: Proceedings of ORE (2012)

27. Zhou, Y., Nenov, Y., Grau, B.C., Horrocks, I.: Pay-as-you-go OWL query answering using a triple store. In: Proceedings of AAAI (2014)

FOOD: FOod in Open Data

Silvio Peroni[1(✉)], Giorgia Lodi[2], Luigi Asprino[2], Aldo Gangemi[2], and Valentina Presutti[2]

[1] DASPLab, DISI, University of Bologna, Bologna, Italy
silvio.peroni@unibo.it
[2] STLab, ISTC, CNR, Rome, Italy
{giorgia.lodi,luigi.asprino,aldo.gangemi,valentina.presutti}@istc.cnr.it

Abstract. This paper describes the outcome of an e-government project named FOOD, FOod in Open Data, which was carried out in the context of a collaboration between the Institute of Cognitive Sciences and Technologies of the Italian National Research Council, the Italian Ministry of Agriculture (MIPAAF) and the Italian Digital Agency (AgID). In particular, we implemented several ontologies for describing protected names of products (wine, pasta, fish, oil, etc.). In addition, we present the process carried out for producing and publishing a LOD dataset containing data extracted from existing Italian policy documents on such products and compliant with the aforementioned ontologies.

Keywords: Linked Open Data · Ontologies · Ontology design patterns

1 Introduction

The recent Open Data Barometer report[1] states that "open data is entering the mainstream" and "the demand is high". Several concrete initiatives and assessments witness that open data, if correctly adopted, can be an extremely powerful driver of innovation for improving different public sectors as well as impacting scientific progresses. Nevertheless, the presence of (linked) open data is not equally distributed in all public sectors. For instance, in the context of the agriculture and food sector in Italy – that has seen some recent development in other European countries, such as Russia [4] – the management of the European Union (EU) quality schemes for agricultural and food products – i.e., PDO (Protected Designation of Origin), PGI (Protected Geographical Indication) and TSG (Traditional Speciality Guaranteed) – is not fully automatized, and no standards are used in the definition of policy documents (or product specifications) that regulate them.

In this context, the Ministry of Agriculture (MIPAAF), the Italian Digital Agency (AgID), and our laboratory conceived and carried out a project named FOod in Open Data (FOOD)[2]. The main goal of FOOD was to extract the data

---

[1] http://opendatabarometer.org/.
[2] https://w3id.org/food.

© Springer International Publishing AG 2016
P. Groth et al. (Eds.): ISWC 2016, Part II, LNCS 9982, pp. 168–176, 2016.
DOI: 10.1007/978-3-319-46547-0_18

contained in the textual content of the policy documents of Italian agricultural PDO, PGI and TSG products that were available online in PDF, and to make them available as LOD. According to informal interviews done by the MIPAAF, the potential impact and the advantages perceived by all the actors of the project are threefold. On the one hand, the availability of interoperable LOD can pave the way to the construction of new applications, targeting for instance food frauds detection, and production and distribution traceability of quality products. On the other hand, it poses the basis for outlining a proper standardization process for the definition of product specifications. Finally, the use of common models for describing such data can improve the interoperability achievable in the data exchange between MIPAAF and other Public Administrations, which may publish the same type of data on a territory base.

In this paper we describe the artefacts (available in [5]) produced as outcomes of the FOOD project, ended in 2015. First, a set of ontologies were developed following a methodology that makes use of well-known Ontology Design Patterns (ODPs). Second, a semi-automatic process for producing and publishing LOD, compliant with the developed ontologies, was carried out on more than 800 policy documents of the Italian quality products.

The rest of paper is structured as follows. Section 2 introduces the ontologies and the methodology that was adopted. Section 3 describes the process we employed to produce and publish the LOD datasets, compliant with the defined ontologies. Section 4 discusses examples of reuse of FOOD's ontologies and data and finally, Sect. 5 concludes the paper.

2   FOOD Ontologies

The process adopted for the development of the ontologies for describing the Italian policy documents was mainly based on the *eXtreme Design* methodology [7], which is a a collaborative, incremental, iterative method for pattern-based ontology design, which we used in several projects in the past. The ontologies developed were also aligned with other existing models, i.e. AGRO-VOC, DOLCE, DBpedia and Wordnet. The development process was organised in few steps, illustrated as follows.

In the first step, two ontology engineers analysed the large set of source policy documents about agricultural products – more than 800 documents available in either PDF or DOCX – involving domain experts of MIPAAF and AgID so as to identify the main high-level concepts characterising the domain. In addition, they also considered the EU schemes for protected names[3] so as to take into account how they are organized at the European level. Three main interlinked classes have been defined as result of this analysis:

- *ProtectedName*, i.e., a trademark label (e.g., "Abruzzo DOC" for a particular Italian wine), issued by an authority granted for certifying agricultural products and foodstuffs, that typically belongs to a certification scheme (i.e., PDO, PGI, TSG);

---

[3] http://ec.europa.eu/agriculture/quality/schemes/index_en.htm.

- *Type*, i.e., a possible type for certain products (e.g., "white" or "red" for wines);
- *Product*, i.e., the agricultural product or foodstuff (e.g., "Abruzzo DOC red wine") which has a type, refers to a protected name disciplined by a certain policy document, and is described in terms of its raw materials (the particular wine varietal used) and physical/chemical/organolectic characteristics (colour, smell, flavour, etc.).

As shown in the Graffoo [1] diagram in Fig. 1 (on the left), in addition to the aforementioned classes, we also modelled several concepts for describing particular features, i.e., the raw material (classes *RawMaterial* and *DescriptionOfRawMaterial*), the characteristics (classes *Characteristic* and *DescriptionOfCharacteristic*), the producer (an *Agent* link with the property *hasProducer*) related to the products. Similarly, other classes and properties defining contextual information of the protected name such the production place (the class *Place* linked by the property *hasProductionPlace*), the logo (the class *Image* linked by the property *hasLogo*), the control authority (another *Agent* linked by the property *hasControlAuthority*) responsible for certifying products of such protected name, and the particular version of the policy document (the class *PolicyDocumentVersion*) source of all the information related.

It is worth noticing that such "upper" ontology reuses several existing ontology design patterns that were considered also during the development, i.e., the pattern Description[4] (for representing raw materials and other product characteristics), the pattern Place[5] (for modelling the production place), the pattern Classification[6] (for expressing both the different characteristics of the products and the quality schemes defined at EU level), and the pattern Information Realization[7] (for relating policy documents with their various versions released during time).

Then, since the ontology had to be appropriately extended according to all the twenty kinds of products[8] described in the policy documents, we decided to involve three more ontology engineers, with different skills and level of expertise, for speeding up the development process. We provided a generic Graffoo template (shown in Fig. 2) created starting from the parts of the upper ontology that had to be extended, so as to guide the development of all the other ontologies in an homogeneous way, independently of the particular ontology engineer assigned as developer of the specific ontologies. Each new ontology was assigned to only one ontology engineer, and the idea was that each engineer had to rename the variables within double square brackets (e.g., "[[product]]") with the appropriate names, thus adding the related labels and comments. The result of this phase

---

[4] http://ontologydesignpatterns.org/wiki/Submissions:Description.
[5] http://ontologydesignpatterns.org/wiki/Submissions:Place.
[6] http://ontologydesignpatterns.org/wiki/Submissions:Classification.
[7] http://ontologydesignpatterns.org/wiki/Submissions:Information_realization.
[8] Bread, cereal, cheese, essential oil, fish, fruit, honey, liquorice, meat, mollusc, oil, pasta, ricotta, saffron, salt, salume, sweet, vegetable, vinegar, wine.

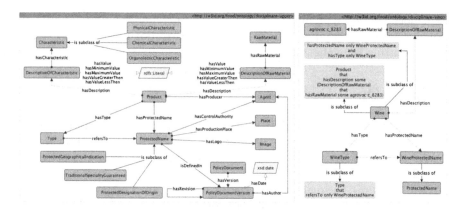

Fig. 1. The Upper Ontology (left) and one of its extension for wines (right). All the classes and properties are here shown with English names, while their IRI are originally in Italian

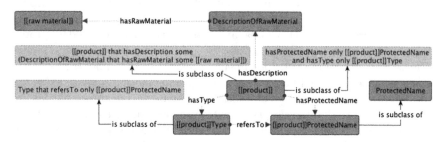

Fig. 2. The template reused by developers for creating the various ontologies for each kind of agricultural product described in the policy documents.

produced twenty new Graffoo diagrams organised as the one for wines shown in Fig. 1 (on the right).

Then, one of the ontology engineers took the responsibility of substituting all the raw materials identified during the development process with the appropriate concepts defined in AGROVOC[9], the multilingual agricultural thesaurus created by the Food and Agriculture Organization (FAO).

Once all the diagrams were finished and stable, we produced all the related OWL files, one for each diagram, by means of DiTTO[10] [3], i.e., a Web application that is able to convert Graffoo diagrams into OWL automatically. Separately, some additional OWL files have been created, so as to map all the developed ontologies with DBpedia, DOLCE, WordNet, and the used ontology design patterns.

---

[9] http://aims.fao.org/standards/agrovoc.
[10] http://www.essepuntato.it/ditto.

The documentation of all the ontologies was generated by using LODE[11] [6], a tool that is able to produce human-readable HTML documents describing an ontology starting from the annotations it contains.

3 Linked Open Datasets

The other main goal of the project aimed at creating LOD datasets containing the data extracted from 847 policy documents on protected names – provided by MIPAAF and publicly available online[12] – according to the ontologies presented in Sect. 2.

At a first reading, the main parts of the documents have a pretty similar organisation: each document consists of at least four articles. The first article usually defines the protected name and the types of related products of such name. The second article defines the raw materials which contribute to the products composition. The third article provides a precise description of the production area. Finally, the rest of the articles usually describe physical/chemical/organoleptic characteristics of the products, introduce historical/geographical evidences supporting the described production area, and describe the methods used for producing the products of that protected name.

The general structure of the documents is shared in principle. In the light of this observation, our initial idea was to develop some mechanisms for extracting all the relevant information within such documents in an automatic fashion. We started trying to develop scripts for extracting relevant data from all the 563 policy documents about wine we had available in DOCX format – the extraction was carried by looking for textual patterns we knew in advance. However, we realized that it was not possible to address this extraction for all the policy documents automatically and to preserve a good and appropriate quality of the data extracted at the same time. There was a rather high degree of heterogeneity among documents describing different kinds of products (e.g., wines vs. bread), as well as among documents describing products of the same type, mainly due to the fact that no predefined templates was actually used for guiding their creation, and that such authorial activity was held across many years and by several authors. In addition, several documents were actually images derived from scanned copies of old paper policy documents, stored as PDF files. Hence, for the remaining 274 PDF documents regarding agricultural products different from wine, we decided to carry out a manual extraction since the beginning.

The pipeline we followed for processing policy documents is shown in the top part of Fig. 3, and was organised in three main steps:

1. extraction – we developed scripts and carried out human-based data extractions from policy documents through the use of Excel files;

---

[11] http://www.essepuntato.it/lode.
[12] https://www.politicheagricole.it/flex/cm/pages/ServeBLOB.php/L/IT/IDPagina/309.

2. validation – the domain experts from MIPAAF and AgID were involved so as to correct and validate the produced Excel documents (that are included in [5] and briefly summarised in the bottom part of Fig. 3);

3. conversion – the validated Excel data were, finally, processed by other scripts we developed for converting such data into RDF according to the FOOD ontologies.

In the latter step we also used TagMe [2] so as to extract DBpedia entities referring to villages, cities and regions from the textual description of the production places. The entities extracted by TagMe were revised by humans as well, so as to remove as more mistakes as possible from the data, and then they were aligned with other existing LOD developed in the past by the Italian Public Administration, i.e., SPCData[13] and the LOD of the Italian National Institute of Statistics[14]. Considering the novelty of these data, we have not found any other relevant dataset to link with.

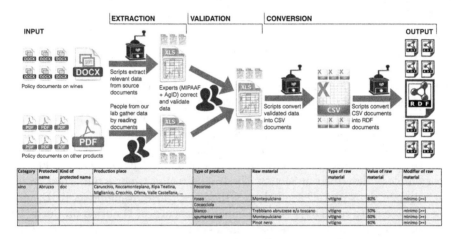

Fig. 3. The pipeline for extracting data from the policy documents (top) and an excerpt of the Excel data validated by the experts (bottom) – table headings in English for the sake of clarity.

The obtained RDF data contain information defining the protected name and various types of related product, the production area of the protected name, the raw materials and the characteristics related with such kinds of products, as shown in the following excerpt (in Turtle with English IRIs for the sake of clarity – while original IRIs are in Italian):

```
product:wine-abruzzo-red a upper:Product , wine:Wine ;
    rdfs:label "Vino 'Abruzzo' Rosso" ;
    upper:hasProtectedName name:wine-abruzzo ;
```

---

[13] http://spcdata.digitpa.gov.it/.
[14] http://datiopen.istat.it/.

```
upper:hasDescription
    descraw:wine-abruzzo-red-raw-material-1 ,
    descraw:wine-abruzzo-red-raw-material-2 ;
upper:hasType type:red-wine .

descraw:wine-abruzzo-red-raw-material-1 a upper:DescriptionOfRawMaterial ;
    rdfs:label "Vitigno 'Montepulciano': minimo (>=) 80%" ;
    upper:hasRawMaterial rawmat:vine-variety-montepulciano ;
    upper:hasMinimumValue "80%" .
```

All the produced resources were defined by means of permanent IRIs via *w3id.org*. The IRI naming convention used for each resource was `http://w3id.org/food/data/[[class]]/[[product]]` (e.g., `http://w3id.org/food/data/prodotto/vino-abruzzo-rosso`). In addition, the datasets also include provenance information (defined using DCAT and PROV-O) about the whole production and publication process. The ontologies, the dataset and the provenance metadata are available in [5] and can be queried by using the SPARQL endpoint[15] provided by the Fuseki 2 triple store of the project. All the data have been released under the license CC BY 4.0, and can be browsed and retrieved in different formats by means of LODView[16].

4   Ontologies and Data Reuse

The LOD paradigm is still not fully adopted in the Italian public sector, although national governmental guidelines encourage public administrations to release their data under the form of LOD for semantic interoperability purposes [8]. Thus, FOOD represents one of a few very recent LOD initiatives, and the first comprehensive project, in the Italian food domain.

In this scenario, early this year we recorded already the reuse of the FOOD upper ontology described in Sect. 2. The Italian region Umbria published, in its regional open data portal, a number of datasets[17] also available as LOD, and some of these concern PDO/PGI/TSG quality schemes of Umbrian products. In particular, they reused the three main classes of the upper ontology, i.e., *Product*, *ProtectedName* and *Type*.

Finally, it is worth mentioning an additional initiative that involved, among the others, the data produced in the context of the FOOD project. In October 2015, AgID carried out a national hackathon[18], where a group of young guys won a price for their idea of an app named *eat it*[19], based on the FOOD LOD datasets. To the best of our knowledge, at the time of this writing, the application is still under development.

---

[15] http://etna.istc.cnr.it/food-sparql/.

[16] https://github.com/dvcama/LodView.

[17] http://dati.umbria.it/dataset/prodotti-dop-igp-umbria/resource/
551c465a-932f-4f10-8b63-8031b15071eb.

[18] http://www.agid.gov.it/notizie/2015/10/12/big-hack-agid-premia-unapp-
valorizzazione-prodotti-dopigp.

[19] http://eatit.emooh.it/index.html.

5   Conclusions

In this paper we have described the outcomes (available in [5]) of an e-government project named FOOD – FOod in Open Data, which was carried out in the context of a collaboration between ISTC-CNR, MIPAAF and AgID, and that concerned the development of ontologies for describing PDO, PGI and TSG products, and the creation of LOD datasets containing data, extracted from existing Italian policy documents on protected names.

The main lesson learnt from the whole process was that the automatic extraction of data from the policy documents – even when they are generally structured similarly and provided in easy-to-process formats (i.e., DOCX), as for wines – it is quite difficult and the results are far from the quality required for being published. In fact, the correction introduced by the experts to the data extracted from such policy documents were quite huge. *A posteriori*, a manual extraction of such data, that could be possible in the same amount of time, would have prevent the introduction of such a large number of mistakes. However, the analysis done on these policy documents about wines have enabled the refactor of existing templates for policy documents so as to write relevant information in an homogeneous manner – that would be a great simplification for guaranteeing some significant automatic processing of such documents in the future. To this end, we are currently discussing with the MIPAAF about possible strategies for integrating our data extraction process in their workflow.

Acknowledgements. The project was partially funded by MIPAAF and AgID in the context of a formal agreement established in accordance to article 15 of the Italian law no. 241/1990. We would like to thank Andrea Giovanni Nuzzolese, Daria Spampinato and the colleagues of IVALSA Department of the Italian National Research Council for their valuable technical contributions.

References

1. Falco, R., Gangemi, A., Peroni, S., Vitali, F.: Modelling OWL ontologies with Graffoo. In: The Semantic Web: ESWC 2014 Satellite Events, pp. 320–325 (2014). http://dx.doi.org/10.1007/978-3-319-11955-7_42
2. Ferragina, P., Scaiella, U.: On-the-fly annotation of short text fragments (by Wikipedia entities). In: Proceedings of the 19th ACM International Conference on Information and Knowledge Management (CIKM 2010), pp. 1625–1628 (2010). http://dx.doi.org/10.1145/1871437.1871689
3. Gangemi, A., Peroni, S.: Diagrams transformation into OWL. In: Proceedings of the ISWC 2013 Posters & Demonstrations Track (2013). http://ceur-ws.org/Vol-1035/iswc2013_demo_2.pdf
4. Kolchin, M., Chistyakov, A., Lapaev, M., Khaydarova, R.: Russian food products as a linked data dataset. In: The Semantic Web: ESWC 2015 Satellite Events, pp. 87–90 (2015). http://dx.doi.org/10.1007/978-3-319-25639-9_17
5. Peroni, S., Lodi, G., Asprino, L., Gangemi, A., Nuzzolese, A.G., Presutti, V., Spampinato, D.: FOOD resources: ontologies and data. In:Figshare. http://dx.doi.org/10.6084/m9.figshare.3187903

6. Peroni, S., Shotton, D., Vitali, F.: The live OWL documentation environment: a tool for the automatic generation of ontology documentation. In: Proceedings of the 18th International Conference on Knowledge Engineering and Knowledge Management (EKAW 2012), pp. 398–412 (2012). http://dx.doi.org/10.1007/978-3-642-33876-2_35

7. Presutti, V., Daga, E., Gangemi, A., Blomqvist, E.: eXtreme design with content ontology design patterns. In: Proceedings of the 1st Workshop on Ontology Patterns (WOP 2009) (2009). http://ceur-ws.org/Vol-516/pap21.pdf

8. SPC Board: An Overview of the Italian "Guidelines for Semantic Interoperability through Linked Open Data" (2012). http://www.agid.gov.it/sites/default/files/documentazione_trasparenza/semanticinteroperabilitylod_en_3.pdf

YAGO: A Multilingual Knowledge Base from Wikipedia, Wordnet, and Geonames

Thomas Rebele[1]([⊠]), Fabian Suchanek[1], Johannes Hoffart[2], Joanna Biega[2], Erdal Kuzey[2], and Gerhard Weikum[2]

[1] Télécom ParisTech, 46 Rue Barrault, 75013 Paris, France
rebele@enst.fr
[2] Max Planck Institute for Informatics, Campus E1 4, 66123 Saarbrücken, Germany
http://www.telecom-paristech.fr,
http://www.mpi-inf.mpg.de/

Abstract. YAGO is a large knowledge base that is built automatically from Wikipedia, WordNet and GeoNames. The project combines information from Wikipedias in 10 different languages into a coherent whole, thus giving the knowledge a multilingual dimension. It also attaches spatial and temporal information to many facts, and thus allows the user to query the data over space and time. YAGO focuses on extraction quality and achieves a manually evaluated precision of 95 %. In this paper, we explain how YAGO is built from its sources, how its quality is evaluated, how a user can access it, and how other projects utilize it.

Keywords: Knowledge base · Wikipedia · WordNet · Geonames

1 Introduction

A knowledge base (KB) is a computer-processable collection of knowledge about the world. A KB usually contains *entities* such as Elvis Presley, Stanford University, or the city of Kobe in Japan. It also contains *facts* about these entities, such as the fact that Elvis Presley plays guitar, that Stanford is a university, or that Kobe is located in Japan. KBs find applications in areas such as machine translation, question answering, and semantic search. Early approaches to create such KBs were mostly manual. With the growth of the Web, more and more approaches constructed KBs automatically by extracting information from Web corpora. Some of the more prominent approaches are YAGO, DBpedia, Wikidata, NELL, and Google's Knowledge Vault. Some of these approaches focus on Wikipedia, the free online encyclopedia.

In this paper, we describe one of the earliest approaches in this direction: The YAGO knowledge base [22]. It was the first academic project to build a KB from Wikipedia, closely followed by the DBpedia project [1]. The particular focus in YAGO has been on precision, i.e., on the correctness of the extracted facts. By sending the extracted facts through a sequence of filters, YAGO achieves a precision of 95 %. Today, YAGO is a larger project at the Max Planck Institute

P. Groth et al. (Eds.): ISWC 2016, Part II, LNCS 9982, pp. 177–185, 2016.
DOI: 10.1007/978-3-319-46547-0_19

for Informatics and Tcom ParisTech University. The KB draws on several sources by now, including WordNet and Geonames, and has grown to 16 million entities and more than 100 million facts. It is part of the Linked Open Data cloud.

This paper is structured as follows. Section 2 gives an overview of YAGO. Section 3 describes the construction of the KB. Section 4 illustrates data formats and tools. Section 5 shows applications of YAGO before Sect. 6 concludes.

2   The YAGO Knowledge Base

2.1   History

The YAGO project started in 2006 from a simple idea: Wikipedia contains a large number of instances, such as singers, movies, or cities. However, its hierarchy of categories is not directly suitable as a taxonomy. WordNet, on the other hand, has a very elaborate taxonomy, but a rather low recall on instances. It thus seemed promising to combine both resources to get the best of the two worlds.

The first version of YAGO [22] extracted facts mainly from the category names of the English Wikipedia. With the first upgrade of YAGO in 2008 [23], the project started extracting also from the infoboxes. In 2010, we started working on the extraction of temporal and geographical meta-facts, which resulted in YAGO2 [11,12]. The system architecture was completely restructured for YAGO2s [2] in 2013. This helped us for YAGO3 [19], which added extraction from 10 different Wikipedia languages in 2015.

The YAGO project shares its goal with other KB projects, most notably DBpedia [1,17], Wikidata [27], and the Google Knowledge Vault [5]. Unlike the Knowledge Vault, YAGO is publicly available for download. Unlike DBpedia and Wikidata, YAGO is not constructed through crowdsourcing, but through information extraction and merging. The YAGO project puts a particular focus on the quality of its data, which is assessed through regular manual evaluations. It also has a rather elaborate taxonomy in comparison to other projects, which it inherits from WordNet [6]. YAGO also integrates several multilingual sources into a single KB. Finally, YAGO pays particular attention to the anchoring of the facts in time and space.

2.2   Content

YAGO facts follows the RDF model [28], where facts are represented by triples of a subject, a predicate, and an object. An example is

$$<\text{Barack\_Obama}><\text{wasBornOnDate}>"1961-08-04"^{\wedge\wedge}\text{xsd:date}.$$

YAGO gives each fact a *fact identifier*. For example, the above fact has the fact identifier <id_1km2mmx_1xk_17y5fnj>. This allows YAGO to state temporal or spatial information, or the origin of facts. We can say, e.g., that the above fact was extracted from the English Wikipedia page about Barack Obama:

<id_1km2mmx_1xk_17y5fnj> <extractionSource>
<http://en.wikipedia.org/wiki/Barack_Obama>.

YAGO covers topics of general interest such as geographical entities, person-
alities of public life or history, movies, and organizations. For this, YAGO uses a
manually predefined set of 76 relations. In total, the KB contains 16 927 153 entities and 1 185 433 982 triples. The triples are partitioned into *themes*, which can
be downloaded separately. YAGO has the following groups of themes (number
of triples in parentheses):

- Taxonomy-related facts (95 m): the class hierarchy (570 k), types (16 m), their
  transitive closure (78 m), and schema information (486).
- A simplified taxonomy with just three layers (17 m). It contains the leaf levels
  of the WordNet taxonomy, the main YAGO branches (person, organization,
  building, artifact, abstraction, physical entity, and geographical entity), and
  the root node owl:Thing.
- The main facts (55 m), i.e., relations between entities (5 m), facts with dates
  (3 m), facts with other literals (1 m), and labels (45 m)
- GeoNames facts (39 m), mainly types, labels, and coordinates of geo-entities.
- Meta-facts (203 m), i.e., facts about the origin of facts (201 m), as well as
  their time and location (2 m)
- Labels for classes in various languages from the Universal WordNet (787 k)
- Links to other KBs (4 m), notably to DBpedia (4 m), GeoNames (117 k) and
  WordNet identifiers (156 k)
- Raw information from Wikipedia in RDF (296 m), which other projects can
  use to avoid parsing of Wikipedia. We provide infobox attributes of enti-
  ties (72 m), the infobox templates that an entity has on its Wikipedia page
  (5 m), the infobox attributes per template (262 k), Wikipedia-links between
  the entities (63 m), and the source facts for all of these.
- Redirect links and hyperlink anchor texts from Wikipedia (471 m).

3   Construction of YAGO

3.1   Sources

Wikipedia. Most of the information in YAGO comes from Wikipedia, the
community-driven online encyclopedia. Wikipedia contains not just textual
material, but also a hierarchical category system and structured data in the
form of *infoboxes*. As a rule of thumb, each Wikipedia page becomes an entity in
YAGO. Facts about these entities are created mainly from Wikipedia Infoboxes,
using a set of manually compiled mappings from Infobox attributes to YAGO
relations. Entity types are extracted from the Wikipedia leaf level categories.
The upper part of the Wikipedia class hierarchy is discarded.

Temporal Knowledge. YAGO extracts the time span of facts by hand-crafted
regular expressions from the Wikipedia infoboxes and categories. For example,
from the infobox excerpt from Cristiano Ronaldo's Wikipedia page

| years2 = 2003 − 2009 | clubs2 = [[Manchester United F.C.]]

YAGO extracts the start time and end time of the fact <Cristiano_Ronaldo> <playsForTeam><Manchester United F.C.>. YAGO stores time points as `xsd:date` literals attached to the fact id of the original fact. If a date contains only the year and month, YAGO uses placeholders, as in "2003-12-##".

WordNet. The WordNet KB [6] is a lexical database of the English language [20]. Among other things, it defines a taxonomy of nouns (e.g. ballet dancer is a hyponym of dancer). YAGO takes the leaves of the Wikipedia category hierarchy and links them to WordNet synsets. This yields, e.g.

    <wikicat_Norwegian_ballet_dancers>

               ⎣⟶ <wordnet_ballet_dancer_109834699>

rdfs:subClassOf

                            ⎣⟶ <wordnet_dancer_109989502>

                 rdfs:subClassOf

YAGO includes WordNet Domains [18], which groups words into 167 thematic domains, and allows, e.g., searching for entities related to "computer science". The Universal WordNet [4] extends WordNet to over 200 languages, and YAGO uses it to add labels in many languages to the WordNet classes in YAGO.

GeoNames. The GeoNames KB[1] contains 7 m geographical entities such as villages, cities, and notable buildings. It contains a class hierarchy and facts such as `locatedIn` facts for cities and countries. GeoNames provides links to Wikipedia, which we use to map the entities to YAGO entities. The GeoNames classes are mapped to WordNet classes by a heuristic defined on the token-overlap of their description. The precision of this matching heuristic is 94.1 %, with a recall of 86.7 % [12].

3.2   Extraction Process

Architecture. In YAGO, an *extractor* is a small code module that is responsible for a single, well-defined extraction subtask. An extractor takes certain themes as input, and produces certain themes as output. Therefore, the architecture of the YAGO extraction system can be represented as a bipartite graph of extractors and themes. This architecture allows for parallelization of the extraction process: Each extractor provides a list of input themes and a list of output themes, and each extractor is started by a scheduler as soon as its input becomes available [2].

Filtering. While the initial extractors are responsible for extracting raw facts from the sources, the following extractors are responsible for cleaning these facts. The facts first undergo redirection, a process where entities are replaced by their canonical versions in Wikipedia. They are then de-duplicated, and sent through various syntactic and semantic checks. Most notably, the facts are checked for compliance with the type signatures of the relations [2,11,12,15,23].

    The modular architecture proved useful when YAGO was made multilingual [19]. Only 3 major new extractors had to be added for the translation of

---

[1] http://www.geonames.org/.

entities. After that, the translated facts later undergo the same procedures as the facts obtained from the English Wikipedia [19].

3.3   Evaluation

Every major release of YAGO is evaluated for quality. Since there is no high quality gold standard of comparable size, this evaluation is done manually. Since the large number of facts in YAGO makes a complete manual evaluation infeasible, we evaluate a randomly chosen sample of facts for every relation. We evaluate only facts obtained by information extraction (not, e.g., imported facts). Facts are evaluated with respect to the extraction source (Wikipedia).

We developed a Web tool that presents a fact with the corresponding Wikipedia pages to a human judge. The judge clicks on "correct", "incorrect" or "ignore", and procedes to the next fact. As YAGO3 extracts facts from Wikipedias in several languages, we extended the tool so as to show the Wikipedia pages of the corresponding language and of the time of the Wikipedia dump.

The last evaluation of YAGO was made in 2015, and took two months. 15 people participated and evaluated 4 412 facts of 76 relations, which contain 60 m facts in total. They judged 98 % of the facts in the sample to be correct. To verify the statistical significance of this result, we calculate the Wilson interval [3]. Weighted by the number of facts, the interval has a center of 95 % and a width of 4.19 %. This means that the true ratio of correct facts in YAGO lies between 91 % and 99 %, with $\alpha = 95\,\%$ probability.[2]

4   Infrastructure

Data format. We provide YAGO in two formats, TTL (Terse RDF Triple Language, also called Turtle)[3] and TSV (Tab Separated Values). The TTL format allows using YAGO with standard Semantic Web software such as Apache Jena. Since TTL does not support fact identifiers directly, we store a fact identifier in a comment that precedes the fact. The TSV format allows users to easily import the facts into a database, or to handle the data programmatically. The format also allows storing fact identifiers as an additional column. We provide a script for importing the TSV files into an SQL database.

Users can download YAGO from the Webpage of the Max-Planck Institute for Informatics[4]. We further published the newest version, YAGO3, to Datahub[5]. The Creative Commons Attribution 3.0 License allows everyone to use YAGO, as long as the origin of the data is credited. YAGO follows the FAIR principles (Findable, Accessible, Interoperable, and Re-usable), thanks to the use of the standard TTL format, its copious metadata, and its open license.

---

[2] See https://w3id.org/yago/statistics for the complete statistics.
[3] See https://www.w3.org/TR/turtle/ for specifications.
[4] https://w3id.org/yago.
[5] https://datahub.io/dataset/yago.

YAGO is an active research project, and the teams at the Max-Planck Institute for Informatics and at Tcom ParisTech provide support and maintenance. Since every major revision of YAGO is evaluated manually, YAGO is updated in the rhythm of months or years.

Tools. We provide several tools to explore the data in YAGO. A graph browser[6] visualizes an entity with its in- and outgoing edges arranged in a star shape. Users can navigate the graph by clicking on an entity. Edges with the same direction and label are grouped together. Flags indicate the origin of the particular fact. The SPOTLX browser (Subject, Predicate, Object, Time, Location, conteXt)[7] allows querying YAGO with spatial and temporal visualizations. Users can ask questions such as "Which politicians born before 1900 were also scientists?". We also provide example queries. The Data Science Center of Paris-Saclay offers a SPARQL endpoint for YAGO[8], together with example SPARQL queries[9].

5   Applications of YAGO

DBPedia. This project [1] is a community effort to extract a KB from Wikipedia. The KB uses two taxonomies in parallel: a hand-crafted one from its contributors, and the YAGO taxonomy. For this purpose, the `type` and `subclassOf` facts from YAGO are imported into a proper namespace in DBpedia.

IBM Watson. The Watson system [7] can answer questions in natural language. It uses several data sources, among them the type hierarchy of YAGO. Watson participated in the TV quizz show *Jeopardy* together with human players, and was awarded the first place.

AIDA. The AIDA system [13] can find names of entities in text documents, and map them to the corresponding YAGO entities. For example, in the sentence "When *Page* played *Kashmir* at *Knebworth*, his *Les Paul* was uniquely tuned.", AIDA recognizes the names in italics using a graph algorithm and entity similarity measures. AIDA can understand that "Page" here refers to *Jimmy Page* of Led Zeppelin fame (and not, e.g., to *Larry Page*), and that "Kashmir" means the song, not the region. YAGO is also used to resolve temporal references such as "the presidency of Obama" or "the second term of Merkel" [16].

Semantic Culturomics. YAGO has been used to annotate articles of the French journal *Le Monde* with entities from the KB [14]. These annotations allow to compute statistics on entities over time, such as: What are the countries where many foreign companies operate (are mentioned)? What is the proportion of women mentioned in Le Monde, and how did it change over time? The

---

[6] https://w3id.org/yago/svgbrowser.
[7] https://w3id.org/yago/demo.
[8] https://w3id.org/yago/sparql.
[9] https://w3id.org/yago/dataset.

combination of structured knowledge (from YAGO) and unstructured knowledge (from Le Monde) illustrates correlations not visible in these resources alone.

6   Conclusions and Future Work

YAGO is a knowledge base that unifies information from Wikipedia, WordNet and GeoNames into a coherent whole. In this paper, we have described the sources, the extraction process, and the applications of YAGO. For future work, we want to extend the knowledge of YAGO along the following dimensions:

Release cycle. Reducing the manual work required for the evaluation could shorten the release cycle. Many relations such as <isLocatedIn> or <wasBornOnDate> will retain almost all facts from the previous version. We could therefore reuse a part of the previously evaluated facts and combined with a manually evaluated proportional sample of the new facts. We will investigate how to assure the validity of the precision estimation.

Textual extension. The textual source of the facts often contains additional subtleties that cannot be captured in triples. We are therefore working on an extended knowledge graph that allows text phrases in the positions of the triples [29]. We are also working on extracting commercial products from the Web [24].

Commonsense knowledge. Properties of everyday objects (e.g. that spiders have eight legs) and general concepts are of importance for text understanding, sentiment analysis, and even object recognition in images and videos. We have started this line of research recently [25, 26].

Intensional knowledge. Commonsense knowledge can also take the form of rules. For example, active sports athletes hardly ever hold political positions. We have already developed methods for mining Horn clauses [8,10], but more general forms of rules need to be tackled [9].

NoRDF. For some information (such as complex events, narratives, or larger contexts), the representation as triples is no longer sufficient. We call this the realm of NoRDF knowledge (in analogy to NoSQL databases), which we want to explore in the near future.

Finally, today's KBs may be correct, but they are hardly ever complete [21].

Acknowledgements. This research was partially supported by Labex DigiCosme (project ANR-11-LABEX-0045-DIGICOSME) operated by ANR as part of the program "Investissement d'Avenir" Idex Paris-Saclay (ANR-11-IDEX-0003-02).

References

1. Auer, S., Bizer, C., Kobilarov, G., Lehmann, J., Cyganiak, R., Ives, Z.: DBpedia: a nucleus for a web of open data. In: Aberer, K., Choi, K.-S., Noy, N., Allemang, D., Lee, K.-I., Nixon, L., Golbeck, J., Mika, P., Maynard, D., Mizoguchi, R., Schreiber, G., Cudré-Mauroux, P. (eds.) ASWC/ISWC -2007. LNCS, vol. 4825, pp. 722–735. Springer, Heidelberg (2007). doi:10.1007/978-3-540-76298-0_52
2. Biega, J., Kuzey, E., Suchanek, F.M.: Inside YAGO2s: a transparent information extraction architecture. In: WWW demo (2013)
3. Brown, L.D., Cai, T.T., DasGupta, A.: Interval estimation for a binomial proportion. Stat. Sci. **16**(2), 101–117 (2001)
4. De Melo, G., Weikum, G.: Towards a universal wordnet by learning from combined evidence. In: CIKM (2009)
5. Dong, X., Gabrilovich, E., Heitz, G., Horn, W., Lao, N., Murphy, K., Strohmann, T., Sun, S., Zhang, W.: Knowledge vault: a web-scale approach to probabilistic knowledge fusion. In: KDD (2014)
6. Fellbaum, C.: WordNet: An Electronic Lexical Database. Language, Speech, and Communication. MIT Press, Cambridge (1998)
7. Ferrucci, D.A., Brown, E.W., Chu-Carroll, J., Fan, J., Gondek, D., Kalyanpur, A., Lally, A., Murdock, J.W., Nyberg, E., Prager, J.M., Schlaefer, N., Welty, C.A.: Building watson: an overview of the deepqa project. AI Magazine **31**(3), 59–79 (2010)
8. Galárraga, L., Symeonidou, D., Moissinac, J.C.: Rule Mining for semantifying wikilinks. In: Linked Open Data Workshop at WWW (2015)
9. Galárraga, L., Suchanek, F.M.: Towards a numerical rule mining language. In: AKBC workshop (2014)
10. Galárraga, L., Teflioudi, C., Hose, K., Suchanek, F.M.: Fast rule mining in ontological knowledge bases with AMIE+. In: VLDBJ (2015)
11. Hoffart, J., Suchanek, F.M., Berberich, K., Lewis-Kelham, E., De Melo, G., Weikum, G.: YAGO2: exploring and querying world knowledge in time, space, context, and many languages. In: WWW (2011)
12. Hoffart, J., Suchanek, F.M., Berberich, K., Weikum, G.: YAGO2: a spatially and temporally enhanced knowledge base from Wikipedia. Artif. Intell. **194**, 28–61 (2013)
13. Hoffart, J., Yosef, M.A., Bordino, I., Fürstenau, H., Pinkal, M., Spaniol, M., Taneva, B., Thater, S., Weikum, G.: Robust disambiguation of named entities in text. In: EMNLP (2011)
14. Huet, T., Biega, J.A., Suchanek, F.M.: Mining history with Le Monde. In: AKBC Workshop (2013)
15. Kasneci, G., Ramanath, M., Suchanek, F., Weikum, G.: The YAGO-NAGA approach to knowledge discovery. ACM SIGMOD Record **37**(4), 41–47 (2009)
16. Kuzey, E., Setty, V., Strötgen, J., Weikum, G.: As time goes by: comprehensive tagging of textual phrases with temporal scopes. In: WWW (2016)
17. Lehmann, J., Isele, R., Jakob, M., Jentzsch, A., Kontokostas, D., Mendes, P.N., Hellmann, S., Morsey, M., van Kleef, P., Auer, S., Bizer, C.: DBpedia - a large-scale, multilingual knowledge base extracted from wikipedia. Semant. Web J. **6**(2), 167–195 (2015)
18. Magnini, B., Cavaglia, G.: Integrating subject field codes into WordNet. In: LREC (2000)

19. Mahdisoltani, F., Biega, J., Suchanek, F.: YAGO3: A knowledge base from multilingual Wikipedias. In: CIDR (2015)
20. Miller, G.A.: WordNet: a lexical database for English. Commun. ACM **38**(11), 39–41 (1995)
21. Razniewski, S., Suchanek, F.M., Nutt, W.: But what do we actually know?. In: AKBC workshop (2016)
22. Suchanek, F.M., Kasneci, G., Weikum, G.: YAGO: a core of semantic knowledge. In: WWW (2007)
23. Suchanek, F.M., Kasneci, G., Weikum, G.: YAGO: a large ontology from wikipedia and wordnet. Web Semant. **6**(3), 203–217 (2008)
24. Talaika, A., Biega, J.A., Amarilli, A., Suchanek, F.M.: IBEX: harvesting entities from the web using unique identifiers. In: WebDB workshop (2015)
25. Tandon, N., de Melo, G., De, A., Weikum, G.: Knowlywood: mining activity knowledge from hollywood narratives. In: CIKM (2015)
26. Tandon, N., de Melo, G., Suchanek, F., Weikum, G.: WebChild: harvesting and organizing commonsense knowledge from the web. In: WSDM (2014)
27. Vrandečić, D., Krtzsch, M.: Wikidata: a free collaborative knowledge base. Communications of the ACM 57, 78–85 (2014)
28. W3C: RDF 1.1 Concepts and Abstract Syntax (2014)
29. Yahya, M., Barbosa, D., Berberich, K., Wang, Q., Weikum, G.: Relationship queries on extended knowledge graphs. In: WSDM (2016)

A Collection of Benchmark Datasets for Systematic Evaluations of Machine Learning on the Semantic Web

Petar Ristoski[1(✉)], Gerben Klaas Dirk de Vries[2], and Heiko Paulheim[1]

[1] Research Group Data and Web Science,
University of Mannheim, Mannheim, Germany
{petar.ristoski,heiko}@informatik.uni-mannheim.de
[2] WizeNoze, Amsterdam, The Netherlands
g.k.d.devries@outlook.com

Abstract. In the recent years, several approaches for machine learning on the Semantic Web have been proposed. However, no extensive comparisons between those approaches have been undertaken, in particular due to a lack of publicly available, acknowledged benchmark datasets. In this paper, we present a collection of 22 benchmark datasets of different sizes. Such a collection of datasets can be used to conduct quantitative performance testing and systematic comparisons of approaches.

Keywords: Linked Open Data · Machine learning · Datasets · Benchmarking

Resource type: Datasets
Permanent URL: http://w3id.org/sw4ml-datasets

1 Introduction

In the recent years, applying machine learning to Semantic Web data has drawn a lot of attention. Many approaches have been proposed for different tasks at hand, ranging from reformulating machine learning problems on the Semantic Web as traditional, propositional machine learning tasks to developing entirely novel algorithms. However, systematic comparative evaluations of different approaches are scarce; approaches are rather evaluated on a handful of often project-specific datasets, and compared to a baseline and/or one or two other systems.

In contrast, evaluations in the machine learning area are often more rigorous. Approaches are usually compared using a larger number of standard datasets, most often from the UCI repository[1]. With a larger set of datasets used in the evaluation, statements about statistical significance are possible as well [3].

At the same time, collections of benchmark datasets have become quite well accepted in other areas of Semantic Web research. Notable examples include the

---

[1] http://archive.ics.uci.edu/ml/.

© Springer International Publishing AG 2016
P. Groth et al. (Eds.): ISWC 2016, Part II, LNCS 9982, pp. 186–194, 2016.
DOI: 10.1007/978-3-319-46547-0_20

Ontology Alignment Evaluation Initiative (OAEI) for ontology matching[2], the *Berlin SPARQL Benchmark*[3] for triple store performance, the Lehigh University Benchmark (LUBM)[4] for reasoning, or the Question Answering over Linked Data (QALD) dataset[5] for natural language query systems.

In this paper, we introduce a collection of datasets for benchmarking machine learning approaches for the Semantic Web. Those datasets are either existing RDF datasets, or external classification or regression problems, for which the instances have been enriched with links to the Linked Open Data cloud [14]. Furthermore, by varying the number of instances for a dataset, scalability evaluations are also made possible.

2   Related Work

Recent surveys on the use of Semantic Web for machine learning organize the proposed approaches in several categories, i.e., approaches that use Semantic Web data for machine learning [16], approaches that perform machine learning on the Semantic Web [11], and approaches that use machine learning techniques to create and improve Semantic Web data [8,16]. Furthermore, there are some challenges, like the *Linked Data Mining Challenge*[6] or the *Semantic-Web enabled Recommender Systems Challenge*[7], which usually focus on only a few datasets and a very specific problem setting.

3   Datasets

Our dataset collection has three categories: (i) existing datasets that are commonly used in machine learning experiments, (ii) datasets that were generated from official observations, and (iii) datasets generated from existing RDF datasets. Each of the datasets in the first two categories are initially linked to DBpedia[8]. This has two main reasons, (1) DBpedia being a cross-domain knowledge base usable in datasets from very different topical domains, and (2) tools like DBpedia Lookup and DBpedia Spotlight making it easy to link external datasets to DBpedia. However, DBpedia can be seen as an entry point to the Web of Linked Data, with many datasets linking to and from DBpedia. In fact, we use the RapidMiner Linked Open Data extension [9], to retrieve external links for each entity to YAGO[9] and Wikidata[10]. Such links could be exploited

---

[2] http://oaei.ontologymatching.org/.

[3] http://wifo5-03.informatik.uni-mannheim.de/bizer/berlinsparqlbenchmark/.

[4] http://swat.cse.lehigh.edu/projects/lubm/.

[5] http://greententacle.techfak.uni-bielefeld.de/~cunger/qald/.

[6] http://knowalod2016.informatik.uni-mannheim.de/en/
linked-data-mining-challenge/.

[7] http://challenges.2014.eswc-conferences.org/index.php/RecSys.

[8] http://dbpedia.org.

[9] http://yago-knowledge.org/.

[10] http://www.wikidata.org.

for systematic evaluation of the relevance of the data of different LOD dataset in different learning tasks.

In the dataset collection, there are four datasets that are commonly used for machine learning. For these datasets, we first enrich the instances with links to LOD datasets, and reuse the already defined target variable to perform machine learning experiments:

- The *Auto MPG* dataset[11] captures different characteristics of cars, and the target is to predict the fuel consumption (MPG) as a regression task.
- The *AAUP* (American Association of University Professors) dataset contains a list of universities, including eight target variables describing the salary of different staff at the universities[12]. We use the average salary as a target variable both for regression and classification, discretizing the target variable into "high", "medium" and "low", using equal frequency binning.
- The *Auto 93* dataset[13] captures different characteristics of cars, and the target is to predict the price of the vehicles as a regression task.
- The *Zoo* dataset captures different characteristics of animals, and the target is to predict the type of the animals as a classification task.

For those datasets, cars, universities, and animals are linked to DBpedia based on their name.

The second category of datasets contains a list of datasets where the target variable is an observation from different real-world domains, as captured by official sources. Again, the instances were enriched with links to LOD datasets. There are thirteen datasets in this category:

- The *Forbes* dataset contains a list of companies including several features of the companies, which was generated from the Forbes list of leading companies 2015[14]. The target is to predict the company's market value as a classification and regression task. To use it for the task of classification we discretize the target variable into "high", "medium", and "low", using equal frequency binning.
- The *Cities* dataset contains a list of cities and their quality of living, as captured by Mercer [7]. We use the dataset both for regression and classification.
- The *Endangered Species* dataset classifies animals into endangered species[15].
- The *Facebook Movies* dataset contains a list of movies and the number of Facebook likes for each movie[16]. We first selected 10,000 movies from DBpedia, which were then linked to the corresponding Facebook page, based on the movie's name and the director. The final dataset contains 1,600 movies, which was created by first ordering the list of movies based on the number of Facebook likes, and then selecting the top 800 movies and the bottom 800 movies. We use the dataset for regression and classification.

---

[11] http://archive.ics.uci.edu/ml/datasets/Auto+MPG.

[12] http://www.amstat.org/publications/jse/jse_data_archive.htm.

[13] http://www.amstat.org/publications/jse/v1n1/datasets.lock.html.

[14] http://www.forbes.com/global2000/list/.

[15] http://a-z-animals.com/.

[16] We use the Facebook Graph API: https://developers.facebook.com/docs/graph-api.

- Similarly, the *Facebook Books* dataset contains a list of books and the number of Facebook likes. Each book was linked to the corresponding Facebook page using the book's title and the book's author. Again, we selected the top 800 books and the bottom 800 books, based on the number of Facebook likes.
- The *Metacritic Movies* dataset is retrieved from Metacritic.com[17], which contains an average rating of all time reviews for a list of movies [12]. The initial dataset contained around 10,000 movies, from which we selected 1,000 movies from the top of the list, and 1,000 movies from the bottom of the list. We use the dataset both for regression and classification.
- Similarly, the *Metacritic Albums* dataset is retrieved from Metacritic.com[18], which contains an average rating of all time reviews for a list of albums [13].
- The *HIV Deaths Country* dataset contains a list of countries with the number of deaths caused by HIV, as captured by the World Health Organization[19]. We use the dataset both for regression and classification.
- Similarly, the *Traffic Accidents Deaths Country* dataset contains a list of countries with the number of deaths caused by traffic accidents[20].
- The *Energy Savings Country* dataset contains a list of countries with the total amount of energy savings of primary energy in 2010[21], which was downloaded from WorldBank[22]. We use the dataset both for regression and classification.
- Similarly, the *Inflation Country* dataset contains a list of countries with the inflation rate for 2011[23].
- The *Scientific Journals Country* dataset contains a list of countries with a number of scientific and technical journal articles published in 2011[24].
- The *Unemployment French Region* dataset contains a list of regions in France with the unemployment rate, used in the SemStats 2013 challenge [10].

Again, for those datasets, the instances (cities, countries, etc.) are linked to DBpedia. For datasets which are used for classification and regression, the regression target was discretized using equal frequency binning, usually into a *high* and a *low* class.

The third, and final, category contains datasets that were generated from existing RDF datasets, where the value of a certain property is used as a classification target. There are five datasets in this category:

- The *Drug-Food Interaction* dataset contains a list of drug-recipe pairs and their interaction, i.e., "negative" and "neutral" [6]. The dataset was retrieved from FinkiLOD[25]. Furthermore, each drug is linked to DrugBank[26]. We drew

---

[17] http://www.metacritic.com/browse/movies/score/metascore/all.
[18] http://www.metacritic.com/browse/albums/score/metascore/all.
[19] http://apps.who.int/gho/data/view.main.HIV1510.
[20] http://apps.who.int/gho/data/view.main.51310.
[21] http://data.worldbank.org/indicator/10.1_ENERGY.SAVINGS.
[22] http://www.worldbank.org/.
[23] http://data.worldbank.org/indicator/NY.GDP.DEFL.KD.ZG.
[24] http://data.worldbank.org/indicator/IP.JRN.ARTC.SC.
[25] http://linkeddata.finki.ukim.mk/.
[26] http://wifo5-03.informatik.uni-mannheim.de/drugbank/.

a stratified random sample of 2,000 instances from the complete dataset. When generating the features, we ignore the `foodInteraction` property in DrugBank, since it highly correlates with the target variable.

- The *AIFB* dataset describes the AIFB research institute in terms of its staff, research group, and publications. In [1] the dataset was first used to predict the affiliation (i.e., research group) for people in the dataset. The dataset contains 178 members of a research group, however the smallest group contains only 4 people, which is removed from the dataset, leaving 4 classes. Also, we remove the `employs` relation, which is the inverse of the *affiliation* relation.

- The *AM* dataset contains information about artifacts in the Amsterdam Museum [2]. Each artifact in the dataset is linked to other artifacts and details about its production, material, and content. It also has an artifact category, which serves as a prediction target. We have drawn a stratified random sample of 1,000 instances from the complete dataset. We also removed the `material` relation, since it highly correlates with the artifact category.

- The *MUTAG* dataset is distributed as an example dataset for the DL-Learner toolkit[27]. It contains information about complex molecules that are potentially carcinogenic, which is given by the `isMutagenic` property.

- The *BGS* dataset was created by the British Geological Survey and describes geological measurements in Great Britain[28]. It was used in [17] to predict the lithogenesis property of named rock units. The dataset contains 146 named rock units with a lithogenesis, from which we use the two largest classes.

An overview of the datasets is given in Tables 1, 2, and 3. For each dataset, we depict the number of instances, the machine learning tasks in which the dataset is used (*C* stands for classification and *R* stands for regression), the source of the dataset, and the LOD datasets to which the dataset is linked. For each dataset, we depict basic statistics of the properties of the LOD datasets, i.e., average, median, maximum and minimum number of *types*, *categories*, *outgoing relations* (rel out), *incoming relations* (rel in), outgoing relations including values (rel-vals out) and incoming relations including values (rel-vals in). The datasets, as well as a detailed description, a link quality evaluation, and licensing information, can be found online[29].

From the given statistics, we can infer the following observations: (i) DBpedia contains significantly less *owl:sameAs* links to YAGO, compared to Wikidata; (ii) DBpedia provides the highest number of types and categories on average per entity; (iii) Wikidata contains the highest number of outgoing and incoming relations for most of the datasets; (iv) YAGO contains the highest number of outgoing and incoming relations values for most of the datasets.

---

[27] http://dl-learner.org.
[28] http://data.bgs.ac.uk/.
[29] http://w3id.org/sw4ml-datasets.

Table 1. Datasets statistics

Name	Source	Task	LOD	#links	types avg	types med	types max	types min	categories avg	categories med	categories max	categories min	rel out avg	rel out med	rel out max	rel out min	rel in avg	rel in med	rel in max	rel in min	rel-vals out avg	rel-vals out med	rel-vals out max	rel-vals out min	rel-vals in avg	rel-vals in med	rel-vals in max	rel-vals in min
Auto MPG	UCI ML	R	DBpedia	371	29.70	31	46	5	11.20	10	25	2	13.48	13	27	3	5.62	5	25	1	16.50	15	70	0	36.65	23	509	0
			YAGO	331	13.99	16	21	0	9.26	9	23	0	8.76	9	18	0	16.96	2	138	1	77.08	70	278	0	3,236.24	60	28,418	0
			Wikidata	371	1.05	1	3	0	0.29	0	3	0	20.20	18	61	9	5.32	5	31	1	13.92	12	54	4	59.33	21	755	3
AAUP	JSE (c=2)	R/C	DBpedia	960	24.40	28	41	0	9.38	9	20	0	12.68	15	28	0	8.20	7	36	0	11.74	11	66	0	62.18	23	2,488	0
			YAGO	889	10.49	11	17	0	3.31	3	11	0	11.37	12	15	0	13.61	3	138	1	85.83	68	446	0	2,455.27	110	28,418	1
			Wikidata	959	2.13	2	5	0	0.88	1	2	0	30.71	29	83	0	8.51	7	44	0	22.38	21	97	0	296.92	20	31,777	0
Auto 93	JSE	R	DBpedia	93	28.76	31	43	5	11.13	10	25	3	12.69	12	22	8	4.92	5	7	2	14.35	11	64	4	22.60	18	64	2
			YAGO	80	13.80	16	19	0	9.09	10	18	0	8.37	10	11	0	21.09	2	138	2	59.33	59	129	4	4,025.90	46	28,418	4
			Wikidata	93	1.00	1	2	0	0.12	0	1	0	17.31	17	26	9	3.56	3	8	1	11.23	11	25	0	19.91	19	57	3
Zoo	UCI ML	C (c=3)	DBpedia	101	8.61	11	26	0	4.67	3	34	0	8.22	9	15	3	3.54	3	8	1	13.26	11	87	1	146.28	24	3,686	2
			YAGO	8	0.74	0	13	0	0.15	0	6	0	0.63	1	8	0	127.23	138	138	2	5.39	1	156	2	26,173.23	28,418	28,418	3
			Wikidata	101	1.00	1	2	0	0.67	1	2	0	29.69	35	57	3	8.28	7	27	0	18.20	21	45	1	125.82	92	785	0
Forbes	Forbes	R/C (c=2)	DBpedia	1,585	14.77	19	62	0	4.87	4	52	0	10.15	11	27	0	2.76	2	27	0	10.44	10	136	0	14.30	4	1,925	0
			YAGO	1,003	7.28	10	33	0	2.35	2	42	0	7.57	12	21	0	52.07	2	138	1	34.42	27	510	0	10,531.37	107	28,418	1
			Wikidata	1,189	0.82	1	4	0	0.22	0	3	0	16.59	16	137	0	5.00	5	52	0	12.69	10	207	0	30.14	8	2,881	0
Cities	Mercer	R/C (c=3)	DBpedia	212	31.28	35	53	0	6.98	7	26	0	18.08	19	38	0	25.66	25	68	0	16.26	13	131	0	1,474.57	678	19,810	0
			YAGO	187	16.66	19	30	0	4.46	4	15	0	13.75	15	32	0	23.56	9	138	2	222.54	214	681	0	8,087.34	3,555	72,320	2
			Wikidata	212	2.11	2	9	1	3.40	4	6	1	69.08	67	153	6	39.99	37	108	6	105.29	89	390	2	5,298.23	1,599	99,865	1
FB Books	Facebook	R/C (c=2)	DBpedia	1,600	19.08	20	42	0	5.15	5	23	0	11.15	11	20	0	1.64	2	7	0	7.04	7	60	0	2.80	8	42	0
			YAGO	1,334	8.37	10	24	0	2.03	2	15	0	8.41	10	13	0	25.32	3	138	1	24.37	22	149	0	4,735.50	8	28,418	1
			Wikidata	1,578	1.00	1	3	0	0.01	0	1	0	21.19	22	55	0	3.15	3	17	0	16.41	16	69	0	7.47	4	165	0
FB Movies	Facebook	R/C (c=2)	DBpedia	1,600	24.90	27	55	0	12.50	11	60	0	12.43	13	21	0	1.46	1	12	0	11.65	12	51	0	4.96	43	110	0
			YAGO	1,339	12.08	14	32	0	6.51	6	27	0	8.39	10	17	0	26.89	6	138	1	55.01	47	280	0	4,682.42	43	28,418	1
			Wikidata	1,585	1.01	1	4	0	0.04	0	1	0	48.75	48	107	0	2.22	1	22	0	56.37	53	372	0	20.75	12	230	0
Metacritic Albums	Metacritic	R/C (c=2)	DBpedia	1,600	17.92	19	36	0	4.27	4	26	0	10.85	12	17	2	2.63	3	7	0	8.92	9	63	0	5.28	3	50	0
			YAGO	1,444	7.22	8	19	0	3.22	3	20	0	8.05	9	10	0	16.02	5	138	1	40.27	32	361	0	2,749.90	10	28,418	1
			Wikidata	1,576	0.99	1	2	0	0.00	0	1	0	17.64	18	45	0	4.00	5	9	1	11.73	12	49	0	8.77	7	54	0

Table 2. Datasets statistics

Name	Source	Task	LOD	#links	types avg	types med	types max	types min	categories avg	categories med	categories max	categories min	rel out avg	rel out med	rel out max	rel out min	rel in avg	rel in med	rel in max	rel in min	rel-vals out avg	rel-vals out med	rel-vals out max	rel-vals out min	rel-vals in avg	rel-vals in med	rel-vals in max	rel-vals in min
Metacritic Movies	Metacritic	R/C (c=2)	DBpedia	2,000	24.38	27	45	0	11.87	11	42	0	12.54	14	19	3	1.35	1	7	1	11.42	12	30	0	3.56	2	31	0
			YAGO	1,588	11.79	14	19	0	6.43	6	28	0	8.34	10	11	0	28.22	6	138	1	48.84	43	216	0	4,960.84	37	28,418	1
			Wikidata	1,981	0.98	1	1	0	0.03	0	1	0	47.86	49	99	0	1.98	1	13	0	52.70	53	237	0	15.77	11	117	0
HIV Deaths Country	WHO	R/C (c=2)	DBpedia	114	35.69	37	52	0	12.61	13	23	0	23.59	24	28	3	34.26	31	89	6	27.75	25	162	10	4,828.36	1,065	70,426	24
			YAGO	108	13.90	15	24	0	9.28	9	18	0	28.41	31	35	0	15.18	9	138	5	302.34	244	1,267	0	12,464.42	4,879	112,032	550
			Wikidata	114	4.12	4	8	0	4.83	5	6	0	120.87	119	173	7	55.68	51	148	2	229.46	210	595	2	45,671.15	4,971	669,273	66
Trafic Accidents Country	WHO	R/C (c=2)	DBpedia	146	36.40	38	53	0	13.12	13	23	0	23.40	24	28	1	37.87	36	94	8	27.44	24	162	8	7,528.18	1,587	218,957	77
			YAGO	139	14.29	15	27	0	9.62	9	16	0	28.44	31	35	0	14.61	9	138	5	345.03	290	2,104	0	17,882.47	6,126	423,559	693
			Wikidata	146	4.42	4	10	0	4.94	5	6	0	124.31	121	191	7	61.68	55	148	2	242.38	213	713	2	85,575.10	7,369	1,557,157	66
Energy Savingss Country	WorldBank	R/C (c=2)	DBpedia	162	36.07	38	53	0	13.12	13	23	0	23.46	24	28	1	36.64	33	94	8	26.72	23	162	8	6,876.80	1,440	218,957	77
			YAGO	152	14.09	15	27	0	9.52	10	16	0	27.82	31	35	0	16.40	9	138	5	329.28	279	2,104	5	16,969.96	5,821	423,559	757
			Wikidata	162	4.41	4	10	1	4.92	5	6	1	123.36	119	191	7	60.02	55	148	2	238.69	210	713	2	77,485.01	5,810	1,557,157	66
Inflation Country	WorldBank	R/C (c=2)	DBpedia	160	36.00	38	53	0	13.11	13	23	0	23.46	24	28	1	36.74	33	94	8	26.85	24	162	8	6,947.59	1,440	218,957	77
			YAGO	150	14.09	15	27	0	9.44	9	16	0	27.80	31	35	0	16.48	9	138	5	331.16	279	2,104	5	17,114.88	5,821	423,559	693
			Wikidata	160	4.39	4	10	1	4.88	5	6	1	123.23	119	191	7	60.12	55	148	2	237.94	210	713	2	78,453.16	5,810	1,557,157	66
Scientific Journals Country	WorldBank	R/C (c=2)	DBpedia	160	36.00	38	53	0	13.11	13	23	0	23.46	24	28	1	36.74	33	94	8	26.85	24	162	8	6,947.59	1,440	218,957	77
			YAGO	150	14.09	15	27	0	9.44	9	16	0	27.80	31	35	0	16.48	9	138	5	331.16	279	2,104	5	17,114.88	5,821	423,559	693
			Wikidata	160	4.39	4	10	1	4.88	5	6	1	123.23	119	191	7	60.12	55	148	2	237.94	210	713	2	78,453.16	5,810	1,557,157	66
Unemployment French Region	SemStats	R/C (c=2)	DBpedia	26	16.38	21	32	0	3.73	3	15	0	7.81	9	10	3	14.19	13	24	7	7.19	7	19	1	975.88	969	2,292	37
			YAGO	26	8.92	8	14	8	2.77	2	8	1	12.42	12	14	12	3.73	4	6	3	81.12	60	299	28	1,793.19	1,424	4,527	88
			Wikidata	26	1.35	1	3	0	2.58	3	4	1	86.23	84	119	74	34.00	33	51	21	83.12	79	157	58	332.69	193	1,464	137
Endangered Species	a-z-animals	R/C (c=2)	DBpedia	301	11.84	12	33	0	6.32	5	34	0	10.77	11	25	0	2.96	3	55	1	12.65	11	87	0	566.25	15	114,742	1
			YAGO	65	2.48	1	16	0	0.76	0	12	0	1.78	0	9	0	108.62	138	138	0	9.53	0	136	0	22,286.36	28,418	28,418	1
			Wikidata	301	1.05	1	4	0	0.44	0	6	0	34.32	37	137	3	6.94	6	78	0	22.04	22	400	1	21,909.94	70	6,460,930	1
Drug-Food Interaction	FinkiLOD	C (c=2)	DBpedia	1,989	8.83	4	38	0	5.46	5	18	0	12.65	14	15	0	1.40	1	5	0	3.63	3	12	0	34.71	24	158	0
			YAGO	588	4.46	4	31	0	0.68	0	6	0	2.15	0	8	0	99.69	138	138	1	7.28	7	61	0	20,427.08	28,418	28,418	0
			Wikidata	1,908	1.96	2	3	0	0.01	0	1	0	45.92	47	79	0	2.78	2	17	0	34.99	27	159	0	32.25	26	487	4
			DrugBank	2,000	2.00	2	2	2	/	/	/	/	61.68	64	71	41	1.70	2	2	0	41.96	41	132	14	62.49	50	211	0
			FinkiLOD	2,000	1.00	1	1	1	/	/	/	/	3.00	3	3	3	0.00	0	0	0	1.00	1	1	1	0.00	0	0	0

Table 3. Datasets statistics

Dataset			types				rel out				rel in				rel-vals out				rel-vals in			
Name	Task	#links	avg	med	max	min	avg	med	max	min	avg	med	max	min	avg	med	max	min	avg	med	max	min
AIFB	C (c=4)	176	1.4	1	2	1	7.1	7	9	5	2.0	2	5	0	18.2	7	219	2	19.8	9	246	0
AM	C (c=11)	1,000	1.0	1	1	1	19.8	20	29	9	0.6	1	3	0	21.9	20	283	7	3.2	1	273	0
MUTAG	C (c=2)	340	1.0	1	1	1	9.8	10	14	5	\	\	\	\	65.8	56	465	4	\	\	\	\
BGS	C (c=2)	146	1.0	1	1	1	29.7	31	36	21	1.4	2	4	0	25.2	24	54	15	2.7	2	12	0

4 Conclusion and Outlook

In this paper, we have introduced a collection of 22 benchmark datasets for
machine learning on the Semantic Web. So far, we have concentrated on classification and regression tasks. There are methods to derive clustering and outlier
detection benchmarks from classification and regression datasets [4,5], so that
extending the dataset collection for such unsupervised tasks is possible as well.
Furthermore, as many datasets on the Semantic Web use extensive hierarchies
in the form of ontologies, building benchmark datasets for tasks like *hierarchical
multi-label classification* [15] would also be an interesting extension.

Acknowledgments. The work presented in this paper has been partly funded by the
German Research Foundation (DFG) under grant number PA 2373/1-1 (Mine@LOD),
and the Dutch national program COMMIT.

References

1. Bloehdorn, S., Sure, Y.: Kernel methods for mining instance data in ontologies. In:
   Aberer, K., et al. (eds.) ASWC/ISWC -2007. LNCS, vol. 4825, pp. 58–71. Springer,
   Heidelberg (2007). doi:10.1007/978-3-540-76298-0_5
2. Boer, V., Wielemaker, J., Gent, J., Hildebrand, M., Isaac, A., Ossenbruggen, J.,
   Schreiber, G.: Supporting linked data production for cultural heritage institutes:
   the Amsterdam museum case study. In: Simperl, E., Cimiano, P., Polleres, A.,
   Corcho, O., Presutti, V. (eds.) ESWC 2012. LNCS, vol. 7295, pp. 733–747.
   Springer, Heidelberg (2012). doi:10.1007/978-3-642-30284-8_56
3. Demšar, J.: Statistical comparisons of classifiers over multiple data sets. J. Mach.
   Learn. Res. **7**, 1–30 (2006)
4. Emmott, A.F., Das, S., Dietterich, T., Fern, A., Wong, W.K.: Systematic construction of anomaly detection benchmarks from real data. In: Proceedings of the ACM
   SIGKDD Workshop on Outlier Detection and Description, pp. 16–21. ACM (2013)
5. Färber, I., Günnemann, S., Kriegel, H.P., Kröger, P., Müller, E., Schubert, E., Seidl,
   T., Zimek, A.: On using class-labels in evaluation of clusterings. In: MultiClust:
   Workshop on Discovering, Summarizing and Using Multiple Clusterings (2010)
6. Jovanovik, M., Bogojeska, A., Trajanov, D., Kocarev, L.: Inferring cuisine-drug
   interactions using the linked data approach. Scientific reports 5 (2015)
7. Paulheim, H.: Generating possible interpretations for statistics from linked open
   data. In: Simperl, E., Cimiano, P., Polleres, A., Corcho, O., Presutti, V. (eds.)
   ESWC 2012. LNCS, vol. 7295, pp. 560–574. Springer, Heidelberg (2012). doi:10.
   1007/978-3-642-30284-8_44

8. Rettinger, A., Lösch, U., Tresp, V., d'Amato, C., Fanizzi, N.: Mining the semantic web. Data Min. Knowl. Disc. **24**(3), 613–662 (2012)
9. Ristoski, P., Bizer, C., Paulheim, H.: Mining the web of linked data with rapidminer. Web Semant. Sci. Serv. Agents WWW **35**, 142–151 (2015)
10. Ristoski, P., Paulheim, H.: Analyzing statistics with background knowledge from linked open data. In: Workshop on Semantic Statistics (2013)
11. Ristoski, P., Paulheim, H.: Semantic web in data mining and knowledge discovery: a comprehensive survey. Web Semant. **36**, 1–22 (2016)
12. Ristoski, P., Paulheim, H., Svátek, V., Zeman, V.: The linked data mining challenge 2015. In: KNOW@ LOD (2015)
13. Ristoski, P., Paulheim, H., Svátek, V., Zeman, V.: The linked data mining challenge 2016. In: KNOW@LOD (2016)
14. Schmachtenberg, M., Bizer, C., Paulheim, H.: Adoption of the linked data best practices in different topical domains. In: Mika, P., et al. (eds.) ISWC 2014. LNCS, vol. 8796, pp. 245–260. Springer, Heidelberg (2014). doi:10.1007/978-3-319-11964-9_16
15. Silla Jr., C.N., Freitas, A.A.: A survey of hierarchical classification across different application domains. Data Min. Knowl. Disc. **22**(1), 31–72 (2011)
16. Tresp, V., Bundschus, M., Rettinger, A., Huang, Y.: Towards machine learning on the semantic web. In: da Costa, P.C.G., et al. (eds.) URSW 2005-2007. LNCS, vol. 5327, pp. 282–314. Springer, Heidelberg (2008)
17. Vries, G.K.D.: A fast approximation of the Weisfeiler-Lehman graph kernel for RDF data. In: Blockeel, H., Kersting, K., Nijssen, S., Železný, F. (eds.) ECML PKDD 2013. LNCS, vol. 8188, pp. 606–621. Springer, Heidelberg (2013)

Enabling Combined Software and Data Engineering at Web-Scale: The ALIGNED Suite of Ontologies

Monika Solanki[1]([✉]), Bojan Božić[2], Markus Freudenberg[3],
Dimitris Kontokostas[3], Christian Dirschl[4], and Rob Brennan[2]

[1] Department of Computer Science, University of Oxford, Oxford, UK
monika.solanki@cs.ox.ac.uk
[2] KDEG, School of Computer Science and Statistics,
Trinity College, Dublin, Dublin, Ireland
[3] AKSW/KILT, University of Leipzig, Leipzig, Germany
[4] Wolters Kluwer, Munich, Germany

Abstract. Effective, collaborative integration of software and big data engineering for Web-scale systems, is now a crucial technical and economic challenge. This requires new combined data and software engineering processes and tools. Semantic metadata standards and linked data principles, provide a technical grounding for such integrated systems given an appropriate model of the domain. In this paper we introduce the ALIGNED suite of ontologies specifically designed to model the information exchange needs of combined software and data engineering. These ontologies are deployed in web-scale, data-intensive, system development environments in both the commercial and academic domains. We exemplify the usage of the suite on a complex collaborative software and data engineering scenario from the legal information system domain.

Resource type: Set of ontologies
Permanent URL: https://github.com/aligned-h2020/ALIGNED_Ontologies

1 Introduction

Recent years have seen a significant increase in the demand for data-intensive applications based on large-scale sources of data. However our engineering techniques for building data-intensive systems are both immature and often partitioned into software engineering and data engineering processes, tasks or teams. There is a need for integrated engineering approaches. The data itself must also be high-quality, which entails a curatorial process to improve and manage data over time. The expressivity of semantic models makes them useful for both addressing data quality [5] and applying model-driven approaches [3] to software engineering. Semantic data, in the form of enterprise linked data is also useful

© Springer International Publishing AG 2016
P. Groth et al. (Eds.): ISWC 2016, Part II, LNCS 9982, pp. 195–203, 2016.
DOI: 10.1007/978-3-319-46547-0_21

for describing, fusing and managing the combined data and software engineering lifecycles to increase productivity, agility and system quality.

In this paper, we present a suite of ontologies developed within the ALIGNED[1] project, that aim to align the divergent processes encapsulating data and software engineering. The key aim of the ALIGNED ontology suite is to support the generation of combined software and data engineering processes and tools for improved productivity, agility and quality. The suite contains linked data ontologies/vocabularies designed to: (1) support semantics-based model driven software engineering, by documenting additional system context and constraints for RDF-based data or knowledge models in the form of design intents, software lifecycle specifications and data lifecycle specifications; (2) support data quality engineering techniques, by documenting data curation tasks, roles, datasets, workflows and data quality reports at each data lifecycle stage in a data intensive system; and (3) support the development of tools for unified views of software and data engineering processes and software/data test case interlinking, by providing the basis for enterprise linked data describing software and data engineering activities (tasks), agents (actors) and entities (artefacts) based on the W3C provenance ontology[2].

This ontology suite has been deployed for validation and incremental improvement in the ALIGNED project on four, large-scale data-intensive systems engineering use cases: the Seshat Global History Databank [10], which is compiling linked data time series relating to all human societies over the past 12,000 years; JURION[3], a legal information platform developed by Wolters Kluwer Germany; PoolParty[4], a semantic technology middleware developed by the Semantic Web Company; and the DBpedia+[5] data quality and release processes.

The paper is structured as follows: Sect. 2 presents an overview of the ALIGNED suite. It provides a brief description of the core ontologies in the suite. Section 3 shows how the vocabularies have been applied to a complex collaborative software and data engineering scenario from the legal information system domain. Section 4 presents an evaluation of the ontologies in the suite. Section 5 briefly discusses related work. Finally, Sect. 6 presents conclusions.

2 Overview of the ALIGNED Suite

Figure 1 illustrates the ALIGNED suite of ontologies split into the provenance, generic, and domain-specific layers. As can be seen from the figure, a high emphasis has been placed on reusing existing, well known and standardised specifications where available. At the top layer, the W3C provenance standard forms the baseline for all our specifications and all our models extend it in some way.

---

[1] http://aligned-project.eu.
[2] http://www.w3.org/ns/prov-o.
[3] https://www.jurion.de/.
[4] https://www.poolparty.biz/.
[5] http://wiki.dbpedia.org/.

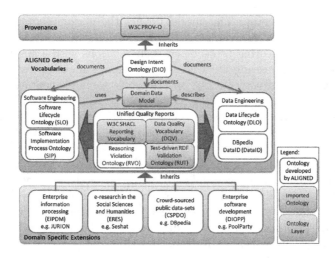

Fig. 1. The ALIGNED suite of ontologies

The split of the ALIGNED ontology suite between a generic layer and a domain specific extensions layer allows rapid evolution of domain-specific extensions for the ALIGNED use cases/trial environments (JURION, Seshat, DBpedia, Pool-Party) based on a stable set of core concepts modelled in the generic layer. As the project progresses these extensions will be evaluated and incorporated into the generic layer if they prove valuable or more widely applicable than a single domain. Within the project the suite of ontologies is known as the "ALIGNED metamodel" due to the links with software engineering practices.

We briefly present here the core ontologies from the suite. Further details of the ontologies including the axiomatisations, graphical representation, serialisations in multiple formats via content negotiation, examples illustrating the usage of the ontologies, typical SPARQL queries that can be formulated using the ontologies as the data model and HTML documentation are available from the individual deployments at their persistent URIs. Due to space constraints we deliberately do not include these in this paper. The ontologies are grouped as follows:

– **Design intent**: This model is used to document the design decisions about data intensive system artefacts such as requirements, designs or datasets. It is based on the design intent ontology (DIO)[6], which allows users to express the design intent or design rationale while undertaking the design of an artefact. DIO [9] is a generic ontology that provides the conceptualisation needed to capture the knowledge generated during various phases of the overall design lifecycle. DIO provides definitions for design artefacts such as requirements, designs, design issues, solutions, justifications and evidence, and relationships between them.

---

[6] https://w3id.org/dio.

- **Software engineering**: This model defines the major agents (e.g. project roles), activities (e.g. lifecycle stages), and entities (design artefacts) involved in a software engineering project and their relations with a special focus on capturing the engineering lifecycle. Two ontologies make up this model: the software process ontology (SPO)[7] and the software implementation processes ontology (SIP)[8].

- **Data engineering**: As software engineering above but with a focus on data engineering and data lifecycles. Two ontologies are used: the data lifecycle ontology (DLO)[9] defined within ALIGNED and the DataID[10] ontology, defined by ALIGNED for the DBpedia association, for describing datasets. DLO provides a set of conceptual entities, agents, activities, and roles to represent the general data engineering process. Furthermore, it is the basis for deriving specific domain ontologies which represent lifecycles of concrete data engineering projects such as DBpedia or Seshat. DataID is a multi-layered meta-data system, which, in its core, describes datasets and their different manifestations, as well as relations to agents like persons or organisations, in regard to their rights and responsibilities. Depending on context, type of data and use case, this core ontology can be augmented by multiple existing extensions (e.g. Linked Data, repository descriptions etc.).

- **Unified quality reports**: Defines a unified reporting representation for data quality metrics, ontology reasoning errors, test cases, and test case results based on the W3C SHACL reporting vocabulary. It is based on four ontologies/vocabularies, three of which are externally developed: W3C SHACL[11], W3C Data Quality[12], and University of Leipzigs test-driven RDF validation ontology [5] (RUT); and one ontology developed within ALIGNED: the reasoning violation ontology (RVO)[13]. RUT is designed to capture the lifecycle of RDF validation with the test driven validation methodology. It is implemented by the RDFUnit tool. RVO describes both ABox and TBox reasoning errors for the integration of reasoners into data lifecycle tool-chains. The ontology covers violations of the OWL 2 direct semantics and syntax detected on both the schema and instance level over the full range of OWL 2 and RDFS language constructs. An overview of RVO and its design, implementation and use cases has been published in [1].

- **Domain data model**: This describes the domain of the data-intensive application being developed and is specific to that application, e.g. the Seshat ontology for historical time-series describing human societies. The lower layer includes the domain-specific extensions to the metamodels. ALIGNED has developed four domain-specific metamodels based on each of our use cases, with a focus on model elements needed for the ALIGNED phase 2 trials.

---

[7] https://w3id.org/slo.

[8] https://w3id.org/sip.

[9] https://w3id.org/dlo.

[10] http://dataid.dbpedia.org/ns/core#.

[11] https://www.w3.org/TR/shacl/.

[12] https://www.w3.org/TR/vocab-dqv/.

[13] https://w3id.org/rvo.

- **Enterprise information processing**: extensions and models for the JURION use case.
- **E-research in the Social Sciences and Humanities**: extensions and models for the Seshat use case.
- **Crowd-sourced public datasets**: extensions and models for the DBpedia use case.
- **Enterprise software development**: extensions and models for the Pool-Party use case.

3   Example Deployment: The ALIGNED Suite in Wolters Kluwer's JURION

JURION is an innovative legal information platform developed by Wolters Kluwer Germany that merges and interlinks over 1 million documents of content and data from diverse sources such as national and European legislation and court judgements, extensive internally authored content and local customer data, as well as social media and semantic web data (e.g. from DBpedia). This data is then presented to users (such as law offices) in the form of highly customised applications for semantic search, annotation, case management and legal information retrieval.

Currently, the software development process and data life cycle are highly independent from each other. Figure 2 illustrates where ontologies from the ALIGNED suite contribute towards facilitating interoperability between the software and data engineering processes and tools used to build and maintain

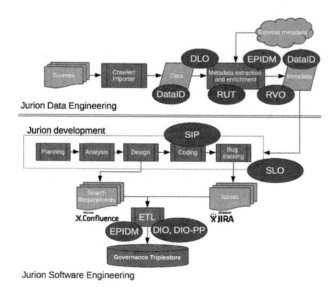

Fig. 2. Usage of the ALIGNED suite of ontologies in the JURION semantics-based legal information system

JURION. The two main uses are tool integration and unified governance. Tool integration includes both cases within a single domain (data or software engineering) and cross-domain tool-chain integration. Unified governance uses ALIGNED provenance records, data extraction and uplift from enterprise engineering tools and data fusion to provide end to end and cross-domain views of the JURION platform engineering processes. We elaborate on the deployment of ALIGNED ontologies for these use cases below.

RUT has been used in JURION for validating & verifying the extraction of metadata [6]. In particular, RDFUnit is used as a data validation tool integrated in JURION's continuous integration (CI) platform (Jenkins). RVO, the reasoning violations ontology, has been used to integrate advanced OWL reasoning-based data quality checks with RDFUnit's triple-query oriented tests to expand the scope of testing possible. DataID descriptors of all the JURION datasets are under evaluation and it is planned to use this to provide consistent meta-data which will be available to all tools thus facilitating further integration. The EIP, enterprise information processing, ontology has been used to describe the JURION environment, systems, artifacts and engineering processes in terms of the ALIGNED software and data lifecycle models.

An upcoming feature in JURION is the integration of search requirements with design issues/software bugs arising during their implementation. The goal is to express integrated requirements and issues as linked data, which is semantically annotated using the DIO and DIO-PP ontologies from the ALIGNED suite. This would further enable the development of customised Confluence interfaces which can be used to provide enhanced query features over the integrated data and produce bespoke reports using visual and statistical analytics.

4   Evaluation

Table 1 presents the evaluation of the ALIGNED suite in accordance to the desired criteria[14].

5   Related Work

SEON[15] is a family of ontologies that describe concepts in the context of software engineering, software evolution and software maintenance. SWO[16] is a resource for describing software tools, their types, tasks, versions and provenance. While they cover some general aspects of software engineering, they do not address the description of design intents and software lifecycles. Representing design intents or design rationales as ontologies have been captured for various specialised domains such as software engineering [2] however there is no generic,

---

[14] https://figshare.com/articles/ISWC2016_Resources_Track_Review_Instructions/2016852.

[15] http://se-on.org/#publications.

[16] http://theswo.sourceforge.net/.

Table 1. Evaluating the ALIGNED suite of ontologies

Generic criteria	Evaluation
Value addition	(1) The ontologies add data and software engineering specific metadata to the process and enrich information about process specific procedures within data and software engineering for a tool, which in return can use this context dependent information for automation and automatic generation purposes. (2) DLO is used to provide details about the data engineering process and SLO details about the software engineering process. (3) RVO helps producing information about reasoning errors in the knowledge base, while DIO enables the mining of design intents from requirements specification as well as the generation of unified governance reports by integrating requirements and design issues
Reuse	(1) Potential reuse across a wider community of content producers, owners of large amounts of data, data managers, ontology engineers of new related ontologies and vocabularies (2) Software development model designers, and developers of human societies datasets (e.g. Seshat Global History Databank). (3) The metamodels are easy to reuse and published on the Web together with detailed documentation. Top level models are general and can be applied for all data and software engineering models. Furthermore, the models are extendable and can be inherited by specialised domain ontologies for specific software and data engineering platforms
Design and technical quality	All ontologies have been designed as OWL DL ontologies, in accordance to ontology engineering principles [7]. Axiomatisations in the ontologies have been defined based on the competency questions identified during requirements scoping
Availability	Ontologies have been made publically available at http://aligned-project.eu/data-and-models/. Further, they have been given persistent w3id URIs, deployed on public facing servers and are content negotiable. DIO has been cited in [9] and RUT in [6]. All ontologies have been licensed under a Creative Commons Attribution License. DIO has also been registered[a] in LOV
Sustainability	All ontologies are deployed on a public Github repositories. Long term sustainability has been assured by the ontology engineers involved in the design
Specific criteria	
Design suitability	Individual ontologies in the suite have been developed in close association with the requirements emerging from corresponding, potential exploiting application.Thus they closely conform to the suitability of the tasks for which they have been designed
Design elegance and quality	Axiomatisation in the ontologies have been developed following Gruber's principles [4] of clarity, coherence, extendability, minimum encoding bias and minimum ontological commitment
Logical correctness	The ontologies have been verified using DL reasoners for satisfiability, incoherency and inconsistencies. Specifically, inconsistencies for DIO has been checked against the instance data in the governance triple store
External resources reuse	External ontologies such as PROV-O, SKOS have been extensively used
Documentation	The ALIGNED public deliverables and publications [6,9] include detailed descriptions of the models. The ontologies have been well documented using rdfs:label and rdfs:comment. HTML documentation via the LODE service has also been enabled. All ontologies have been graphically illustrated

[a] http://lov.okfn.org/dataset/lov/vocabs/dio.

domain-independent design intent capture model available as a design pattern. OOPS! [8] is a tool with a catalogue for validating ontologies by spotting common pitfalls, however it detects design flaws rather than logical errors and does not use an ontology for error reporting. The DCAT vocabulary includes the special class Distribution for the representation of the available materialisations of a dataset. These distributions cannot be described further within DCAT. The Asset

Description Metadata Schema[17] (ADMS) is a profile of DCAT, which only describes a specialised class of datasets: so-called Semantic Assets.

6   Conclusions

Combining data and software engineering processes to increase productivity and agility, is a challenge being faced by several organisations aiming to exploit the benefits of big data. Ontologies and vocabularies developed in accordance to competency questions, objective criteria and ontology engineering principles can provide useful support to data scientists and software engineers undertaking the challenge. In this paper we have proposed the ALIGNED suite of ontologies that provide semantic models of design intents, domain specific datasets, software engineering processes, quality heuristics and error handling mechanisms. The suite contributes immensely towards enabling interoperability and alleviating some of the complexities involved. We have exemplified the usage of the suite on a real-world use case from the legal domain and evaluated it against the desired criteria. As ontologies from the suite are now in various stages of adoption by the ALIGNED use cases, the next steps would incorporate their empirical evaluation.

Acknowledgment. This research has received funding from the European Unions Horizon 2020 research and innovation programme under grant agreement No 644055, the ALIGNED project (www.aligned-project.eu).

References

1. Bozic, B., Brennan, R., Feeney, K., Mendel-Gleason, G.: Describing reasoning results with RVO, the reasoning violations ontology. In: ESWC 2016 (2016, to appear)
2. de Medeiros, A.P., Schwabe, D., Fejjó, B.: Kuaba ontology: design rationale representation and reuse in model-based designs. In: Delcambre, L., Kop, L., Mayr, H.C., Mylopoulos, J., Pastor, O. (eds.) Conceptual Modeling - ER 2005. LNCS, vol. 3716, pp. 241–255. Springer, Heidelberg (2005)
3. Gasevic, D., Djuric, D., Devedzic, V.: Model Driven Engineering and Ontology Development, 2nd edn. Springer, Heidelberg (2009)
4. Gruber, T.R.: Toward principles for the design of ontologies used for knowledge sharing. Int. J. Hum. Comput. Stud. **43**(5–6), 907–928 (1995)
5. Kontokostas, D., Brüummer, M., Hellmann, S., Lehmann, J., Ioannidis, L.: NLP data cleansing based on linguistic ontology constraints. In: Presutti, V., d'Amato, C., Gandon, F., d'Aquin, M., Staab, S., Tordai, A. (eds.) ESWC 2014. LNCS, vol. 8465, pp. 224–239. Springer, Heidelberg (2014). doi:10.1007/978-3-319-07443-6_16
6. Kontokostas, D., Mader, C., Dirschl, C., Eck, K., Leuthold, M., Lehmann, J., Hellmann, S.: Semantically enhanced quality assurance in the jurion business use case. In: Sack, H., Blomqvist, E., d'Aquin, M., Ghidini, C., Ponzetto, S.P., Lange, C. (eds.) ESWC 2016. LNCS, vol. 9678, pp. 661–676. Springer, Heidelberg (2016). doi:10.1007/978-3-319-34129-3_40

---

[17] https://www.w3.org/TR/vocab-adms/.

7. Noy, N.F., Mcguinness, D.L.: Ontology development 101: a guide to creating your first ontology. Technical report, Stanford Center for Biomedical Informatics Research (BMIR) (2001)
8. Poveda-Villalón, M., Suárez-Figueroa, M.C., Gómez-Pérez, A.: Validating ontologies with OOPS!. In: Teije, A., et al. (eds.) EKAW 2012. LNCS (LNAI), vol. 7603, pp. 267–281. Springer, Heidelberg (2012). doi:10.1007/978-3-642-33876-2_24
9. Solanki, M.: A pattern for capturing the intents underlying designs. In: Proceedings of the 6th Workshop on Ontology and Semantic Web Patterns (WOP 2015), vol. 1461. CEUR-WS.org (2015)
10. Turchin, P., Brennan, R., Currie, T., Feeney, K., Francois, P., Hoyer, D., Manning, J., Marciniak, A., Mullins, D., Palmisano, A., Peregrine, P., Turner, E.A., Whitehouse, H.: Seshat: the global history databank. Cliodynamics **6**(1), 77–107 (2015)

A Replication Study of the Top Performing Systems in SemEval Twitter Sentiment Analysis

Efstratios Sygkounas[1], Giuseppe Rizzo[2], and Raphaël Troncy[1($\boxtimes$)]

[1] EURECOM, Sophia Antipolis, France
{sygkouna,troncy}@eurecom.fr
[2] ISMB, Turin, Italy
giuseppe.rizzo@ismb.it

Abstract. We performed a thorough replicate study of the top performing systems in the yearly SemEval Twitter Sentiment Analysis task. We highlight and discuss differences among the results obtained by those systems that have been officially published and the ones we are able to compute. Learning from the studies being made on the systems, we also propose SentiME, an ensemble system composed of five state-of-the-art sentiment classifiers. SentiME trains the different classifiers using the Bootstrap Aggregating Algorithm. The classification results are then aggregated using a linear function that averages the classification distributions of the different classifiers. SentiME has also been evaluated over the SemEval2015 test set, properly trained with the SemEval2015 train test. We show that SentiME would outperform the best ranked system of the challenge.

1 Introduction

Replication studies are a core element of scientific research. They play a crucial role, either during a peer-review process, or *a posteriori*, for validating results and approaches and enabling further scientific progress. They aim to generate the same overall conclusions rather than producing the same exact figures [2–4]. We observe a steady rise of challenges organized by particular scientific communities, that aim to share common datasets, tasks and scorers to statistically evaluate results and enable comparison of approaches. There is also a strong encouragement from the research community to publish software source code, scripts and models as citable resources alongside traditional papers describing a particular approach and its evaluation.

In this paper, we propose to perform a thorough replication and reproduction study of the top systems that have competed to the yearly SemEval Twitter Sentiment Analysis tasks [17]. Specifically, we replicated the Webis system [6], an ensemble system of four state-of-art sub-classifiers that ranked first in SemEval 2015 among forty different systems. These four individual sub-classifiers have themselves participated during previous SemEval years where they were also among the top performing systems. The ensemble approach adopted by the

© Springer International Publishing AG 2016
P. Groth et al. (Eds.): ISWC 2016, Part II, LNCS 9982, pp. 204–219, 2016.
DOI: 10.1007/978-3-319-46547-0_22

Webis system has also inspired us to propose the SentiME system that adds another classifier on top of the Webis system.

Similar to [8], we adopt the following definitions that have been proposed during the recent SIGIR 2015 workshop on Reproducibility, Inexplicability, and Generalizability of Results (RIGOR)[1]:

- **Replicability**: "Repeating a previous result under the original conditions (e.g. same system configuration and datasets)";
- **Reproducibility**: "Reproducing a previous result under different, but comparable conditions";
- **Generalizability**: "Applying an existing, empirically validated technique to a different task/domain than the original one".

We aim to perform a replication study using the Webis source code and the pre-trained models provided by the authors. We also perform a reproducible study using again the Webis system but this time, training ourselves the models using the same features reported by the authors. Finally, our generalizability study leads to the creation of the SentiME system, an ensemble approach developed on top of the Webis system where a fifth classifier, namely the off-the-shelves Stanford Sentiment System, is added. We have also evaluated SentiME on a different dataset composed of one million Amazon reviews of products [21].

The remainder of this paper is structured as follows: in Sect. 2, we describe the particular SemEval task, datasets and systems we aim to replicate. In Sect. 3, we present our replicate study, where we first tried to replicate the results of the Webis system using the models provided by the authors, and then, by re-training ourselves those models. We present SentiME, our own ensemble system in Sect. 4 and we show that it would outperform the best performing systems in SemEval. We provide some lessons learned during this replication study (Sect. 5) before concluding and outlining future work (Sect. 6).

2   SemEval Twitter Sentiment Analysis Task

SemEval (Semantic Evaluation) is an ongoing series of evaluations of computational semantic analysis systems, where "semantic analysis" refers to a formal analysis of meaning, and "computational" refers to approaches that in principle support effective implementation [1]. While the series of evaluation has first focused on word sense disambiguation, it has evolved to investigate the interrelationships among the elements in a sentence (e.g., semantic role labeling), relations between sentences (e.g., coreference), and the nature of what we are saying in sentences (semantic relations and sentiment analysis). Since 2012, SemEval is part of the *SEM Conference. In 2013, the organizers have created a new task about Sentiment analysis in Twitter which was identified as the Task 2 in SemEval-2013, the Task 9 in SemEval-2014 and the Task 10 in SemEval-2015.

The general goal of this task is to better understand how the sentiment is expressed in short text messages such as tweets which are still considered

---

[1] https://sites.google.com/site/sigirrigor/.

as constrained 140 characters message. Each year, the task is actually divided into two sub-tasks: Subtask A about "Contextual Polarity Disambiguation" and Subtask B about "Message Polarity Classification". In 2015, the organizers have added three additional sub-tasks: Subtask C about "Topic-Based Message Polarity Classification", Subtask D about "Detecting Trends Towards a Topic" and Subtask E about "Determining strength of association of Twitter terms with positive sentiment (or, degree of prior polarity)". This replication study focuses on systems participating and ranked in the top positions for the Subtask B since the inception of the competition.

2.1 SemEval Subtask B in 2013–2015: Task and Corpus

The Subtask B asks participants to classify a given sentence (message or tweet) in three possible categories according to the overall sentiment which is conveyed: negative, neutral or positive. For tweets that transmit both positive and negative sentiment, the stronger should be chosen. SemEval provides each participation team a common training and test datasets, which have been annotated by the organizers. This allows different teams to compete on one controlled environment and to compare the performance of different algorithms and approaches in a fair way.

In 2013, the SemEval organizers have collected tweets to compose the training dataset over a one-year period. The test dataset corresponds to a three-months period collection realized before the competition. The raw tweets being heavily skew towards neutral, the SemEval organizers have filtered out large amount of neutral tweets and organize them in topics. Table 1 shows the total number of tweets and their distribution following the classification positive, negative and neutral.

Table 1. Datasets statistics for the Subtask B of SemEval-2015

Corpus	Positive	Negative	Neutral	Total
Twitter2013-train	3, 662	1, 466	4, 600	9, 728
Twitter2013-dev	575	340	739	1, 654
Twitter2013-test	1, 572	601	1, 640	3, 813
Twitter2014-test	982	202	669	1, 853
Twitter2014-sarcasm	33	40	13	86
LiveJournal2014-test	427	304	411	1, 142
Twitter2015-test	1040	365	987	2392

Annotations have been performed by Amazon Mechanical Turk (Turkers) resulting ultimately in a gold standard file composed of four fields: the original tweet's ID, the tweet's gold standard ID, the tweet's polarity and the textual

Table 2. SemEval's datasets structure

Tweet ID	Gold standard ID	Polarity	Tweet content
522931511323275264	T15111115	Positive	Catch Rainbow Valley at ...
522838326126989314	T15111137	Neutral	I wonder if Billy Joe ...
520829332525441024	T15111111	Negative	Saturday without Leeds ...

content of the tweet. These fields are stored into column separated tsv files. Table 2 provides some examples following this schema.

Systems participating in SemEval2015 must generate a output file which contains for each tweet, its classification result: Positive, Neutral or Negative. A system output is then compared with the gold standard file by the organizers who compute the Precision of positive, the Recall of positive, the Precision of negative and the Recall of negative tweets classification. The F-scores for both positive (Eq. 1) and negative (Eq. 2) are computed to finally generate the general F-score (Eq. 3) which is used to rank the various participating systems.

$$F_{pos} = 2 * \left(\frac{P_{pos} * R_{pos}}{P_{pos} + R_{pos}} \right) \tag{1}$$

$$F_{neg} = 2 * \left(\frac{P_{neg} * R_{neg}}{P_{neg} + R_{neg}} \right) \tag{2}$$

$$F = \frac{P_{pos} + R_{neg}}{2} \tag{3}$$

2.2 Top Scoring Systems in 2013–2015

NRC-Canada. The NRC-Canada team ranked 1st in SemEval 2013, using a SVM classifier to extract the sentiment from tweets [12]. They used different lexicons such as lists of words assigned with either a positive or a negative sentiment, the NRC Emotion Lexicon [13,14], the MPQA Lexicon [22], and the Bing Liu Lexicon [9]. They also used specific Twitter-based lexicon such as the NRC Hashtag Sentiment Lexicon [11] and the Sentiment140 Lexicon [12]. The system is trained with a linear kernel of a state-of-the-art Support Vector Machine (SVM) algorithm. A pre-processing phase enables to make the tweets easier to be processed. Each tweet is then represented by a feature vector composed of: N-grams, ALLCAPS, POS, Polarity Dictionaries, Punctuation Marks, Emoticons, Word Lengthening, Clusters and Negation.

GU-MLT-LT. The GU-MLT-LT team ranked 2nd in SemEval 2013, using a linear classifier trained by stochastic gradient descent with hinge loss and elastic net regularization for their predictions [5]. They also perform a pre-processing phase for the tweets where they included a variety of linguistics and lexical features such as: Normalized Uni-grams, Stems, Clustering, Polarity Dictionary and Negation.

KLUE. The KLUE team ranked 5th in SemEval 2013, using a simple bag-of-words models with three different features unigrams, unigrams and bigrams, and an extended unigram model that includes a simple treatment of negation [16]. They also pre-process the tweets and they used features based on a sentiment dictionary such as SentiStrength and an extended version of AFINN-111. Large-vocabulary distributional semantic models (DSM) have been used in order to obtain better word coverage, constructed from a version of the English Wikipedia[2] and the Google Web 1 T 5-Grams databases[3]. Finally, they included features based on emoticons and slang abbreviations mostly used on the Web and manually classified by themselves.

TeamX. The TeamX team ranked 1st in SemEval 2014, using a variety of pre-processors and features [10]. More specifically, the TeamX system used a large variety of lexicons categorized into "FORMAL" and "INFORMAL" such as AFINN-111 [15], Bing Lius Opinion Lexicon1 [9], General Inquirer [20], MPQA Subjectivity Lexicon [22], NRC Hashtag Sentiment Lexicon [11], Sentiment140 Lexicon Lexicon [12] and SentiWordNetBaccianella2010. Furthermore, they used additional features such as word ngrams, character ngrams, clusters and word senses. Eventually, they fed these features to a supervised machine learning algorithm which utilizes Logistic Regression (LIBLINEAR).

Webis. The Webis team ranked 1st in SemEval 2015, using an ensemble method over the four state-of-the-art systems previously described: NRC-Canada, GU-MLT-LT, KLUE and TeamX [6]. Initially, they selected the wining system of SemEval-2013, namely NRC-CANADA, and they manually choose the remaining three systems having as one and only criterion the level of dissimilarity of these systems with respect to NRC-CANADA. Having dissimilar systems in an ensemble is very important since it ultimately leads to a bigger diversity of features and lexicons being used. In other words, each one of the sub-classifiers complement each other which makes the ensemble method special and particularly effective on such a challenge.

In the Webis ensemble system, the authors did not use the classification results of each of the four sub-classifiers but instead, they used their confidence scores. Hence, if two sub-classifiers are not confident enough to provide a classification, the final sentiment will only depend on the other two remaining sub-classifiers providing a higher confidence. The authors also preferred not to build a weighting schema but to use a linear function which averages the classification distributions provided by the four sub-classifiers and produce the final classification according to the maximum value of the labels in the average classification distribution. In summary, the Webis system works as follows: the sub-classifiers are trained individually; the ensemble ignores the individual classification results

---

[2] The pre-processed and linguistically annotated Wackypedia corpus they used is from http://wacky.sslmit.unibo.it/doku.php?id=corpora.

[3] http://googleresearch.blogspot.fr/2006/08/all-our-n-gram-are-belong-to-you.html.

coming the four sub-classifiers but it considers the confidence scores (possibilities) for each class (positive, neutral and negative). The final classification is done by averaging the confidence scores for each class, the highest confidence score providing the final classification.

2.3 Stanford Sentiment System

The Stanford Sentiment System [19] is one of the sub-systems of the Stanford NLP Core toolkit. It contains the Stanford Tree Parser, a machine-learning model which can parse the input text into Stanford Tree format and use some existing models, some of them are trained especially for parsing tweets. The Stanford Sentiment Classifier is at the heart of the system. This classifier takes as input Stanford Trees and outputs the classification results for Stanford Trees. The Stanford Sentiment Classifier provides also useful detailed results such as classification label and classification distribution on all the nodes in the Stanford Tree.

The Stanford Sentiment System is a Recursive Neural Tensor Network trained on the Stanford Sentiment TreeBank that is the first corpus with fully labeled parse trees which makes possible training a model with large and labeled dataset. This model store the information for compositional vector representations, its size of parameters is not very large and the computation cost is reasonable. Moreover, the Stanford Sentiment System can capture the meaning of longer phrases and shows a great strength in classifying negations. It beats the bag of word approaches when predicting fine-grained sentiment labels.

While this system has never participated in previous SemEval years, we have decided to include it as an off-the-shelves classifier in a new ensemble system called SentiME (Sect. 4). In particular, we will demonstrate that its addition enables to outperform previous system on the sarcasm corpus.

3 Replication Study

3.1 Methodology

The initial goal of a replication study is to identify and to select state-of-the-art approaches that have performed relatively well in a given settings and to compare results with the ones published by the original authors. In this replication study, we first focused our research on the three top performing systems of the SemEval-2015: Webis [6] (1st), UNITN [18] (2nd) and Lsislif [7] (3rd). We decided to replicate the Webis system for three reasons: first, it is the best performing system in SemEval-2015; second, it is a state-of-the-art implementation of an ensemble method combining four sub-classifiers which themselves participated in previous years of SemEval thus giving a broader scope of this replicate study. Furthermore, the ensemble method give us more flexibility to achieve the generalizability we are looking for, in particular when including new classifiers to improve the overall system; third, but not least, the Webis source code has been released openly. While replicating a system without having access to the source code is not impossible, it is much harder, a paper or a technical report being rarely self-sufficient to re-implement a system from scratch. We have contacted the UNITN team, letting them know about our research, and asked if they were willing to share with us their source code but we did not receive any response. Similarly, the Lsislif paper description leaves too many open questions for deciding to re-implement this particular system.

The Webis team has released both the source code of the system and the models they have trained for the SemEval challenge at https://github.com/webis-de/ECIR-2015-and-SEMEVAL-2015. We verified that this version corresponds to the system which has being used to report official results at SemEval.

We manage to download and cleanse all the datasets through a download script from the SemEval's official website[4] via the Twitter API. A small number of tweets are not accessible anymore via the API. Consequently, the training and test sets are slightly different, which might have affected the performance of our replicate system. Table 3 reports the differences in terms of downloaded tweets across all datasets compared to the number of tweets SemEval-2015 organizers presented in [17].

Table 3. Comparison of datasets: some tweets are unfortunately not available anymore

Corpus	Nb tweets collected	Nb tweets originally in SemEval	Multiple tweets
SemEval2013-train+dev-B	11, 338	11,382	50
SemEval2013-test-gold-B	3, 813	3,813	3
SemEval2014-test-gold-B	1, 853	1,853	0
SemEval2014-test-sarcasm-B	86	86	0
SemEval2015-gold-B	2, 390	2,392	11
SemEval2015-test-sarcasm-B	60	N/A	0

We observe that we can almost collect all the data that has been officially provided by the SemEval organizers. The main differences, in the SemEval2013-train+dev-B and SemEval2015-gold-B datasets, are either due to the fact that some tweets have been deleted or made inaccessible, or because they are out of the time window that SemEval used to publish its data. Due to the Twitter TOS, nobody can publicly publish the original tweets content. However, the small volume of missing tweets do not hinder the validity of our replicate experiment.

The Multiple Tweets column shows a very interesting phenomenon in the dataset. Each count in Multiple Tweets represents one true identical tweet (exact

---

[4] http://alt.qcri.org/semeval2015/task10/index.php?id=data-and-tools.

same tweet ID) appearing more than once in the dataset. This is not due to multiple users publishing the same tweet content (or using the RT functionality) but because SemEval does not filter out some multiple tweets either intentionally or unintentionally. Although we are not sure what was the purpose of the SemEval organizers, we decided to not filter out the multiple tweets because the other participating teams such as Webis did not claim to have performed this operation.

3.2   Replicating Webis Using Pre-trained and Re-Trained Models

Due to the fact that the Webis system is the ensemble of four sub-classifiers and that each sub-classifier is built using the classifier API of WEKA, the performance of the system is based on some external libraries. Versioning (of software libraries and dependencies) is an important aspect to be considered in any replication study. In this study, we removed the old external libraries which are related to Stanford NLP Core from the Webis system and we added the newest version of Stanford NLP Core libraries.

We have created two separate workbenches. The first replicate system use the Webis system as well as the models provided by the authors on their github repository. In the second replicate system, we re-train ourselves the various sub-classifiers composing the Webis system: NRC-CANADA, GU-MLT-LT, KLUE and TEAMX. Our replicate system can train each one of the four sub-classifiers individually and test them individually or all together as an ensemble system using any of the dataset we have. We compared the classification results using either the models provided by Webis or the models we were able to train, using the same Webis ensemble configuration (Table 4).

Table 4. F-scores for pre-trained and re-trained models in comparison with the results reported in the original papers

Dataset	Claimed in papers [6,17]	Webis's models	Our models
Replicate Webis system on test 2013	68.49	69.62	70.06
Replicate Webis system without TeamX on test 2013	N/A	69.04	70.34
Replicate Webis system on test 2014	70.86	66.65	69.31
Replicate Webis system without TeamX on test 2014	N/A	66.51	68.56
Replicate Webis system on test 2015	64.84	66.17	66.57
Replicate Webis system without TeamX on test 2015	N/A	65.58	66.19

Concerning the training workflow, the sub-classifiers are trained individually and one attribute file is generated for storing the model for each one of the sub-classifiers. We also provide one function in our replicate system to train the four sub-classifiers all together. The training of each individual sub-classifier involves four steps. First, a feature extractor processes all tweets in the training dataset to generate feature vectors from the tweets. This step is different among the four sub-classifiers due to the different features each sub-classifier uses. Second, feature vectors are passed to the classifiers. Third, each sub-classifier is trained to generate the parameters of the classifier. The fourth and final step is to store all those information into an attribute file that represents the model.

Regarding the testing process, each sub-classifier is also processing independently the dataset. The first step is to load all the parameters into the feature extractor and the classifier. Then we pre-process the tweets similar to the training process. The next step is to extract the feature vectors from the cleansed tweet texts and to pass them to the classifier.

Each sub-classifier will give a classification result and a the confidence scores for each of the three classes (positive, neutral and negative). When we aggregate the classification results of the four sub-classifiers, we just use a simple linear function which averages the classification distributions of the four sub-classifiers and classify the tweet according to the label which holds the maximal value in the average classification distribution.

After building our replicate system, we launch several experiments to test whether our replicate system achieve exact or similar results than the ones reported by the Webis team in their paper [6]. Consequently, we perform experiments using the individual sub-classifiers and the ensemble system on the SemEval2013-test, SemEval2014-test and SemEval2015-test datatset using either the models provided by the Webis team or the models we re-trained using the SemEval2013-train+dev corpus. Those results are reported in Table 4.

4 SentiME: Generalizing the Webis System

We originally aim to replicate and to reproduce the Webis system in order to see if we can get comparable results. Our investigations have lead us to generalize the system and to propose SentiME, a new ensemble system that add a fifth sentiment classifier to the Webis system, the Stanford Sentiment System that is used as an off-the-shelf classifier[5]. We also propose to use bagging to boost the training of the ensemble system.

The Stanford Sentiment System is a recursive neural tensor network parsed by the Stanford Tree Bank. It is significantly different from all the other classifiers used on tweets polarity prediction and it shows great performance on negative classification. Hence, the negative recall of the sole Stanford Sentiment System is over 90 % on average which makes it trustworthy to detect negation (Table 5). We want to investigate whether the addition of this new sub-classifier would improve our Webis replicate system.

---

[5] The Stanford Sentiment System will not be trained with the task datasets.

Table 5. Negative recall of the sole Stanford Sentiment System on SemEval datasets

Corpus	Negative recall
SemEval2014-test-gold-B	0.9108910891089109
SemEval2015-gold-B	0.8980716253443526

The classification distribution provided by the Stanford Sentiment Classifier consists of five labels: very positive, positive, neutral, negative and very negative. Consequently, we need to map these five labels into the three classes expected by the SemEval challenge for a consistent integration with our replicate system. We only extract the root classification distribution because it represents the classification distribution of the whole tweet text. We have tested different configurations for mapping the Stanford Sentiment System classification to the three classes modem. According to the results of these tests, we decide to use the following mapping algorithm: very positive and positive are mapped to Positive, neutral are mapped to Neutral and negative and very negative are mapped to Negative.

There are multiple ways to do an ensemble of different systems. In the case of SentiME, we propose to use the Stanford Sentiment System as an off-the-shelve classifier that will not be re-trained. We also propose to use a bagging algorithm for boosting the training of the four other sub-classifiers. To perform bagging, we generate new training data set for each sub-classifier using uniformly sampling with replacement. In order to simulate this procedure, we use the pseudo random number generator which is provided by the Java Util package to generate random tweet indexes used to sample from the loaded tweet list. Afterwards, we provide the system time as seed to Random Class to increase the randomness between different sub-classifiers and we carefully pick and test the size of the bootstrap samples. After training the four sub-classifiers, we store their resulting attribute files for further usage. When we test these models on the test dataset, the aggregating function is just one simple linear function which averages the classification results from the four sub-classifiers as done by Webis.

Due to the fact that bagging introduces some randomness into the training process, and that the size of the bootstrap samples are not fixed, we decide to perform multiple experiments with different sizes ranging from 33 % to 175 %. We perform the training process three times and get three models to test for each size. We observe that doing bagging with 150 % of the initial dataset size leads to the best performance in terms of F1 score (Table 6).

We choose the linear averaging function as the aggregating function because this is the simplest method and the performance of the ensemble system is easy to predict. To be more precise, the linear aggregating function averages the classification distributions of each sub-classifiers and choose the polarity which holds the maximum value among the average classification distributions as the final classification.

Table 6. Experiments performed with different bagging sizes on SemEval2013-train+dev-B (11,338 original size) training dataset

Model	19,842(175 %)	17,007(150 %)	14,173(125 %)	11,338(100 %)
Model 1	64.76	65.45	65.26	65.05
Model 2	64.65	65.81	64.40	64.15
Model 3	64.50	65.71	64.29	65.25
	9,000(80 %)	7,525(66 %)	5,644(50 %)	3,780(33 %)
Model 1	63.37	63.80	64.64	62.81
Model 2	64.54	64.92	62.93	62.85
Model 3	63.54	64.67	63.85	61.65

We already mention the problem of multiple identical tweets in the dataset, where the same tweet ID is present several times with different gold standard ID but also different gold standard polarity. In this experiment, we filter out those multiple tweets to evaluate the SentiME system.

We performed four different experiments to evaluate the performance of SentiME compare to our previous replicate of the Webis system:

1. Webis replicate system: this is the replicate of the Webis system using re-trained models as explained in the Sect. 3;
2. SentiME system: this is the ensemble system composed of the four sub-classifiers used in by Webis plus the Stanford Sentiment System. The ensemble uses a bagging approach for the training phase;
3. Webis replicate system without TeamX;
4. SentiME system without TeamX;

The experiments 3 and 4 are variations of the experiments 1 and 2 where we simply remove the TeamX sub-classifier, based on the observation that this particular sub-classifier plays a similar role than the Stanford Sentiment system.

The Table 7 reports the F-scores for the four different set ups we described above on four different datasets: two datasets contain regular tweets and two datasets contain sarcasm tweets. We evaluated the four sub-classifiers on SemEval2014 test data set, SemEval2014 sarcasm data set, SemEval2015 test data set and SemEval2015 sarcasm data set in order to figure out whether the Stanford Sentiment System has significant impacts on the performance of our ensemble system. The last row of the Table 7 presents the F scores of the Webis system as reported in the authors' paper [6].

We observe that the SentiME system outperforms the Webis Replicate system on all datasets except on the SemEval2014-test, in which the SentiME system without the TeamX sub-classifier has almost the same performance than the Webis Replicate system. Concerning the performance on both sarcasm datasets, it is clear that SentiME system improves the F score by respectively 2,5 % and 6,5 % on SemEval2014-sarcasm and SemEval2015-sarcasm datasets. However, it is unclear why we observe a significant difference of performance on the

Table 7. F-scores for the four systems on four different datasets (highest scores are in bold)

System	SemEval2014-test	SemEval2014-sarcasm	SemEval2015-test	SemEval2015-sarcasm
Webis replicate system	**69.31**	60.00	66.57	54.19
SentiME system	68.27	**62.57**	**67.39**	**60.92**
Webis replicate system without TeamX	68.56	62.04	66.19	56.86
SentiME system without TeamX	69.27	62.04	66.38	58.92
Webis	70.86	49.33	64.84	53.59

SemEval2014-sarcasm dataset between the original Webis system (49.33 %) and our replicate (60 %). The fact that the dataset is extremely small (only 86 tweets) prevents us to draw any conclusion.

We notice that some features used in TeamX come from the Stanford NLP Core package and we assume that TeamX shares some common characteristics with the Stanford Sentiment System. Since the idea of our experiments was to figure out what benefits the Stanford Sentiment System can bring to our replicate system, we consider it is reasonable to exclude the TeamX sub-classifier from our replicate system.

The complete workflow of the SentiME system is depicted in the Fig. 1. The SentiME system trains the four sub-classifiers independently and the results are stored into four different attribute files. Then, concerning the test process, each sub-classifier reads the attribute files as well as the test dataset. The Stanford Sentiment System just reads the test dataset. The final step is to average the five different classification results of the sub-classifiers in order to derive the final classification result.

5   Lessons Learned

A few aspects emerged from this comprehensive replication study:

Paper and source code: We started from replicating the experimental setup of the Webis system by studying and implementing the approach reported in the paper. We found the paper not self contained for the settings: we believe that this is partially due do the page limits authors have to respect for complying with publication rules. The source code of the system has helped the replication study presented in this paper. We encounter a few minor issues related to the existence of libraries not included in the source code. We can conservatively state that the availability of the source code has significantly helped to pursue this study and to reproduce the results.

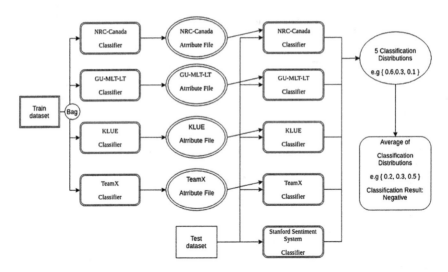

Fig. 1. Workflow of training and test for SentiME

Differences in the performance results due to the training data: The pre-trained models provided by Webis are not exactly the same as the re-trained models we have created from the data at disposal. This is significant for the SemEval2014-test (Table 4) and can be explained by the loss we had in the collection of tweets. We have also to consider that older tweets are more likely to be removed or made inaccessible because of the tweet persistence that is dependent of the Twitter platform. This is how we can explain the under-performance of the SentiME system on the same dataset (Table 7).

Differences in performance results due to the differences in the source code: Another topic of discussion is the differences between the F1 scores claimed in SemEval's 2015 and Webis's papers [6,17] with the F1 scores of the pre-trained models we computed. These differences indicate that our replicate Webis system is not exactly the same than the original Webis system. This is reasonable because the Webis system contains a lot of libraries which have been updated between the time their experiments and our experiments have been performed. The differences we noticed in the datasets organizers provided for the SemEval competition played also an important role for the final results.

Differences in performance results due to missing features: In addition, another factor that we need to consider about the re-trained models we computed and why they differ is the possibility that the Webis's authors did not detail the full set of features they have used. Feature engineering is an art and the devil is in the details.

6   Conclusion and Future Work

The replication of a prior study is not an easy task since one has always to anticipate possible differences from the original study that may lead to different results. When achieving a thorough replication study, a natural evolution is to propose a generalization for improving a system. The number of tweets we can collect for each dataset is not strictly identical with the numbers reported by the SemEval-2015 organizers [17]. Nevertheless, we manage to replicate the Webis system and to reproduce its functioning by re-training ourselves the models being used. We observe that the Stanford Sentiment System is heavily skew towards negative classification and share a lot of commonalities with the TeamX sub-classifier which is being used by the Webis systems. We also demonstrate that the Stanford Sentiment System improves the performance of a sentiment detection system on a sarcasm dataset.

We manage to improve the Webis system by 1 % in the general case by introducing a fifth sub-classifier (the Stanford Sentiment System) and by boosting the training with bagging 150 % of the original training dataset while filtering out the multiple tweets. The SentiME system also outperforms the Webis system by 6,5 % on the particular and more difficult sarcasm dataset. Additional experiments performed on product reviews confirm that the use of bagging during the training phase is the main driver for improving significantly the performance [21]. The SentiME system is available at https://github.com/MultimediaSemantics/sentime and is itself fully replicable.

We suggest, for future work, to improve the aggregating algorithm used in our experiments which, so far, is a simple linear function which averages the classification distributions of each sub-classifier. This is the most basic aggregating function and there is consequently a lot of space for improvement. We suggest to use some weighted aggregating functions and perform some related experiments in order to find out the best possible system set up. Moreover, it is worth trying to train the Stanford Sentiment System with the SemEval training datasets. This will require to convert the training dataset into the Stanford Tree Bank format.

Concerning the training process, finding the best size of bootstrap samples is a real challenge. One should not only look at the performance improvement in terms of the F scores, but should also consider whether a stable training process can be established since bagging introduces some randomness to the training process. Consequently, a series of very fine-grained experiments that may take a long time to run must be performed. The aggregating algorithm we use in our bagging process is the linear function which averages the classification distributions of each sub-classifier. Since this linear function is very simple and does not involve any careful consideration, it is possible that the performance of the system could be improved by replacing the aggregating algorithm by a new technique.

Acknowledgments. This work was partially supported by the innovation activity 3cixty (14523) of EIT Digital and by the European Union's H2020 Framework Programme via the FREME Project (644771).

References

1. Blackburn, P., Bos, J.: Representation and Inference for Natural Language: A First Course in Computational Semantics. CSLI, Stanford (2005)
2. Buchert, T., Nussbaum, L.: Leveraging business workflows in distributed systems research for the orchestration of reproducible and scalable experiments. In: 9th French Conference on MAnifestation des JEunes Chercheurs en Sciences et Technologies de l'Information et de la Communication (2012)
3. Dalle, O.: On reproducibility and traceability of simulations. In: WSC - Winter Simulation Conference (2012)
4. Drummond, C.: Replicability is not reproducibility: nor is it good science. In: Workshop on Evaluation Methods for Machine Learning (2009)
5. Günther, T., Furrer, L.: GU-MLT-LT: sentiment analysis of short messages using linguistic features and stochastic gradient descent. In: 7th International Workshop on Semantic Evaluation (SemEval-2013) (2013)
6. Hagen, M., Potthast, M., Büchner, M., Stein, B.: Webis: an ensemble for Twitter sentiment detection. In: 9th International Workshop on Semantic Evaluation (SemEval-2015) (2015)
7. Hamdan, H., Bellot, P., Bechet, F.: Lsislif: feature extraction and label weighting for sentiment analysis in Twitter. In: 9th International Workshop on Semantic Evaluation (SemEval-2015) (2015)
8. Hasibi, F., Balog, K., Bratsberg, S.E.: On the reproducibility of the TAGME entity linking system. In: Ferro, N., et al. (eds.) ECIR 2016. LNCS, vol. 9626, pp. 436–449. Springer, Heidelberg (2016). doi:10.1007/978-3-319-30671-1_32
9. Hu, M., Liu, B.: Mining opinion features in customer reviews. In: 19th National Conference on Artificial Intellgience (2004)
10. Miura, Y., Sakaki, S., Hattori, K., Ohkuma, T.: TeamX: a sentiment analyzer with enhanced lexicon mapping and weighting scheme for unbalanced data. In: 8th International Workshop on Semantic Evaluation (SemEval-2014) (2014)
11. Mohammad, S., Kiritchenko, S.: Using hashtags to capture fine emotion categories from tweets. In: Special Issue on Semantic Analysis in Social Media, Computational Intelligence (2012)
12. Mohammad, S.M., Kiritchenko, S., Zhu, X.: NRC-Canada: building the state-of-the-art in sentiment analysis of tweets. In: 7th International Workshop on Semantic Evaluation (SemEval-2013) (2013)
13. Mohammad, S., Turney, P.: Emotions evoked by common words and phrases: using mechanical turk to create an emotion lexicon. In: NAACL-HLT 2010 Workshop on Computational Approaches to Analysis and Generation of Emotion in Text (2010)
14. Mohammad, S., Turney, P.: Crowdsourcing a word emotion association lexicon. Comput. Intell. **29**(3), 436–465 (2013)
15. Nielsen, F.: AFINN. Tech. rep., Informatics and Mathematical Modelling, Technical University of Denmark (2011). http://www2.imm.dtu.dk/pubdb/p.php?6010
16. Proisl, T., Greiner, P., Evert, S., Kabashi, B.: KLUE: simple and robust methods for polarity classification. In: 7th International Workshop on Semantic Evaluation (SemEval-2013) (2013)

17. Rosenthal, S., Nakov, P., Kiritchenko, S., Mohammad, S., Ritter, A., Stoyanovm, V.: SemEval-2015 task 10: sentiment analysis in Twitter. In: 9th International Workshop on Semantic Evaluation (SemEval-2015) (2015)
18. Severyn, A., Moschitti, A.: UNITN: training deep convolutional neural network for Twitter sentiment classification. In: 9th International Workshop on Semantic Evaluation (SemEval-2015) (2015)
19. Socher, R., Perelygin, A., Wu, J.Y., Chuang, J., Manning, C., Ng, A., Potts, C.: Recursive deep models for semantic compositionality over a sentiment treebank. In: Conference on Empirical Methods in Natural Language Processing (EMNLP) (2013)
20. Stone, P., Dunphy, D., Smith, S.: The general inquirer: a computer approach to content analysis. J. Reg. Sci. **8**(1), 113116 (1968)
21. Sygkounas, E., Rizzo, G., Troncy, R.: Sentiment polarity detection from amazon reviews: an experimental study. In: Semantic Web Evaluation Challenges (2016)
22. Wilson, T., Wiebe, J., Hoffmann, P.: Recognizing contextual polarity in phrase-level sentiment analysis. In: Human Language Technology Conference and Conference on Empirical Methods in Natural Language Processing (2005)

VoldemortKG: Mapping schema.org and Web Entities to Linked Open Data

Alberto Tonon[(✉)], Victor Felder, Djellel Eddine Difallah,
and Philippe Cudré-Mauroux

eXascale Infolab, University of Fribourg, Fribourg, Switzerland
{alberto,victor,ded,pcm}@exascale.info, philippe.cudre-mauroux@unifr.ch

Abstract. Increasingly, webpages mix entities coming from various sources and represented in different ways. It can thus happen that the same entity is both described by using schema.org annotations and by creating a text anchor pointing to its Wikipedia page. Often, those representations provide complementary information which is not exploited since those entities are disjoint. We explored the extent to which entities represented in different ways repeat on the Web, how they are related, and how they complement (or link) to each other. Our initial experiments showed that we can unveil a previously unexploited knowledge graph by applying simple instance matching techniques on a large collection of schema.org annotations and Wikipedia. The resulting knowledge graph aggregates entities (often tail entities) scattered across several webpages, and complements existing Wikipedia entities with new facts and properties. In order to facilitate further investigation in how to mine such information, we are releasing (i) an excerpt of all Common Crawl webpages containing both Wikipedia and schema.org annotations, (ii) the toolset to extract this information and perform knowledge graph construction and mapping onto DBpedia, as well as (iii) the resulting knowledge graph (VoldemortKG) obtained via label matching techniques.

Keywords: Knowledge graphs · schema.org · Instance matching · Data integration · Dataset

1 Introduction

Annotating webpages with structured data allows webmasters to enrich their HTML pages by including machine-readable content describing what we call *Web Entities*, along with their properties and the relationships that might exist among them. Such machine-readable content is embodied into the HTML markup by using specific formats like microdata or RDFa, and vocabularies coming from different ontologies. According to Bizer et al. [1], in 2013 the ontologies that were most widely used to describe Web Entities were: schema.org, a schema designed and promoted by several technology companies including Google, Microsoft, Pinterest, Yahoo! and Yandex; the Facebook Open Graph Protocol (OGP), which helps web editors integrating their content to the social networking platform; and

P. Groth et al. (Eds.): ISWC 2016, Part II, LNCS 9982, pp. 220–228, 2016.
DOI: 10.1007/978-3-319-46547-0_23

the GoodRelation vocabulary, which defines classes and properties to describe e-commerce concepts. As a result, the Web is now a prime source of structured data describing self-defined entities.

We argue that there is an underlying unexploited knowledge graph formed by such data, which overlaps and possibly complements other knowledge graphs in the Linked Open Data (LOD) cloud[1]. More specifically, we are interested in identifying connections between entities represented through annotations in webpages and entities belonging to further datasets, as well as discovering new entities that are potentially missing from well-known knowledge bases.

To extract such information, some challenges must first be overcome:

- Due to the decentralized nature of the Web, this knowledge graph is scattered across billions of webpages, with no central authority governing the creation and indexing of Web Entities;
- The markups are added by a crowd of non-experts driven by Search Engine Optimization (SEO) goals, hence the quality of the data is generally-speaking questionable;
- In order to produce high-quality links, one needs to extract supporting evidence from the annotated webpages, track provenance, clean and parse text and identify additional named entities.

In this context, we propose to help the Semantic Web research community tackle the open research problem of mapping Web Entities across webpages and finding their counterparts in other knowledge bases. To that end, we construct and release a dataset containing all webpages extracted from the Common Crawl dump[2] containing both Web Entities and links to Wikipedia. This data structure is designed to disseminate enough contextual information (full HTML content) and prior ground (Wikipedia links) to effectively perform the task of instance matching. In addition to the raw dataset of webpages and triples that we publish, we also showcase the generation of a proof-of-concept knowledge graph (VoldemortKG[3]). Our technique performs instance matching of microdata triples to their DBpedia counterparts via simple label matching. The resulting graph is also available as a downloadable (and browsable) resource and can serve as a baseline for more advanced methods.

2   Related Work

Extracting and leveraging online structured data has been of interest to many companies and was the core of a number of Web services. Sindice [7], was a search engine that indexed LOD data and provided a keyword search interface over RDF. The Sig.ma project [12] was an application built on top of Sindice that allowed browsing the Web of data by providing tools to query and mashup

---

[1] http://linkeddata.org/.
[2] http://commoncrawl.org/.
[3] The knowledge graph that everyone knows exists, but no one talks about.

the retrieved data[4]. While potential applications of VoldemortKG could overlap with these projects, our present endeavor aims at providing key building blocks to perform data integration on the Web of data.

The Web Data Commons (WDC) initiative [5] extracts and publishes structured data available on the Web. The project makes available two important resources (i) Datasets, namely: RDFa, Microdata and Microformat, Web tables, Web hyperlinks, and IsA relations extracted from webpages, and (ii) the toolset for processing the Common Crawl dataset. Similarly, we build on top of the WDC Framework, and in addition extract and organize both structured data and HTML contents encompassing links pointing to Wikipedia. In contrast to the Web Data Commons, our objective is not only to collect and distribute the triples, but also the context in which they appear.

Instance Matching and Ontology Alignment. The process of matching entities across knowledge graphs is usually referred to as *Instance Matching*. The challenge is to automatically identify the same real world object described in different vocabularies, with slightly different labels and partially overlapping properties. Advanced techniques for instance matching compare groups of records to find matches [6] or use semantic information in the form of dependency graphs [3]. A task that often goes hand in hand with instance matching is *Ontology Alignment*. This task requires to map concepts belonging to an ontology to concepts of another ontology; for example, one can align the schema.org classes to their equivalent classes in the DBpedia ontology. The Ontology Alignment Evaluation Initiative (OAEI)[5] aims at stimulating and comparing research on ontology alignment. We point the reader to the work by Otero-Cerdeira et al. [8] for a detailed survey of the state of the art on the subject.

Our dataset will pose new challenges for both the instance matching and the ontology alignment community given the complexity of automatically mapping embedded structured data onto other LOD datasets. New methods need to be investigated in order to leverage the webpage contents for these tasks.

Entity Linking/Typing and AOR. Another relevant task in our context is *Entity Linking*, where the goal is to detect named entities appearing in text, and identify their corresponding entities in a knowledge base. Similarly, the dataset we release can be exploited for designing new methods for Ad-hoc Object Retrieval (AOR) [9], that is, building a ranked list of entities to answer keyword queries. Recent approaches for AOR make use of the literals connected to the entities in some knowledge base in order to use language modeling techniques to retrieve an initial ranked list of results that can be then refined by exploiting different kinds of connections among entities [2,11]. Lastly, Entity Typing (ET) is the task of finding the types relevant to a named entity. For instance, some ET systems focus on the following types Organization, Person and Location [4]. More recent work try to map named entities to fine-grained types [10]. Our dataset will challenge

---

[4] We note that both http://sindice.com and http://sig.ma are defunct as of April 2016.

[5] http://oaei.ontologymatching.org/.

these tasks by providing novel use-cases, where the extracted entities together with their types will then be used to verify and match against structured data embedded in the document.

3   The Dataset

As pointed out above in the introduction, it is worth exploring multiple representations of entities and connections among them. To foster investigations on this subject, we gathered a dataset that guarantees the presence of at least two sources of entities: i) DBpedia (as wikipedia anchors), and ii) structured data describing Web Entities. The dataset is created starting from the Common Crawl dated from November 2015[6], a collection of more than 1.8 billion pages crawled from the World Wide Web.

Data Extraction. We slightly modified the Web Data Commons Framework [5] to extract both the semantic annotations contained in the pages and the source code of all pages containing anchors pointing to any Wikipedia page. To lower the computational complexity during the extraction, we first test for the presence of Wikipedia anchors by matching against a set of simple regular expressions.

Even though we designed these regular expressions to achieve high recall—thus accepting the possibility of having many false positive pages—this simple filtering process significantly reduced the number of pages that we had to parse in order to extract the triples.

The whole process ran on 100 c3.4xlarge Amazon AWS spot instances featuring 16 cores and 30 GiB of memory each. The instances ran for about 30 h and produced 752 GiB of data, out of which 407 GiB contained compressed raw webpages and 345 GiB contained compressed semantic annotations in the form of 5-ple (subject, predicate, object, page url, markup format). In the rest of this document, we use the word "triple" to refer to the first three components of such extracted 5-ple. We release our modified version of WDC together with the dataset.[7]

Data Processing. In this step we process the webpages we extracted previously to build the final datasets we release. To create the datasets we used Apache Spark[8] and stored the pages, the 5-ples, and the anchors using the Parquet storage format[9] combined with the Snappy compression library:[10] this allows selective and fast reads of the data. We then used SparkSQL methods to discard semantic annotations extracted from pages not containing Wikipedia anchors, to determine the Pay Level Domains of the pages, to compute statistics, and to generate the final data we release. Together with the data, we also provide a

---

[6] http://commoncrawl.org/2015/12/.
[7] https://github.com/XI-lab/WDCFramework.
[8] https://spark.apache.org/.
[9] https://parquet.apache.org/.
[10] https://github.com/google/snappy.

Table 1. (left) Markup formats for including structured data in webpages and their popularity in number of annotations and number of webpages. (right) Top-10 vocabularies used to denote properties of structured data.

Format	N. Triples	N. Pages		Vocabulary	N. Triples	N. Pages
μformats-hcard	317,636,734	4,190,649		www.w3.org	363'103'085	7'113'775
μdata	84,073,194	2,539,539		schema.org	47'211'476	2'202'504
RDFa	17,451,754	1,675,747		vocab.sindice.com	7'886'539	396'102
μformats-xfn	9,567,666	396,102		data-vocabulary.org	6'213'512	243'531
μformats-adr	4,333,580	165,601		purl.org	5'408'960	2'625'015
μformats-geo	778,343	114,092		ogp.me	2'231'434	398'927
μformats-hcalendar	4,802,572	70,174		opengraphprotocol.org	1'927'743	313'734
μformats-hreview	234,221	17,210		historical-data.org	1'213'531	43'026
μformats-hrecipe	144,836	4,237		rdfs.org	1'041'408	28'943
μformats-species	18,190	2,282		www.facebook.com	1'005'237	665'205
μformats-hresume	1,385	75				
μformats-hlisting	2,610	20				

framework written in Scala allowing researchers to easily run their own instance matching methods on each webpage of the provided dataset.[11]

Key Statistics. Out of the 21,104,756 pages with Wikipedia anchors, 7,818,341 contain structured data. Table 1 (left) shows the distribution of the markup formats used to include structured data on webpages. As can be seen, 54 % of the webpages of our dataset are annotated by using some type of Microformats, 28 % of the pages contain Microdata, and 18 % contain RDFa annotations. This gives an idea of the diversity of sources one could tap into in order to extract entities and, possibly, connect them to other knowledge bases. In addition, we notice that more than one million pages contained in the dataset feature more than one type of markup format; detecting when the same entity is represented using different formats is an interesting open topic.

Table 1 (right) lists the top-10 vocabularies used in our dataset. The most wildely used is "www.w3.org" since the tool we used to extract structured data uses properties defined in that domain to encode is-a relations (e.g., all the `itemtype` Microdata annotations are translated into triples featuring the http://www.w3.org/1999/02/22-rdf-syntax-ns#type predicate). We observe that more than 3.3 million pages feature properties coming from more than one vocabulary and, more interestingly, almost 2.5 million pages feature properties selected from more than three vocabularies.

Distribution of the datasets. The dataset and the tool-chain used throughout this project is duly described on our website, for which we created a permanent URL: https://w3id.org/voldemortkg/. The extracted data is provided according the same terms of use, disclaimer of warranties and limitation of liabilities that apply to the Common Crawl corpus.[12]

---

[11] https://github.com/XI-lab/WDCTools.

[12] http://commoncrawl.org/terms-of-use/.

4   The VoldemortKG Knowledge Graph

To demonstrate the potential of the dataset we release, we built a proof of concept knowledge graph called VoldemortKG. VoldemortKG integrates schema.org annotations and DBpedia entities by exploiting the Wikipedia links embedded in the webpages we share. Equivalence between schema.org entities and DBpedia entities in VoldemortKG is based on string matching between the name of the former and the labels of the latter. Specifically, given a webpage P containing a DBpedia entity w, and a schema.org entity s, we say that w and s denote the same entity if the name of s, extracted from p by using the http://schema.org/name property, is also a label of w. A string s is a label of a DBpedia entity w if there is either a triple $(w, \text{rdfs:label}, s)$ in DBpedia, or if in some webpage there is an anchor enclosing the text s and pointing to the Wikipedia page of w. We also exploit transitivity to generate equivalences among entities. Figure 1 shows our matching algorithm applied to a simple example.

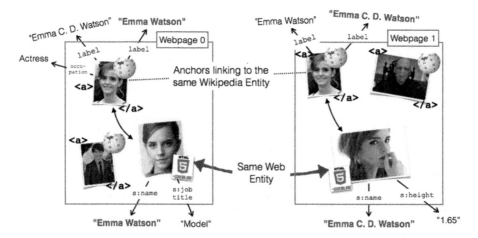

Fig. 1. Instance matching method used to build VoldemortKG. On the left-hand side, the DBpedia entry for Emma Watson is considered equivalent to a schema.org entity as its name is also a label of Emma Watson in DBpedia. On the right-hand side, a similar situation takes place for the same DBpedia entry and another Web entity. We thus conclude that all the mentioned entities refer to the same actress.

VoldemortKG is composed by 2.8 millions triples and contains information about 55,869 entities of 134 different types extracted from 202,923 webpages. Table 2 shows the top-15 entity types ordered by the number of instances (left), number of pages (center), and number of Pay Level Domains (PLDs, right) in which one of their instances appear. It is interesting to observe how top types change: notice that the top ranked type is different depending on the statistics taken into consideration. For example, the top ranked type in the right table is WebSite, with a count that is much higher than the number of Voldemor-tKG entities. This is due to the fact that entity E13418[13] appears in 132,616 webpages. This shows how one can get compelling results by leveraging simple string matching techniques in order to connect schema.org entities mentioned in different pages. Nevertheless, relying on such a simple method may result in many false positives, such as entity E13140[14], which is a person in VoldemortKG but an organization in DBpedia. This calls for further research on the topic.

Entity Fragmentation. It often happens that information about the same entity is scattered across several webpages. During the construction of VoldemortKG we extracted data about entities from 4 pages on average per entity (min. 1, max 132,616). As expected, there were cases in which the same (entity, prop-erty) pair was found in more than one webpage. For example, the properties `s:alternateName` and `owl:sameAs` appear, on average, in 367 and 11 pages per entity. Deciding which values of the property should be assigned to the entity taken into consideration is out of the scope of this work and is an interesting subject for future research.

Table 2. Top-15 entity types ordered by the number of instances (left), number of pages (center), and number of Pay Level Domains (PLDs, right) in which one of their instances appear. The prefix "hd" refers to http://historical-data.org.

Entity Type	N. Instances	Entity Type	N. Pages	Entity Type	N. PLDs
s:Person	36,370	s:WebSite	132,666	s:Article	295
s:RadioStation	3,524	s:Person	33,457	s:Person	203
hd:HistoricalPerson	1,963	s:Store	5,709	s:SiteNavigationElement	72
s:Movie	1,734	s:RadioStation	4,393	s:WebPage	70
hd:Person	1,496	s:AccountablePerson	3,645	s:Recipe	56
s:CollegeOrUniversity	1,308	s:CollegeOrUniversity	2,591	s:ListItem	53
s:AdministrativeArea	1,230	s:Movie	2,456	s:Product	40
s:School	1,118	s:Recipe	2,313	s:BlogPosting	36
s:Article	1,020	hd:Person	2,070	s:Organization	33
s:TVSeries	906	hd:HistoricalPerson	1,997	s:Place	25
s:AccountablePerson	844	s:Article	1,451	s:LocalBusiness	25
s:Blog	578	s:AdministrativeArea	1,252	s:Blog	24
s:Place	569	s:School	1,156	s:Thing	20
s:Landmarks...	518	s:TVSeries	948	s:MusicGroup	19
s:Book	428	s:QAPage	677	s:Book	17

5   Conclusions and Open Challenges

Taking advantage of the growing amount of structured data produced on the Web is critical for a number of tasks, from identifying tail entities to enriching existing knowledge bases with new properties as they emerge on the Web. While this information has been essentially exploited by commercial companies, it remains

---

[13] http://voldemort.exascale.info/resource/E13418.
[14] http://voldemort.exascale.info/resource/E13140.

an under-explored ground for the research community where several fundamental research challenges arise.

In this paper, we proposed a new dataset composed of webpages containing both Web Entities and Wikipedia links. Our goal was to extract and match structured pieces of data with high confidence in addition to provenance data, which constitutes a playground for researchers interested in a number of tasks including entity disambiguation & linking, entity typing, ad-hoc object retrieval or provenance management.

To demonstrate the usefulness of this dataset, we built a proof-of-concept knowledge graph (VoldemortKG) by label-matching triples to corresponding Wikipedia entities found on the same webpage. The resulting data was also made available in a browsable and downloadable format, and can be used as a baseline for further extraction and linking efforts.

Acknowledgements. This work was supported by the Swiss National Science Foundation under grant number PP00P2 153023. The authors would like to thank DAPLAB (http://daplab.ch/) for providing the resources necessary to create the datasets and, in particular, Benoit Perroud and Christophe Bovigny for their precious help.

References

1. Bizer, C., Eckert, K., Meusel, R., Mühleisen, H., Schuhmacher, M., Völker, J.: Deployment of RDFa, microdata, and microformats on the web - a quantitative analysis. In: ISWC 2013, pp. 17–32 (2013)
2. Bron, M., Balog, K., Rijke, M.: Example based entity search in the web of data. In: Serdyukov, P., Braslavski, P., Kuznetsov, S.O., Kamps, J., Rüger, S., Agichtein, E., Segalovich, I., Yilmaz, E. (eds.) ECIR 2013. LNCS, vol. 7814, pp. 392–403. Springer, Heidelberg (2013). doi:10.1007/978-3-642-36973-5_33
3. Dong, X., Halevy, A., Madhavan, J.: Reference reconciliation in complex information spaces. In: SIGMOD 2005, pp. 85–96 (2005)
4. Finkel, J.R., Grenager, T., Manning, C.: Incorporating non-local information into information extraction systems by Gibbs sampling. In: ACL 2005, pp. 363–370 (2005)
5. Meusel, R., Petrovski, P., Bizer, C.: The Webdatacommons microdata, RDFa and microformat dataset series. In: Mika, P., Tudorache, T., Bernstein, A., Welty, C., Knoblock, C., Vrandečić, D., Groth, P., Noy, N., Janowicz, K., Goble, C. (eds.) ISWC 2014. LNCS, vol. 8796, pp. 277–292. Springer, Heidelberg (2014). doi:10.1007/978-3-319-11964-9_18
6. On, B., Koudas, N., Lee, D., Srivastava, D.: Group linkage. In: ICDE 2007, pp. 496–505 (2007)
7. Oren, E., Delbru, R., Catasta, M., Cyganiak, R., Stenzhorn, H., Tummarello, G.: Sindice.com: a document-oriented lookup index for open linked data. IJMSO **3**(1), 37–52 (2008)
8. Otero-Cerdeira, L., Rodríguez-Martínez, F.J., Gómez-Rodríguez, A.: Ontology matching: a literature review. Expert Syst. Appl. **42**(2), 949–971 (2015)
9. Pound, J., Mika, P., Zaragoza, H.: Ad-hoc object retrieval in the web of data. In: WWW 2010, pp. 771–780 (2010)

10. Tonon, A., Catasta, M., Demartini, G., Cudré-Mauroux, P., Aberer, K.: Trank: ranking entity types using the web of data. In: ISWC 2013, pp. 640–656 (2013)
11. Tonon, A., Demartini, G., Cudré-Mauroux, P.: Combining inverted indices and structured search for ad-hoc object retrieval. In: Proceedings of the 35th International ACM SIGIR Conference on Research and Development in Information Retrieval, SIGIR 2012, pp. 125–134. ACM, New York (2012)
12. Tummarello, G., Cyganiak, R., Catasta, M., Danielczyk, S., Delbru, R., Decker, S.: Sig.ma: live views on the web of data. JWS 8(4), 355–364 (2010)

AUFX-O: Novel Methods for the Representation of Audio Processing Workflows

Thomas Wilmering[✉], György Fazekas, and Mark B. Sandler

Centre for Digital Music (C4DM),
Queen Mary University of London, London E1 4NS, UK
{t.wilmering,g.fazekas,mark.sandler}@qmul.ac.uk

Abstract. This paper introduces the Audio Effect Ontology (AUFX-O) building on previous theoretical models describing audio processing units and workflows in the context of music production. We discuss important conceptualisations of different abstraction layers, their necessity to successfully model audio effects, and their application method. We present use cases concerning the use of effects in music production projects and the creation of audio effect metadata facilitating a linked data service exposing information about effect implementations. By doing so, we show how our model facilitates knowledge sharing, reproducibility and analysis of audio production workflows.

1 Introduction

Audio effect devices are essential tools in contemporary music production and composition and are used to enhance the perceived quality of an audio signal, as well as for creative sound design. They can be defined as signal processing techniques used to modify an audio signal in a controlled way. Audio effect devices can take the form of hardware units or software applications, both of which find widespread use in professional and home studios. Effects usually exhibit an interface that allows setting parameters of the implementation, thereby adjusting the behaviour of the signal processing mechanism. Detailed descriptions of effect types and their implementation can be found in [11]. In light of the popularity of Web services for music production and consumption, significant added value may be provided by a thorough description of production workflows using well defined concepts and properties. Furthermore, multimedia and audio specifically constitutes a significant part of the Web. This should be made better accessible and represented on the Semantic Web. Recommendation systems and cultural archives consuming Linked Data about music production will require the transparent, interoperable provenance metadata this ontology enables.

This paper introduces the Audio Effect Ontology (AUFX-O)[1] for the description of audio signal processing units and workflows in the context of music production [8]. The ontology builds on previous work, such as the Music Ontology framework [6] and borrows ideas from the Vamp Plugins Ontology for

---

[1] Specification at https://w3id.org/aufx/ontology/1.0#.

© Springer International Publishing AG 2016
P. Groth et al. (Eds.): ISWC 2016, Part II, LNCS 9982, pp. 229–237, 2016.
DOI: 10.1007/978-3-319-46547-0_24

the description of feature extraction plugins and transformations[2]. It extends
the Studio Ontology which is designed to capture the work of the audio engineer and producer in the studio [2]. It enables a detailed description of audio
effects, audio effects implementations and their application in music production
projects, including domain-specific provenance information based on the generalised model of the Provenance Ontology (PROV-O) [4]. This present work
extends and refines previous theoretical models [10] and describes the first published version of the ontology. The need for layered conceptualisation of audio
effects to support the description and analysis audio production workflows is
discussed, followed by use cases concerning the application of effects in music
production and their description in linked data services. Finally we conclude and
present future directions.

2   The Audio Effect Ontology

2.1   Conceptual Model of Audio Effects

The representation of effects and effect implementations in the Audio Effect
Ontology closely follows the conceptualisation of devices in the Device Ontology
within the Studio Ontology framework [2]. An audio effect is conceptualised as
some perceived effect of a physical process. For example, the reflection of sound
waves (echo) produces an audio effect. This physical phenomenon is the most
abstract element of our model. Its level of abstraction similar to the abstract
concept of intellectual works in the Functional Requirements for Bibliographic
Records (FRBR) [5]. An algorithm that reproduces the effect can be described as
a model that approximates the perceived effect. An echo effect, for instance, can
be modelled by various delay line arrangements, comparable with how a work
may be realised through several expressions. We define the level of abstraction of
a model's implementation as being analogous with that of manifestations. Different implementations of a model in our domain may be an algorithm implemented
in different programming languages, or a circuit design implemented using discrete components. On the most concrete level, an instance of an implementation
represents a concrete device, for example an audio effect plugin running on a
specific computer. Using this model we can describe concrete instances of audio
effect devices, linked to an actual implementation of a model or algorithm that
represents a physical phenomenon. Figure 1 shows these main concepts and the
four levels of abstractions.

AUFX-O introduces additional concepts to describe a product for the distribution of a given audio effect implementation published by an individual or a
software company. As opposed to the solution previously employed in the Device
Ontology where product names are expressed as string literals linked to individual devices, in AUFX-O specific concepts for products and product families can
be linked to implementations. This offers increased expressivity and allows the
description of audio effects in a database that assigns URIs to products and

---

[2] http://vamp-plugins.org/rdf/.

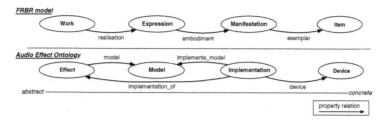

Fig. 1. Entities and relationships in the model of audio effect devices in the Audio Effect Ontology. The levels of abstraction are comparable to those of the FRBR model.

different versions of a product unified in a product family, independently from instantiations of devices in a music production scenario. Implementations and products need to be clearly distinguished – different versions of a *product*, for instance, updates of audio effect software or versions targeting different operating systems may exhibit different characteristics and functionality, but are nevertheless marketed under the same name.

2.2   Modelling Details

AUFX-O is designed with two main areas of application in mind: *(1)* the description of audio effects on the implementation level and *(2)* the description of audio signal transformations on the device level. Modelling decisions draw from the authors' expertise in the domain and specific use cases, including collecting, sharing and querying production metadata in digital audio workstations (DAWs) and web-based tools for linking and retrieving such data. Data-driven evaluation [1] of the ontology was presented in our earlier work [10] showing good lexical fit using a text corpus consisting of product descriptions of over 3000 effect implementations.

Figure 2 shows an overview of how implementations and transformations are modelled in AUFX-O. The `Implementation` is linked directly to the `Effect`, skipping the `Model` layer with the term `implementation_of`. This property has been introduced to enable efficient querying. In many cases, especially for commercial effect implementations, the model or algorithm employed is not made public. An important factor in the description of an effect implementations is the precise representation of the variable parameters exposed to the user. These parameters allow the user to alter the behaviour of a signal processing device, i.e. to adjust the sound output within the constraints imposed by the parameter range. In AUFX-O, an `Implementation` can be linked to a `Parameter` using the property `has_parameter`. To describe a parameter's value range, we reuse terms from the Quantities, Units, Dimensions and Data Types Ontology (QUDT) [3]. The terms `minimum_value` and `maximum_value` are subsumed by the property `qudt:value`, enabling the explicit, unambiguous description of parameter values.

The `Transform` class represents a signal transformation performed by a `Device` which stands for a concrete instance of an `Implementation`. To describe the device state at the time of transformation, we reuse the device:State concept of the Device ontology. The settings of variable parameters are represented by a ParameterSetting linked to a given Parameter. The value of the setting is specified using the qudt:value property of QUDT. An audio effect performs a signal transformation, i.e., it takes an input signal and produces a new signal, the output signal, by applying the effect in a controlled way. The terms input_signal and output_signal link the Transform to instances of mo:Signal of the Music Ontology.

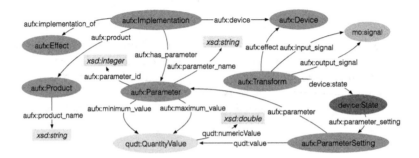

Fig. 2. Selected concepts for the description of implementation characteristics and signal transformations, reusing concepts of the Music Ontology (mo), Device Ontology (device) and QUDT (qudt).

In addition to the properties specifying parameter values, multiple other terms of AUFX-O are subsumed by concepts of existing ontologies. The four abstraction layers for the conceptualisation of audio effects from physical phenomenon to concrete device is based on the Device ontology (see Sect. 2.1). The AUFX-O classes Effect, Model, Implementation and Device are subsumed by respective terms of the Device Ontology. To increase flexibility of AUFX-O, it

Table 1. Core classes and properties of AUFX-O subsumed by concepts of the Studio/Device Ontology (studio/device), QUDT (qudt) and PROV-O (prov).

AUFX class	subclass_of	AUFX property	subproperty_ of
Effect	device:Phenomenon	device	device:exemplar
Model	device:Model	effect	studio:effect,
Implementation	device:Implementation		prov:wasAssociatedWith
Device	device:Device,	input_signal	device:consumed_signal,
	prov:Agent		prov:used
Transform	prov:Activity	output_signal	device:generated_signal,
			prov:generated
		minimum_value	qudt:value
		maximum_value	qudt:value

defines inverse property relations to link instances of these classes. Several concepts are linked to PROV-O to enable interoperable provenance metadata. For instance, a `Device` is defined as a `prov:Agent` while `Transform` is a subclass of `prov:Activity`. Table 1 provides an overview of some core terms and their relations to existing ontologies. Data-driven evaluation of AUFX-O measuring "fit" to a corpus of audio effect related documents with 246544 stemmed words and 8023 unique stems is given in [10].

3   Use Cases

3.1   Audio Effects Product Database on the Semantic Web

Linked data about audio effects may benefit music production professionals and amateurs alike. For instance, linking effect implementations by their available parameters or sonic properties can help finding the right effect for a production scenario. Linking effects and parameter settings to released audio has several applications too, for example in music education. Albeit there are already existing online databases of audio effects, these do not provide standard query end points and do not rely on a common, interoperable conceptualisation and machine readable representation of effect data.

A linked data service using AUFX-O exposes metadata about audio effect implementations and a Web application for data entry and retrieval [9]. An overview of the database and its use is shown in Fig. 3. The data is exposed via a triple store that can be accessed by software agents. Content of the database is initially obtained by parsing Web resources and audio effect binaries that expose information about their structure. This is translated to RDF. For instance, unstructured data retrieved from the KVR audio effect database[3] is mapped onto AUFX-O. The KVR database is accessible through a Website and contains descriptions of over 3500 digital audio effect products, including developer and product name, plugin APIs, operating systems, pricing, and product information in text form. Effects are classified by user-generated tags indicating the effect type, which can be mapped to AUFX-O using the `kvr_tag` property. Listing 1 shows the partial description of an effect implementation. It includes details such as product name and versioning information, as well as the plugin API and parameter description. AUFX-O is capable of representing the information in KVR, however, it also allows for more fine-grained semantic descriptions using the terms discussed in Sect. 2.2.

This linked data service enables novel queries that automate manually performed tasks. An audio engineer may want run simple queries finding audio effects with a given plugin API, in order to identify effect implementations compatible with a given studio setup. Furthermore, users and software agents can find similar audio effects based on their parameter setup and specific tags for the automatic classification of effects following specified implementation characteristics. The detailed description of effect parameters enables more complex

---

[3] http://www.kvraudio.com/plugins/effects/.

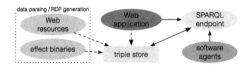

Fig. 3. High level overview of the audio effects database system

```
:ex_implementation a :Implementation ; :plugin_api [ a :AudioUnit ] ;
:ex_product [ :product_name "exDelayFx"^^xsd:string ;
 :kvr_tag "Delay / Echo"^^xsd:string ; version "1.0"^^xsd:string ] ;
:has_parameter [ a :NumericParameter ;
 rdfs:label "delay time"^^xsd:string ;
 :maximum_value [ qudt:numericValue "18.0" ; qudt:unit unit:Decibel ] ;
 :minimum_value [ qudt:numericValue "-18.0" ; qudt:unit unit:Decibel ] ;
 :parameter_id "12"^^xsd:integer ] .
```

Listing 1. Partial Description of an audio effect implementation.

queries with regards to the capabilities of effect implementations. For instance, the SPARQL query in Listing 2 retrieves Echo effect implementations that are compatible with Audio Unit hosts and allow effect settings delaying the incoming signal by at least 1000 ms. In this particular example, the returned data contains the publisher and product name. Integrated in a DAW, this database can be the basis for an effect recommender system. This may return implementations that are available to the user, i.e., plugins that are actually installed on the system, or may refer the user to online sources, e.g., companies offering similar effects.

```
SELECT ?publisher ?name WHERE {
  ?x a :Implementation ; :plugin_api [ a :AudioUnit ] ;
    :has_parameter [ rdfs:label"delay time";
      :maximum_value [ qudt:numericValue ?max ; qudt:unit unit:Millisecond ] ;
      :product [ dc:publisher [ fc:name ?publisher ] ;
      :product_name ?name ; :kvr_tag"Delay / Echo"^^xsd:string ] .
  FILTER ( ?max >= 1000 ) }
```

Listing 2. SPARQL query retrieving echo effect implementations with a delay time parameter that ranges above 1000 ms.

3.2   Semantic Metadata in Digital Audio Workstations

Audio effects used in music production are usually controlled by parameter settings linked to variables in an effect algorithm that alter low-level properties of the audio signal. Configuring an audio effect to achieve a desired outcome often requires in-depth technical knowledge about the algorithms. Music producers and musicians however often describe sound transformations using semantic descriptors such as *warm, bright* or *harsh*. These descriptors cannot always be mapped to effect parameters and may have a nontrivial relationship to the settings. The Semantic Audio Feature Extraction (SAFE) project[4] aims to address

---

[4] http://semanticaudio.co.uk.

this by interpreting the relationship between parameter settings, audio signal features and semantic descriptors to provide a queryable database of signal transformations associating descriptors with effect settings [7].

Specialised audio effect plugins have been developed for the project to gather data including effect parameters, changes in audio features, details about the music as well as the user including the intended sonic outcome (e.g. warm). A triple store is used on the server side to collect this information using ontologies including AUFX-O (see Fig. 4). Here, the `Transform` becomes a core concept to link crucial metadata elements. Querying the SAFE data facilitates novel ways for finding effect settings using semantic descriptors. A producer for instance can find an equaliser setting for processing a guitar track with the intention of giving it a warmer sound. The SPARQL query in Listing 3 retrieves equaliser settings associated with this term, as well as the instrument and music genre, enabling effect parameters to be set solely on the basis of high-level descriptions. Effect settings may also be retrieved by specifying expected change in the audio features.

At the time of writing, the publicly available effect plugins have been downloaded over 4500 times. A triple store collecting data from effect application enables sharing data among a community of users, both music producers and researchers. A dataset of over 240 million triples describing several thousand effect transforms and audio analysis results will be published in the near future.

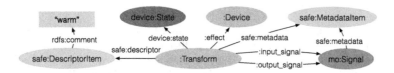

Fig. 4. Data model for the description of signal transformations in the SAFE project.

```
SELECT ?descriptor ?paramName ?paramValue WHERE {
  ?transform :effect [ :device_of
    aufxdb:implementation_SFEQ_VST_1.32 ] ;
      :input_signal ?inputSignal ; device:state ?state ;
      safe:Descriptor [ rdfs:comment ?descriptor ] .
  FILTER REGEX(?descriptor, "warm", "i").
  ?inputSignal safe:Metadata [ rdfs:label "genre" ; rdfs:comment ?genre ] ;
    safe:Metadata [ rdfs:label "instrument" ; rdfs:comment ?instrument ] .
  FILTER REGEX(?genre, "reggae", "i") .
  FILTER REGEX(?instrument, "guitar", "i") .
  ?state :parameter_setting [ :parameter [ qudt:value ?paramValueNode ;
                                           :parameter_id ?paramId ] ] .
  ?paramValueNode qudt:numericValue ?paramValue .
  aufxdb:implementation_SFEQ_VST_1.32 :has_parameter [
    :parameter_id ?paramId ; rdfs:label ?paramName ] .
} ORDER BY ?transform ?paramId
```

Listing 3. SPARQL query to retrieve equaliser parameter settings being used to alter a guitar sound in the reggae genre, semantically described as *warm*.

4   Conclusions and Future Work

This paper discusses the newly published Audio Effect Ontology for describing of audio effects, their implementations and their use in music production. Through several use cases we have shown that detailed metadata about audio effects provided as Linked Data facilitates novel applications for the description and analysis music production workflows, including the search and retrieval of audio effect implementations. Educational tools focusing on audio production and engineering may also benefit from AUFX-O. The modelling principles for effect parameters and settings are applicable in other domains where device settings for specific processes are of interest, e.g. for signal processing devices in other media domains or laboratory devices in scientific experiments.

Future work includes the publication of effect classification vocabularies extending AUFX-O, both for audio effect types and parameter types. These vocabularies combine low-level descriptors for technical classification targeted to audio engineers, and high-level descriptors and auditory perceptual attributes which may be preferred by musicians and composers. Well-defined conceptualisations of effect and parameter types can further improve audio effect related semantic descriptions and enable tasks such as querying for similar audio effects based on their types or parameters. The Studio Ontology and AUFX-O may also be used to support intelligent music production tools, such as Web-based audio workstations capable of storing and interpreting semantic metadata to inform the production process.

Acknowledgments. This paper is supported by EPSRC Grant EP/ L019981/1, Fusing Audio and Semantic Technologies for Intelligent Music Production and Consumption and the European Commission H2020 research and innovation grant AudioCommons (688382). Mark B. Sandler acknowledges the support of the Royal Society as a recipient of a Wolfson Research Merit Award.

References

1. Brewster, C., Alani, H., Dasmahapatra, S., Wilks, Y.: Data driven ontology evaluation. In: Proceedings of the International Conference on Language Resources and Evaluation, Lisbon, Portugal (2004)
2. Fazekas, G., Sandler, M.B.: The studio ontology framework. In: Proceeding of the 12th International Society for Music Information Retrieval Conference (ISMIR) (2011)
3. Hodgson, R., Keller, P.J., Hodges, J., Spivak, J.: QUDT - quantities, units, dimensions and data types ontologies (2014). http://www.qudt.org/
4. Lebo, T., Sahoo, S., McGuiness, D.: PROV-O: the PROV ontology. W3C recommendation, 30 April 2013. World Wide Web Consortium (2013)
5. Plassard, M.F. (ed.): Functional Requirements For Bibliographic Records : final report/IFLA Study Group on FRBR, vol. 19. K.G. Saur (1998)
6. Raimond, Y., Abdallah, S., Sandler, M., Giasson, F.: The music ontology. In: Proceeding of the International Conference on Music Information Retrieval (ISMIR), Vienna, Austria (2007)

7. Stables, R., Enderby, S., Man, B.D., Fazekas, G., Reiss, J.G.: SAFE: a system for the extraction and retrieval of semantic audio descriptors. In: 15th International Society for Music Information Retrieval Conference (ISMIR), Taipei, Taiwan (2014)
8. Wilmering, T., Fazekas, G.: AUFX-O: the audio effect ontology (2016). http:// isophonics.net/content/aufx
9. Wilmering, T., Fazekas, G., Allik, A., Sandler, M.B.: Audio effects data on the semantic web. In: 139th Audio Engineering Society Convention, New York, USA (2015)
10. Wilmering, T., Fazekas, G., Sandler, M.B.: Semantic metadata for music production projects. In: Proceeding of the 1st International Workshop on Semantic Music and Media (SMAM), Sydney, Australia (2013)
11. Zölzer, U. (ed.): DAFX - Digital Audio Effects, 2nd edn. Wiley, New York (2011)

Applications

Translating Ontologies in Real-World Settings

Mihael Arcan[1], Mauro Dragoni[2(✉)], and Paul Buitelaar[1]

[1] Insight Centre for Data Analytics, National University of Ireland, Galway, Ireland
{Mihael.Arcan,Paul.Buitelaar}@insight-centre.org
[2] FBK- Fondazione Bruno Kessler, Via Sommarive 18, 38123 Trento, Italy
dragoni@fbk.eu

Abstract. To enable knowledge access across languages, ontologies that are often represented only in English, need to be translated into different languages. The main challenge in translating ontologies is to disambiguate an ontology label with respect to the domain modelled by ontology itself. Machine translation services may help in this task; however, a crucial requirement is to have translations validated by experts before the ontologies are deployed. For this reason, real-world applications must implement a support system addressing this task to relieve experts in validating all translations. In this paper we present the Expert Supporting System for Ontology Translation, called ESSOT, which exploits the semantic information of the label's context for improving the quality of label translations. The system has been tested within the Organic.Lingua project by translating the ontology labels in three languages. In order to evaluate further the effectiveness of the system on handling different domains, additional ontologies were translated and evaluated. The results have been compared with translations provided by the Microsoft Translator API and the improvements demonstrate a better performance of the proposed approach for automatic ontology translation.

1 Introduction

Nowadays, most of the semantically structured data, i.e. ontologies or taxonomies, have labels stored in English only. Although the increasing amount of ontologies offers an excellent opportunity to link this knowledge together, non-English users may encounter difficulties when using the ontological knowledge represented in English only [1]. Furthermore, applications in information retrieval or knowledge management, using monolingual ontologies are limited to the language in which the ontology labels are stored. Therefore, to make ontological knowledge accessible beyond language borders, these monolingual resources need to be enhanced with multilingual information [2].

Another important reason to translate ontologies is that they may already exist in different languages, but without aligning the concepts across languages we are not able to combine, compare or extend them. Furthermore, government institutions may be obliged to publish their ontologies or other structured data in their native language, e.g. financial reports need to be written in the language in which the financial institution operates. Therefore, performing ontology based data analytics would fail on providing reports in German, Spanish,

© Springer International Publishing AG 2016
P. Groth et al. (Eds.): ISWC 2016, Part II, LNCS 9982, pp. 241–256, 2016.
DOI: 10.1007/978-3-319-46547-0_25

or other official European languages, if we would use only existing ontologies in English. Additionally, medical ontologies, e.g. the ICD Ontology[1] can be used to standardize medical reports, but physicians, researchers or patient organizations will access these reports in their native language; therefore, only a cross-lingual aligned ontology may give an appropriate overview. Another example is the Europeana project,[2] in the heritage domain, where preservation of the cultural treasure of Europe shows the need of cross-lingual alignment of different resources.

Since manual multilingual enhancement of domain-specific ontologies is very time consuming and expensive, we engage a domain-aware statistical machine translation (SMT) system to automatically translate the ontology labels. As ontologies may change over time; having in place an SMT system adaptable to an ontology can therefore be very beneficial. Nevertheless, the quality of the SMT generated translations relies strongly on the translation model learned from the information stored in parallel corpora. In most cases, the inference of translation candidates cannot always be learned accurately when specific vocabulary, like ontology labels, appears infrequent in a parallel corpus. Additionally, ambiguous labels built out of only a few words do not always express enough semantic information to guide the SMT system in translating a label correctly in regards to the targeted domain. This can be observed in domain-unadapted SMT systems, e.g. Microsoft Translator,[3] where an ambiguous expression, like *vessel* stored in a medical ontology, is translated as *Schiff*[4] (en. *ship*) in German, but not into the targeted medical domain as *Gefäß*.

In this paper, we present ESSOT, a collaborative knowledge management platform with a domain-aware SMT system for supporting language experts in the task of translating ontologies. The benefits of such a platform are *(i)* the possibility of having an all-in-one solution, containing both an environment for modelling ontologies which enables the collaboration between different type of experts and *(ii)* a pluggable domain-adaptable service for supporting ontology translations. The proposed solution has been validated in two different settings: (i) in a real-world context, namely Organic.Lingua,[5] from quantitative and qualitative points of view, and, quantitatively only, (ii) on a set of ontologies aiming to evaluate the effectiveness of the SMT service in different domains.

The paper is structured as follow. Section 2 introduces the context of our main use case, the Organic.Lingua project. Section 3 provides an overview of the ESSOT architecture; while, Sect. 4 focuses on user facilities implemented for supporting the ontology translation task. In Sect. 5 we reported the evaluation conducted on both the translations suggested by the service and the usability of the platform. Finally, Sect. 6 provides a general overview about ontology translation and knowledge management tool; while, Sect. 7 concludes the paper.

---

[1] http://www.who.int/classifications/icd/en/.

[2] http://www.europeana.eu/portal/.

[3] http://www.bing.com/translator/.

[4] Translation performed on 16.04.2016.

[5] http://www.organic-lingua.eu.

2   The Organic.Lingua Project

Organic.Lingua is an EU-funded project that aims at providing automated multilingual services and tools facilitating the discovery, retrieval, exploitation and extension of digital educational content related to Organic Agriculture and AgroEcology. More concretely, the project aims at providing, on top of a web portal, cross-lingual facility services enabling users to *(i)* find resources in languages different from the ones in which the query has been formulated and/or the resource described (e.g., providing services for cross-lingual retrieval); *(ii)* manage meta-data information for resources in different languages (e.g., offering automated meta-data translation services); and *(iii)* contribute to evolving content (e.g., providing services supporting the users in content generation).

These objectives are reached in the Organic.Lingua project by means of two components: on the one hand, a web portal offering software components and linguistic resources able to provide multilingual services and, on the other hand, a conceptual model (formalized in the "Organic.Lingua ontology") used for managing information associated with the resources provided to the final users and shared with other components deployed on the Organic.Lingua platform. In a nutshell, the usage of the Organic.Lingua ontology is twofold:

- Resource annotation: each time a content provider inserts a resource in the repository, the resource is annotated with one or more concepts extracted from the ontology. The list of available concepts is retrieved by using an ontology service deployed in the ontology management component. Then, this list is exploited for annotating the learning resources published on the Web portal.
- Resource retrieval: when web users perform queries on the system, the ontology is used, by the back-end information retrieval system, to perform advanced searches based on semantic techniques. Moreover, the Web portal is equipped with a graphical semantic tree that exploits the content of the ontology for facilitating the browsing of the resource repository classification. Finally, the ontology is used also by the Cross-Language Information Retrieval component for query expansion purposes.

Due to this intensive use of the ontology in the entire Organic.Lingua portal, the accuracy of the linguistic layer, represented by the set of translated labels, is crucial for supporting the annotation and retrieval functionalities. The maintenance of such an accuracy requires a precise methodology, and dedicated tools, for avoiding the loss of effectiveness of the components deployed on the platform.

3   Platform Architecture

In this Section, we present a general overview of the platform for managing the life-cycle of translating ontological entities. Figure 1 shows the architecture diagram, where we distinguish two main blocks:

- the service-side: containing the components for the machine translation models used for suggesting translations when requests are performed by users; and,

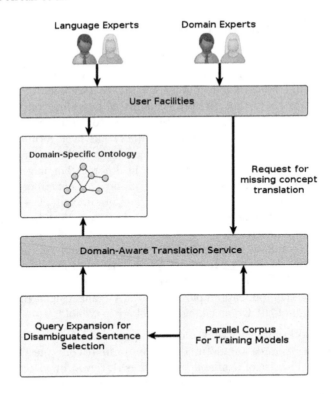

Fig. 1. Diagram of the overall architecture of the ESSOT platform

– the user-side: containing the facilities implemented for supporting experts in managing the multilingual layer of ontologies.

The service side contains the components used for creating and updating the model used by the domain-aware machine translation service. Such components are described below, while in Sect. 4 we provide a description of the facilities implemented for supporting users in managing ontologies.

Statistical Machine Translation. Our approach is based on statistical machine translation, where we wish to find the best translation **e** of a source string **f**, given by a log-linear model combining a set of features. The translation that maximizes the score of the log-linear model is obtained by searching all possible translations candidates. The decoder, which functions as a search procedure, provides the most probable translation based on a statistical translation model learned from sentence aligned corpora.

For a broader domain coverage of datasets necessary to train an SMT system, we merged several parallel corpora, e.g. JRC-Acquis [3], Europarl [4], DGT (translation memories generated by the *Directorate-General for Translation*) [5], MultiUN corpus [6] and TED talks [7] among others, into one parallel dataset. For the translation approach, we engage the widely used Moses toolkit [8]. Word alignments were built with GIZA++ [9] and a 5-gram language model was build with KenLM [10].

Query Expansion for Sentence Selection. Due to the shortness of ontology labels, there is a lack of contextual information, which can otherwise help disambiguating short or ambiguous expressions. Therefore, our goal is to translate the identified ontology labels within the textual context of the targeted domain, rather than in isolation. With this selection approach, we aim to retain relevant sentences, where the English label *vessel* or *injection* belongs to the medical domain, but not to the technical domain. This process reduces the semantic noise in the translation process, since we try to avoid contextual information that does not belong to the domain of the targeted ontology.

Due to the specificity of the ontology labels, just an *n-gram overlap* approach is not sufficient to select all the useful sentences. For this reason, we follow the idea of [11], where the authors extend the semantic information of ontology labels using Word2Vec[6] for computing distributed representations of words. The technique is based on a neural network that analyses the textual data provided as input and outputs a list of semantically related words [12]. Each input string, in our experiment ontology labels or source sentences, is vectorized using the surrounding context and compared to other vectorized sets of words in a multi-dimensional vector space. Word relatedness is measured through the cosine similarity between two word vectors. A score of 1 would represent a perfect word similarity; e.g. *cholera* equals *cholera*, while the medical expression *medicine* has a cosine distance of 0.678 to *cholera*. Since words, which occur in similar contexts tend to have similar meanings [13], this approach enables to group related words together.

The usage of the ontology hierarchy allows us to further improve the disambiguation of short labels, i.e., the related words of a label are concatenated with the related words of its direct parent. Given a label and a source sentence from the used concatenated corpus, related words and their weights are extracted from both of them, and used as entries of the vectors to calculate the cosine similarity. Finally, the most similar source sentence and the label should share the largest number of related words.

4   Supporting Users in the Ontology Translation Activity

The ESSOT system has been equipped with facilities supporting the collaborative translation of domain-specific ontologies in order to satisfy the requirements of the ontology translation task from a user perspective.

Concerning users, we identified two distinct groups: the Domain Experts and the Language Experts. Domain Experts are in charge of the modelling aspect of ontologies (i.e. creation of concepts, individuals, properties, and the relationships between them); while Language Experts are responsible of managing the labels associated with each entity by evaluating their correctness and, eventually, by

---

[6] https://code.google.com/p/word2vec/.

providing a more fine-grained adaptation with respect to the domain described by the ontology.

Below, we present the list of the implemented facilities specifically designed for supporting the management of the multilingual layer of ontologies. Here, we focused on the interface that have been specifically implemented for managing the Organic.Lingua ontology.[7] However, such facilities can be adopted, in general, for managing the multilingual aspect of any ontology.

Domain And Language Experts View. The page dedicated to the management of an ontology label, specifically designed for the Domain and Language Experts, has been equipped with functionalities that permits revisions of the linguistic layer. This set of functionalities permits to revise translations of names and descriptions of each entity (concepts, individuals, and properties).

For facilitating the browsing and the editing of the translations, a quick view box has been inserted into the mask (as shown in Fig. 2); in this way, language experts are able to navigate through the available translations and, eventually, invoke the translation service for retrieving a suggestion or, alternatively, to edit the translation by themselves (Fig. 3).

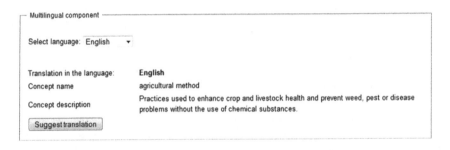

Fig. 2. Multilingual box for facilitating the entity translation

Approval And Discussion Facilities. Given the complexity of translating domain specific ontologies, translations often need to be checked and agreed upon by a community of experts. This is especially true when ontologies are used to represent terminological standards which need to be carefully discussed and evaluated. To support this collaborative activity we foresee the usage of a wiki-style paradigm [14], expanded with the possibility of assigning specific translations of ontology labels to specific experts who need to monitor, check, and approve the suggested translations. This customization promotes the management of the changes carried out on the ontology (in both layers) by providing the facilities necessary to manage the life-cycle of each change.

These facilities may be split in two different sets of features. The first group may be considered as a monitor of the activities performed on each entity page.

---

[7] A read-only version, but with all functionalities available, of the MoKi instance described in this paper is available at https://dkmtools.fbk.eu/moki/3_5/essot/.

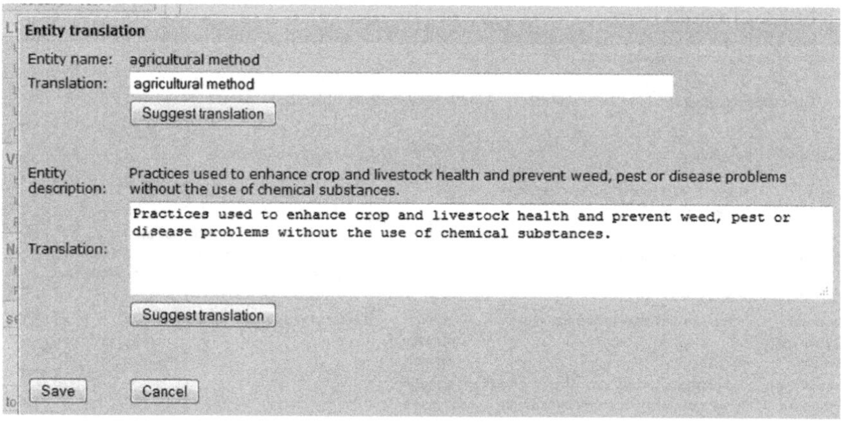

Fig. 3. Quick translation box for editing label translations

When changes are committed, approval requests are created. They contain the identification of the expert in charge of approving the change, the date on which the change has been performed, and a natural language description of the change. Moreover, a mechanism for managing the approvals and for maintaining the history of all approval requests for each entity is provided. Instead, the second set contains the facilities for managing the discussions associated with each entity page. A user interface for creating the discussions has been implemented together with a notification procedure that alerts users when new topics/replies, related to the discussions that they are following, have been posted.

"Quick" Translation Feature. For facilitating the work of language experts, we have implemented the possibility of comparing side-by-side two lists of translations. This way, the language expert in charge of revising the translations, avoiding to navigate among the entity pages, is able to speed-up the revision process.

Figure 4 shows such a view, by presenting the list of English concepts with their translations into Italian. At the right of each element of the table a link is placed allowing to invoke a quick translation box (as shown in Fig. 3) that gives the opportunity to quickly modify information without opening the entity page. Finally, in the last column, a flag is placed indicating that changes have been performed on that concept, and a revision/approval is requested.

5   Evaluation

Our goal is evaluating the usage and the usefulness of the ESSOT user facilities and the underlying service for suggesting domain-adapted translations.

In detail, we are interested in answering two main research questions:

RQ1 Does the proposed system provide an *effective* support, in terms of the quality of suggested translations, to the management of multilingual ontologies?

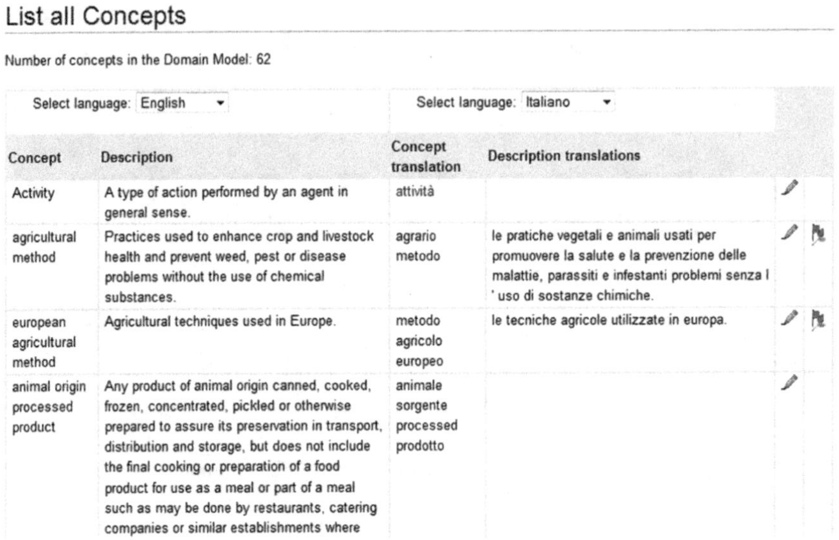

List all Concepts

Number of concepts in the Domain Model: 62

Select language: English ▾   Select language: Italiano ▾

Concept	Description	Concept translation	Description translations	
Activity	A type of action performed by an agent in general sense.	attività		✎
agricultural method	Practices used to enhance crop and livestock health and prevent weed, pest or disease problems without the use of chemical substances.	agrario metodo	le pratiche vegetali e animali usati per promuovere la salute e la prevenzione delle malattie, parassiti e infestanti problemi senza l' uso di sostanze chimiche.	✎ 🖉
european agricultural method	Agricultural techniques used in Europe.	metodo agricolo europeo	le tecniche agricole utilizzate in europa.	✎ 🖉
animal origin processed product	Any product of animal origin canned, cooked, frozen, concentrated, pickled or otherwise prepared to assure its preservation in transport, distribution and storage, but does not include the final cooking or preparation of a food product for use as a meal or part of a meal such as may be done by restaurants, catering companies or similar establishments where	animale sorgente processed prodotto		✎

Fig. 4. View for comparing label translations

RQ2 Do the ESSOT functionalities provide an *effective* support to the collaborative management of a multilingual ontology?

In order to answer these questions, we performed two types of analysis:

1. Qualitative: the tool has been validated in the context of the Organic.Lingua project where we collected subjective judgements from the language experts. They have been involved in the evaluation of the tool on the general usability of the components and by providing feedback for future improvements.
2. Quantitative: beside the user evaluation, we collected objective measures concerning the effectiveness of the translations suggested by the embedded machine translation service. This information allows to have an estimation of the effort needed for adapting all translations by the language experts.

5.1   User Evaluation Context

Eleven language experts have been involved in the evaluation of the proposed platform for translating the Organic.Lingua ontology in three different languages: German, Spanish, and Italian. They were all experts of the agricultural domain, therefore, labels used by them have to be considered as a gold standard from the domain point of view. From the mother tongue perspective, the evaluation was performed by three German, four Spanish and four Italian native speaking experts. Most of them had no previous knowledge of the tool, hence an initial phase of training was necessary. The training was organized according to the following steps:

- A one-day overall introduction to the tool.
- A few short, on-line, training sessions with the ESSOT tool guided by ontology and tool experts, targeted to help domain experts to better understand the capabilities of the tool.
- Hands-on usage of the tool: language experts were left to "play" with ESSOT in order to become familiar with the functionalities that they would use during the revision process. This exercise also had the secondary objective to collect doubts and problems encountered by experts.

After the initial training, experts were asked to translate the ontology in the three languages mentioned above. The experts used ESSOT facilities for completing the translation task and, at the end, they provided feedback on the tool support for accomplishing the task. A summary of these findings and lessons learned are presented in Sect. 5.2.

5.2 Qualitative Evaluation Results

To investigate the subjective perception of the eleven experts about the support provided for translating ontologies, we analysed the data collected through a questionnaire. For each functionality described in Sect. 4, we provided the information how often each aspect has been raised by the language experts.

Language Experts View
Pros: Easy to use for managing translations (9)
Usable interface for showing concept translations (3)

Approval And Discussion
Pros: Pending approvals give a clear situation about concept status (4)
Cons: Discussion masks are not very useful (8)

Quick Translation Feature
Pros: Best facility for translating concepts (8)
Cons: Interface design improvable (3)

The results show, in general, a good perception of the implemented functionalities, in particular concerning the procedure of translating a concept by exploiting the quick translation feature. Indeed, 9 out of 11 experts reported advantages on using this capability. Similar opinions have been collected about the language expert view, where the users perceived such a facility as a usable reference for having the big picture about the status of concept translations.

Results concerning the approach and discussion facility are inconclusive. On the one hand, the experts perceived positively the solution of listing approval requests on top of each concept page. This fact is connected with a personalization that we embed into the ESSOT home page. Indeed, after the login, users are able to see the list of pending approvals require their action. This way, it is more easy for them to locate the translations that have to be evaluated and, eventually, to approve or to modify them. On the other hand, we received negative opinions by almost all experts (8 out of 11) about the usability of discussion forms. This result shows us to focus future effort in improving this aspect of the tool.

Finally, concerning the "quick" translation facility, 8 out of 11 experts judged this facility as the most usable way for translating a concept. The main characteristic that has been highlighted is the possibility of performing a "masstranslation" activity without opening the page of each concept, with the positive consequence of saving a lot of time.

5.3 Quantitative Evaluation Results

The automatic evaluation on label translations provided by ESSOT is based on the correspondence between the automatically generated output and reference translations (gold standard), provided by domain and language experts. For the automatic evaluation we used the BLEU [15], METEOR [16] and TER [17] algorithms.

BLEU is calculated for individual translated segments (n-grams) by comparing them with reference translations. Those scores, between 0 and 100 (perfect translation), are then averaged over the whole *evaluation dataset* to reach an estimate the automatically generated translation's overall quality. METEOR is based on the harmonic mean of precision and recall, whereby recall is weighted higher than precision. Along with standard exact word (or phrase) matching it has additional features, i.e. stemming, paraphrasing and synonymy matching. Differently to BLEU, the metric produces good correlation with human judgement at the sentence or segment level. TER is an error metric (lowers scores are better) for machine translation measuring the number of edits required to change a system output into one of the references.

Datasets. To demonstrate the performance of the proposed framework, we use several ontologies coming from different domains:

- The Organic.Lingua ontology contains 291 concepts in the agricultural domain. All concepts within the ontology have been translated into 16 languages. In addition, mappings to Agrovoc and Eurovoc have also been defined.
- The DOAP (Description of a Project) Ontology[8] defines the vocabulary to describe software projects. It was created to convey semantic information associated with free and open source software projects. It holds translations of labels into 6 languages,[9] whereby we use German and Spanish translations as the gold standard, which is compared with the automatically generated labels.
- The Geoskills ontology[10] holds the competencies, topics and educational contexts in five different languages, i.e., English, German, French, Spanish and Dutch.

---

[8] https://github.com/edumbill/doap.
[9] French, Spanish, German, Czech, Portuguese, Japanese.
[10] https://github.com/i2geo/GeoSkills.

- The STW Thesaurus for Economics [18] provides the vocabulary of more than 6,000 standardized subject headings (in English and German) and 20,000 additional entry terms (keywords) belonging to the economical domain. In addition to that, the entries are richly interconnected by 16,000 broader/narrower and 10,000 related relations.
- The Thesaurus for the Social Sciences (TheSoz) [19] enables indexing documents and research information in the social sciences. In overall it stores about 8,000 standardized subject headings in English, German and French.

Table 1. Automatic translation evaluation of the targeted ontologies by the Microsoft Translator API and our proposed system (bold results = best performance)

	English → German			English → Italian			English → Spanish		
Organic.Lingua	BLEU	METEOR	TER	BLEU	METEOR	TER	BLEU	METEOR	TER
Microsoft	3.7	19.6	**95.1**	**13.5**	28.6	87.1	21.0	36.9	73.3
ESSOT	**7.4**	**31.0**	99.1	13.0	**34.2**	**78.8**	**25.7**	**44.4**	**66.7**
DOAPOntology	BLEU	METEOR	TER	BLEU	METEOR	TER	BLEU	METEOR	TER
Microsoft	6.0	24.3	93.3	/	/	/	**20.2**	32.7	81.6
ESSOT	**6.4**	**31.3**	**91.8**	/	/	/	20.1	**35.3**	**72.5**
GeoSkills	BLEU	METEOR	TER	BLEU	METEOR	TER	BLEU	METEOR	TER
Microsoft	11.8	22.3	97.5	/	/	/	16.3	30.1	88.9
ESSOT	**14.0**	**30.7**	**94.6**	/	/	/	**16.7**	**35.9**	**83.0**
STW	BLEU	METEOR	TER	BLEU	METEOR	TER	BLEU	METEOR	TER
Microsoft	6.5	15.1	89.9	/	/	/	/	/	/
ESSOT	**6.8**	**21.7**	**98.8**	/	/	/	/	/	/
TheSoz	BLEU	METEOR	TER	BLEU	METEOR	TER	BLEU	METEOR	TER
Microsoft	12.1	37.0	**88.1**	/	/	/	/	/	/
ESSOT	**20.4**	**39.8**	103.5	/	/	/	/	/	/

We evaluate the automatically generated translations into German, Italian and Spanish provided within ESSOT and the Microsoft Translator API. Since reference translations are needed to evaluate automatically generated translations, we use the translated labels provided by the domain experts as the gold standard.

The Organic.Lingua ontology provides 274 German, 354 Italian and 355 Spanish existing translations out of 404 English labels. As seen in Table 1, the contextual information for label translation used in ESSOT, significantly outperforms[11] the Microsoft Translator API. When translating English labels into German we gain a 51.3% averaged improvement over the commercial system and 51.7% for Spanish. In addition to that, it produces comparable results when translating into Italian (10.5% improvement).

---

[11] The approximate randomization approach in MultEval [20] is used to test whether differences among system performances are statistically significant with a p-value < 0.05.

Besides the evaluation on the Organic.Lingua ontology labels, the multilingual gold standard within the DOAP and GeoSkills ontology enables the evaluation of automatically translated ontology labels into German and Spanish. In detail, the results for the DOAP ontology show similar performance between both translation systems. On the other hand, the results of the GeoSkills ontology labels show statistically significant (p-value < 0.05) improvements over the Microsoft Translator API. For the STW and TheSoz ontology, which enables automatic evaluation of German language, only translations of the TheSoz labels show significant improvements of our system. Although the evaluation metrics show slight improvements when translating the STW ontology, the improvements are not significant.

As a final evaluation, we manually analysed the TheSoz translated labels regarding the most frequent errors of the both translation systems. The first observation is related to compound words, a frequent error class when translating into German. We observed that Microsoft often provided a non-compound translation in German. As an example, labels like *company takeover, working week* or *crime fighting* were translated word by word into German, i.e., *Übernahme der Firma, wöchentliche Arbeitszeit* or *Bekämpfung von Kriminalität*. Although these translations can be seen as correct translations, the provided gold standard in the ontology preferred German compounds *Unternehmensübernahme, Wochenarbeitszeit, Verbrechensbekämpfung*. Besides a small amount of wrong translations (*partnership* into *marriege, translator* into *translator, young worker* into *Junge Arbeitnehmer*), Microsoft's system showed expected problems in disambiguating short expressions. Due to the shortness of the labels, the ontology label *driver* was translated as *Treiber*, which is correct in the IT domain (as *hardware driver*), but not in the targeted domain. Similarly, *stroke*, without contextual information, was translated as *Strich* (en. *line, dash*), although *Schlaganfall* would be the correct translation into German. For these ambiguous labels, our proposed system, which used a disambiguated contextual information, provided correct translations, i.e. *Fahrer* from *driver* or *Schlaganfall* from the English label *stroke*. On the other hand also our system did not always perform best. The largest observed error class were out-of-vocabulary issues, i.e. alignments between source and target language, which were not learned during the SMT training step. For example, TheSoz labels *bonapartism, shamanism, patriciate* or *praxeology*, which are no stored in our translations models, were provided as untranslated words on the target side.

5.4   Findings and Lessons Learned

The quantitative and qualitative results demonstrate the viability of the proposed platform in real-world scenarios and, in particular, its effectiveness in the proposed use case. Therefore, we can positively answer to both research questions, **RQ1**: the back-end component provides helpful suggestions for performing the ontology translation task, and **RQ2**: the provided interfaces are usable and useful for supporting the language experts in the translation activity.

Besides these, there were other insights, either positive and negative, that emerged during the subjective evaluation that we conducted.

The main positive aspect highlighted by the experts was related to the easy and quick way of translating a concept with respect to other available knowledge management tools (see details in Sect. 6), which do not enable specific support for translation. The suggestion-based service allowed effective suggestions and reduced the effort required for finalizing the translation of the ontology. As example, we may consider the Organic.Lingua specific use case, where the time for translating the ontology was reduced from 3.5 h (completely manual translation) to 2.1 h (translation performed with ESSOT). This point confirms the capability of the domain-aware translation service of providing translations adapted to the specific topic of the ontology experts are going to model. However, even if on one hand, the experts perceived such a service very helpful from the point of view of domain experts (i.e. experts that are generally in charge of modeling ontologies but that might not have enough linguistic expertise for translating label properly with respect to the domain), facilities supporting the direct interaction with language experts (i.e. discussion form) should be more intuitive, for instance as the approval one.

The criticism concerning the interface design was reported also about the quick translation feature, where some of the experts commented that the comparative view might be improved from the graphical point of view. In particular, they suggested (i) to highlight translations that have to be revised, instead of using a flag, and (ii) to publish only the concept label instead of putting also the full description in order to avoid misalignments in the visualization of information.

Connected to the quick translation facility, experts judged it as the easiest way for executing a first round of translations. Indeed, by using the provided translation box, experts are able to translate concept information without navigating to the concept page and by avoiding a reload of the concepts list after the storing of each change carried out by the concept translation.

Finally, we can judge the proposed platform as a useful service for supporting the ontology translation task, especially in a collaborative environment when the multilingual ontology is created by two different types of experts: domain experts and language experts. Future work in this direction will focus on the usability aspects of the tool and on the improvement of the semantic model used for suggesting translations in order to further reduce the effort of the language experts. We plan also to extend the evaluation on other use cases.

6   Related Work

In this section, we want to summarize approaches related to the pure ontology translation task and to present a brief review of the most known ontology management tools current available by emphasizing their capabilities in supporting language experts for translating ontologies.

The task of ontology translation involves generating an appropriate translation for the lexical layer, i.e. labels stored in the ontology. Most of the previous related work focused on accessing existing multilingual lexical resources, like EuroWordNet or IATE [21,22]. Their work focused on the identification of the lexical overlap between the ontology and the multilingual resources, which guarantees a high precision but a low recall. Consequently, external translation services like BabelFish, SDL FreeTranslation tool or Google Translate were used to overcome this issue [23, 24]. Additionally, [23, 25] performed ontology label disambiguation, where the ontology structure is used to annotate the labels with their semantic senses. Similarly, [26] show positive effect of different domain adaptation techniques, i.e., using web resources as additional bilingual knowledge, re-scoring translations with Explicit Semantic Analysis, language model adaptation) for automatic ontology translation. Differently to the aforementioned approaches, which rely on external knowledge or services, the machinery implemented in ESSOT is supported by a domain-aware SMT system, which provides adequate translations using the ontology hierarchy and the contextual information of labels in domain-relevant text data. Current frameworks for ontology label translation are accessing directly commercial systems, such as Google Translate or Microsoft Translate, whereby both systems are unable to detect the domain when translating short ambiguous expression, e.g. *vessel, injection, track, head, equity*. In this paper, we demonstrate a platform supporting a machine translation system to translate ontology labels in a domain-specific context.

If we perform a "skimming" of the systems available for ontology management, we identified four of them that may be compared with the capabilities provided by ESSOT : *Neon* [27], *VocBench* [28], *Protégé* [29], and *Knoodl*[12]. However, they do not fully support experts in the specific task of translating ontologies. While the first two, *Neon* and *VocBench*, are the ones more oriented for supporting the management of multilinguality in ontologies by including dedicated mechanisms for modelling the multilingual fashion of each concept; the support for multilinguality provided by *Protégé* and *Knoodl* is restricted to the sole description of the labels. Finally, none of them implements the capability of connecting the tool to an external machine translation system for suggesting translations automatically.

7   Conclusions

This paper presents ESSOT, an Expert Supporting System for Ontology Translation implementing an automatic translation approach based on the enrichment of the text to translate with semantically structured data, i.e. ontologies or taxonomies. ESSOT system integrates a domain-adaptable semantic translation component and a collaborative knowledge management facilities for supporting language experts in the ontology translation activity. The platform has been concretely used in the context of the Organic.Lingua EU project and on a set of multilingual ontologies coming from different domains by demonstrating the effectiveness in the quality of the suggested translations and in the usefulness from the language experts point of view.

Acknowledgments. This publication has emanated from research conducted with the financial support of Science Foundation Ireland (SFI) under Grant Number SFI/12/RC/2289.

---

[12] http://www.knoodl.com.

References

1. Gómez-Pérez, A., Vila-Suero, D., Montiel-Ponsoda, E., Gracia, J., Aguado-de Cea, G.: Guidelines for multilingual linked data. In: Proceedings of the 3rd International Conference on Web Intelligence, Mining and Semantics. ACM (2013)
2. Gracia, J., Montiel-Ponsoda, E., Cimiano, P., Gómez-Pérez, A., Buitelaar, P., McCrae, J.: Challenges for the multilingual web of data. Web Semantics: Science, Services and Agents on the World Wide Web 11 (2012)
3. Steinberger, R., Pouliquen, B., Widiger, A., Ignat, C., Erjavec, T., Tufis, D., Varga, D.: The JRC-Acquis: a multilingual aligned parallel corpus with 20+ languages. In: Proceedings of the 5th International Conference on Language Resources and Evaluation (LREC) (2006)
4. Koehn, P.: Europarl: a parallel corpus for statistical machine translation. In: Conference Proceedings: The Tenth Machine Translation Summit, AAMT (2005)
5. Steinberger, R., Ebrahim, M., Poulis, A., Carrasco-Benitez, M., Schlüter, P., Przybyszewski, M., Gilbro, S.: An overview of the European union's highly multilingual parallel corpora. Lang. Res. Eval. 48(4), 679–707 (2014)
6. Eisele, A., Chen, Y.: Multiun: A multilingual corpus from united nation documents. In: Tapias, D., Rosner, M., Piperidis, S., Odjik, J., Mariani, J., Maegaard, B., Choukri, K. (Chair), Calzolari, N.C., (eds.) Proceedings of the Seventh conference on International Language Resources and Evaluation, European Language Resources Association (ELRA), vol. 5, pp. 2868–2872 (2010)
7. Cettolo, M., Girardi, C., Federico, M.: Wit³: Web inventory of transcribed and translated talks. In: Proceedings of the 16th Conference of the European Association for Machine Translation (EAMT), Trento, Italy, pp. 261–268, May 2012
8. Koehn, P., Hoang, H., Birch, A., Callison-Burch, C., Federico, M., Bertoldi, N., Cowan, B., Shen, W., Moran, C., Zens, R., et al.: Moses: open source toolkit for statistical machine translation. In: Proceedings of the 45th Annual Meeting of the ACL on Interactive Poster and Demonstration Sessions, Association for Computational Linguistics, pp. 177–180 (2007)
9. Och, F.J., Ney, H.: A systematic comparison of various statistical alignment models. Comput. Linguist. 29, 19–51 (2003)
10. Heafield, K.: KenLM: faster and smaller language model queries. In: Proceedings of the EMNLP 2011 Sixth Workshop on Statistical Machine Translation, Edinburgh, pp. 187–197, July 2011
11. Arcan, M., Turchi, M., Buitelaar, P.: Knowledge portability with semantic expansion of ontology labels. In: Proceedings of the 53rd Annual Meeting of the Association for Computational Linguistics, Beijing, China, July 2015
12. Mikolov, T., Chen, K., Corrado, G., Dean, J.: Efficient estimation of word representations in vector space. ICLR Workshop (2013)
13. Harris, Z.: Distributional structure. Word 10(23), 146–162 (1954)
14. Dragoni, M., Bosca, A., Casu, M., Rexha, A.: Modeling, managing, exposing, and linking ontologies with a wiki-based tool. In: Calzolari, N., Choukri, K., Declerck, T., Loftsson, H., Maegaard, B., Mariani, J., Moreno, A., Odijk, J., Piperidis, S. (eds.) Proceedings of the Ninth International Conference on Language Resources and Evaluation (LREC-2014), Reykjavik, Iceland, 26–31 May 2014, European Language Resources Association (ELRA), pp. 1668–1675 (2014)
15. Papineni, K., Roukos, S., Ward, T., Zhu, W.J.: BLEU: a method for automatic evaluation of machine translation. In: Proceedings of the 40th Annual Meeting on Association for Computational Linguistics. ACL 2002, pp. 311–318 (2002)

16. Denkowski, M., Lavie, A.: Meteor universal: language specific translation evaluation for any target language. In: Proceedings of the EACL 2014 Workshop on Statistical Machine Translation (2014)
17. Snover, M., Dorr, B., Schwartz, R., Micciulla, L., Makhoul, J.: A study of translation edit rate with targeted human annotation. In: Proceedings of Association for Machine Translation in the Americas (2006)
18. Borst, T., Neubert, J.: Case study: Publishing stw thesaurus for economics as linked open data. Case study (2009)
19. Zapilko, B., Schaible, J., Mayr, P., Mathiak, B.: TheSoz: a SKOS representation of the thesaurus for the social sciences. Semantic Web 4(3), 257–263 (2013)
20. Clark, J., Dyer, C., Lavie, A., Smith, N.: Better hypothesis testing for statistical machine translation: controlling for optimizer instability. In: Proceedings of the Association for Computational Lingustics (2011)
21. Cimiano, P., Montiel-Ponsoda, E., Buitelaar, P., Espinoza, M., Gómez-Pérez, A.: A note on ontology localization. Appl. Ontol. 5(2), 127–137 (2010)
22. Declerck, T., Pérez, A.G., Vela, O., Gantner, Z., Manzano, D.: Multilingual lexical semantic resources for ontology translation. In. In Proceedings of the 5th International Conference on Language Resources and Evaluation, Saarbrücken, Germany (2006)
23. Espinoza, M., Montiel-Ponsoda, E., Gómez-Pérez, A.: Ontology localization. In: Proceedings of the Fifth International Conference on Knowledge Capture, New York (2009)
24. Fu, B., Brennan, R., O'Sullivan, D.: Cross-lingual ontology mapping – an investigation of the impact of machine translation. In: Gómez-Pérez, A., Yu, Y., Ding, Y. (eds.) ASWC 2009. LNCS, vol. 5926, pp. 1–15. Springer, Heidelberg (2009). doi:10.1007/978-3-642-10871-6_1
25. McCrae, J., Espinoza, M., Montiel-Ponsoda, E., Aguado-de Cea, G., Cimiano, P.: Combining statistical and semantic approaches to the translation of ontologies and taxonomies. In: Fifth Workshop on Syntax, Structure and Semantics in Statistical Translation (SSST-5) (2011)
26. McCrae, J.P., Arcan, M., Asooja, K., Gracia, J., Buitelaar, P., Cimiano, P.: Domain adaptation for ontology localization. Science, Services and Agents on the World Wide Web, Web Semantics (2015)
27. Espinoza, M., Gómez-Pérez, A., Mena, E.: Enriching an ontology with multilingual information. In: Bechhofer, S., Hauswirth, M., Hoffmann, J., Koubarakis, M. (eds.) ESWC 2008. LNCS, vol. 5021, pp. 333–347. Springer, Heidelberg (2008). doi:10.1007/978-3-540-68234-9_26
28. Stellato, A., Rajbhandari, S., Turbati, A., Fiorelli, M., Caracciolo, C., Lorenzetti, T., Keizer, J., Pazienza, M.T.: VocBench: a web application for collaborative development of multilingual thesauri. In: Gandon, F., Sabou, M., Sack, H., d'Amato, C., Cudré-Mauroux, P., Zimmermann, A. (eds.) ESWC 2015. LNCS, vol. 9088, pp. 38–53. Springer, Heidelberg (2015). doi:10.1007/978-3-319-18818-8_3
29. Gennari, J., Musen, M., Fergerson, R., Grosso, W., Crubézy, M., Eriksson, H., Noy, N., Tu, S.: The evolution of protégé: an environment for knowledge-based systems development. Int. J. Hum. Comput. Stud. 58(1), 89–123 (2003)

EnergyUse - A Collective Semantic Platform for Monitoring and Discussing Energy Consumption

Grégoire Burel$^{(\boxtimes)}$, Lara S.G. Piccolo, and Harith Alani

Knowledge Media Institute (KMi), The Open University, Milton Keynes, UK
{g.burel,lara.piccolo,h.alani}@open.ac.uk

Abstract. Conserving fossil-based energy to reduce carbon emissions is key to slowing down global warming. The 2015 Paris agreement on climate change emphasised the importance of raising public awareness and participation to address this societal challenge. In this paper we introduce EnergyUse; a collective platform for raising awareness on climate change, by enabling users to view and compare the actual energy consumption of various appliances, and to share and discuss energy conservation tips in an open and social environment. The platform collects data from smart plugs, and exports appliance consumption information and community generated energy tips as linked data. In this paper we report on the system design, data modelling, platform usage and early deployment with a set of 58 initial participants. We also discuss the challenges, lessons learnt, and future platform developments.

Keywords: Energy consumption · Climate change · Semantic collective platforms · Energy monitors

1 Introduction

Global warming is one of the biggest current threats to lives, livelihoods, and economies. If unmitigated, the cost of global warming in the US alone is estimated to reach over \$500 billion by 2025 [2], and to cause an average drop in global income of 23 % by 2100 [22]. The World Heath Organisation predicts that climate change will cause around 250 K additional deaths per year by 2030 due to malnutrition, disease, and heat stress.[1] The 2015 Paris Agreement on climate change emphasised the importance of raising public awareness and participation to address climate change [13]. Although most citizens are aware of the general threats of climate change, they tend to be less aware of the concrete actions that they can take to reduce carbon emissions in their homes, to more actively participate in the global fight against climate change [8,19].

Engaging people with energy conservation is a complex task [1,20], where lack of adequate consumption feedback, habitual aspects, appliances design, and

---

[1] WHO Climate Change and Health Fact Sheet, http://www.who.int/mediacentre/factsheets/fs266/en.

© Springer International Publishing AG 2016
P. Groth et al. (Eds.): ISWC 2016, Part II, LNCS 9982, pp. 257–272, 2016.
DOI: 10.1007/978-3-319-46547-0_26

the choice of energy suppliers, are some of many factors that influence daily consumption of energy. Energy consumption is generally perceived at a high level [14], where the majority of people are unaware of the consumption levels of their various appliances [19]. Energy monitors could ease these issues, and have shown to lead to energy savings of 5–15% [9]. However, studies showed that energy monitors rarely attract user's attention for more than a few weeks, unless combined with other interventions, such as providing tips, motivations, and social engagement [10]

The European Environmental Agency stresses the role of technology, and of community-based initiatives, in engaging citizens and in achieving long-term behaviour change [8]. Bringing citizens together, and enabling them to compare their energy consumption levels and habits, have been found to be effective in these contexts [18].

In this paper, we introduce EnergyUse;[2] a collective awareness platform that aims to leverage the social power to engage citizens and to influence behavioural change, by encouraging people to discover, share, and discuss tips for conserving energy, and as a consequence, learn how to reduce carbon emission and help slowing climate change. With the help of electricity monitors, EnergyUse enables users to view and compare the electricity consumption of their entire households, or of specific devices and appliances. Semantics are used in EnergyUse for content augmentation from DBpedia, for environment-related tag extraction, and for Linked Data exports.

Requirements are described in the next section. Section 3 introduces EnergyUse, and in Sect. 4 we describe the usage of semantics in EnergyUse. Various evaluations are provided in Sect. 5. We end with a discussion and conclusions (Sects. 6 and 7 respectively).

2   Platform Design Requirements

Our design of EnergyUse followed a set of requirements extracted from (a) the literature, which consist of several general capabilities that were associated with successful energy saving initiatives, and from (b) community representatives, expressing their needs and interests. The following lists the recommendations gleaned from the literature:

1. **Personal approach.** The belief that climate change is a distant and non-personal threat is widespread [11]. More personal communication models are therefore needed [17,21,25], to replace generic-information based climate change campaigns, with actionable and personal experiences.
2. **Pragmatic emotions.** Rational, monetary, guilt, fear, and environmental benefits are common campaigns that tend to be ineffective in influencing citizen's behaviour towards energy and climate [15,17,23,25]. Instead, citizens should be encouraged to evaluate and understand the trade offs between environmental options and their individual values (e.g., comfort, preferences).

---

[2] EnergyUse, http://energyuse.eu.

3. **Social engagement.** Community initiatives tend be more successful than others in influencing energy-consumption behaviour [24]. Being a community not only encourages information exchange, it also apply more pressure to adopt, long lasting, greener social norms [8].

4. **Open discussions.** Online discussions is a promising engagement strategy for environment related topics [19]. Sharing and debating energy saving tips can provide citizens with direct feedback and concrete actions, which tend to be a highly effective intervention strategy [8].

5. **Direct feedback.** In-house energy monitors can provide households with real-time feedback on the impact of their energy consumption behaviours, thus raising their interest, understanding and awareness [6,7].

6. **Comparisons and Competitions.** Peer comparisons and competitions tend to be effective interventions in nudging towards greener behaviour [3,10].

In addition to the above, we organised 3 workshops with a group of 9 local community leaders who are active in disseminating energy conservation practices. Participants held non-technical occupations, aged between 21–72 (average age: 49.3), with an average household size of 2. In these workshops, participants discussed values, motivations and barriers related to energy savings, and shared tips and discussed energy usage in their daily routines.

The need to tackle the energy conservation subject with **personal** and **emotional** approaches constantly emerged in the discussions. For example: *"There are many things going around global warming and carbon emissions. And it scares people for discussing energy. Politicians and scientists arguing, people can have this perception. . . . But if you do in a light way, small actions, and in the end you say - "you saved 3 penguins", then it is ok.".* Also, against providing general hints *"If you say in kWh and carbon usage, people say, what does it mean to me? It's my hygiene standards, my time. . . (. . .) I have my lifestyle and expectations. Imagine saying to someone: you shouldn't be hoovering more than once a week.".*

In terms of **social engagement**: *"Rather than giving out leaflets or sheets of information, it is better to have a [community member] to give simple messages"*. Participants also mentioned asking neighbours for help since they are likely to have similar appliances. The need for **direct feedback** to build knowledge was frequently mentioned; *"I assume that if I use the [washing machine] eco mode I save energy, but I am not sure. It takes 3 h to wash!"*, and *"We all have this mythology to say this or that is more expensive. It is not always true"*. Beyond understanding appliances consumption, the interest in aggregating the cost of a domestic tasks also emerged: *"I would like to know how much energy I use for my breakfast"*. Participants also expressed interest in comparing the consumption of different appliances to perform similar activities, such as cooking with gas oven or microwave, or evaluating time vs efficiency: *"Sometimes less power takes longer"*, as vacuum cleaners or hair dryers.

One community leader also stated the advantages of adopting external motivations to bring people online, such as **comparison and competition**: *"some people have a natural desire to learn more about, but something to put people in,*

Table 1. EnergyUse (EU) design principles, following common guidelines from energy-related literature.

Guideline	EnergyUse design
Personal Approach	Focus on energy consumption of personal devices and appliances around the house
Pragmatic Emotions	Citizens are free to propose and discuss energy saving options in accordance with their own values and preferences
Social Engagement	Registered users are members of the EnergyUse online community site, where they can engage, interact, and influence each other
Open Discussions	Online forum for members to share and rate questions, answers, and energy saving tips
Direct Feedback	Users are equipped with electricity monitoring plugs and displays, to enable them to directly gauge their consumption, which is automatically fed into the EnergyUse platform
Comparisons and Competitions	Users can compare their own consumption against the aggregated energy consumption of appliances. Highly rated content and a high number of contributions are praised to inject an element of competition

like vouchers, might help". Table 1 shows how we designed EnergyUse to meet the requirements and guidelines described above.

3   EnergyUse Architecture and Components

EnergyUse (energyuse.eu) is a web platform designed for addressing the needs described in Sect. 2. Its development is heavily based on a customised version of BioStar[3] [16]; a Python and Django[4] based Question Answering (Q&A) software, with additional EnergyUse specific features such as concept pages, consumption data from energy monitoring devices, and linked data publication of actual energy usage of various equipment and appliances. The general architecture of EnergyUse is shown in Fig. 1. As shown in the figure, EnergyUse consists of four main modules:

– **Automatic Semantic Tagging:** This module is designed for increasing the number of concepts associated with user posts, to expand linking between conversations, and enrich the browsing experience. Third party semantic annotation tools are used by this module.

---

[3] BioStar, http://github.com/ialbert/biostar-central.
[4] Django, http://www.djangoproject.com.

- **Automatic Semantic Description Generator:** Topic and appliance pages on EnergyUse can be automatically populated with descriptions and background images using this module, which locates and retrieves this information from DBpedia.[5]

- **Energy Consumption Processing:** This module deals with the automatic collection of energy consumption data from user's energy monitoring devices.

- **Ontology Mapper:** Public data on EnergyUse is made available as linked data for third party tools, by mapping the data to the EnergyUse ontology.[6]

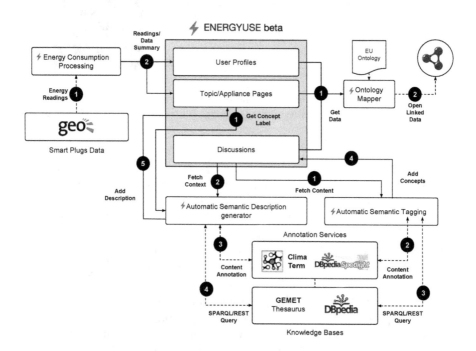

Fig. 1. Architecture and main components of the EnergyUse platform. Numbers and arrows indicate the data flow and process order for each platform modules. The modules are indicated with a *lightning bolt* symbol.

3.1 Platform Design

Figure 2 shows the front page of EnergyUse. The platform consists of three main components areas: (1) Energy related community discussions; (2) Appliance pages, and; (3) Personal energy consumption readings.

---

[5] DBpedia, http://dbpedia.org.
[6] EnergyUse Ontology, http://socsem.open.ac.uk/ontologies/eu.

Fig. 2. The EnergyUse (energyuse.eu) homepage, displaying recent conversations, top contributors and posts, community average energy consumption, and featured topics and information.

Discussion Pages: The discussions follow the Q&A structure inherited from the Biostar platform. Each user can create discussions and post comments, "*upvote*" good posts and bookmark interesting discussions. They can also follow discussions and topics and receive email notifications when new posts are contributed. All visitors to EnergyUse can browse and read all existing discussions and public summary information about the consumption of appliances. However, only EnergyUse registered users can initiate and contribute to discussions.

Profile Pages: Profile pages contain diverse information about individual users such as their profile picture, their username, reputation score, and recent contributions. If the user connected their energy monitoring account to EnergyUse (Sect. 3.2), they can see the electric consumption of the appliances they connected to their monitoring plugs. The interface allows users to filter the display of data according to time period, type of device, and day of the week.

Appliance and Topic Pages: Posts can be tagged with keywords, which could either refer to topics (e.g., breakfast, lightening) or appliances (e.g., kettle, light bulb). Such tags are also linked with corresponding semantic concepts from DBpedia. For each topic or appliance, a dedicated page is created, containing a description, image, icon, and the list of discussions that mention that particular tag. The aim of these pages is to improve content access and topic understanding whereas the images and icons are displayed so that users feel less intimidated by the technical aspects of particular topics. If available, average energy consumption data for a given appliance will be displayed. Finally, users

Fig. 3. EnergyUse (EU) page header of the *Microwave* appliance automatically generated using DBPedia.

can subscribe to the topic or appliance feed to receive updates of new posts. Figure 3 shows an example of a dedicated appliance page.

Tags and Users List Pages: Besides the pages above, users can view the list of all community members (registered users), to see their profile, reputations, contributions, etc. Users can also browse the list of all tags created in EnergyUse, which are linked to relevant discussion.

3.2 Connecting with Energy Monitoring Devices

One main novelty of EnergyUse is its ability to collect, visualise and publish actual appliance energy consumption information. Visitors to the platform are prompted to create an account, and to connect their energy monitoring accounts if they have one. EnergyUse currently supports energy monitoring devices from Green Energy Options (GEO)[7]. GEO enables users of their devices to select one of 41 different type of appliances for each monitoring plug they own. Examples of appliances include TV, Kettle, PC, other, etc. Energy readings from GEO devices are read by EnergyUse for registered users, thus providing them with direct, secure, and private access to their data via the platform. For privacy reasons, users are only able to view aggregated energy consumptions from all other community members, for any given appliance.

Energy consumption data is collected from GEO every 15 min,[8] and thus for each user, EnergyUse accumulates 96 readings per day per plug. When aggregating community consumption readings, outliers are removed. Outlier user-data may exist when a user moves the monitoring plug to another appliance, without changing the setup on GEO.

For each appliance where consumption data is available, EnergyUse calculates the following measurements:

---

[7] Green Energy Options, http://geotogether.com.
[8] GEO energy monitors send data readings once every 15 min to GEO servers.

- *Number of Observations:* We report the number of data points used for calculating the summary statistics.
- *Number of Providers:* The number of users providing the readings.
- *Number of Removed Observations:* Number of data observations found as outliers, and removed from all subsequent measurements.
- *Min Consumption:* The minimum electricity consumption recorded when the appliance is switched on (i.e. when consumption is higher than 0 kWh).
- *Max Consumption:* The maximum electricity usage observed.
- *Mean Consumption:* The average electricity consumption observed when the appliance is switched on.
- *Mean Daily Consumption:* The daily average electricity usage observed.

4 Semantic Descriptions and Modelling

Semantics are used in EnergyUse in 3 areas; (1) concept extraction from posts; (2) appliance and topic description generation from an external knowledge base, and; (3) publishing of aggregated energy consumption data as Linked Open Data (LOD). These are detailed in the following sections.

4.1 Semantic Tagging

In order to find discussions and energy consumption information more easily, user are expected to add tags when creating a discussion, to describe the appliance in question and related topics. For example, a discussion about dishwashers could be tagged with *dishwasher* appliance, *dishes, washing topics*, etc. To ensure that all relevant tags are generated, EnergyUse automatically identifies relevant topics and appliances from the post content, using DBpedia Spotlight[9] [5], and ClimaTerm [12].[10] entity recognition tools. With such annotation tools, EnergyUse is able to extracts concepts automatically from existing discussions and posts.

These services are designed for recognising terms in plain text and linking them with relevant concepts from different knowledge bases. In the case of DBpedia Spotlight, the knowledge is provided by DBpedia, while ClimaTerm uses GEMET[11] and REEGLE[12] as data sources. ClimaTerm is especially designed for identifying environment related terms whereas DBpedia has a more generic and higher coverage.

Since ClimaTerm specifically recognises climate-related terms, we directly use the extracted terms as additional tags. In the case of DBpedia Spotlight, we only focus on the detection of appliance related terms by adding a restriction to the type of entities identified by the tools. More specifically, we use the following SPARQL restriction to only select entities linked with the home appliance category:

---

[9] DBpedia Spotlight, http://github.com/dbpedia-spotlight/dbpedia-spotlight.
[10] ClimaTerm, http://services.gate.ac.uk/decarbonet/term-recognition.
[11] GEMET Thesaurus, https://www.eionet.europa.eu/geme.
[12] REEGLE, http://www.reegle.info/glossary.

```
SELECT DISTINCT ?appliance WHERE {
?appliance ?related
<http://dbpedia.org/resource/Category:Home_appliances>
}
```

4.2 Semantic Descriptions

As discussed earlier, each appliance or topic page contains a description, list of relevant tags, and a representative background image and icon. When such a page is first created (when the appropriate tag is used for the first time), there will be no description or image to describe it. This information is inserted manually by the EnergyUse administrators. As a result, such pages could remain relatively empty for a little while, especially when several of such pages are created in a short period of time. To populate such descriptive pages automatically and avoiding missing descriptions, EnergyUse retrieves relevant content from DBpedia using an approach similar to the semantic tagging method described in the previous section (Sect. 4.1).

DBpedia uses `dbo:abstract` from the DBpedia Ontology[13] for providing concept descriptions. Many DBpedia concepts also have a `dbo:thumbnail` which link to an image resource. When such information is available, we use the linked image as the background image displayed behind the concept description. When the page is automatically generated, a default *"plug"* icon is used temporarily, since no icons are available from DBpedia.

Instead of simply matching tag labels to knowledge bases directly, we use the post's text as contextual information to help the tools to disambiguate the corresponding concepts. If the annotation tools identify any concept related to the given labels, we retrieve the information mentioned above and populate the appliance or topic descriptions directly from DBPedia. To automatically verify whether or not the given page/tag is referring to a home appliance, EnergyUse detects if the corresponding DBpedia entity is associated with `dbc:Home_appliances`.

4.3 Linked Open Data Publishing

EnergyUse aggregated and anonymised energy consumption are published in JSON-LD,[14] along with all other public information on the platform. This roughly falls into four components: (1) User profiles; (2) Content data; (3) Appliances and topic information, and; (4) Energy measurements. Energy consumption of individual users is not published.

In general we reuse six different ontologies to fully represent all this data, and use `owl:equivalentClass`, `owl:equivalentProperty`, and `rdfs:sameAs` to connect EnergyUse instances with existing resources from DBpedia and GEMET. The ontology is represented in Fig. 4.

---

[13] DBpedia Ontology, http://dbpedia.org/ontology.

[14] JSON-LD, http://json-ld.org.

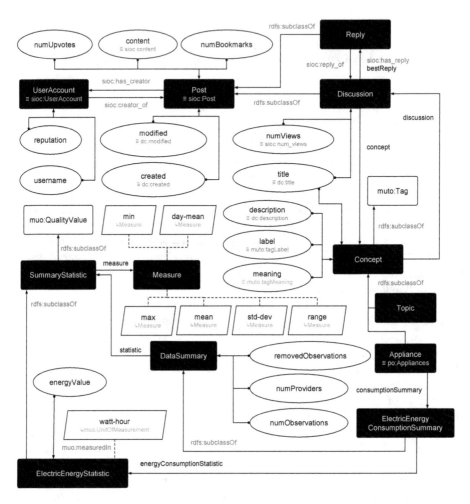

Fig. 4. EnergyUse ontology, which imports properties and classes from 6 ontologies: SIOC [4] (`sioc`), PowerOnt (`po`), MUTO (`muto`), DC Terms (`dc`), FOAF (`foaf`) and MUO (`muo`). Rectangles denote classes, ellipses properties, and parallelograms class instances.

SIOC is used to represent users and their contributions. Users' reputation is represented via a subclass of `sioc:UserAccount`, extended with `eu:reputation` and `eu:username` properties. EnergyUse discussion are not confined to Q&A, and hence to link discussions with concepts, we use `eu:Post` with an equivalence relation with `sioc:Post`. This is then extended with the `eu:Discussion` and `eu:Reply` concepts that respectively represent the initial post of a discussion and its replies. SIOC is used to represent replies (`sioc:has_reply/reply_of`) and their creators (`sioc:has_creator/creator_of`), and we add (`eu:bestReply`) as a sub-property of `sioc:has_reply` to represent best replies. New properties are used for the content, titles, upvotes, creation dates, modification dates, views and bookmarks, all appropriately linked to equivalent properties in SIOC, Dublin Core and FOAF. Each `eu:Concept` has a description, title, label and meaning, which are linked with corresponding properties in MUTO and Dublin Core. `eu:Concept` is subclassed into `eu:Appliance` and `eu:Topic`. To model tags (representing appliances and topics), we reuse and extend the Modular Unified Tagging Ontology[15] (MUTO) tag representation. FOAF `foaf:topic/page` properties are used to link topics to `eu:Discussion`. To represent the measurements in Sect. 3.2, we define `eu:DataSummary` and `eu:SummaryStatistic`, associated with multiple `eu:Measure` instances.

5   Evaluation and Analysis

In the following sections we provide various evaluations and analysis of EnergyUse.

5.1   EnergyUse User Community

Users of EnergyUse were selected as part of a general energy trial. Over 400 expression-of-interest were submitted via a dedicated online form. We then randomly selected 150 UK participants (average household size: 2.6, average age: 40.9), acquired their acceptance to our Terms & Conditions, and supplied them with energy monitoring kits, and provided installation support when needed. Few months later, we invited the participants to register with EnergyUse. 58 participants registered, and 29 of them linked their GEO account (Sect. 3.1) with the EnergyUse platform, thus allowing their energy consumption data to be read by EnergyUse. According to Google Analytics, EnergyUse so far received 520 unique visitors, generating 1,142 sessions with an average duration of 4:48 minutes, and 4,655 pageviews, with 54.4 % returning visitors.

5.2   Semantic Tag Generation

Automatic tagging (Sect. 4.1) is performed using two distinct methods: (1) ClimTerm, for adding climate related terms from the GEMET Thesaurus and REEGLE to a given discussion, and; (2) DBpedia spotlight annotations for identifying potential appliances.

In this section we focus our analysis on how the automatic tagging improved cross-concept browsing by increasing the tagging network between EnergyUse discussions. EnergyUse contained 48 tags manually created by the posts' authors. Using ClimaTerm, an additional 17 tags (concepts) were automatically generated. A full evaluation of ClimaTerm on Twitter can be found in [12]. Although ClimaTerm was evaluated on a different dataset, we do not evaluate the accuracy of ClimaTerm for our data preferring to focus on how additional tags improve user navigation between discussions.

---

[15] MUTO, http://muto.socialtagging.org.

With DBpedia spotlight on the other hand only found 4 concept tags, 2 of which were already found by ClimaTerm. Reason for the limited performance of DBpedia Spotlight is the enforcement of the SPARQL restriction that was presented in Sect. 4.1, which strictly limits new concept suggestions to dbc:Home_appliances.

In order to show that the additional tags may improve user navigation between discussions, we compute the discussion network density by considering that documents sharing the same concept tags are linked. The network density measure is the ratio between the number of observed connection between graph nodes and the number of potential node connections. The higher the value, the more likely a node can be reached from any other node. For our evaluation, we compute this measure before and after adding the new concepts through the semantic tagging approach described in Sect. 4.1. As a result, we observe that before adding concepts, the network density is 0.061 and when additional concepts are added automatically, the density rise to 0.065. This shows that although the density increase is minimal, automatic tagging may improve discussion navigation.

Such results could be enhanced further by relaxing the restrictions on the DBPedia Spotlight annotator. For example, we would obtain more concepts from discussions if we were removing the dbc:Home_appliances restriction. However, this could retrieve many generic concepts that might disrupt navigation in EnergyUse.

5.3   Semantic Description Generation

EnergyUse currently contains 67 concept labels, referring to topic or appliances. Remember that a page is created automatically for every new concept label. Only 25 of the 67 pages have been manually populated with descriptions and images. Here we evaluate the ability of the description generation module (Sect. 4.2) to associate DBpedia concepts to our existing labels. First we created a gold standard by manually linking all 67 labels to DBpedia. We found that only two could not be linked to any appropriate concepts. Those were *vampires* (devices that consume considerable energy even when on standby), and *baseload* (household electricity baseload). Secondly, we compare the automatic association of tags to the gold standard. Out of 67 tags, 46 were automatically linked to the exact concept as in the gold standard (69 %).

Thirdly, for the remaining 21, we check the semantic distance between the gold standard concept, and the one automatically chosen by EnergyUse. We use this to estimate how off-target the concept linking is. For different link distances $Link_d = \{0, 1, 2, 3\}$, the accuracies are respectively $Acc = \{0.69, 0.84, 0.89, 0.94\}$. In other words, 15 % of the tags were linked to DBpedia concepts that were 1 link away from the ones in our gold standard, 6 % were 2 links away, and 4 % were 3 links away. We observed that in many cases the selected concepts were more general to the gold standard. For example, the tag "*gas*" was associate with dbo:Natural_gas in our gold standard, and automatically linked to the broader concept of dbo:Gas.

Finally, we check how well did EnergyUse distinguish between topic tags and appliance tags. Out of the correctly linked 46 tags, 16 actually refer to appliances, and the 30 to topics. We found that 65 % of these tags were accurately linked to topics or to appliances. This shows that not all appliances were actually identified with `dbc:Home_appliances`. Note that such misclassifications are not impactful and directly visible to the users.

5.4   Linked Data Publishing

As discussed in Sect. 4.3, public data in EnergyUse are published in JSON-LD using the EnergyUse ontology (Fig. 4). Currently, the EnergyUse platform publishes 58 user profiles, 67 concepts, 121 posts from 38 discussions, and summarised energy consumption data for 37 appliances. Although it is early to evaluate the impact of publishing this data, we could draw insights with regards to its potential reach from the current usages of the ontologies imported in EnergyUse.

According to statistics computed in 2014 about the LOD cloud,[16] the FOAF and Dublin Core Terms were respectively used in 69.13 % and 56.02 % of the 2014 datasets crawled for computing the LOD statistics. SIOC was also highly popular with 17.65 % of the datasets using it. The other ontologies used by EnergyUse were not reported or they were not used significantly in the data crawl.

In spite of the vast amount of research on climate change related topics, there are very few energy related ontologies and datasets. This shows the potential value of the EnergyUse datasets to fill this gap and to support this rapidly growing field of research.

5.5   Scalability Assessment

As mentioned earlier, EnergyUse platform is built over the Biostar platform. This platform currently hosts the Biostars.org website which receives more than 600 user visits per hour, has nearly 26 K registered users and over 181 K posts. Different deployment modes can be used depending on the performance required. For instance, the platform has a *high traffic deployment mode* that enables asset compression and caching and a *low traffic mode* that allows for easier debugging.

Currently, the average loading time page of EnergyUse is 4.02 seconds which is sub-optimal. This is largely due to the current environment in which the server runs. Since EnergyUse is currently in still emerging and being constantly improved, the focus has been mostly on getting user feedback and rapid prototyping of necessary functionalities. For instance, features such as compression are not activated yet, as well as the cache mechanisms available in Biostar. Although we are not currently focusing on raw performance, the current platform can be easily scaled up as the large scale deployments of Biostars.org shows.

---

[16] State of the LOD Cloud 2014, http://linkeddatacatalog.dws.informatik.uni-mann heim.de/state.

6   Discussions and Future Work

In this section we shed light on a number of relevant issues to be addressed in near future versions of EnergyUse. Energy monitoring devices tend to be proprietary and their data are not easily accessible. Currently, EnergyUse can automatically retrieve consumption data from GEO energy monitoring devices, knowing that GEO is a major supplier of monitoring devices to British Gas;[17] the largest energy supplier in the UK. Extending this architecture to include other monitoring devices might not be easily achievable, since it would require special agreements with various manufacturers of such devices. As an alternative, in the next version of EnergyUse we will enable users to manually enter their energy readings, which they could acquire from the energy monitoring device of their choice.

The broad use of semantic technologies in the EnergyUse platform shows how entity linking can be used for integrating external information without requiring constant human supervision. This is particularly important for websites that have a small amount of administrators and when the velocity of user contributions is outpacing manual information assessment and management (e.g. when creating topic and appliance description for newly created keywords).

LOD export currently only publishes entire discussion posts, in addition to various other public data from EnergyUse. However, some posts tend to contain concrete energy saving tips, which could be extracted in a more detail and structured fashion, thus potentially creating a valuable linked dataset. We plan to test ClimateMeasure[18] to fulfil this ambition. With regards to improving appliance-tag identification and concept-linking, beside using DBpedia Spotlight to drive this linking step, we plan to test other tools such as alchemyapi.com and textrazer.com.

EnergyUse aims to influence citizen behaviour towards greener lifestyle and habits. Next, we will study the evolution of behaviour of all members, while taking their actual energy consumption data into account, as well as the quality and quantity of their engagement and contributions on the platform. We also plan to run a user survey and usability evaluation of the platform this summer.

EnergyUse is an emerging platform and we are currently testing various interventions to increase engagement, and plan to expand the energy trials to the Netherlands which was a source of many expressions of interest (Sect. 5.1).

7   Conclusions

EnergyUse is a collective platform that targets the critical societal challenge of climate change, by raising awareness and engagement of citizens in energy-consumption discoveries and debates. One of the unique features of EnergyUse is its integration with energy monitoring devices, thus enabling users to view their

---

[17] GEO Press Release, http://www.mynewsdesk.com/uk/geo/pressreleases/british-gas -chooses-geo-for-next-generation-energy-displays-883182.

[18] ClimateMeasure, http://services.gate.ac.uk/decarbonet/indicators.

actual consumption levels, and to compare against community averages. Another feature is the focus on household devices and appliances, and thus bringing the topic of climate change to a personal level, and facilitating the identification of concrete and actionable energy-saving practices. In this paper, we described the rational behind EnergyUse, its user and design requirements, and detailed and evaluated the main components on the platform.

Acknowledgement. This work is funded by the EC-FP7 project DecarboNet (grant number 610829).

References

1. Abrahamse, W., et al.: A review of intervention studies aimed at household energy conservation. J. Environ. Psychol. **25**(3), 273–291 (2005)
2. Ackerman, F., Stanton, E.A.: The cost of climate change: what well pay if global warming continues unchecked (2008). https://www.nrdc.org/sites/default/files/cost.pdf
3. Allcott, H.: Social norms and energy conservation. J. Public Econ. **95**(9), 1082–1095 (2011)
4. Breslin, J.G., Harth, A., Bojars, U., Decker, S.: Towards semantically-interlinked online communities. In: Gómez-Pérez, A., Euzenat, J. (eds.) ESWC 2005. LNCS, vol. 3532, pp. 500–514. Springer, Heidelberg (2005). doi:10.1007/11431053_34
5. Daiber, J., Jakob, M., Hokamp, C., Mendes, P.N.: Improving efficiency and accuracy in multilingual entity extraction. In: Proceedings of the 9th International Conference on Semantic Systems (I-Semantics) (2013)
6. Darby, S.: Literature review for the energy demand research project. Ofgem (Office of Gas and Electricity Markets), London (2010)
7. Erhardt-Martinez, K., Donnelly, K., Laitner, J.: Advanced metering initiatives, residential feedback programs: a metareview for household energy-saving opportunities. Report no. E105. American Council for an Energy-Efficient Economy, Washington, DC (2010)
8. Agency, E.E.: Achieving energy efficiency through behaviour change: what does it that? Technical report N5/2013 (2013)
9. Hargreaves, T., Nye, M., Burgess, J.: Making energy visible: a qualitative field study of how householders interact with feedback from smart energy monitors. Energ. Policy **38**(10), 6111–6119 (2010)
10. Letwin, O., Barker, G., Stunell, A.: Behaviour change and energy use. Cabinet Office: Behavioural Insights Team (2011)
11. Lorenzoni, I., Langford, I.H.: Climate change now, in the future: a mixed methodological study of public perceptions in norwich (UK) (2001). http://www.cserge.ac.uk/sites/default/files/ecm_2001_05.pdf
12. Maynard, D., Bontcheva, K.: Understanding climate change tweets: an open source toolkit for social media analysis. In: 29th International Conference on Informatics for Environmental Protection (EnviroInfo 2015), Copenhagen (2015)
13. United Nations: Adoption of the paris agreement (2015). https://unfccc.int/resource/docs/2015/cop21/eng/l09r01.pdf
14. Neustaedter, C., Bartram, L., Mah, A.: Everyday activities, energy consumption: how families understand the relationship. In: Proceedings of CHI 2013, Paris, France (2013)

15. O'Neill, S., Nicholson-Cole, S.: Fear wont do it: promoting positive engagement with climate change through visual and iconic representations. Sci. Commun. **30**(3), 355–379 (2009)
16. Parnell, L.D., Lindenbaum, P., Shameer, K., Dall'Olio, G.M., Swan, D.C., Jensen, L.J., Cockell, S.J., Pedersen, B.S., Mangan, M.E., Miller, C.A., Albert, I.: Biostar: an online question & answer resource for the bioinformatics community. PLoS Comput. Biol. **7**(10), 1–5 (2011)
17. Petkov, P., Goswami, S., Köbler, F., Krcmar, H.: Personalised eco-feedback as a design technique for motivating energy saving behaviour at home. In: Proceedings of the 7th Nordic Conference on Human-Computer Interaction: Making Sense Through Design, NordiCHI 2012, Copenhagen, Denmark. ACM (2012)
18. Petkov, P., Köbler, F., Foth, M., Krcmar, H.: Motivating domestic energy conservation through comparative, community-based feedback in mobile and social media. In: 5th International Conference on Communities & Technologies (C&T 2011), Brisbane, Australia, June 2011
19. Piccolo, L., Baranauskas, C., Fernandez, M., Alani, H., de Liddo, A.: Energy consumption awareness in the workplace: technical artefacts and practices. In: Proceedings of the 13th Brazilian Symposium on Human Factors in Computing Systems (IHC 2014), Parana, Brazil (2014)
20. Pierce, J., Paulos, E.: Beyond energy monitors: interaction, energy, and emerging energy systems. In: Proceedings of CHI 2012, Texas, US (2012)
21. Stern, P.C., Dietz, T., Abel, T., Guagnano, G.A., Kalof, L.: A value belief norm theory of support for social movements: the case of environmental concern. Hum. Ecol. Rev. **6**(8), 81–97 (1999)
22. Sterner, T.: Economics: higher costs of climate change. Nature **527**, 177–178 (2015)
23. Sweney, M.: Government's 6m climate change ads cleared. Guardian (2010). http://www.guardian.co.uk/media/2010/oct/11/governmentclimate-change-ad
24. Umpfenbach, K.: Influences on consumer behaviour. Policy implications beyond nudging. Final report. Ecologic institute, Berlin (2014)
25. van der Linden, S.: Towards a new model for communicating climate change. In: Cohen, S., Higham, J., Peeters, P., Gossling, S. (eds.) Understanding and Governing Sustainable Tourism Mobility: Psychological and Behavioural Approaches, pp. 243–275. Routledge, Taylor and Francis (2014)

Extracting Semantic Information for e-Commerce

Bruno Charron, Yu Hirate, David Purcell, and Martin Rezk[✉]

Rakuten Inc., Tokyo, Japan
{bruno.charron,yu.hirate,david.purcell,martin.rezk}@rakuten.com

Abstract. Rakuten Ichiba uses a taxonomy to organize the items it sells. Currently, the taxonomy classes that are relevant in terms of profit generation and difficulty of exploration are being manually extended with data properties deemed helpful to create pages that improve the user search experience and ultimately the conversion rate. In this paper we present a scalable approach that aims to automate this process, automatically selecting the relevant and semantically homogenous subtrees in the taxonomy, extracting from semi-structured text in items descriptions a core set of properties and a popular subset of their ranges, then extending the covered range using relational similarities in free text. Additionally, our process automatically tags the items with the new semantic information and exposes them as RDF triples. We present a set of experiments showing the effectiveness of our approach in this business context.

1 Introduction

Semantic technologies are not new in the e-commerce business. In particular, ontologies (or ontology-like artifacts) have been used for a number of different tasks that go from data integration [11] to items classification and search support [7]. Rakuten Ichiba (the largest e-commerce site in Japan) uses a large legacy taxonomy with around 40,000 classes to organize the items and provide the users with discovery axes relevant to their searches. In order to assist the user with her search, the catalog team is manually extending the taxonomy and the items with new semantic information. They start by selecting the classes in this taxonomy that need to be improved based on profitability and difficulty of exploration. For each of these classes, they study the domain together with the customers shopping behavior, such as search keywords, browsed items, etc. Based on that, they extend the taxonomy with a small set of properties deemed helpful to the customers along with the most representative values (not necessarily all) in the range of these properties. Finally, they use these properties/values to create pages that help the users to explore the items in the class.

Being completely manual, this taxonomy extension effort currently requires a massive amount of time and human effort. Acquiring domain knowledge and modeling the properties alone for a single class can take days, the operation is error-prone, and the result cannot be easily ported to other Rakuten market places across the globe (Taiwan, France, US, Germany, etc.). As such, there is

© Springer International Publishing AG 2016
P. Groth et al. (Eds.): ISWC 2016, Part II, LNCS 9982, pp. 273–290, 2016.
DOI: 10.1007/978-3-319-46547-0_27

an important business value in automating this process, both to speed up the completion of the project and to drastically reduce the cost of this task.

This work raises a number of questions: (i) how to automatically select relevant and semantically homogenous subtrees in the taxonomy which would most benefit from being extended with data properties; (ii) how to extract a core set of properties from existing datasets that can effectively assist the users in their product searches; (iii) how to extract a representative set of range values to be displayed together with these properties; (iv) how to find for each item in each class the corresponding property values; (v) how to provide a generalizable solution (to arbitrary taxonomies, languages, and datasets) that scales to run over tens of thousands of classes and TB of data.

In this paper we propose answers to these questions adopting and extending techniques from ontology-based information extraction and triple extraction, and proposing new techniques when needed. We present an end-to-end scalable unsupervised approach that automatizes the process of extracting semantic information from item descriptions and shopping behaviors to improve the user experience. This process adds new properties to the relevant subtrees of the taxonomy and tags the items with this new information, exposing the outcome as RDF triples. The pipeline of our approach is depicted in Fig. 1. In this work when we say taxonomy/ontology we do not necessary mean OWL ones, and when we say triples do not necessary mean RDF ones. However, we plan to use this project to push for the use of these semantic web standards inside Rakuten.

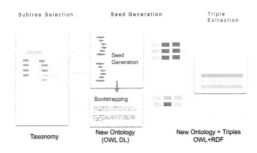

Fig. 1. Expected outcome

Our main contributions can be summarized as follows:

- A novel technique to select relevant subtrees in the taxonomy in terms of overall profitability, ease of product discovery and similarity of users' shopping behavior.
- An extension to the work presented in [23] to extract, for the relevant subtrees, an initial set (seed) of popular property-value list from semi-structured text. The extension includes a new technique for the aggregation of synonymous property names, as well as an improvement in the precision of the result by exploiting users' shopping data.

- An approach, based on neural networks [16], to extend the initial seed with new relevant values using semantic similarity.
- A set of experiments to measure the precision, the coverage, and the scalability of these techniques.

The remainder of the paper is organized as follows: Sect. 2 describes the business need, and highlights the problems with the current approach. In Sect. 3 we briefly survey other related works. In Sect. 4 we describe our approach towards data property extraction, range extraction, and triple generation. Section 5 gives an overview of the architecture of our current implementation of the proposed approach. In Sect. 6 we provide a number of experiments evaluating the correctness and scalability of our framework. Section 7 concludes the paper.

2   Business Case

Rakuten Group is one of the world's leading Internet service companies, providing a variety of consumer and business-focused services including e-commerce, eBooks, travel, banking, securities, credit card, e-money, portal and media, online marketing, professional sports, etc. Currently, around 40[1] of the total operating income comes from Rakuten Ichiba, the leading e-commerce website in Japan, which is Rakuten Group's online shopping mall where third-party merchants can set up shops and sell their products. Rakuten Ichiba offers around 200 million items classified in a large legacy taxonomy (there are only classes and subclass relations between them) with around 40,000 classes. The first version of this taxonomy was developed around 2001 (three years before the standardization of OWL) and has been evolving since then, although it was never moved to a well-known ontology language. Currently there are several internal projects to shrink and improve the quality of this taxonomy, aiming to enhance the user experience, increase the conversion rate[2], and as a consequence increase the *gross merchandise sales* (abbr. GMS) per class. In this work we focus on the project that aims to extend the relevant taxonomy classes and the items with new semantic information deemed helpful to the user.

An appeal of Rakuten Ichiba which contributed to its initial popularity is the ease and freedom with which the merchants can register products on the platform, allowing them to freely design their item descriptions in order to construct a shop identity and a connection with the customers. As a result, any information about the items must be found in their titles and descriptions specified by the merchants, that is, HTML code mostly consisting of free-form text, semi-structured text (table-like free-form text), structured text (tables) and images. This data is therefore the primary source to extract item-specific information in this project but it is also used to automatically acquire domain-specific knowledge as it reflects the knowledge of the merchants.

---

[1] http://global.rakuten.com/corp/investors/documents/results/.
[2] Proportion of customers making a purchase within a given browsing session.

When extending the taxonomy, it is necessary to avoid extracting properties for *every* single class. First, because of the legacy nature of the taxonomy, some classes are either obsolete, or in an unintuitive branch, or too specific to be interesting, therefore displaying data properties for those classes will not improve the user experience (abbr. UX). Recall that here we are not trying to build a complete model of the domain, like the work in [17], but use the ontology to improve the UX. Second, extending the whole taxonomy would not only dramatically increase its already massive size but also lead to unintuitive results if we extract properties of semantically inhomogeneous classes. For instance, the class Wine has among its subclasses Red Wine, White Wine, and Wine Accessories. Clearly the properties for Wine and Wine Accessories are very different. Thus, trying to extract property-values from *all* the items in the Wine category directly leads to unintuitive (although correct) results, such as "yellow" for the property "color".

Currently, the extension of the taxonomy is done manually and we aim to reduce the human effort needed for this task, although some human work might still be needed. A production-level system should have a precision above 80 %, and the cost of the deployment and maintenance of the proposed solution should not exceed the benefits from the use of the system.

3   Related Work

This work intersects mainly with two fields: Ontology-based information extraction and triple extraction. In this section we will briefly survey some of the works in these areas and their relation with our approach. It is worth noting that none of the articles mentioned here tackle the problem of selecting subtrees as described in Sects. 4 and 5.

Several studies mentioned in this section rely heavily on a number of linguistic resources (such as Yago, DBpedia, Wikipedia) and complex NLP tools (such as entity/predicate disambiguation and matching [24]). These resources and tools for languages different than English (we work with Japanese texts) are often not as accurate, and the inaccuracies tend to accumulate and propagate through layers. In addition, performing such analysis over terabytes of data can be computationally expensive. Thus, in this work we only use a tokenizer to prepare the datasets and we use data mining and machine learning to extract relations and triples. In the future we plan to use NLP techniques to improve the results when our techniques perform sub-optimally.

OBIE: In Ontology-based information extraction (abbr. OBIE) [25] the ontology is used to drive the information extraction process. In [21], the authors describe an OBIE application (GATE) for e-business. In this approach, the properties are already given and the concepts are selected manually with the help of domain experts. The work in [1] extracts properties names using an NLP approach, but the extraction process relies on a predefined (manually by experts) semantic lexicon, and on a manually created set of rules. Another NLP approach for OBIE is the one in [2], where the extracted relations have to be mapped to an existing (manually created) list of properties. [22] presents a system called RelExt for relation extraction. They use semi-structured text to map terms to concepts, and linguistic analysis to automatically find relations between those terms. In addition, they use a number of statistical metrics to score the popularity of the triples. [19] presents a system called DOGMA for relation extraction from free text. This system first finds verbs-objects in the text, then builds semantics sets by clustering the verbs and the terms, and finally finds the relations between those sets of terms. Unlike our work (and [22]) they make no use of semi-structure data, but the precision reported is rather low (around 11 %). [9] introduces *opal*, an ontology-based web pattern analysis approach that extracts semantic information (such as properties, values, and classes) from web forms exploiting a number of prolog rules. The classification step (that maps terms to concepts) requires terms in the domain to be annotated to create a set of ground facts, such as, City(London). [13] tackles the same problem but instead of having a logic-based approach, they use a machine learning approach (they use Bayesian classifiers). Some manual labor is also required to build the training sets of the classifiers. [17] presents the never-ending learning paradigm for machine learning, that endlessly extends a small original ontology (Tbox) with new classes, properties, and facts (Abox) from the web.

It is worth noticing that in the works mentioned above, the authors start from text, and then they populate the classes/relation in the ontology. In our scenario, we start from the ontology, and look for its instances in text.

We also include under the OBIE discussion the extraction of attributes and values such as the works presented in [4,6,12,14,23]. Although they do not constitute an ontology, the same techniques can be applied to extract ontological properties (in fact, that is what we do). These studies propose similar unsupervised frameworks that extract a set of properties and values from HTML semi-structure data. After this core set has been extracted, in [4,23], they use machine learning (Conditional Random Fields) to expand the set of values by extracting values from free text. [4] takes into account the popularity of the properties, as we do, but they use user reviews instead of user's queries. [12] extracts first an initial seed of property-values from user queries, and then it uses distant supervision to extend it. [23,26] also present methods to aggregate semantically equivalent property names, in Sect. 5 we discuss the differences with our approach. [18] uses supervised learning to classify HTML fragments, and then combines two techniques to extract the property-values: the first strategy is similar to the one in [23], and the second one is based on the outcome of a number of noisy annotators that rely on manually labeled data.

Triple Extraction from Text: Works in this field aim to detect semantic relations between instances in the text. [24] presents two complementary approaches to extract relations: a rule-based approach and a machine learning approach. Their rule-based approach provides high precision but it requires manual work to build the rules. The ML based approach relies heavily on a number of NLP tools. The systems presented in [8,14] follow a similar approach to extract triples from unstructured text. In particular, [14] is a plugin for [21]. Observe that their goal, in all the cases above, is different from ours. They aim to discover as many relations as they can (+7000 in [24]). Here we aim to extract a small set of "helpful" properties for and from an automatically selected taxonomy subtree.

4   Overview of the Solution

In this section we describe the different steps of our approach. These steps are depicted in Fig. 2. Observe that this solution is general, and it does not depend on any Rakuten-specific artifact, therefore in this section we only explain *what* each step should do. In Sect. 5 we describe *how* we implemented each of these steps in the case of Rakuten Ichiba. Also note that the production of the formal OWL ontology and RDF triples is not mandatory in this process. Although in our implementation we do output the ontology in this format (c.f. Sect. 5.2), the adoption of W3C standards for these artifacts in Rakuten is still under consideration, and this work represents the first steps towards it. We will make use of Description Logics (DL) terminology, but if the reader is not familiar with this formalism s/he can skip these parts.

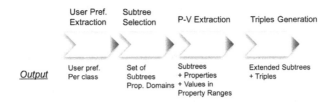

Fig. 2. Extracting semantic information to extend an ontology

User Preference Extraction: Since the end goal is to assist the users in their purchases, the first step is to extract relevant information about their shopping behavior. What data is needed here depends on the particular implementation of the following steps, but query logs and browsed items are some examples of the type of data that can be used.

Subtree Selection: Intuitively, relevant subtrees are those that: *(i)* have *high business impact*; *(ii)* can benefit from *new discovery axes*, i.e. from a UI perspective, they need some navigational assistance; and *(iii)* are *semantically homogeneous* from the user perspective. A class has a high need for navigational assistance (NNA) if it is "sufficiently far" from purchases, that is, the purchases occur in one of its non-immediate subclasses, or the diversity of the class in terms of the number of popular items is high. A class has a high business impact (BI) if the items in the class contribute significantly to the GMS. For instance, Water is a class with high BI but with low NNA because it contains few popular items which amount to a large sales volume. On the other hand, Red Wine has both high BI and high NNA because the large sales volume of the class is distributed over a high number of moderately popular items. A subtree of the taxonomy is semantically homogeneous from the user perspective (UH) if users associate the classes in the subtree with similar terms. For instance, although American whisky and wine are semantically similar (alcoholic drinks), these classes are reached in different ways. Most searches leading to American whisky (in Japan) are very specific, such as "Old Crow"; on the other hand, searches leading to red wine, white wine, and even Champagne tend to be more general and often overlap, for instance "French wine" (non-alcoholic wine would be in a different subtree). Observe that this way of splitting the tree does not only consider the semantics of the class, but also the behavior of the shoppers. In our example, shoppers buying American whisky will tend to have (statistically) a better background in that domain, while shoppers buying wine are more diverse. This process is illustrated in Fig. 3.

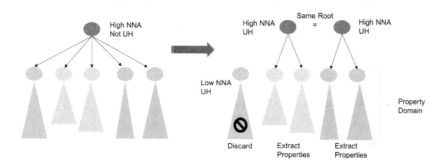

Fig. 3. Subtree selection process

Property-Value Extraction: This step consists of extending a subtree, T, selected in the previous step, with a *relevant* set of data property. The definition of relevant depends on the particular goal, but in any case it must align with the requirements of the particular use-case. The domain of these new properties is the union of the classes in $T \setminus root(T)$, that is, all the classes in T minus the root. Formally, let C be the root of the tree representing the subclass relation, $D_1 \ldots D_n$ the subclasses of C in T, and R a new relation, then the ontology (Tbox) is extended with the following OWL 2 DL axiom (in DL notation for succinctness): $\exists R \equiv D_1 \sqcup \cdots \sqcup D_n$.

Once the ontology for the subtree T has been extended with the set of relevant properties $R_1 \ldots R_n$, the next step is to find a fragment of the range of these properties (literals) that is popular among the users that browse the tree T, and that will be used to generate the triples. In this work we focus on data properties, but this work can be extended to object properties using the approaches described in [9,13,17].

Triples Generation: The final step is to link the items in T with the properties and values which have been extracted. Observe that depending on the scenario, a confidence score might be attached to each triple (s, p, o) indicating how certain the system is that the item s has the property p with value v. In such case, one would need to reify the tuples using the standard technique [3].

5 Architecture

In Sect. 4 we described the general flow of the approach. In this section we present how we implemented the different steps in Rakuten. The architecture, which is illustrated in Fig. 4, can be divided in four different layers: (i) the inputs, i.e., artifacts such as the Rakuten taxonomy, the search logs, etc.; (ii) the computational framework on which we run the different modules of the system; (iii) the processing layer, in charge of generating the intermediate artifacts from the input: sets of subtrees, tokenized item descriptions, word2vec models, etc.; (iv) the extension layer, in charge of extending the taxonomy using the intermediate artifacts and exposing the output as RDF. First we briefly describe the input layer and the computational framework and then we give closer looks to the processing and extension layers.

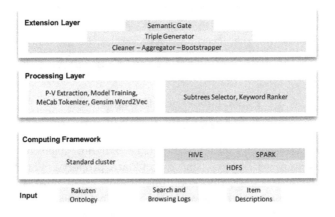

Fig. 4. System architecture

Input Layer: The system takes the following artifacts as input:

- *Rakuten Taxonomy:* The Rakuten taxonomy contains 38,167 classes: 35 with depth 1, 405 with depth 2, 3,790 with depth 3, 15,849 with depth 4, and the remaining classes with depth 5. Each class has exactly one parent except for the root. Unfortunately it is not formalized in a well-known ontology language. The intended semantics behind the parent-child relation is the same one as `owl:subclass`.

- *Search and Browsing Logs:* This dataset (2TB per year) contains information about all the search queries executed by users of Rakuten Ichiba: keywords, class being browsed, items browsed after the query, etc.
- *Item Descriptions:* This dataset (800GB per year) contains the HTML pages describing each item sold in Rakuten Ichiba, together with the class in which they belong and other metadata irrelevant for this project.

Computing Framework: Given the size of the datasets, the system relies on distributed computing frameworks to perform computations at the level of the full taxonomy, i.e. computations which require considering all the classes. A Hive cluster with 1000 nodes is used to pre-process and analyze the query logs then score the search keywords. A Spark cluster with 200 nodes is used to select the subtrees for which the taxonomy needs to be extended. On the other hand, the processing of the individual subtrees is independent and bounded in memory and computational requirements for each instance. Therefore, the subtree-level pipeline, which consists of a series of Python scripts, is run on a standard cluster and does not require a particular computing framework to scale.

5.1   Processing Layer

Subtree Selector. This modules selects the subtrees in the Rakuten taxonomy that will be extended. Next we give concrete definition of the different abstract concepts listed in Sect. 4.

Need for Navigational Assistance: To measure the need for new discovery axes of a subtree, we compute its GMS *diversity*, defined as $\exp(-\sum_i p_i \ln p_i)$ where the sum is over the items in the subtree and p_i is the proportion of the total GMS of the subtree which is due to the item i. This is the exponential of the Shannon entropy of the subtree's item-level GMS which intuitively represents the effective number of items in a subtree making up its GMS. A subtree is said to have a high need for navigational assistance (NNA) if its effective number of items is more than $Z_1 = 2^{15}$. It is said to have a low NNA, and is therefore discarded, if its effective number of items is less than $Z_2 = 2^7$. These values for Z_1 and Z_2 are based on an initial exploration of the data and the final values will be decided by the catalog team.

Business Impact: The business impact is not used in addition to the NNA requirement as we found in practice that subtrees not discarded by the previous requirement have high enough business impact. Indeed, a counter-example would necessitate a large number of items with very low and almost evenly distributed sales volumes, which is not found in our datasets.

Semantic Homogeneity: We use search query logs to measure how semantically homogeneous a given subtree t_1 is. For each node with depth 1 in t_1, we compute the set of search keywords leading to that node (class). A keyword is said to lead to a class if a user searching for this keyword clicked on an item of this class

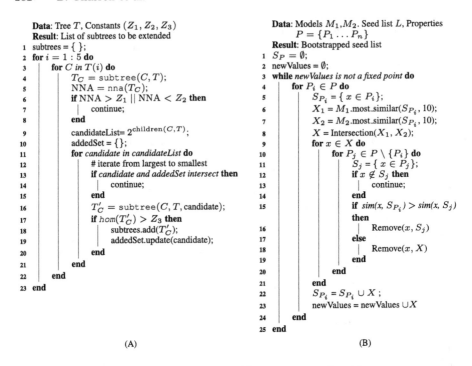

Fig. 5. Subtree selection and bootstrapping algorithms

immediately after the query. Then the subtree is said to be homogeneous if the number of such keywords is larger than $Z_3 = 30$, a value determined empirically.

In Fig. 5(A) we show the algorithm that we use to find the subtrees in the taxonomy. For the sake of clarity we use a number of functions that we will informally define next. Let T a tree, C a class in R, and M a model built as explained above, and S a set of classes. Then $T(i)$ returns the set of classes in T with depth i; children(C, T) returns the children of C in T; subtree(C, T) returns the subtree of T "hanging" from C; subtree(C, T, S) returns the subtree T_1 of T hanging from C but restricting the nodes with depth 1 in T_1 to those in S; nna(T) returns the effective number of items in T; hom(S) returns the homogeneity of T.

Property-Value Seed Extraction. We extract the initial set of properties and values (PV) from HTML tables and semi-structured text inputted by merchants in the items descriptions as these are easy to parse, quite accurate and reflect the domain knowledge of the merchants. There are several approaches in the literature to extract information from HTML tables [6,10,23], here we adopt the one used in [23] (Sect. 4.1.1) and use a slightly modified version of their implementation. Intuitively, the property names are first extracted from the headers of HTML tables in the item descriptions, and the associated values are found as the adjacent keywords either in the tables or in semi-structured text.

Model Training. The initial PV list obtained only from HTML tables and semi-structured text can lack a number of popular values, depending on the class. To increase the coverage of the property range we bootstrap the list using neural models of context similarity (word2vec). This module is in charge of training these models on the set of item descriptions within a given subtree.

We use word2vec's CBOW model with negative sampling to find words which appear in item descriptions within similar contexts. The item descriptions are first stripped from non-text features such as HTML tags and URLs. Then, they are tokenized using Mecab, a Japanese language morphological analyzer of which only the tokenizing part is used. Two models are trained on the resulting data. The first one is directly trained on the tokenized descriptions seen as bags of words. The second one is trained on the tokenized descriptions after performing collocation using popular search keywords extracted previously (by the Keyword Ranker). Collocation is done in two steps due to the specificities of the Japanese language. The first step is to join adjacent tokens into popular "words". Indeed, as the words are usually not separated by spaces or punctuation in Japanese, the tokenizer may cut words into several tokens. The second step is to join the resulting words into popular ngrams (up to trigrams).

After training we obtain two models trained on slightly different representations of the item descriptions. This module relies on the word2vec implementation of the library gensim.

5.2   Extension Layer

Seed Cleaning. The initial PV list extracted in the processing layer has usually a fairly high precision but is not usable as is. The first issue is the existence of redundant property names, meaning that the merchants use different words to identify the same concept. Redundant property names can either be *(i)* different terms, such as (manufacturer) and (maker), or *(ii)* the same term written in different alphabets or combinations thereof such as and (grape variety)[3]. Another issue is the existence of values that are not useful as discovery axes, such as expiration date, model numbers or long ingredient lists. It is critical to point out that this information might be accurate but is not deemed relevant for the purpose of this project. This is why we do not aim to obtain a complete model of the domain, but to extract the core fragment that is relevant to the users.

Properties Aggregation: We first remove redundant property names. For this we develop the following score function. Let P_1 and P_2 be two properties in the seed, m_1 and m_2 their respective range sizes and n the size of the intersection of their ranges. Their similarity score function is defined as:

$$L(P_1, P_2) = L_{\text{conf}}\left(\frac{n}{\min(m_1, m_2)}\right) \times L_{\text{size}}\left(\frac{\min(m_1, m_2)}{\max(m_1, m_2)}\right) - L_{\text{error}}\left(\frac{1}{n}\right)$$

---

[3] The Japanese language has three different alphabets: Hiragana, Katakana, and Kanji.

where L_{conf} is an increasing function representing the naive confidence that two properties are similar if they share many values respective to the maximum number of shared values, L_{size} is a decreasing function which tempers that confidence if the properties have comparable range sizes and L_{error} is an increasing function, with value 0 at 0, that increases as the number of shared values decrease, modelling the uncertainty over the score computed by L_{conf} and L_{size}. In practice we use $L_{conf}(x) = x$, $L_{size}(x) = \exp(-ax)$ and $L_{error}(x) = bx$, with the ad-hoc parameters $a = 0.33$ and $b = 0.1$. Two properties are considered similar if their similarity score is larger than 0.1 and the equivalence classes for this relation are computed. For each of these classes, we pick the representative that occurs more often in item descriptions as final property name.

Observe that [23,26] also performs an aggregation step. Intuitively, in [23] two properties are aggregated into a vector if they have a popular value (among merchants) in common. Two vectors are aggregated if the cosine similarity is above a given threshold. We empirically observed that the score function presented here can more accurately single out synonym properties since it does not depend on a single value occurring multiple times, but on the set of shared values and on the property sizes. In [26] they use cosine similarity the aggregate property names, again, in our experiments (using word2vec) we obtained better results using the score function presented here. As in [23], two properties are not aggregated if both are found to appear in the same item description during the seed extraction.

Values Filtering: The next step is to clean the properties' ranges by discarding any non-popular value, measured by their frequency in the search queries logs (obtained by the Keyword Ranker). The result of these two steps is a small list of property-values pairs with low redundancy and high precision which is representative of the interests of the users.

Bootstrapping. We then expand the coverage of the PV seed obtained so far. The bootstrapping algorithm, simplified for the sake of presentation, is shown in Fig. 5(B). We use two models, described in the previous subsection, to mitigate spurious high similarity between words that are not semantically similar as caused by the text pre-processing and tokenizing errors (particularly relevant for Japanese). Another use of the two similar models is to introduce a natural stopping condition to the bootstrapping algorithm. For each property, we only consider the 10 words most similar to the current range for each model then intersect the two outputs. This overcomes the problem of setting a meaningful threshold on similarities provided by word2vec.

The algorithm iterates over the property list, P, adding new values to the range of each property, S_{P_i}, until no more new values are found (newValues reaches a fixed point). For a new keyword x to be added to the property range S_{P_i} two conditions must hold: *(i)* all the models (two in this case) must agree that x is similar to S_{P_i} (lines 8–9); *(ii)* there should not be another property P_j such that x is more similar to S_{P_j} than to S_{P_i} (line 12). Observe that if x is added to S_{P_i} and also belongs to a less similar S_{P_j}, then x is removed from S_{P_j} enforcing the disjointness of values in the property ranges.

Intuitively, the function most_similar finds the top n words in the vocabulary that are most similar to the words in the range of the property P. More specifically, it finds the top n vectors that maximizes the multiplicative combination of the cosine similarities (originally proposed in [15]) between the given set of vectors, S_{P_i}, and the candidate vector in the model vocabulary $V \setminus S_{P_i}$:

$$score_{M_i}^{S_P}(candidate) = \Pi_{v \in S_P} \cos(\texttt{candidate}, \texttt{v})$$

Triple Generator. This modules takes all the items titles/descriptions in a given subtree t_1, and the bootstrapped property-value list for t_1. For every item I and every property P, it first look for the values of P in the HTML tables and semi-structured text of the description of I, if it cannot find it looks for the values in the title of I, and finally it looks for the value in the free text in the description of I. If two different values for P appear together in one of the three steps above, it ignores them and moves to the next step. Once it finds a value v, for P in I, it generates the triple (I, P, v).

Semantic Gate. The semantic gate is in charge of exposing the triples through a SPARQL end-point, and moving the new extended ontology into OWL 2 (DL). Recall that the existing taxonomy is not available in any well-known ontology language. For the SPARQL end-point we use Sesame Workbench[4] and Ontop[5]. Ontop implements an ontology-based data access (OBDA) approach. Interested readers can look at [5,20]. We decided to use OBDA since it is a non-invasive way to introduce semantic standards (RDF/OWL/SPARQL) to the different business units in Rakuten, and it still allows the different departments to access the data through standard SQL, in which they are already proficient.

5.3   Limitations

The current implementation has two major limitations that we will work on in the future. The first one is that it only handles words as property values. Thus, alphanumeric values such as 100 ml or 2 kg cannot be handled at the moment, and therefore properties such as size are discarded. Extending our approach to handle this does not present any technical challenge but it requires time to implement it in such a way that it is not detrimental for performance. The second limitation of this implementation is that we only consider subtrees with a root with depth at most 3.

---

[4] http://rdf4j.org/sesame/2.8/docs/articles/workbench.docbook?view.
[5] http://ontop.inf.unibz.it/.

6 Experiments

At a high level, the system divides the taxonomy extension into the independent extension of subtrees. As such, we will use a two-fold approach to assess the system capabilities. First we look into the division of the taxonomy in relevant subtrees. Second we analyze the precision and coverage of the extension of the subtrees.

6.1 Subtree Selection

Our system selected 1,251 subtrees from the taxonomy. There is no way to quantitatively evaluate the correctness of the subtree extraction without putting the results in production and A/B testing to check the response of the users. Thus we will only briefly discuss the results of this process looking at the selected subtrees with a non-empty intersection with the subtrees hanging from either of these two dissimilar classes: wine and small fashion items.

In both cases the selected subtrees are found to hang from classes at depth 2, meaning that any class above that was too diverse. Our algorithm selects 4 subtrees containing classes related to wine. The largest subtree, denoted T_w below, includes red wine, white wine, sparkling wine, wine sets and "other wines". The other wine-related classes at depth 3 (rosé wine, non-alcoholic wine, wine accessories) are separated in the remaining 3 subtrees. The height of these subtrees varies between 2 and 3, therefore containing classes of depth up to 4 or 5 in the original tree. For classes related to small fashion items, all classes at depth 3 (neckties, belts, handkerchiefs, key cases, etc.) are separated in different subtrees. The subtree containing neckties is denoted T_n below.

In the case of the wine-related classes, the fact that non-alcoholic wines and wine accessories are separated from the others is consistent with the fact that they are semantically different. However, it could be argued that rosé should belong to the same subtree as red wine, etc. An explanation would be that shoppers searching for rosé wine are more knowledgeable of the domain and therefore use more specific terms in their searches, compared to more mainstream red, white or sparkling wines. Therefore, this separation allows to provide more specific properties to the shoppers browsing the rosé wine class. On the other hand, the classes related to small fashion items are very semantically diverse and would not benefit from being aggregated.

Subtree Extension. The extension of the taxonomy at the subtree-level consists of first extracting core properties and values (PV), then linking these to the items to generate triples. We evaluate these two steps separately.

Comparison to manual work: The PV extraction is currently done manually by domain experts, which we aim to supersede with our automated approach. A first quality criterion to compare both approaches is the recall, in this case the ability to discover all the relevant properties for a given subtree. The catalog team provided us with the outcome of the manual work on rice and beef, corresponding respectively to subtrees T_r and T_b below, which we used to summarize

Subtree	M \ A	M ∩ A	A \ M
T_r	1	3	3
T_b	2	3	6

(A)

Subtree	count	overall	max	median	mean	min
T_r	6	0.92	1.00	0.97	0.81	0.20
T_b	9	0.86	1.00	0.88	0.83	0.50
T_w	9	0.91	1.00	0.81	0.77	0.20
T_n	8	0.88	1.00	0.85	0.70	0.00

(B)

Fig. 6. Results for the property-value extraction

the overlaps of the extracted properties for both approaches in the Table (A) in Fig. 6.

The properties missed by the automated process (M\A) are: for rice, Size, which was discarded due to the alphanumeric values; for beef they are Brand, which was discarded due to the infrequency of the values, and Intended Use, which did not appear in the semi-structured text and is arguably not a property of the items. The automated process, on the other hand, found several properties that where missed by humans (A\M). In the case of rice, Shipping fee, Country of production and Composition. In the case of beef, Shipping fee, Ingredient, Product Name, Processing area, Country of production and Allergens.

For the properties which were extracted by both processes (M ∩ A), the total number of values for rice was 47 for the manual process and 102 for the automated process, while it was respectively 18 and 43 for beef. This, combined with the accuracy discussed below shows that the automated process increases the coverage by a large margin.

Accuracy of PV pairs: A second quality criterion for the PV extraction is the accuracy of the property-value pairs. Assuming naturally that the manual extraction by domain experts has a satisfying accuracy, we independently assess the accuracy of our automated process by manually annotating its results and checking a business minimal requirement of 80 %. In the Table (B) in Fig. 6 we analyze the accuracy in PV pairs extracted for the four subtrees T_w, T_n, T_r and T_b described previously. The table shows for each subtree: the number of properties, the overall accuracy of the pairs, and the distribution of the accuracies by property (max, median, mean, min). Observe that the overall accuracy is above 80 % in each subtree and the median property accuracies are also high. The properties with minimal accuracies in T_r, T_b, and T_w are properties with small initial ranges which were erroneously extended in the bootstrapping process. In T_n there is an ambiguous property that we could not evaluate its correctness and therefore assumed that all its values were wrong.

Analysis of triples: The triples are evaluated by annotating 50 randomly drawn triples for each subtree. Each triple was annotated with one of the three following labels: (i) "Correct" if the property and value are correct for the item; (ii) "Wrong pair" if the property and value would be incorrect for any item in the subtree; (iii) "Wrong linking" if the property and value would be correct for some other items in the subtree but not for the current item. We summarize the proportions of these labels in the Table (A) in Fig. 7. As we can see, the accuracy is not consistently above our target of 80 %.

However, in the context of the business implementation of this work, a manual step will ultimately be added to identify the incorrect pairs in the PV extraction step (which is a limited amount of work, due to the moderate number of pairs), so that the final accuracy of the triples would be above 80 %.

Subtree	correct	wrong pair	wrong linking
T_r	0.74	0.08	0.18
T_b	0.68	0.16	0.16
T_w	0.86	0.12	0.02
T_n	0.78	0.10	0.12

Property	#values	eff. #values	coverage
Place of Origin	60	26.4	0.75
Country of Origin	15	2.3	0.24
Rice Variety	35	9.6	0.79
Composition	5	3.4	0.65
Category	7	3.9	0.57
Shipping Fee	4	2.5	0.53

(A)   (B)

Fig. 7. Results for the triples generation

We then compute the coverage for each property as the proportion of the items in the subtree with a triple containing this property. Another interesting measure is the effective number of values for a property. It is computed as $\exp(-\sum_v p_v \ln p_v)$ where the sum is over the values of the property (plus a catch-all "Unknown" value) and p_v is the proportion of the items linked to this property and the value v (for the "Unknown" value it is the proportion of items not covered by the property). In Fig. 7(B) we show these measures for rice (T_r) as well as the number of value per property. The coverage is found to be substantial with several properties over 50 %. Note how the property Country of Origin has 15 values but only an effective number of values of 2.3 as most of rice sold in Rakuten Ichiba is from Japan; the effective values being Japan and "Unknown" as most merchants do not feel the need to mention Japan, which is reflected in the low coverage of the property.

7   Conclusion

In this work we propose an end-to-end unsupervised approach that automatizes the process of extracting semantic information from text to: (i) extend the *relevant* fragments of a taxonomy with data properties deemed helpful to the user; and (ii) tag the items with this new information by generating the corresponding triples. We presented a novel technique to select relevant subtrees in the taxonomy in terms of overall profitability, ease of product discovery and similarity of users' shopping behavior, as well as a number of techniques to clean the text sources, aggregate equivalent properties and extend the range of the properties through text mining of semantic similarity. The approach presented here can be easily ported to other scenarios beyond Rakuten Ichiba since it is language/data/technology independent. We provided a number of experiments showing the effectiveness of our approach in terms of precision and coverage.

In the future we plan to work on extracting object properties and axioms, lift the limitations discussed in Sect. 5, extract information from images, and trying to make publicly available an SPARQL endpoint with information about Rakuten Ichiba products.

References

1. Anantharangachar, R., Ramani, S., Rajagopalan, S.: Ontology guided information extraction from unstructured text. CoRR abs/1302.1335 (2013)
2. Barkschat, K.: Semantic information extraction on domain specific data sheets. In: Presutti, V., dAmato, C., Gandon, F., dAquin, M., Staab, S., Tordai, A. (eds.) ESWC 2014. LNCS, vol. 8465, pp. 864–873. Springer, Heidelberg (2014). doi:10. 1007/978-3-319-07443-6_60
3. Berardi, D., Calvanese, D., Giacomo, G.D.: Reasoning on UML class diagrams. Artif. Intell. **168**(1–2), 70–118 (2005)
4. Bing, L., Wong, T.L., Lam, W.: Unsupervised extraction of popular product attributes from e-commerce web sites by considering customer reviews. ACM Trans. Internet Technol. **16**(2), 12:1–12:17 (2016)
5. Calvanese, D., Cogrel, B., Ebri, S.K., Kontchakov, R., Lanti, D., M.R., Rodriguez-Muro, M., Xiao, G.: Ontop: Answering SPARQL queries over relational databases. Semant. Web J. (2016)
6. Chen, H.H., Tsai, S.C., Tsai, J.H.: Mining tables from large scale html texts. In: Proceedings of the 18th Conference on Computational Linguistics, pp. 166–172. Association for Computational Linguistics (2000)
7. Ding, Y., Fensel, D., Klein, M.C.A., Omelayenko, B., Schulten, E.: The role of ontologies in ecommerce. In: Handbook on Ontologies, pp. 593–616. Springer (2004)
8. Exner, P., Nugues, P.: Entity extraction: from unstructured text to dbpedia rdf triples. In: International the Semantic Web Conference, ISWC 2012, pp. 58–69. CEUR (2012)
9. Furche, T., Gottlob, G., Grasso, G., Guo, X., Orsi, G., Schallhart, C.: Real understanding of real estate forms. In: WIMS, p. 13 (2011)
10. Gatterbauer, W., Bohunsky, P., Herzog, M., Krüpl, B., Pollak, B.: Towards domain-independent information extraction from web tables. In: Proceedings of the 16th International Conference on World Wide Web, pp. 71–80. ACM (2007)
11. Giese, M., Soylu, A., Vega-Gorgojo, G., Waaler, A., Haase, P., Jiménez-Ruiz, E., Lanti, D., Rezk, M., Xiao, G., Özçep, Ö.L., Rosati, R.: Optique: zooming in on big data. IEEE Comput. **48**(3), 60–67 (2015)
12. Gupta, R., Halevy, A., Wang, X., Whang, S., Wu, F.: Biperpedia: an ontology forsearch applications. In: Proceedings of the 40th International Conference on Very Large Data Bases (PVLDB) (2014)
13. He, H., Meng, W., Lu, Y., Yu, C., Wu, Z.: Towards deeper understanding of the search interfaces of the deep web. World Wide Web **10**(2), 133–155 (2007)
14. Krestel, R., Witte, R., Bergler, S.: Predicate-Argument EXtractor (PAX). In: New Challenges for NLP Frameworks, pp. 51–54. ELRA (2010)

15. Levy, O., Goldberg, Y.: Linguistic regularities in sparse and explicit word representations. In: Conference on Computational Natural Language Learning, pp. 171–180. Ann Arbor, Michigan, June 2014

16. Mikolov, T., Chen, K., Corrado, G., Dean, J.: Efficient estimation of word representations in vector space. CoRR abs/1301.3781 (2013)

17. Mitchell, T.M., Cohen Jr., W.W., Hruschka, E.R., Talukdar, P.P., Betteridge, J., Carlson, A., Mishra, B.D., Gardner, M., Kisiel, B., Krishnamurthy, J., Lao, N., Mazaitis, K., Mohamed, T., Nakashole, N., Platanios, E.A., Ritter, A., Samadi, M., Settles, B., Wang, R.C., Wijaya, D.T., Gupta, A., Chen, X., Saparov, A., Greaves, M., Welling, J.: Never-ending learning. In: AAAI, Texas, USA, pp. 2302–2310 (2015)

18. Qiu, D., Barbosa, L., Dong, X.L., Shen, Y., Srivastava, D.: DEXTER: large-scale discovery and extraction of product specifications on the web. PVLDB 8(13), 2194–2205 (2015)

19. Reinberger, M.L., Spyns, P.: Discovering knowledge in texts for the learning of dogma-inspired ontologies. In: ECAI 2004 Workshop on Ontology Learning and Population (2004)

20. Rodriguez-Muro, M., Rezk, M.: Efficient SPARQL-to-SQL with R2RML mappings. J. Web Sem. 33, 141–169 (2015)

21. Saggion, H., Funk, A., Maynard, D., Bontcheva, K.: Ontology-based information extraction for business intelligence. In: Aberer, K., et al. (eds.) ASWC/ISWC - 2007. LNCS, vol. 4825, pp. 843–856. Springer, Heidelberg (2007). doi:10.1007/978-3-540-76298-0_61

22. Schutz, A., Buitelaar, P.: RelExt: a tool for relation extraction from text in ontology extension. In: Gil, Y., Motta, E., Benjamins, V.R., Musen, M.A. (eds.) ISWC 2005. LNCS, vol. 3729, pp. 593–606. Springer, Heidelberg (2005). doi:10.1007/11574620_43

23. Shinzato, K., Sekine, S.: Unsupervised extraction of attributes and their values from product description. In: Sixth International Joint Conference on Natural Language Processing, IJCNLP 2013, pp. 1339–1347 (2013)

24. Wang, C., Kalyanpur, A., Fan, J., Boguraev, B., Gondek, D.: Relation extraction and scoring in deepqa. IBM J. Res. Dev. 56(3), 9 (2012)

25. Wimalasuriya, D.C., Dou, D.: Ontology-based information extraction: an introduction and a survey of current approaches. J. Inf. Sci. 36(3), 306–323 (2010)

26. Yoshida, M., Torisawa, K.: A method to integrate tables of the world wide web. In: International Workshop on Web Document Analysis (WDA 2001), pp. 31–34 (2001)

Building Urban LOD for Solving Illegally Parked Bicycles in Tokyo

Shusaku Egami[1]([✉]), Takahiro Kawamura[1,2], and Akihiko Ohsuga[1]

[1] Graduate School of Informatics and Engineering,
The University of Electro-Communications, Tokyo, Japan
egami.shusaku@ohsuga.lab.uec.ac.jp, ohsuga@uec.ac.jp
[2] Japan Science and Technology Agency, Tokyo, Japan
takahiro.kawamura@jst.go.jp

Abstract. The illegal parking of bicycles is an urban problem in Tokyo and other urban areas. The purpose of this study was to sustainably build Linked Open Data (LOD) for the illegally parked bicycles and to support the problem solving by raising social awareness, in cooperation with the Bureau of General Affairs of Tokyo. We first extracted information on the problem factors and designed LOD schema for illegally parked bicycles. Then we collected pieces of data from Social Networking Service (SNS) and websites of municipalities to build the illegally parked bicycle LOD (IPBLOD) with more than 200,000 triples. We then estimated the missing data in the LOD based on the causal relations from the problem factors. As a result, the number of illegally parked bicycles can be inferred with 70.9 % accuracy. Finally, we published the complemented LOD and a Web application to visualize the distribution of illegally parked bicycles in the city. We hope this raises social attention on this issue.

Keywords: Linked Open Data · Urban problem · Illegally parked bicycles

1 Introduction

An increased awareness of health problems [1] and energy conservation [2] led to a 2.6-fold increase in bicycle ownership in Japan from 1970 to 2013. Consequently, illegally parked bicycles around railway stations have become an urban problem in Tokyo and other urban areas. In addition to the insufficient availability of bicycle parking spaces, inadequate public knowledge on bicycle parking laws has contributed to this urban problem. Illegally parked bicycles obstruct vehicles, cause road accidents, encourage theft, and disfigure streets.

In order to address this problem, we believe it would be useful to publish the distribution of illegally parked bicycles as Linked Open Data (LOD). For example, it would serve to visualize illegally parked bicycles, suggest locations for optimal bicycle parking spaces, and assist with the removal of illegally parked bicycles. However, Open Data sets available for illegally parked bicycles are

© Springer International Publishing AG 2016
P. Groth et al. (Eds.): ISWC 2016, Part II, LNCS 9982, pp. 291–307, 2016.
DOI: 10.1007/978-3-319-46547-0_28

currently distorted, and it is difficult for services to utilize the data. In addition, other data concerning issues such as bicycle parking spaces and government statistics, have been published in a variety of formats. Hence, unification of the data formats and definition of schema for data storage are important issues that need to be addressed. Bischof et al. [3] proposed a method for integrating Open City Data as Linked Data and proposed methods for the complementation of missing values. The study improved the utilization of unreusable Open Data. However, more spatio-temporal data and factor data are necessary to develop services for combating illegally parked bicycles.

In this study, we first extracted domain requirements of illegally parked bicycles from articles on the Web and design LOD schema. Next, we collected data about illegally parked bicycles from Twitter and the data describing factors that affect the number of illegally parked bicycles. In order to reuse these data sets, which have different formats, we unify the data formats based on designed schema and publish the data on the Web as LOD. Moreover, we estimate the missing data (the number of illegally parked bicycles) based on the causal relations from the factors. Our predictions take into consideration factors such as time, weather, nearby bicycle parking information, and nearby points of interest (POIs). However, since there are cases that lack these factor values, the missing factor values are also complemented by searching similar observation data. We thus use Bayesian networks to estimate the number of illegally parked bicycles for data sets after complementation of the factors. These results are also incorporated to build LOD with a specified property. In addition, we develop a service that visualizes the illegally parked bicycles using the constructed LOD. This visualization service raises the awareness of the issue in local residents and prompts users to provide more information about illegally parked bicycles. Therefore, our contributions are as follows.

1. Proposal of a methodology for designing LOD schema for an urban problem
2. Collection of data from SNS and municipalities of Tokyo and other urban areas, and the building of illegally parked bicycle LOD (IPBLOD)
3. Development and evaluation of an approach for complementing the missing factor values and estimating the missing values
4. Development and publishing of a Web application for visualizing illegally parked bicycles in Tokyo and other urban areas

The remainder of this paper is organized as follows. In Sect. 2, related works of data collection and urban LOD are described. In Sect. 3, the methodology for designing the LOD schema and IPBLOD are presented. In Sect. 4, two approaches that complement the missing factors, and estimate the illegally parked bicycles using Bayesian networks, are described. Also, we evaluate our results and our findings. In Sect. 5, visualization of the IPBLOD is described. Finally, Sect. 6 concludes this paper with future works.

2   Related Work

In most cases, LOD sets have been built based on existing databases. However, there is little LOD available so far that describes urban problems. Thus, methods for collecting new data to build urban LOD are required. Data collection methods for building Open Data include crowdsourcing and gamification. A number of projects have employed these techniques. OpenStreetMap [4] is a project that creates an open map using crowdsourced data. Anyone can edit the map, and the data are published as Open Data. Similarly, FixMyStreet [5] is a platform for reporting regional problems such as road conditions and illegal dumping. Crowdsourcing to collect information in FixMyStreet has meant that regional problems are able to be solved more quickly than ever before. Zook et al. [6] reported a case, where crowdsourcing was used to link published satellite images with OpenStreetMap after the Haitian earthquake. A map of the relief efforts was created, and the data were published as Open Data. Celino et al. [8] proposed an approach for editing and adding Linked Data using a game with a purpose (GWAP) [7] and human computation. However, since the data concerning illegally parked bicycles are time-series data, it is difficult to collect data using these approaches. Therefore, new techniques are required. We propose a method to build urban LOD while complementing the missing data.

Also, there have been studies about building Linked Data for cities. Lopez et al. [9] proposed a platform that publishes sensor data as Linked Data. The platform collects streamed data from sensor and publishes Resource Description Framework (RDF) in real time using IBM InfoSphere Stream and C-SPARQL [10]. The system is used in Dublinked2[1], which is a data portal of Dublin, Ireland, that publishes information of bus routes, delays, and congestion updated every 20 s. However, since embedding sensors is costly, this approach is not suitable for our study.

Furthermore, Bischof et al. [3] proposed a method for the collection, complementation, and republishing of data as Linked Data, as with our study. This method collects data from DBpedia [13], Urban Audit[2], United Nations Statistics Division (UNSD)[3], and U.S. Census[4] and then utilizes the similarity among such large Open Data sets on the Web. However, we could not find the corresponding data sets and thus could not apply the same approach to our study.

3   Building LOD

In this study, we propose a method for sustainably building urban LOD and applying them to Tokyo and other urban areas. Managing urban problem data *joining* multiple tables in (distributed) RDBs is troublesome from the aspect of data interoperability and maintenance, since the urban problem is closely related

---

[1] http://www.dublinked.ie/.
[2] http://ec.europa.eu/eurostat/web/cities.
[3] http://unstats.un.org/unsd/default.htm.
[4] http://www.census.gov/.

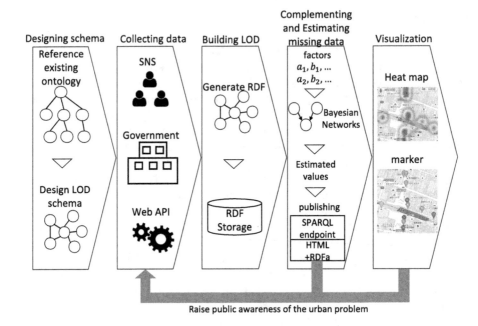

Fig. 1. Overview of this study

to multiple domains, such as government data, legal data, and social data as we already incorporated POIs and weather data in this application, and also those have different schemata. Thus, Linked Data is a suitable format as the data infrastructure of not only illegally parked bicycles, but also urban problems in general, since Linked Data can have advantages of flexible linkability and schema.

Figure 1 provides an overview of this study. This study is divided into the following five steps. Steps (2) to (5) are executed repeatedly as more input data become available.

1. Designing LOD schema
2. Collecting observation data and factor data
3. Building the LOD based on schema
4. Using Bayesian networks to estimate the missing number of illegally parked bicycles at each location
5. Visualizing illegally parked bicycles using LOD

3.1   A Methodology for Designing LOD Schema

Illegally parked bicycles can be observed by social sensors, as it is difficult to install physical sensors in the streets. In our previous work [11], the schema for IPBLOD was based on the Semantic Sensor Network ontology. However, in order to address this urban problem using LOD, the LOD should not only

Table 1. Results of clustered keywords

Property	Category	Keyword
Place and factor	POI	shopping street, large-scale retail stores, public office, school, library, public hall, department store, supermarket, bank, amusement, pachinko, commercial building
	bicycle parking	usability, safety, conforts, supply, fee
	train	train, distance from home to the station
	storage space	capacity
Time and factor	time	weekday, holiday, hour
Product	accident	congestion, blocking vegicle traffic
Factor	objective	commutation, destination
	weather	rainy
	bicycle	price reduction, deterioration, cost of maintenance, variety

have the number of illegally parked bicycles, location and time information but also contain the factors related to illegally parked bicycles, such as POIs and weather. In this paper, we present an LOD schema, including the factors related to the illegally parked bicycles, and we propose a methodology for designing LOD schema for urban problems, such as illegally parked bicycles.

In the ontology study, the methodology for building ontology has been discussed. We propose the methodology for designing practical LOD schema in reference to Activity-First Method [12]. The schema for IPBLOD is designed based on this methodology, which consists of two steps as follows:

1. Extraction of domain requirements
   a. Select an ontology that models the urban problem
   b. Search for articles on the urban problem using a search engine
   c. Extract keywords from the articles based on properties of the ontology
   d. Cluster the keywords
2. Designing schema
   a. Design classes based on the ontology
   b. Design instances and properties based on the result of the clustering

First, the existing ontology is selected in order to build LOD based on the ontology. To serve as a source of this problem, it is necessary to consider the accessibility of the LOD as well as the semantic consistency. Thus, we select Event Ontology (EO)[5] as a practical and intuitive structure wherein illegally parked bicycles can be considered as an event. In the EO, an event class has properties for place, time, agent, factor, and product.

---

[5] http://purl.org/NET/c4dm/event.owl.

Next we search for articles on illegally parked bicycles using Google. Then we investigate the top 10 articles and their references and then manually extract the keywords based on the properties of the existing ontology. Specifically, keywords are extracted from sentences that describe the place, time, agent, factor, and product. Even if keywords that are not defined in the ontology appear to be important in the article, the keywords are also extracted.

The extracted keywords are clustered manually as in Table 1. Then the classes are designed based on the EO. The expression of Description Logic (DL) is as follows:

$$IllegallyParkedBicycles \sqsubseteq Event$$
$$IllegallyParkedBicycles \sqsubseteq \exists place.SpatialThing$$
$$IllegallyParkedBicycles \sqsubseteq \exists time.TemporalEntity$$
$$IllegallyParkedBicycles \sqsubseteq \exists weather.WeatherState$$
$$IllegallyParkedBicycles \sqsubseteq \exists factor.Thing$$
$$IllegallyParkedBicycles \sqsubseteq \exists agent.Agent$$
$$IllegallyParkedBicycles \sqsubseteq \exists product.Thing$$
$$IllegallyParkedBicycles \sqsubseteq \exists value.Integer$$

The IllegallyParkedBicycles class refers to a set of illegally parked bicycles, and it is a subclass of the Event class. IllegallyParkedBicycles contains the place, time, weather, agent, factor, product, and number of illegally parked bicycles. Weather and the number of illegally parked bicycles are not defined in the EO, but since these are considered to be important in the domain of illegally parked bicycles, we add them to the LOD schema.

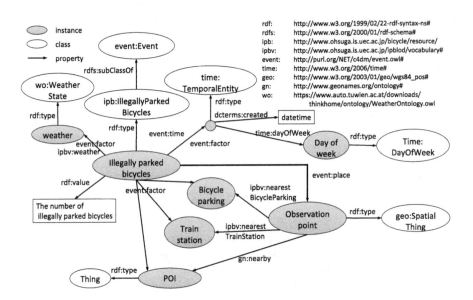

Fig. 2. LOD schema containing instances

Then we design their instances in reference to Table 1. Figure 2 shows an overview of LOD schema with the instances. In Table 1 the column of category refers to the instance, and the instance is linked with the EO property. Also, some instances are linked to other instances. However, it is difficult to obtain data on individuals, who park their bicycles illegally. Therefore, we omit the `event:agent` property. However, we use population statistics data instead of the data on individuals. We obtain population statistics from the portal site of official statistics of Japan called e-stat[6], and we use "population density per 1 km^2 of habitable area" and "the number of commuters who use trains." In fact, there are a large number of illegally parked bicycles near the stations, which are located in densely populated ateas in Japan. Many bicycles are illegally parked during the morning-commuting hours. Also, we omit the `event:product` property, since it is difficult to obtain data on accidents caused by illegally parked bicycles. In the same way the storage space, the objective of the person who parked the bicycle illegally, and the price of the bicycle are also omitted. Moreover, the POI, nearest bicycle parking, nearest train station, time, and weather are added to the LOD schema as factors related to illegally parked bicycles. The nearest train station is a resource of DBpedia Japanese[7].

Fig. 3. Screenshot of the tweet application

---

[6] http://www.e-stat.go.jp/SG1/estat/eStatTopPortalE.do.

[7] http://ja.dbpedia.org.

3.2   Collection of Observation Data

We started this study by collecting tweets containing location information, pictures, hash-tags, and the number of illegally parked bicycles. However, obtaining the correct locations from Twitter was difficult, since mobile phones often attach incorrect location information. Mobile phones are equipped with inexpensive GPS chips, and it is known that the accuracy is often low due to weather conditions and GPS interference area [15]. To address this problem, we developed a Web application that enables users to post tweets on Twitter after correcting their location information, and we made an announcement asking public users to post tweets of illegally parked bicycles using our application. Figure 3 shows a screen shot of this application. After OAuth authentication, a form and buttons are shown. When the location button is pressed, a marker is displayed at the user's current location on a map. The marker is draggable, thus allowing users to correct their location information. When users add their location information, enter the number of illegally parked bicycles, take pictures, and submit them, then tweets including this information with a hashtag are posted.

Furthermore, we collected information on POI using Google Places API[8] and Foursquare API[9]. Also, we obtained bicycle parking information from websites of municipalities and in cooperation with the Bureau of General Affair of Tokyo[10]. The Bureau of General Affairs of Tokyo publishes Open Data on bicycle parking areas as CSV. The data contain names, latitudes, longitudes, addresses, capacities, and business hours. More information was collected from municipalities, for example, monthly parking fees and daily parking fees. Also, we retrieved weather information from the website of the Japanese Meteorological Agency (JMA)[11].

3.3   Building LOD Based on Designed Schema

The collected data on illegally parked bicycles are converted to LOD based on the designed schema. Figure 4 shows the process of building IPBLOD. First, the server program collects tweets containing the particular hash-tags, the location information, and the number of illegally parked bicycles in real time. The number of illegally parked bicycles is extracted from the text of tweets using regular expressions.

Next, the server program checks whether there is an existing observation point within a radius of less than 30 m by querying our endpoint[12] using the SPARQL query. If there is no observation point on the IPBLOD, the point is added as a new observation point. In order to add new observation points, the nearest POI information is obtained using Google Places API and Foursquare API. The new observation point is generated based on the name of the nearest POI. It is possible to obtain the types of the POI from Google Places API and

---

[8] https://developers.google.com/places/?hl=en.
[9] https://developer.foursquare.com/.
[10] http://www.soumu.metro.tokyo.jp/30english/index-en.htm.
[11] http://www.jma.go.jp/jma/indexe.html.
[12] http://www.ohsuga.is.uec.ac.jp/sparql.

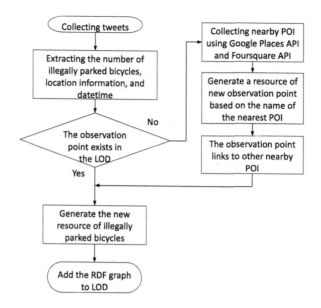

Fig. 4. Process of LOD building

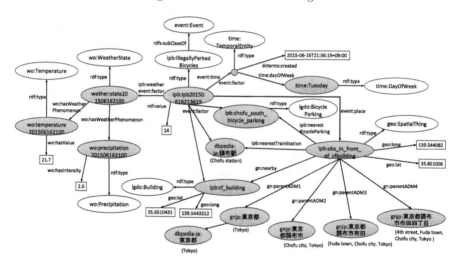

Fig. 5. Part of the integrated LOD

Foursquare API. We map the types of POI to classes in LinkedGeoData [16]. Thus, the POI is an instance of classes in LinkedGeoData. However, some POIs do not have a recognized types. Therefore, their types are decided by a keyword search with the name of the POI.

Then the address, prefecture's name, city name, and land lot name are obtained using Yahoo! reverse geocoder API, and then the links to GeoNames.jp[13] are generated based on the obtained information. GeoNames.jp is a Japanese geographical database. Thus, data are collected and added to the IPBLOD using Web APIs in real time. Figure 5 shows part of the IPBLOD. The LOD are stored in Virtuoso[14] Open-Source Edition. Also, the RDF data set is published with CC-BY license on our website[15].

4   Complementing and Estimating Missing Values

Since we rely on the public to observe illegally parked bicycles, we do not have round-the-clock data for every place, and thus, missing data in the IPBLOD are inevitable. However, the number of the illegally parked bicycles should be influenced by several factors, thus we try to estimate these missing data using Bayesian networks. If the data is expanded in density through the estimation, it will serve, for example, as the suitable location of bicycle parking spaces, the decision on variable prices of the parking fee and efficient timing of removal of illegally parked bicycles by the city, and part of the references for future urban design.

4.1   Complementing Missing Factors

As the factors (attributes), we use observation points, day of week, hours, precipitation, temperature, daily fee for the nearest bicycle parking, monthly fee for the nearest bicycle parking, "population density per $1\,km^2$ habitable are," "the number of commuters who use trains," and types of POIs. We selected Building, Bank, Games, DepartmentStore, Supermarket, Library, Police, and School as the types of POIs based on Table 1. However, there are also missing factor values. We assume that the missing factor values are similar to the corresponding factor value in the similar observation data. Therefore, we used the factor values of the similar observation data as substitutes for actual values which cannot be obtained. The similar observation data are found using Jaccard coefficient. Suppose the aggregates of each factor are given by Location, Day={sun, mon,...,sat}, Hour={0,1,...,23}, Precipitation={0,1,...}, Temperature={...,-1,0,1,...}, DailyFee={0,1,...}, MonthlyFee={0,1,...}, Density={0,1,...}, Commuters={0,1,...}, Building, Bank, DepartmentStore, Games, Supermarket, Library, Police, School={0,1}, and Number (of illegally parked bicycles)={1,...,4}, then the observation data are stored as an aggregate O of vectors $o \in Location \times Day \times Hour \times Precipitation \times Temperature \times DailyFee \times MonthlyFee \times Density \times Commuters \times Building \times Bank \times DepartmentStore \times Games \times Supermarket \times Library \times Police \times School \times Number$. The number of illegally parked bicycles is classified into four classes by Jenks natural breaks [14], which are often used in Geographic Information Systems (GISs). The range is 0 to 6, 7 to 17, 18 to 35, and 36 to 100. Therefore, the similarity of observation data o_1 and o_2 is $sim(o_1, o_2) = |o_1 \cap o_2|/|o_1 \cup o_2|$.

---

[13] http://geonames.jp/.

[14] http://virtuoso.openlinksw.com/.

[15] http://www.ohsuga.is.uec.ac.jp/bicycle/dataset.html.

4.2   Estimating the Number of Illegally Parked Bicycles Using Bayesian Network

We then estimate the number of illegally parked bicycles, at observation points, where the number data are missing. The input dataset is the dataset complemented using the method described in Sect. 4.1. We use the Bayesian network tool Weka[16] to estimate the unknown numbers of illegally parked bicycles. There are 897 observation data. The input data is a set O that consists of vectors with eight elements at first. We used HillClimber as a search algorithm, and also used Markov blanket classifier. The maximum number of parent nodes was two. As a result of 10-fold cross validation, we got 65.2 % accuracy.

To raise the accuracy, we focused on types of POIs. We did not restrict the types of POIs when building the IPBLOD, but we restricted types to the POIs contained in Table 1 when estimating the number of illegally parked bicycles using Bayesian networks. However, other POIs could become factors related to illegally parked bicycles. Hence, we first used all POI types as factors, and the number of POI types became 68. However, the accuracy became relatively low due to too many factors. Thus, we used super classes in LinkedGeoData ontology for clustering those types. Since we mapped the POI types to classes of LinkedGeoData, it was possible to obtain their super classes by querying the LinkedGeoData. As the result, the number of POI types became 46, as follows.

```
Pharmacy, Park, Retail, Restaurant, Police, University,
FastFood, BusStation, Gym, Parking, Church, Florist, Cafe,
Supermarket, Hospital, Nightclub, Sport, Advertising,
Casino, Hairdresser, Doctor, Bar, Bakery, Bank, TakeAway,
Amenity, Dentist, EmergencyThing, Hall, Office, Residential
, PlaceOfWorship, School, CommunityCentre, Building, Spa,
CarRental, VideoRental, Hotel, Cinema, CoffeShop,
Construction, Lawyer, HighwayThing, Shop,
PublicTransportThing
```

Therefore, an observation datum became a vector $o \in Location \times Day \times Hour \times Precipitation \times Temperature \times DailyFee \times MonthlyFee \times Density \times Commuters \times Pharmacy \times ... \times PublicTransportThing \times Number$, which resulted in 56 possible elements. Finaly, the average estimation accuracy of ten times 10-fold cross validation became 70.9 %. The maximum number of parent nodes was seven, after random sampling with a 90 % rate. We estimated the number of illegally parked bicycles on unobserved dates using the above parameters. Specifically, we examined the observation data in each observation point from the first observation date to the last observation date. If there are no data at 9 am or 9 pm, we estimated and complemented the number of illegally parked bicycles. Then, we added the estimated number and its probability to IPBLOD as follows.

---

[16] http://www.cs.waikato.ac.nz/ml/weka/.

```
@prefix ipb: <http://www.ohsuga.is.uec.ac.jp/ipblod/
    vocabulary#>
@prefix bicycle: <http://www.ohsuga.is.uec.ac.jp/bicycle/
    resource/>
bicycle:ipb_{observation point}_{datetime}
    ipb:estimatedValue [ rdf:value "0--7" ;
        ipb:probability  "0.772"^^xsd:double ]  .
```

4.3 Evaluation and Discussion

The observational data were collected from January 2015 to April 2016. Eighteen users posted data on Twitter using our application in Fig. 3. The number of triples included in the IPBLOD became more than 200,000. Table 2 shows statistics about the observation data. Observation points are places at which someone observed the bicycles, and the amount of observation data is the total number of submitted data (tweets), in which the same observation points may appear several times. As the result, Chofu City in Tokyo had the largest amount of the observation data. Since we posted promotional tweets on our Twitter accounts, contributors from our university increased.

There are 219 pieces of observation data that have missing factor values, and these values have been complemented using the method discussed in Sect. 4.1. The missing factors are found in the daily monthly fees of the nearest bicycle parking since the municipalities publish information on bicycle parking in different details. Also, the missing factors are found in precipitation and temperature values since there are not the source data in JMA.

Also, we found that LOD is also useful for constructing probabilistic networks for the estimation since possible nodes in the network can be obtained by following the properties like event:factor and ontological hierarchies. As the result of 10-fold cross validation repeated ten times, the precision became 69.9 %, the recall became 70.9 %, and the F-measure became 69.7 %. The precision is the ratio of correct data in the estimated data. The recall (accuracy) is the correctly estimated data divided by correct data. The training data are 897 observation data and their attributes. The accuracy of the estimated data in this study was low for the following reasons. The number of observations was not very many, and it was also unbalanced. Also, Table 3 shows a confusion matrix. The amount of 36–100 data is few; thus, this class of data is not correctly estimated.

Also, since we did not define the range of observation points, there were differences of the range decisions for each person. It was found that some people tweeted many illegally parked bicycles at one times, while some people divided the illegally parked bicycles and tweeted individually. Thus, data for 0–6 and 7–17 were higher, and this fact affected the estimation accuracy. We plan to visualize a specified range of circles which indicate observation points in the tweet application. Also, we plan to add a selection button such as "low (less than 10)" and "high (greater than 30)" in order to reduce the work burden.

Table 2. Statistics for observation data

Area	# of observation points	Amount of observation data
Chofu City, Tokyo	19	673
Nerima City, Tokyo	4	96
Naka-ku, Yokohama City, Kanagawa	5	39
Fussa City, Tokyo	2	23
Fuchu City, Tokyo	7	20
Musashino City, Tokyo	4	16
Chuo-ku, Sapporo City, Hokkaido	9	14
Shinjuku City, Tokyo	3	4
Isogo-ku, Yokohama City, Kanagawa	2	3
Kokubunji City, Tokyo	2	3
Shibuya City, Tokyo	3	3
Kita-ku, Sapporo City, Hokkaido	2	2
Shinagawa City, Tokyo	1	1

Table 3. Estimated results

		0–6	7–17	18–35	36–100	Total
Estimated data	0–6	339	38	10	2	389
	7–17	62	158	19	2	241
	18–35	22	34	89	2	147
	36–100	2	16	5	5	28
TP		0.871	0.656	0.605	0.179	

5 Visualization of LOD

Data visualization enables people to intuitively understand data contents. Thus, it can possibly raise the awareness of an issue among local residents. Furthermore, it is expected that we shall collect more urban data. In this section our visualization application of the IPBLOD is described.

The IPBLOD are published on the Web, and a SPARQL endpoint[17] is also available. Consequently, anyone can download and use IPBLOD as APIs via the SPARQL endpoint. As an example of the use of these data, we developed a Web application that visualizes illegally parked bicycles. The application can display time-series changes in the distribution of illegally parked bicycles on a map. Also, the application has a responsive design, so it is possible to use it on various devices such as PCs, smartphones, and tablets. When the start and end times are selected, and the play button is pressed, the time series changes of the

---

[17] http://www.ohsuga.is.uec.ac.jp/sparql.

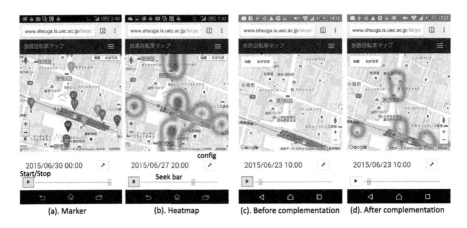

(a). Marker	(b). Heatmap	(c). Before complementation	(d). After complementation

Fig. 6. Screenshots of the visualization application

distribution of the illegally parked bicycles are displayed. Figure 6(a) and (b) show screenshots of an Android smartphone, on which the Web application is displaying such an animation near Chofu Station in Tokyo using a heatmap and a marker UI. This visualization application and the tweet application in Fig. 3 are hosted on the same website, so it is possible to see the visualized information just after tweeting. Thus, users obtain the instant feedback on posting new data.

The IPBLOD contain not only the data collected from Twitter, but also the data estimated by Bayesian networks. Therefore, time-series changes in the distribution of illegally parked bicycles become smoother than before estimating the missing values. Figure 6(c) and (d) show the comparison between the before and after complementation. The time-series changes after complementation are successive, whereas the time-series changes before complementation are intermittent.

As another example, we can see the average number of illegally parked bicycles per hour using the short SPARQL query. Figure 7 shows a visualization of the result. We found from the result that there are more illegally parked bicycles at night rather than in the morning. In general, many bicycles are thought to be illegally parked during morning commuting hours. However, the opposite result was shown in this study.

The number of page views of the visualization application increased from January to April 2016 (in Fig. 8). The number of page views of the visualization application was 187 in January 2016, and the number of page views gradually increased to 705 in April 2016. Also, the average session duration is 2 min and 32 s. Therefore, it was found that visitors are increasing and tend to use our application for longer periods of time.

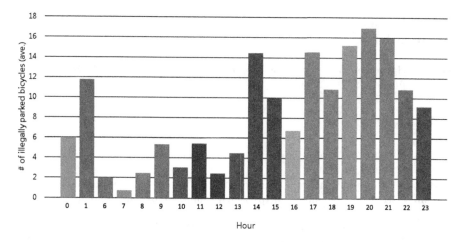

Fig. 7. The average number of illegally-parked bicycles per hour

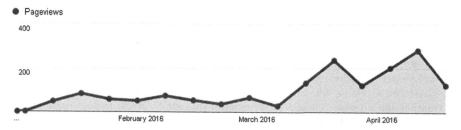

Fig. 8. Pageview of visualization application

6 Conclusion and Future Work

In this paper, the building and visualization of urban LOD was described as a solution for an illegally parked bicycles problem. The techniques proposed were a methodology for designing LOD schema from pages about an urban problem, data collection from Twitter with the exact location, a schema design of the illegally parked bicycles, complementation and estimation of the missing data, and then visualization of the LOD. We expect that this will increases the public awareness of local residents regarding the problem and encourages them to post more data.

In the future we will increase the amount of observation data and factors in order to improve the accuracy of the estimation. Also, we will collect more bicycle parking information and illegally parked bicycle data in cooperation with the Bureau of General Affair of Tokyo and NPOs. Moreover, we will visualize more statistics of the IPBLOD and clarify the problems caused by illegally parked bicycles in cooperation with local residents. We estimated and complemented the temporally missing values in this paper. However, there are also missing spatial data where bicycles might be illegally parked. In the future we will also estimate these missing spatial values. Then we will measure the growth rate of IPBLOD in Tokyo.

Acknowledgments. This work was supported by JSPS KAKENHI Grant Numbers 16K12411, 16K00419, 16K12533.

References

1. Nishi, N.: The 2nd health Japan 21: goals and challenges. J. Fed. Am. Soc. Exp. Biol. 28(1), 632.19 (2014)
2. Ministry of Internal Affairs, Communications: Current bicycle usage and bicycle-related accident (Japanese). http://www.soumu.go.jp/main_content/000354710.pdf. Accessed 10 Sept 2015
3. Bischof, S., Martin, C., Polleres, A., Schneider, P.: Collecting, integrating, enriching and republishing open city data as linked data. In: Arenas, M., et al. (eds.) ISWC 2015. LNCS, vol. 9367, pp. 57–75. Springer, Heidelberg (2015). doi:10.1007/978-3-319-25010-6_4
4. Haklay, M., Weber, P.: Openstreetmap: user-generated street maps. IEEE Pervasive Comput. 7(4), 12–18 (2008)
5. King, S.F., Brown, P.: Fix my street or else: using the internet to voice local public service concerns. In: Proceedings of the 1st International Conference on Theory and Practice of Electronic Governance, pp. 72–80 (2007)
6. Zook, M., Graham, M., Shelton, T., Gorman, S.: Volunteered geographic information and crowdsourcing disaster relief: a case study of the haitian earthquake. World Med. Health Policy 2(2), 7–33 (2010)
7. Ahn, L.V.: Games with a purpose. IEEE Comput. 39(6), 92–94 (2006)
8. Celino, I., Cerizza, D., Contessa, S., Corubolo, M., DellAglio. D., Valle, E.D., Fumeo, S., Piccinini, F.: Urbanopoly: collection and quality assesment of geospatial linked data via a human computation game. In: Proceedings of the 10th Semantic Web Challenge (2012)
9. Lopez, V., Kotoulas, S., Sbodio, M.L., Stephenson, M., Gkoulalas-Divanis, A., Aonghusa, P.M.: QuerioCity: a linked data platform for urban information management. In: Cudré-Mauroux, P., et al. (eds.) ISWC 2012. LNCS, vol. 7650, pp. 148–163. Springer, Heidelberg (2012). doi:10.1007/978-3-642-35173-0_10
10. Barbieri, D.F., Ceri, S.: C-SPARQL: SPARQL for continuous querying. In: Proceedings of the 18th International Conference on World Wide Web, pp. 1061–1062 (2012)
11. Egami, S., Kawamura, T., Sei, Y., Tahara, Y., Ohsuga, A.: Visualization of open urban data for illegally parked bicycles. In: CompleXity: Technology for Complex Urban Systems in the 49th Hawaii International Conference on System Sciences (2016)
12. Mizoguchi, R., Ikeda, M., Seta, K., Vanwelkenhuysen, J.: Ontology for modeling the world from problem solving perspectives. In: Proceedings of the 1995 International Joint Conference on AI (IJCAI) Workshop on Basic Ontological Issues in Knowledge Sharing, pp. 1–12 (1995)
13. Auer, S., Bizer, C., Kobilarov, G., Lehmann, J., Cyganiak, R., Ives, Z.: DBpedia: a nucleus for a web of open data. In: Aberer, K., et al. (eds.) ASWC/ISWC - 2007. LNCS, vol. 4825, pp. 722–735. Springer, Heidelberg (2007). doi:10.1007/978-3-540-76298-0_52

14. Jenks, G.F.: The data model concept in statistical mapping. Int. Yearb. Cartography **7**(1), 186–190 (1967)
15. Hwang, S., Yu, D.: GPS location improvement of smartphones using built in sensors. Int. J. Smart Home **6**(3), 1–8 (2012)
16. Stadler, C., Lehmann, J., Höffner, K., Auer, S.: LinkedGeoData: a core for a web of SpatialOpen data. Seman. Web J. **3**(4), 333–354 (2012)

Ontology-Based Design of Space Systems

Christian Hennig[1(✉)], Alexander Viehl[2], Benedikt Kämpgen[2], and Harald Eisenmann[1]

[1] Space Systems, Airbus Defence and Space, Friedrichshafen, Germany
{christian.hennig,harald.eisenmann}@airbus.com
[2] FZI Research Center for Information Technology, Karlsruhe, Germany
{viehl,kaempgen}@fzi.de

Abstract. In model-based systems engineering a model specifying the system's design is shared across a variety of disciplines and used to ensure the consistency and quality of the overall design. Existing implementations for describing these system models exhibit a number of shortcomings regarding their approach to data management. In this emerging applications paper, we present the application of an ontology for space system design that provides increased semantic soundness of the underlying standardized data specification, enables reasoners to identify problems in the system, and allows the application of operational knowledge collected over past projects to the system to be designed. Based on a qualitative evaluation driven by data derived from an actual satellite design project, a reflection on the applicability of ontologies in the overall model-based systems engineering approach is pursued.

Keywords: Space systems · Systems engineering · MBSE · ECSS-E-TM-10-23 · Conceptual data model · OWL · Reasoning

1 Introduction

The industrial setting for producing systems to be deployed in space, such as satellites, launch vehicles, or science spacecraft, involves a multitude of engineering disciplines. Each involved discipline has its own view on the system to be built, along with its own models, based on its own model semantics. For forming a consistent picture of the system, information from all relevant discipline-specific models is integrated towards an interdisciplinary system model, forming the practice of model-based systems engineering (MBSE).

Usually the data used to describe these system models is specified in technologies focused on software specification, such as UML or Ecore. However, these means of specification fail or neglect to address a number of important aspects. Semantic accuracy is often sacrificed for efficient implementation, the interdisciplinary nature of the engineering process is neglected in its data specification, and mechanisms for deriving knowledge from the information stored in the system models do not exist.

This emerging applications paper demonstrates how using OWL 2 ontologies for model specification addresses existing shortcomings in describing system models, making the following contributions:

© Springer International Publishing AG 2016
P. Groth et al. (Eds.): ISWC 2016, Part II, LNCS 9982, pp. 308–324, 2016.
DOI: 10.1007/978-3-319-46547-0_29

- Outlining of shortcomings of the established industrial data management practice used in space system engineering.
- Delivery and application of an ontology for model-based space systems engineering for supporting numerous activities involved in a satellite's design.
- Demonstration of the utility of an OWL 2 conceptual data model for solving current problems in the context of MBSE.
- Critical discussion on how OWL 2 can fit into current industrial data management and model-based systems engineering settings.

The ontologies developed for this paper are available online under Creative Commons licensing. Details on how to access the ontologies can be found in Sect. 4.2.

2 Model-Based Systems Engineering at Airbus DS

2.1 The Practice of Systems Engineering

In many industrial domains today, and especially in the space community, a multitude of disciplines is involved in creating a product. For space projects such as satellites, launch vehicles, or resupply spacecraft these disciplines involve, only to name a few, mechanical engineering, electrical engineering, thermal engineering, requirements engineering, software engineering, verification engineering, and their respective sub-disciplines. Each of these disciplines specifies, designs, and verifies specific parts or aspects of the system. In order to provide an all-encompassing understanding of the system of interest, the unique, yet complementary, views from every involved discipline are combined. The science and art of integrating different views on one system towards system thinking is called Systems Engineering.

As NASA [1] elegantly puts it: "Systems engineering is a holistic, integrative discipline, wherein the contributions of structural engineers, electrical engineers, mechanism designers, power engineers, human factors engineers, and many more disciplines are evaluated and balanced, one against an-other, to produce a coherent whole that is not dominated by the perspective of a single discipline."

2.2 Model-Based Systems Engineering (MBSE)

Many of the engineering activities performed within these disciplines are already well supported by computer-based models. Mechanical design models built with tools such as CATIA V5, mechanical analysis models built with tools such as PATRAN and thermal analysis models built with tools such as ESATAN-TMS are well established in the space engineering community today. Furthermore, requirements models based on DOORS play an important role, while also "traditional" tools such as Excel or Visio are used on a regular basis for specifying models.

Some tools provide an automated data exchange interface to other tools inside their own domain, e.g. in the case of mechanical design and mechanical analysis. However, classically, most engineering tools have been regarded in isolation from each other. Personal communication and documents are usually the main mechanism for bringing information from one domain into another one. Such a manual information management process is a pragmatic approach that can easily be established as it does not involve a lot of technical prerequisites, but is generally regarded as error-prone, producing inconsistencies, and involving a high amount of coordination overhead.

In order to cope with the shortcomings imposed by manual data integration, a further integration of these engineering tools is being pushed by the systems engineering community in numerous industrial sectors, including the space business unit of Airbus DS [2, 3]. The approach of MBSE focuses on employing a truly interdisciplinary representation of a system that incorporates design information of relevance to all domains. A common approach to achieve this kind of system-level data integration is employing a system model that serves as a central hub integrating the different modeling paradigms, and model semantics. This architecture is visualized in Fig. 1.

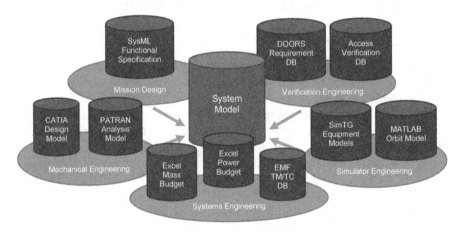

Fig. 1. System model as hub for integrating information from different engineering tools

For the purpose of designing space systems at Airbus DS, a product line of system engineering tools [4] has been developed that manages the system model. This model contains artefacts of system-wide relevance such as requirements, the product structure, operational modes, geometric properties, or verification approaches and traces between those artefacts. For the purpose of easing integration across engineering disciplines, and easing integration across the whole customer-supplier chain along multiple companies, the specification of these artefacts is done in a conceptual data model (CDM) that is shared between many stakeholders from the European space community, including the European Space Agency (ESA) and Airbus DS. This common data model, forming the meta-model to the system model, has been defined in a specification called ECSS-E-TM-10-23A [2] (abbreviated 10-23) and is specified using UML. It has been implemented in a number of tools, such as those employed in the Virtual Spacecraft Design Project [5], the SVTLC and FSS ESA TRP activities [6], in the EGS-CC project [7], and in Airbus DS' internal system engineering product line [4].

2.3 Shortcomings of Data Management Approach in Use by Current Tools

The current approach to describe system models neglects to address a number of important aspects that are of essential relevance to the MBSE domain. These points originate from shortcomings (SCs) identified in the Airbus DS data management approach, but are generalizable for other modeling approaches based on similar specification technologies.

SC1 - Semantics sacrificed for ease of implementation: The necessity for cost-efficient deployment of a system modeling application often leads to using generic structures for defining central concepts, as is the case for the Product Structure of 10-23. These generic structures ease implementation and applicability to a wide range of systems, but their logical consistency is not necessarily ensured. For instance, a system element might be a piece of software, but also exhibit mechanical properties, such as mass, being intuitively incorrect, but nevertheless possible in the system model.

SC2 - Inadequate tailoring support: In the space engineering community, the practice of tailoring is very dominant, meaning that selected engineering standards are adapted (albeit in a limited way) for each individual project. This implies that specific data structures have to be changed for every individual project, leading to high adaption cost.

SC3 - Missing discipline context information on elements: Although the need for having a system model comes from an interdisciplinary problem, this interdisciplinary nature is not directly reflected in the CDM. Consequently there is no logical link that allows a derivation of which engineering discipline is involved with which system element, falling short of improving discipline management and coordination.

SC4 - No exploitation of existing system design data: As the main purpose of current databases is data storage, versioning, and exchange, using system description data for other purposes, such as logical soundness checking, is currently not done.

SC5 - No existing knowledge capture and application mechanism: There is no mechanism for continuously gathering and formalizing knowledge across numerous projects, and making it available for application to future projects, steadily improving system quality as more knowledge becomes available.

2.4 Related Work on Describing System Models

Describing a system model in the MBSE context can also be done using other, yet similar, means. On the one hand, the System Modeling Language SysML [8] has been gaining a lot of traction and can be employed for the functional specification of a system using tools such as Papyrus [9]. On the other hand, a tool named Capella [10] surfaced, implementing the Arcadia meta-model, and supporting the Arcadia method [11]. In the case of SysML, being a profile to UML, the CDM's specification language is MOF [12] while the Arcadia data model is described using the modeling language of EMF, Ecore [13]. Further approaches include the Open Concurrent Design Tool (OCDT) [14] and the Concurrent Design Platform (CDP) [15].

3   Improving the System Modeling Approach Using OWL 2

3.1   Improvement of System Model Through Employment of OWL 2

For overcoming the shortcomings associated with system models based on object-oriented specification languages, describing the system model and the CDM with an ontology description language is proposed. The motivation behind this approach is based upon the following assumptions: Improvement of system model semantic accuracy (SC1), support of tailoring (SC2), classification of elements according to their engineering discipline (SC3), and further improvement of system model utility by inferring new information with help of a knowledge base (SC4 and SC5).

"Re-hosting" of the conceptual data model started with a manual transformation of the original 10-23 UML meta-model according to pre-defined transformation rules, transforming entities and relations. Subsequently operational knowledge from engineering domains contained in system design documents was formalized and discussed with discipline experts. A validation followed in order to ensure that data from both documents and the 10-23 meta-model is able to accurately be represented by the MBSE ontology. This approach, along with further information on the evaluation approach, is illustrated in Fig. 3.

The decision to use OWL 2 was reached by determining the semantic expressivity required by the MBSE ontology. One evaluation case relies heavily on using disjoints between classes, eliminating RDF(S). The MBSE ontology also exploits concepts such as reflexive properties and sub-property chains that are covered well by OWL 2. This leads to the question if an OWL 2 profile can be used in order to benefit from a performance advantage. The MBSE ontology uses inverses of object properties, ruling out OWL 2 EL. OWL 2 QL is ruled out since, as already mentioned, property chains are required. Some classes state information about yet unknown individuals in their superclass expressions, ruling out OWL 2 RL. Consequently, required expressivity can only be supplied using OWL 2.

3.2   Related Work on Ontological Modeling of Engineering Data

Numerous authors have been considering the role of ontologies in the systems engineering context [16–18]. While these considerations contain valuable remarks and highlight issues, the works do not provide usable solutions.

An approach regularly pursued is modeling engineering vocabularies using OWL or other ontology languages, instead of their native description language. This was done, for example, with ISO 15926 [19], and ISO 15288 [20]. The main purpose of these activities is to provide better communication and more accurate specification of data. The idea of integrating SysML with OWL has been proposed [21, 22] in order to improve modeling semantics. However, due to the remaining implicit semantics of the engineering domains to be distinguished, this approach does not fulfill the need for semantically driven consistency within the engineering domains and between them.

Another approach is specifically constructing engineering ontologies, as Borst [23] does for the purpose of better facilitating knowledge sharing and reuse. Van Renssen [24] even coins his own ontology and vocabulary, Gellish, for describing engineering facts. The goal of both approaches also does not go further than modeling, sharing, and reusing, however.

Some authors do apply reasoners to their engineering ontologies. Jenkins [25] uses inference along the taxonomic structure of system elements in order to query the system model more easily. Graves [26] suggests using inference in order to evaluate design consistency in terms of violated constraints. Takker et al. [27] employ OWL and inferencing for coming to diagnoses of tunnel disorders by examining the modeled set of symptoms observed in tunnels. Abele et al. [28] use the same technologies for validating design consistency of industrial plant models. Feldmann et al. [29] use SPARQL queries for specifying and finding design problems. Wende et al. [30] also check basic consistency of domain models using reasoning.

This paper picks up at consistency checking, but goes beyond the use cases described by the works above. On the one hand, we approach the multi-disciplinary problem of MBSE by determining the responsibility of engineering disciplines in specific system building blocks using inference. Furthermore, we perform engineering tasks automatically that were entirely manual processes before. Also, we enable new engineering activities by making information explicit that was only implicitly scattered across system models before, and introduce a knowledge base to engineering activities.

4   Evaluation of Ontological System Modeling Approach

4.1   Definition of Evaluation Scenario

For evaluating the improvements enabled by re-hosting the system model using ontology modeling technologies, the approach outlined in Sect. 3.1 will be demonstrated using examples based on data derived from an actual satellite design, resulting in the satellite dataset. This mission is representative for a mission in Phase B of the ESA project cycle [31], close to the Preliminary Design Review (PDR), utilizing data derived from a real project wherever possible. Data subject to intellectual property ownership issues was deliberately left out. For providing the necessary data to demonstrate this research, the ontologies are extended in a representative manner at required points with data not covered by the established engineering processes. This approach allows using project data not subject to intellectual property issues whenever possible, and to extend this data at specific points when required for demonstration purposes.

The goal of the spacecraft's mission is to measure the Earth's magnetic field using a set of specialized instruments, along with other scientific experiments. More specifically, the mission consists of a constellation of numerous identical satellites which are approximately 5 m long, 1.4 m in width, 0.8 m in height, and weighing 600 kg. The mechanical design of these spacecraft is illustrated in Fig. 2.

Structure
Electric Power Subsystem
Data Handling Subsystem
Telemetry, Tracking, and Telecommand Subsystem
Attitude and Orbit Control Subsystem
Instruments

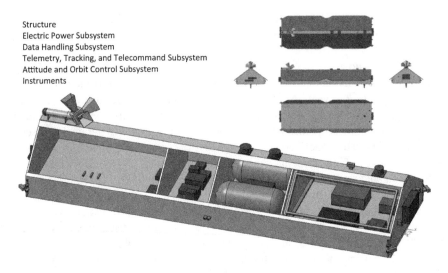

Fig. 2. Isometric view and side-views of the satellite

This object-oriented satellite system model, based upon a derivation of 10-23 data specified in UML, is re-hosted in an ontology specified in OWL 2, consisting of a concept ontology (MBSE Ontology) and an ontology containing instance data of the satellite (satellite dataset). The transformation for object-oriented satellite instances to OWL individuals can principally be done in an automated, generic fashion, since it is a very direct instance mapping, but is done manually in the scope of this paper for pragmatic reasons.

A number of evaluation cases are provided that rely on usage of both satellite dataset and MBSE ontology. These cases pick up on the shortcomings identified in Sect. 2.3, giving a proposal on how to solve them with OWL 2, resulting in the provision of

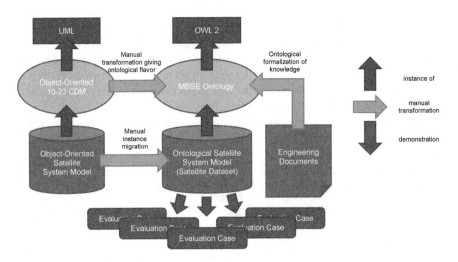

Fig. 3. Illustration of evaluation approach

objective evidence that the shortcomings have indeed been solved using an OWL 2 ontology. This evaluation approach is illustrated in Fig. 3.

4.2 Availability of Employed Ontologies

Both MBSE ontology and satellite dataset are available at Zenodo under Creative Commons Attribution-ShareAlike (CC BY-SA) license and may be accessed via the following DOIs:

http://dx.doi.org/10.5281/zenodo.57955 (MBSE Ontology)

http://dx.doi.org/10.5281/zenodo.57957 (Satellite Dataset)

These ontologies have been spawned through research on using ontologies in the model-based systems engineering context. Being research on the future functionality of system models, these ontologies have not yet been rolled out to actual end-users in an industrial deployment process. The team involved in ontology design and design of the infrastructure around them consists of three researchers, with a significant contribution from domain experts. Both ontologies will continue to be maintained and extended.

The evaluation scenario is based upon the MBSE ontology in version 1.1, and the satellite dataset in version 1.1. Modeling was performed using Protégé 5.0.0. Reasoning was done using PelleT 2.2.0. Inferring all statements on the MagSat dataset takes an average of 24 s on a Core i7 4th generation quad core CPU at 2.8 GHz with 16 GB of RAM.

4.3 Evaluation Cases

The five selected evaluation cases together encompass activities from the beginning of a project in Phase 0/A up to system production in Phase D. Their allocation to different system design phases is illustrated in Fig. 4.

Evaluation and Insurance of Property Consistency. A significant part of describing space systems is done with physical properties. Based on these properties, assertions can be made, such as that a Star Sensor has a mass with the value 0.380, based on the

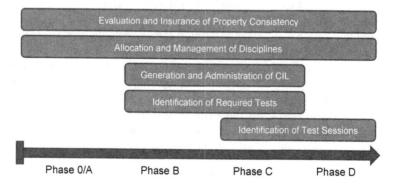

Fig. 4. Positioning of evaluation cases along the system design cycle

physical quantity mass, and the unit Kilogram. The assertion of such properties is based upon the SysML QUDV standard [8] and is employed as reusable library that can be loaded into multiple projects and adapted accordingly. In order to be applicable to virtually any data model element, the implementation of these properties is a generic structure as outlined in SC1. However, there exist certain data model elements that are not applicable to any properties, and combinations of properties that do not make sense to occur together. More specifically, through employing such a generic structure, the possibility to specify inconsistent model populations emerges.

For example, a property of *powerConsumption* in Watts could be attributed to a screw, a software could exhibit dimensions in length, width, and height, or a battery could have a data rate. While these assertions are illogical at first glance, modeling mistakes do happen. In other cases, errors are more subtle. For example a component should get the mass property for items on component level, while subsystems have a mass property that is composed of *mass*, *margin*, and *nominalMass*, whereas the satellite on system level exhibits *nominalMass*, *margin*, *totalDryMass*, and *totalWetMass* (including propellant and other fluids required for full operation).

In order to deal with the possibility for specifying illogical models, the *categories* (essentially containers for properties) of 10-23 are transformed to OWL classes, and disjoints are added to them. Using the disjoint axioms, logical exclusions between categories and specific equipments, or between categories themselves, can be modeled and evaluated using a reasoner. This essentially tackles SC1. Selected concepts and disjoint relations are shown in Fig. 5.

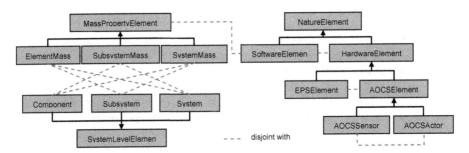

Fig. 5. Disjoints between model concepts

The satellite's design was validated through extensive verification, validation, and reviewing and does not exhibit any inconsistencies in terms of disjoints. However, as intermediate design baselines may very well contain such inconsistencies, a number of logical violations were provoked in order to evaluate how the reasoner behaves for inconsistent system models. As such, inconsistent types were defined and successfully found for the *On-Board Software* (ComponentMass and SoftwareElement), *Attitude and Orbit Control System* (SystemMass and Subsystem), and *Attitude Control Thruster* (AttitudeControlThruster and OrbitControlThruster).

Allocation and Management of Disciplines. In the multi-disciplinary environment of space system engineering, each discipline is a stakeholder in several system elements.

Consequently, every system element has an involvement in numerous disciplines. For example, a star sensor may involve Optical Engineering (due to its sensor), System Engineering (mass budget, power budget, ...), Harness Engineering (the star sensor has electrical connectors), Mechanical Engineering (geometric accommodation), Operational Engineering (modes, services), Requirements Engineering, etc. Due to the amount of combinations between system elements and disciplines, the exact nature of discipline involvement is often obscured by sheer mass and complexity.

For overcoming this problem, a reasoner is employed in order to assert what disciplines are involved in each system element. Depending on the nature of discipline involvement, the inference might be fairly simple (every system element with a requirement involves Requirements Engineering) or more complex (a *SystemElement* involves *SystemsEngineering* with the *PowerBudget* if it has a *DiscreteModel* with a *DiscreteState* that has some *powerConsumption*).

This approach enables the automated assertion of discipline involvement on all system elements. Based on the current elaboration of the system, the discipline allocation is automatically adjusted as the system progresses along its life-cycle. Furthermore, the reasoner is able to provide an explanation as to why a specific discipline is a stakeholder in a specific element. This addresses SC3. Additionally, due to the fact that the ontology can be adapted during runtime, SC2 is also tackled.

For evaluation, five system elements are chosen and selected disciplines allocated by the reasoner (Table 1). A direct comparison to the original 10-23 satellite cannot be performed due to the fact that there is no explicit overview on discipline allocation. Nevertheless, the asserted properties of both models are very similar, with the OWL model able to make the required inferences. The satellite itself is involved in many disciplines due to the fact that it contains an element from every discipline. The Star Tracker Electronics (STRE) also has an involvement from almost every discipline, except Instruments Engineering, since it is not an instrument per se. The On-Board Software (OBSW) involves mainly Software Engineering, and requirements-related disciplines. The Orbit Control Thruster (OCT) and the Battery (BATT) both involve a limited number of disciplines.

Table 1. Allocation of disciplines to system elements

	Satellite	STRE	OBSW	OCT	BATT
Requirements Engineering	•	•	•	•	•
Operational Engineering	•	•	•		
Mass Budget	•	•		•	•
Mechanical Engineering	•	•		•	•
Electrical Engineering	•	•		•	•
Thermal Engineering	•	•			•
Instruments Engineering	•				
Control Engineering	•	•		•	
Software Engineering	•	•	•		
Verification Engineering	•	•	•	•	•

Generation and Administration of Critical Items List. A central part in space system engineering is played by the Critical Items List (CIL). Its purpose is to identify elements of the system that have major difficulties or uncertainties associated and expected in their design. This list identifies the exact risks associated with these items and provides risk reduction measures accordingly. An item might be critical for a variety of reasons, e.g. it may be a single point of failure, be related to personnel safety, exhibit low reliability, be sensitive to radiation, or limited in its lifespan. One element in the system may also belong to multiple categories, such as a battery, which can result in personnel injury in case of failure during spacecraft assembly (safety critical item), but also be a life limited item due to a limited number of charge and discharge cycles. Generating and maintaining the CIL is very formal work involving discussion of every single system element, with a potential of missing out on important item-category-assertions. The original 10-23 data model does not contain any aspect of the CIL. The idea is to formalize the knowledge about all critical items originally in the CIL document of the satellite in order to specify the definition of numerous critical item categories in the MBSE ontology.

As an example, the following specification defines the nature of a battery sensitive that introduces some criticality to spacecraft with magnetic instruments on board:

```
Class: fdir:MagneticCriticalBattery
    EquivalentTo:
        mbse:Battery
        and (mbse:isContainedByElement some
            (mbse:System
                and (mbse:containsElement some mbse:MagneticInstrument)))
    SubClassOf:
        fdir:MagneticCleanlinessElement,
        fdir:hasFailureEffect value fdir:MagFieldCausesMagInstrPerfDegr,
        fdir:hasRiskReductionMeasure value fdir:CompensationInDataProc,
        fdir:severityLevel value fdir:Major_3
```

Based on the specification above, the reasoner concludes that the battery aboard the satellite is a magnetically critical item, since it exhibits its own magnetic field and is aboard a spacecraft that contains magnetic instruments. In order to avoid performance degradation of scientific measurements (failure effect), data processing routines have to be calibrated accordingly (compensatory provision).

Furthermore, the spacecraft itself is also inferred as being a magnetically critical item, resulting in the need to use demagnetized tools, and to instate the working instruction to keep demagnetized tools away from conventional tools. The benefits for managing large parts of the CIL with help of a reasoner are the automated classification of all system elements according to their criticality, the automated inference of failure effects and of compensatory measures. In the event of architectural change of the system, the reasoner will update the CIL accordingly. The approach of using the CIL as a knowledge base, extending it throughout multiple projects, and applying this knowledge to every subsequent project, improves SC4 and SC5.

Table 2 contains the evaluation results of modeling the satellite's CIL. Across four categories, from the original definition of 56 different critical item definitions,

37 definitions of critical items that are applicable to a wide variety of different missions were formalized in the ontology. 19 definitions were deemed not worth formalizing because they were based upon highly specialized mechanisms with a low likelihood of reuse for future missions. The amount of critical item assertions is higher than the amount of definitions, since the satellite has different kinds of thrusters (attitude and orbit control thrusters), different kinds of electronics, etc. aboard. A total of 67 failure effects across all system elements have automatically been derived, along with 91 risk reductions measures to be incorporated in the system design.

Table 2. Inference of selected critical items, failure effects, and compensatory measures

	Classes in document	Applicable classes	Inferred critical item assertions	Inferred total effects	Inferred total measures
Contamination critical items	3	3	4	4	4
Life limited items	12	5	6	7	13
Magnetic cleanliness items	10	7	11	11	10
Safety critical items	31	22	27	45	64
Total	56	37	48	67	91

Identification of Required Tests. Across the whole design cycle of a satellite, a variety of tests are performed. While some tests are performed in a simulated environment (e.g. for control algorithms), others are performed in hybrid environments, while yet others are performed on real flight hardware. For instance, every On-board Control Procedure (OBCP) requires an Integrated System Test (IST). Every electronic equipment, as well as the spacecraft's Central Software, requires a hardware IST. Every item that has the spacecraft harness connected to it requires an Electric Integration Test (ELI), etc.

```
Class: fv:ElementWithIST
    EquivalentTo:
        mbse:CentralSoftware or mbse:Equipment
    SubClassOf:
        fv:ElementWithTest,
        fv:requiresTest value fv:IST

Class: fv:ElementWithELI
    EquivalentTo:
        mbse:ElementDefinition
         and (el:hasFunctionalPort some el:FunctionalPort)
    SubClassOf:
        fv:ElementWithTest,
        fv:requiresTest value fv:ELI
```

Similarly to the CIL, deriving all necessary tests is also very formal work with a potential to not achieve full coverage, that is currently performed manually. The idea behind handling test management with OWL is that the definition of each required test is formalized in OWL and each relation to a test to be performed is inferred by the reasoner, helping in doing manual work and being an aid in achieving exhaustiveness. The definition of tests also behaves as a kind of knowledge base, addressing SC4 and SC5. Due to the fact that this structure can be adapted during system model runtime, SC2 is also addressed.

In the example at hand, the reasoner concludes that the GPS receiver aboard is defined to be an equipment and thus has to undergo an IST. Furthermore, the reasoner will conclude that the GPS receiver electronics has a functional electrical port, resulting in the necessity to perform an ELI.

Table 3 contains evaluation results for the test derivation case. Across all considered categories, a total of 79 out of 84 tests have been identified by the reasoner. Differences occur for the ISTs where non-standard ISTs are performed for the spacecraft's Thermal Control System, and its OBC's Mass Memory Unit, splitting the original OBC IST into two parts. The difference for OBCP ISTs results from the introduction of tests for specific failure cases, e.g. where the nominal execution of the OBC's power-up procedure is tested normally, but also with intentionally provoked errors.

Table 3. Inferred tests on the satellite per category

	OO 10-23	OWL 10-23	Missing tests
IST	12	10	2
ELI	16	16	0
AFT	1	1	0
SFT	1	1	0
CLT	1	1	0
EMCFT	1	1	0
RFCFT	1	1	0
TVFT	1	1	0
OBCP IST	48	45	3
Total	84	79	5

Identification of Possible Test Sessions. During the integration of all satellite components towards a fully functional system, the majority of tests take place. Every single test requires a specific configuration in order to be performed. For example, the Star Tracker Integrated System Test (STR IST) requires all three Star Tracker sensors (one for each axis) to be integrated, and at least one of the two electronics boxes. Furthermore, the on-board computer is required for data processing and the power control and distribution unit (PCDU) for providing power. Furthermore, the spacecraft harness is required for wiring all of these components together. Additionally, external test equipment is required, e.g. the test caps that simulated a given view of a star field on ground. This configuration of a satellite during integration changes very frequently, usually several times per day. This is due to the fact that components are only procured once in the beginning of test campaigns, there might only be a limited number of external test equipment available, or items might be taken out of the spacecraft for further analysis and improvement. This situation is outlined in Fig. 6.

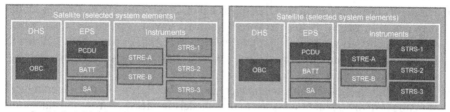

Possible Test: Abbreviated Functional Test (AFT)   Possible Tests: Star Tracker IST, AFT

Fig. 6. Configuration required for executing selected tests

This frequent change in configuration makes it difficult to keep track of what tests can currently be performed on the available configuration. Determining what tests are to be performed several days ahead is usually not feasible due to the random occurrence of issues in test campaigns. Also, evaluating what tests can be performed on the current configuration is a manual effort involving a significant amount of work.

For tackling this problem, the required configuration for a given test is specified in the OWL model. Depending on the current spacecraft configuration, the reasoner then evaluates what tests can be performed. This definition serves as a kind of knowledge base as well, tackling SC4 and SC5. For the satellite example, a selected number of tests were modeled with their configuration prerequisites specified as equivalent class axioms and successfully tested as a proof of concept.

5   Lessons Learned

OWL 2 provides valuable functionality that can be used to significant benefit in the context of MBSE. In particular the following capabilities have been proven helpful to the overall MBSE process:

- The ability to apply operational knowledge from past projects to a current design.
- The ability to ensure the semantic coherence of a multi-classification environment.
- The ability to infer new knowledge on a system.
- The ability to explain made inferences.

Computational performance for the mentioned reasoning tasks was found to be not an issue, as all required statements can be inferred on the given dataset in reasonable time. It is expected that for significantly larger datasets that require longer this will remain to not be an issue, as approaches such as nightly consistency checking have been deemed a viable solution in the MBSE context.

However, there is also functionality required by the current data management approach that was identified as being difficult to realize, or incompatible with, the employment of OWL 2 ontologies. This includes:

- Lack of capability to apply closed world checks and negation as failure: Many consistency checks in the MBSE context rely on evaluating if values are present in the model or not. Lack of this functionality imposes significant restrictions and has to be worked around accordingly.
- Possibility to model data in an ABox that is out of scope of the TBox: The conceptual data model in MBSE is used to specify allowed model populations, upon which further functionality, such as importers and exporters, relies upon. OWL 2 allows to model data on instance level that is not scoped by the data's specification, imposing significant challenges.
- Lack of built-in part-of/containment/aggregation relationships: Explicit hierarchical structures are of central importance to data models in MBSE, but are not supported by OWL. This essentially does not allow modeling these relationships with sound semantics without resorting to workarounds.
- Shortcomings when reasoning with numeric values: Reasoning activities described in this paper could be significantly improved by the possibility to base inferences on numerical values. Although some rule-based approaches exist, their reasoner support is currently inadequate.

Due to the fact that system databases in the MBSE context require functions currently not addressed by OWL 2 ontologies, a replacement of the system model by an OWL 2 ontology as a standalone solution cannot be recommended. Instead, a solution where the main data management activities are performed by a traditional system database on the one hand, and a knowledge-based enhancement of the system model relying on OWL 2 on the other hand, is seen as viable solution in the industrial context. Research is currently being pursued in exploring such hybrid architectures in more detail. Alternatives could include the re-hosting of ontological functions in the object-oriented context, or a synchronized coexistence of a system model in both the ontology world and the object-oriented world and are currently subject of investigation at Airbus DS.

6   Conclusions and Future Outlook

This research demonstrates that ontologies are able to solve problems that current system databases neglect to address. Using the reasoner to infer new knowledge about the system enables higher efficiency and better scaling of established system engineering activities, an improved quality of the system and system model in terms of consistency, and an improved information consolidation and reuse process. Having a knowledge base in the background that grows from project to project improves system quality steadily and ensures exhaustiveness in performing specific design steps. Furthermore, the practice of tailoring, predominant in space engineering, is supported by the ability of OWL 2 ontologies to adapt the data model during runtime.

However, due to their non-conformance to existing data management requirements in the MBSE context, ontologies cannot directly take the place of "traditional" system databases in the near future. Instead, a solution where the main data management activities are performed by a traditional system database on the one hand, and a knowledge-based enhancement of the system model based on OWL 2 on the other hand, is seen as viable solution in the industrial context. Our research continues towards intersecting object-oriented and ontological approaches, addressing core requirements for effective data management, as well as those newly identified for semantic models and knowledge management activities.

References

1. NASA, NASA Systems Engineering Handbook (NASA-SP-2007-6105) Rev1 (2007)
2. ESA, Space engineering - Space system data repository. ESA Technical Memorandum ECSS-E-TM-10-23A (2011)
3. INCOSE, Systems Engineering Vision 2025 (2014). http://www.incose.org/docs/default-source/aboutse/se-vision-2025.pdf. Accessed 4 Dec 2015
4. Eisenmann, H., Cazenave, C.: Evolving a classic SRDB into an engineering database. In: 6th International Workshop on Systems and Concurrent Engineering for Space Applications (SECESA), 8–10 October 2014 (2014)
5. ESA, The Virtual Spacecraft Design Project (2012). http://vsd.esa.int/
6. Fischer, P.M., Eisenmann , H., Fuchs, J.: Functional verification by simulation based on preliminary system design data. In: 6th International Workshop on Systems and Concurrent Engineering for Space Applications (SECESA), 8–10 October 2014 (2014)
7. ESA, EGS - CC - European Ground Systems - Common Core (2013). http://www.egscc.esa.int/
8. OMG, OMG Systems Modeling Language (OMG SysML), Version 1.3 (2012)
9. Eclipse Foundation, Papyrus (2015). https://eclipse.org/papyrus/. Accessed 4 Dec 2015
10. Eclipse Foundation, Capella (2014). https://www.polarsys.org/capella/. Accessed 4 Dec 2015
11. Voirin, J.-L.: Method & tools to secure and support collaborative architecting of constrained systems. In: 27th International Congress of the Aeronautical Sciences, 19–24 September 2010 (2010)
12. OMG, Meta Object Facility (MOF) Version 2.5 (2015)
13. Eclipse Foundation, org.eclipse.emf.ecore (EMF Documentation) (2014). http://download.eclipse.org/modeling/emf/emf/javadoc/2.9.0/org/eclipse/emf/ecore/package-summary.html
14. ESA, Open Concurrent Design Tool (OCDT) Community Portal (2014). https://ocdt.esa.int/. Accessed 4 Dec 2015
15. RHEA System, CDP - Concurrent Design Platform (2015). http://www.rheagroup.com/products/cdp/. Accessed 4 Dec 2015
16. Ferreira, S., Sarder, B.: Developing systems engineering ontologies. In: 2007 IEEE International Conference on System of Systems Engineering, San Antonio (2007)
17. Chourabi, O., Pollet, Y., Ben Ahmed, M.: Ontology based knowledge modeling for System Engineering projects. In: Second International Conference on Research Challenges in Information Science, Marrakech (2008)
18. Ernadote, D.: An ontology mindset for system engineering. In: 2015 IEEE International Symposium on Systems Engineering (ISSE), Italy, Rome (2015)
19. Klüwer, J.W., Skjæveland, M.G., Valen-Sendstad, M.: ISO 15926 templates and the Semantic Web. In: W3C Workshop on Semantic Web in Energy Industries; Part I: Oil & Gas, Houston (2008)
20. Van Ruijven, L.: Ontology for systems engineering as a base for MBSE. In: 25th Annual INCOSE International Symposium (IS 2015), Seattle (2015)
21. Graves, H.: Integrating SysML and OWL. In: Proceedings of OWL: Experiences and Directions (OWLED 2009) (2009)

22. Wagner, D., Bennett, M., Karban, R., Rouquette, N., Jenkins, S., Ingham, M.: An ontology for state analysis: formalizing the mapping to SysML. In: IEEE Aerospace Conference, 3–10 March 2012, pp. 1–16 (2012)
23. Borst, W.N.: Construction of Engineering Ontologies for Knowledge Sharing and Reuse. University of Twente, Twente (1997)
24. Van Renssen, A.S.H.P.: Gellish - A Generic Extensible Ontological Language. Technische Universiteit Delft, Delft (2005)
25. Jenkins, S.: Ontologies and model-based systems engineering. In: INCOSE IW 2010 MBSE Workshop, 7–10 February 2010 (2010)
26. Graves, H.: Integrating reasoning with SysML. In: INCOSE International Symposium, vol. 22(1), pp. 2228–2242 (2012)
27. Thakker, D., Dimitrova, V., Cohn, A.G., Valdes, J.: PADTUN - using semantic technologies in tunnel diagnosis and maintenance domain. In: Gandon, F., Sabou, M., Sack, H., d'Amato, C., Cudré-Mauroux, P., Zimmermann, A. (eds.) ESWC 2015. LNCS, vol. 9088, pp. 683–698. Springer, Heidelberg (2015)
28. Abele, L., Legat, C., Grimm, S., Müller, A.W.: Ontology-based validation of plant models. In: 11th IEEE International Conference on Industrial Informatics (INDIN), Bochum, Germany (2013)
29. Feldmann, S. Herzig, S.J.I., Kernschmidt, K., Wolfenstetter, T., Kammerl, D., Qamar, A., Lindemann, U., Krcmar, H., Paredis, C. J. J., Vogel-Heuser, B.: Towards effective management of inconsistencies in model-based engineering of automated production systems. In: 15th IFAC Symposium on Information Control Problems in Manufacturing (INCOM 2015), Ottawa (2015)
30. Wende, C., Siegemund, K., Thomas, E., Zhao, Y., Pan, J.Z., Silva Parreiras, F., Walter, T., Miksa, K., Sabina, P., Aßmann, U.: Ontology reasoning for consistency-preserving structural modeling. In: Pan, J.Z., Staab, S., Aßmann, U., Ebert, J., Zhao, Y. (eds.) Ontology-Driven Software Development, pp. 193–218. Springer, Heidelberg (2013)
31. ESA, ECSS-M-ST-10C: Space project management - Project planning and implementation (2009)

Capturing Industrial Information Models with Ontologies and Constraints

Evgeny Kharlamov[1]([⊠]), Bernardo Cuenca Grau[1], Ernesto Jiménez-Ruiz[1],
Steffen Lamparter[2], Gulnar Mehdi[2], Martin Ringsquandl[2], Yavor Nenov[1],
Stephan Grimm[2], Mikhail Roshchin[2], and Ian Horrocks[1]

[1] University of Oxford, Oxford, UK
evgeny.kharlamov@cs.ox.ac.uk
[2] Siemens AG, Corporate Technology, Munich, Germany

Abstract. This paper describes the outcomes of an ongoing collaboration between Siemens and the University of Oxford, with the goal of facilitating the design of ontologies and their deployment in applications. Ontologies are often used in industry to capture the conceptual information models underpinning applications. We start by describing the role that such models play in two use cases in the manufacturing and energy production sectors. Then, we discuss the formalisation of information models using ontologies, and the relevant reasoning services. Finally, we present SOMM—a tool that supports engineers with little background on semantic technologies in the creation of ontology-based models and in populating them with data. SOMM implements a fragment of OWL 2 RL extended with a form of integrity constraints for data validation, and it comes with support for schema and data reasoning, as well as for model integration. Our preliminary evaluation demonstrates the adequacy of SOMM's functionality and performance.

1 Introduction

Software systems in the domain of industrial manufacturing have become increasingly important in recent years. Production machines, such as assembly line robots or industrial turbines, are equipped with and controlled by complex and costly pieces of software; according to a recent survey, over 40 % of the total production cost of such machines is due to software development and the trend is for this number only to continue growing [35]. Additionally, many critical tasks within business, engineering, and production departments (e.g., control of production processes, resource allocation, reporting, business decision making) have also become increasingly dependent on complex software systems.

Recent global initiatives such as Industry 4.0 [9,18,34] aim at the development of *smart factories* based on fully computerised, software-driven, automation of production processes and enterprise-wide integration of software components. In smart factories, software systems monitor and control physical

This work was partially funded by the Royal Society under a University Research Fellowship, the EU project Optique [24] (FP7-ICT-318338), and the EPSRC projects MaSI3, DBOnto, ED3.

P. Groth et al. (Eds.): ISWC 2016, Part II, LNCS 9982, pp. 325–343, 2016.
DOI: 10.1007/978-3-319-46547-0_30

processes, effectively communicate and cooperate with each other as well as with humans, and are in charge of making decentralised decisions. The success of such ambitious initiatives relies on the seamless (re)development and integration of software components and services. This poses major challenges to an industry where software systems have historically been developed independently from each other.

There has been a great deal of research in recent years investigating key aspects of software development in industrial manufacturing domains, including life-cycle costs, dependability, compatibility, integration, and performance (e.g., see [41] for a survey). This research has highlighted the need for enterprise-wide *information models*—machine-readable conceptualisations describing the functionality of and information flow between different assets in a plant, such as equipment and production processes. The development information models based on ISA and IEC standards[1] has now become a common practice in modern companies [30] and Siemens is not an exception in this trend.

In practice, however, many types of models co-exist, and applications typically access data from different kinds of machines and processes designed according to different models. These information models have been independently developed in different (often incompatible) formats using different types of proprietary software; furthermore, they may not come with a well-defined semantics, and their specification can be ambiguous. As a result, model development, maintenance, and integration, as well as data exchange and sharing pose major challenges in practice.

Adoption of semantic technologies has been a recent development in many large companies such as IBM [11], the steel manufacturer Arcelor Mittal [2], the oil and gas company Statoil [21], and Siemens [1, 4, 19, 20, 22, 25, 32]. An important application of these technologies has been the formalisation of information models using OWL 2 ontologies and the use of RDF for storing application data. OWL 2 provides a rich and flexible modelling language that seems well-suited for describing industrial information models: it not only comes with an unambiguous, standardised, semantics, but also with a wide range of tools that can be used to develop, validate, integrate, and reason with such models. In turn, RDF data can not only be seamlessly accessed and exchanged, but also stored directly in highly scalable RDF triple stores and effectively queried in conjunction with the available ontologies. Moreover, legacy and other data that must remain in its original format and cannot be transformed into RDF can be virtualised as RDF using ontologies following the Ontology-Based Data Access (OBDA) approach [21, 23, 29].

In this paper, we describe the outcomes of an ongoing collaboration between Siemens Corporate Technology in Munich and the University of Oxford, with the goal of facilitating deployment of ontology-based industrial information models. We start by describing the key role that information models play in two use cases in the manufacturing and energy production sectors. Then, we present industrial information models that are used for describing manufacturing and

---

[1] International Society of Automation and International Electrotechnical Commission.

energy plants, and discuss how they can be captured using ontologies. In our discussion, we stress the modelling choices made when formalising these models as ontologies and identify the key OWL constructs required in this setting. Our analysis revealed the need for integrity constraints for data validation [27,37], which are not available in OWL 2. Hence, we discuss in detail what kinds of constraints are needed in industrial use cases and how to incorporate them. We then illustrate the use of reasoning services, such as concept satisfiability, data constraint validation, and query answering for addressing Siemens' application requirements.

Ontologies are currently being created and maintained in Siemens by qualified R&D personnel with expertise in ontology languages and ontology engineering. In order to widen the scope of application of semantic technologies in the company it is crucial to make ontology development accessible to other teams of engineers. To this end, we have developed the Siemens-Oxford Model Manager (SOMM)—a tool that has been designed to fulfil industrial requirements and which supports engineers with little background on semantic technologies in the creation and use of ontologies. SOMM provides a simple interface for ontology development and enables the introduction of instance data via automatically generated forms that are driven by the ontology and which help minimising errors in data entry. SOMM implements a fragment of the OWL 2 RL profile [26] extended with database integrity constraints for data validation; the supported language is sufficient to capture the main features of ISA and ICE based information models used by Siemens. SOMM is built on top of Web-Protégé [40], which provides built-in functionality for ontology versioning and collaborative development. It relies on HermiT [10] for ontology classification and LogMap [16] to support model alignment and merging. For query answering and constraint validation, SOMM requires a connection to a triple store or a rule inference system that supports Datalog reasoning and stratified negation-as-failure.

We showcase the practical benefits of our tool using two ontologies in the manufacturing and power generation domains. Both ontologies have been developed using SOMM by Siemens engineers to capture information models currently in use. Based on these ontologies, we conducted an empirical evaluation of SOMM's performance in supporting constraint validation and query answering over realistic manufacturing and gas turbine data. In our experiments, we coupled SOMM with the rule inference engine IRIS [3], which is available under the LGPL license.[2] Our evaluation demonstrates the adequacy of SOMM's functionality and performance for industrial applications.

2   Industrial Information Models

Conceptual information models can be exploited in a wide range of manufacturing and energy production applications. In this Section, we discuss two concrete use cases and describe the underpinning models and their limitations.

---

[2] http://www.iris-reasoner.org/.

2.1 Applications in Manufacturing and Energy Production

In manufacturing and energy production plants it is essential that all processes
and equipment run smoothly and without interruptions.

In a typical manufacturing plant, data is generated and stored whenever a
piece of equipment consumes material or completes a task. This data is then
accessed by plant operators using *manufacturing execution systems (MES)*—
software programs that steer the production in a manufacturing plant. MESs
are responsible for keeping track of the material inventory and tracing their consumption, thus ensuring that equipment and materials needed for each process
are available at the relevant time [30]. Similarly, turbines in energy plants are
equipped with sensors that are continuously generating data. This data is consumed by *remote monitoring systems (RMS)*, which analyse turbine data to
prevent faults, report anomalies and ensure that the turbines operate without
interruption. In both application scenarios, the use of information models is
twofold.

1. Models are used to provide machine-readable specifications for the data generated by equipment and processes, and for the data flow across assets and
   processes in a plant.
2. Models provide a schema for constructing and executing complex queries. In
   particular, monitoring tasks in MESs are realised by means of queries issued
   to production machines and data hubs; similarly, anomaly detection in an
   RMS relies on queries spanning the structure of the turbines, the readings of
   their sensors, and the configuration of turbines within a plant.

2.2 Information Models Based on Industrial Standards

We next describe the information models in Siemens relevant to the aforementioned applications. These models have been developed in compliance with ISA,
IEC, and ISO/TS international standards.

Manufacturing Models. For many manufacturing applications it is a common practice to rely on information models that are based on the international
standard ISA-88/95.

The ISA-88/95 standard provides general guidelines for specifying the functionality of and interface between manufacturing software systems. The standard consists of UML-like diagrammatic descriptions accompanied with tables
and unstructured text, which are used to extend the diagrams with additional
information and examples. Figure 1 presents an excerpt of the ISA-88/95 standard modelling materials, equipment, personnel, and processes in a plant. For
instance, one of these diagrams establishes that pieces of equipment can be
composed by other pieces of equipment and are described by a number of specified 'equipment properties'. The table complementing this diagram indicates
that each piece of equipment must have a numeric ID and may have a textual

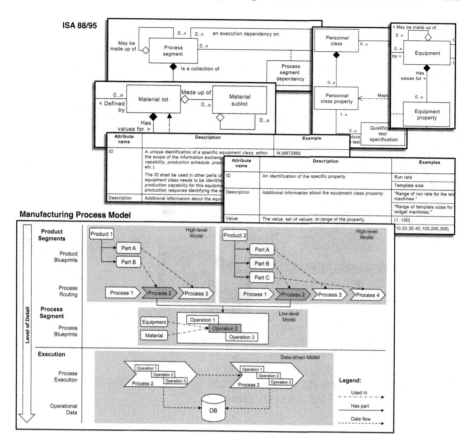

Fig. 1. Fragment of ISA 88/95 and an example model based on it.

description; additional properties of equipment can be introduced by providing an ID, a textual description of the property, and a value range.

Figure 1 provides a simplified version of an information model based on the standard ISA-88/95. The model is organised in three layers: *product*, *process*, and *execution*. On the product level, we can see the specification of two products and their relationship to production processes; for instance, *Product1* consists of *PartA* and *PartB*, which are manufactured by two consecutive processes. The process segment level provides more fine-grained specifications of the structure of each process; for instance, *Process2* consists of three operations, where the second one relies on specific kinds of materials and equipment. Finally, at the execution level, we can see how data is stored and accessed by individual processes.

Energy Plant Models. Information models for energy plants are often based on the *Reference Designation System for Power Plants (RDS-PP)* and *Kraftwerk-*

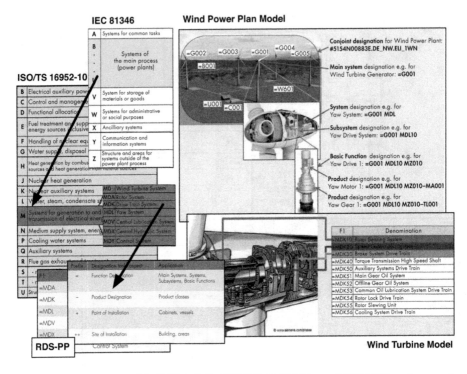

Fig. 2. Designation models IEC 81346, ISO/TS 16952-10, and RDS-PP and example energy information model for an energy plant [31].

Kennzeichensysten (KKS) standards, which are in turn extensions for the energy sector of the IEC 81346 and ISO/TS 16952-10 international standards.

IEC 81346 and ISO/TS 16952-10 provide a generic dictionary of codes for designating and classifying industrial equipment. Figure 2 provides an except of these standards and their dependencies. For instance, in IEC-81346 letters 'B' to 'U' are used for generically designating systems in power plants. ISO/TS 16952-10 makes this specification more precise by indicating, for example, that letter 'M' refers to systems for generating and transmitting electricity, and that we can append 'D' to 'M' to refer to a wind turbine system. RDS PP and KKS provide a more extensive vocabulary of codes for equipment, their functionality and locations, as well as a system for combining such codes.

A typical energy plant model describes the structure of a plant by providing the functionality and location of each equipment component using RDS PP and KKS codes. Having this information in a machine-readable format is important for planning and construction, as well as for the software-driven operation and maintenance of the plant. Figure 2 shows how a specific plant is represented in a model; for instance, code *=G001 MDL10* denotes that the yaw drive system number 10 of type *MDL* is located in the wind turbine generator number 001.

2.3   Technical Challenges

The development and use of information models in practice poses major challenges.

1. Model development is costly, as it requires specialised training and proprietary tools; as a result, model development often cannot keep up with the arrival of new equipment and introduction of new processes.
2. Models are difficult to integrate and share since they are often independently developed using different types of proprietary software and they are based on incompatible data formats.
3. Monitoring queries are difficult to compose and execute on top of information models: they must comply with the requirements of the models (e.g., refer to specific codes in the energy use case), and their execution requires access to heterogeneous data from different machines and processes.

In order to overcome these challenges Siemens has recently applied semantic technologies in a number of applications [13,15,19,22,32]. In particular, OWL 2 has been used for describing information models. The choice of OWL 2 is not surprising since it provides a rich and flexible modelling language that is well suited for addressing the aforementioned challenges: it comes with an unambiguous, standardised semantics, and a wide range of tools and infrastructure. Moreover, RDF provides a unified data exchange format, which can be used to seamlessly access and exchange data, and hence facilitate monitoring tasks based on complex queries.

3   From Information Models to Ontologies and Constraints

In this section we describe the ontologies that we have developed to capture manufacturing and energy production models presented in Sect. 2. The goal of our ontologies is to eventually replace their underpinning models in applications. Thus, their design has been driven towards fulfilling the same purposes as the models they originate from; that is, to act as schema-level templates for data generation and exchange, and to enable the formulation and execution of monitoring queries.

The representation of industrial information models and standards using ontologies has been widely acknowledged as a non-trivial task [5,12,14,36]. In Sect. 3.1 we discuss the modelling choices underpinning the design of our ontologies and identify a fragment of OWL 2 RL that is sufficient to capture the basic aspects of the information models. Our analysis of the models, however, also revealed the need to incorporate database integrity constraints for data validation, which are not supported in OWL 2 [27,37]. Thus, we also discuss the kinds of constraints that are relevant to our applications.

Finally, in Sect. 3.2 we discuss how the OWL 2 RL axioms and integrity constraints can be captured by means of rules with stratified negation for the purpose of data validation and query answering. We assume basic familiarity with Datalog—the rule language underpinning OWL 2 RL and SWRL—as well as with stratified negation-as-failure (see [6] for an excellent survey on Logic Programming).

3.1   Modelling

From an ontological point of view, most building blocks of the the typical industrial information models are rather standard in conceptual design and naturally correspond to OWL 2 classes (e.g., *Turbine, Process, Product*), object properties (e.g., *hasPart, hasFunction, locatedIn*) and data properties (e.g., *ID, hasRotorSpeed*).

The main challenge that we encountered was to capture the constraints of the models using ontological axioms. We next describe how this was accomplished using a combination of OWL 2 RL axioms and integrity constraints.

Standard OWL 2 RL Axioms. The specification of the models suggests the arrangement of classes and properties according to subsumption hierarchies, which represent the skeleton of the model and establish the basic relationships between their components. For instance, in the energy plant model a *Turbine* is specified as a kind of *Equipment*, whereas *hasRotorSpeed* is seen as a more specific relation than *hasSpeed*. The models also suggest that certain properties must be declared as transitive, such as *hasPart* and *locatedIn*. Similarly, certain properties are naturally seen as inverse of each other (e.g., *hasPart* and *partOf*). These requirements are easily modelled in OWL 2 using the following axioms written in functional-style syntax:

$$\text{SubClassOf}(\textit{Turbine Equipment}) \tag{1}$$

$$\text{SubDataPropertyOf}(\textit{hasRotorSpeed hasSpeed}) \tag{2}$$

$$\text{TransitiveObjectProperty}(\textit{hasPart}) \tag{3}$$

$$\text{InverseObjectProperties}(\textit{hasPart partOf}) \tag{4}$$

These axioms can be readily exploited by reasoners to support query answering; e.g., when asking for all equipment with a rotor, one would expect to see all turbines that contain a rotor as a part (either directly or indirectly).

Additionally, the models describe *optional relationships* between entities. In the manufacturing model certain materials are optional to certain processes, i.e., they are compatible with the process but they are not always required. Similarly, certain processes can optionally be followed by other processes (e.g., conveying may be followed by packaging). Universal (i.e., *AllValuesFrom*) restrictions are well-suited for attaching an optional property to a class. For instance, the axiom

$$\text{SubClassOf}(\textit{Conveying} \; \text{ObjectAllValuesFrom}(\textit{followedBy Packaging})) \tag{5}$$

states that only packaging processes can follow conveying processes; that is, a conveying process can be either terminal (i.e., not followed by any other process) or it is followed by a packaging process. As a result, when introducing a new conveying process we are not forced to provide a follow-up process, but if we do so it must be an instance of *Packaging*.

All the aforementioned types of axioms are included in the OWL 2 RL profile. This has many practical advantages for reasoning since OWL 2 RL is amenable to efficient implementation using rule-based technologies.

Constraint Axioms. In addition to optional relationships, the information models from Sect. 2 also describe relationships that are inherently *mandatory*, e.g., when introducing a new turbine, the energy model requires that we also provide its rotors.

This behaviour is naturally captured by an integrity constraint: whenever a turbine is added and its rotors are not provided, the application should flag an error. Integrity constraints are not supported in OWL 2; for instance, the axiom

$$\text{SubClassOf}(\textit{Turbine } \text{ObjectSomeValuesFrom}(\textit{hasPart Rotor})) \qquad (6)$$

states that every turbine must contain a rotor as a part; such rotor, however, can be possibly unknown or unspecified.

The information models also impose cardinality restrictions on relationships. For instance, each double rotor turbine in the energy plant model is specified as having exactly two rotors. This can be modelled in OWL 2 using the axioms

$$\text{SubClassOf}(\textit{TwoRotorTurbine } \text{ObjectMinCardinality}(2 \textit{ hasPart Rotor})) \qquad (7)$$

$$\text{SubClassOf}(\textit{TwoRotorTurbine } \text{ObjectMaxCardinality}(2 \textit{ hasPart Rotor})) \qquad (8)$$

Such cardinality restrictions are interpreted as integrity constraints in many applications: when introducing a specific double rotor turbine, the model requires that we also provide its two rotors. The semantics of axioms (7) and (8) is not well-suited for this purpose: on the one hand, (7) does not enforce a double rotor turbine to explicitly contain any rotors at all; on the other hand, if more than two rotors are provided, then (8) non-deterministically enforces at least two of them to be equal.

There have been several proposals to extend OWL 2 with integrity constraints [27,37]. In these approaches, the ontology developer explicitly designates a subset of the OWL 2 axioms as constraints. Similarly to constraints in databases, these axioms are used as checks over the given data and do not participate in query answering once the data has been validated. The specifics of how this is accomplished semantically differ amongst each of the proposals; however, all approaches largely coincide if the standard axioms are in OWL 2 RL.

3.2 Data Validation and Query Answering

Our approach to data validation and query answering follows the standard approaches in the literature [27,37]: given a query Q, dataset $\mathcal{D}$, and OWL 2 ontology $\mathcal{O}$ consisting of a set $\mathcal{S}$ of standard OWL 2 RL axioms and a set $\mathcal{C}$ of axioms marked as constraints, we proceed according to Steps 1–4 given next.

Table 1. OWL 2 RL axioms as rules. All entities mentioned in the axioms are named. By abuse of notation, we use SubPropertyOf and AllValuesFrom to refer to both their Object and Data versions in functional syntax.

OWL 2 Axiom	Datalog Rules
SubClassOf(A B)	$B(?x) \leftarrow A(?x)$
SubPropertyOf(P_1 P_2)	$P_2(?x, ?y) \leftarrow P_1(?x, ?y)$
TransitiveObjectProperty(P)	$P(?x, ?z) \leftarrow P(?x, ?y) \land P(?y, ?z)$
InverseObjectProperties(P_1, P_2)	$P_2(?y, ?x) \leftarrow P_1(?x, ?y)$ and
	$P_1(?y, ?x) \leftarrow P_2(?x, ?y)$
SubClassOf(A AllValuesFrom(P B))	$B(?y) \leftarrow P(?x, ?y) \land A(?x)$

1. Translate the standard axioms $\mathcal{S}$ into a Datalog program Π_S using the well-known correspondence between OWL 2 RL and Datalog.
2. Translate the integrity constraints $\mathcal{C}$ into a Datalog program Π_C with stratified negation-as-failure containing a distinguished binary predicate *Violation* for recording the individuals and axioms involved in a constraint violation.
3. Retrieve and flag all integrity constraint violations. This can be done by computing the extension of the *Violation* predicate.
4. If no constraints are violated, answer the user's query Q using the query answering facilities provided by the reasoner.

Steps 3 and 4 can be implemented on top of RDF triple stores with support for OWL 2 RL and stratified negation (e.g., [28]), as well as on top of generic rule inference systems (e.g., [3]). In the remainder of this Section we illustrate Steps 1 and 2, where standard axioms and constraints are translated into rules. **Standard Axioms.** Table 1 provides the standard OWL 2 RL axioms needed to capture the information models of Sect. 2 and their translation into negation-free rules. In particular, the axioms (1)–(5) are equivalent to the following rules:

$$Equipment(?x) \leftarrow Turbine(?x) \tag{9}$$

$$hasSpeed(?x, ?y) \leftarrow hasRotorSpeed(?x, ?y) \tag{10}$$

$$hasPart(?x, ?z) \leftarrow hasPart(?x, ?y) \land hasPart(?y, ?z) \tag{11}$$

$$Packaging(?y) \leftarrow Conveying(?x) \land followedBy(?x, ?y) \tag{12}$$

Constraint Axioms. Table 2 provides the constraint axioms required to capture the models of Sect. 2 together with their translation into rules with negation. Our translation assigns a unique id to each individual axiom marked as an integrity constraint in the ontology, and it introduces predicates not occurring in the ontology in the heads of all rules. Constraint violations are recorded using the fresh predicate *Violation* relating individuals to constraint axiom ids.

The constraint (6) from Sect. 3.1 is captured by the following rules:

$$hasPart\_Rotor(?x) \leftarrow hasPart(?x, ?y) \land Rotor(?y) \tag{13}$$

$$Violation(?x, \alpha) \leftarrow Turbine(?x) \land \textbf{not}\ hasPart\_Rotor(?x) \tag{14}$$

Table 2. Constraints axioms as rules. All entities are named, $n \geq 1$, and α is the unique id for the given constraint. *SomeValuesFrom, HasValue, FunctionalProperty, MaxCardinality* and *MinCardinality* denote both their Object and Data versions.

OWL Axiom	Datalog rules
SubClassOf(A SomeValuesFrom(R B))	$R\_B(?x) \leftarrow R(?x, ?y) \wedge B(?y)$ and $Violation(?x, \alpha) \leftarrow A(?x) \wedge$ **not** $R\_B(?x)$
SubClassOf(A HasValue(R b))	$Violation(?x, \alpha) \leftarrow A(?x) \wedge$ **not** $R(?x, b)$
FunctionalProperty(R)	$R\_2(?x) \leftarrow R(?x, ?y_1) \wedge R(?x, ?y_2) \wedge$ **not** $owl{:}sameAs(?y_1, ?y_2)$ and $Violation(?x, \alpha) \leftarrow R\_2(?x)$
SubClassOf(A MaxCardinality(n R B))	$R\_(n{+}1)\_B(?x) \leftarrow \bigwedge\limits_{1 \leq i \leq n+1} (R(?x, ?y_i) \wedge B(?y_i))$ $\bigwedge\limits_{1 \leq i < j \leq n+1}$ (**not** $owl{:}sameAs(?y_i, ?y_j)$) and $Violation(?x, \alpha) \leftarrow A(?x) \wedge R\_(n{+}1)\_B(?x)$
SubClassOf(A MinCardinality(n R B))	$R\_n\_B(?x) \leftarrow \bigwedge\limits_{1 \leq i \leq n} (R(?x, ?y_i) \wedge B(?y_i))$ $\bigwedge\limits_{1 \leq i < j \leq n}$ (**not** $owl{:}sameAs(?y_i, ?y_j)$) and $Violation(?x, \alpha) \leftarrow A(?x) \wedge$ **not** $R\_n\_B(?x)$

Rule (13) identifies all individuals with a rotor as a part, and stores them as instances of the auxiliary predicate *hasPart_Rotor*. In turn, Rule (14) identifies all turbines that are not known to be instances of *hasPart_Rotor* (i.e., those with no known rotor as a part) and links them to the constraint α they violate.

Integrity constraints based on cardinalities require the use of the OWL 2 equality predicate *owl:sameAs*. For instance, the constraint axiom (7) from Sect. 3.1, to which we assign the id β_1, is translated into the following rules:

$$hasPart\_2\_Rotor(?x) \leftarrow \bigwedge_{1 \leq i \leq 2} (hasPart(?x, ?y_i) \wedge Rotor(?y_i)) \wedge$$

$$\wedge \, (\textbf{not} \; owl{:}sameAs(?y_1, ?y_2))$$

$$Violation(?x, \beta_1) \leftarrow TwoRotorTurbine(?x) \wedge \textbf{not} \; hasPart\_2\_Rotor(?x)$$

The first rule infers an instance of the auxiliary predicate *hasPart_2_Rotor* if it is connected to two instances of *Rotor* that are not known to be equal; in turn, the second rule infers that all instances of *TwoRotorTurbine* that are not known to be instances of the auxiliary predicate violate the constraint (7). Similarly, axiom (8), to which we assign the id β_2, is translated as follows:

$$hasPart\_3\_Rotor(?x) \leftarrow \bigwedge_{1 \leq i \leq 3} (hasPart(?x, ?y_i) \wedge Rotor(?y_i)) \wedge$$

$$\wedge \bigwedge_{1 \leq i < j \leq 3} (\textbf{not} \; owl{:}sameAs(?y_i, ?y_j))$$

$$Violation(?x, \beta_2) \leftarrow TwoRotorTurbine(?x) \wedge hasPart\_3\_Rotor(?x)$$

Analogously to the previous case, the first rule infers that an individual is an instance of *hasPart_3_Rotor* if it is connected to three instances of *Rotor* that are not known to be equal; in turn, the second rule infers that every such individual that is also an instance of *TwoRotorTurbine* violates the constraint axiom (8).

To conclude this section, we note that our translation in Table 2 yields a stratified program for any set $\mathcal{C}$ of constraints. We can always define a stratification where the lowest stratum consists of the predicates in $\mathcal{C}$ and *owl:sameAs*, the intermediate stratum contains all predicates of the form R_B, R_n_B, and R_n, and the uppermost stratum contains the special *Violation* predicate.

4 SOMM: An Industrial Ontology Management System

We have developed the *Siemens-Oxford Ontology Management (SOMM)* tool[3] to support engineers in building ontologies and inserting data based on their information models. The interface of SOMM is restricted to support only the kinds of standard OWL 2 RL axioms and constraints discussed in Sect. 3.

SOMM is built on top of the Web-Protégé platform [40] by extending its front-end with new visual components and its back-end to access a Datalog-based triple store or a generic rule inference system for query answering and constraint validation, the OWL 2 reasoner HermiT [33] for ontology classification, and LogMap [16] to support ontology alignment and merging. Our choice of WebProtégé was based on Siemens' requirements for the platform underpinning SOMM, namely that it *(i)* can be used as a Web application; *(ii)* is under active development; *(iii)* is open-source and modular; *(iv)* includes built-in functionality for ontology versioning and collaborative development; *(v)* provides a form-based and end-user oriented interface; and *(vi)* enables the automatic generation of forms to insert instance data. Although we considered other alternatives such as Protégé-desktop [39], NeON toolkit [8], OBO-Edit [7], and TopBraid Composer [38], we found that only WebProtégé satisfied all the aforementioned requirements.

In the remainder of this section, we describe the main features of SOMM.

Insertion of axioms and constraints. We have implemented a form-based interface for editing standard axioms and constraints. Figure 3 shows a screenshot of the SOMM class editor representing the following axioms about *SteamTurbine* (abbreviated below as *ST*), where all but the last axiom represent constraints.

$$\mathrm{SubClassOf}(ST \ \mathrm{ObjectSomeValuesFrom}(hasState \ State))$$
$$\mathrm{SubClassOf}(ST \ \mathrm{DataSomeValuesFrom}(hasId \ xsd{:}string))$$
$$\mathrm{SubClassOf}(ST \ \mathrm{ObjectMinCardinality}(1 \ hasConfig \ STConfig))$$
$$\mathrm{SubClassOf}(ST \ \mathrm{ObjectMaxCardinality}(3 \ hasConfig \ STConfig))$$
$$\mathrm{SubClassOf}(ST \ \mathrm{ObjectAllValuesFrom}(hasProductLine \ ProductLine))$$

The interface shows that the class *SteamTurbine* has three mandatory properties (*hasState*, *hasID* and *hasConfig*) marked as 'Required' and interpreted as constraints, and an optional property (*hasProductLine*) interpreted as a standard

---

[3] http://www.cs.ox.ac.uk/isg/tools/SOMM/.

axiom. Object and data properties are indicated by blue and green rectangles, respectively. For each property we can specify their filler using a WebProtégé autocompletion field. Finally, the fields 'Min' and 'Max' are used to represent cardinality constraints on mandatory properties.

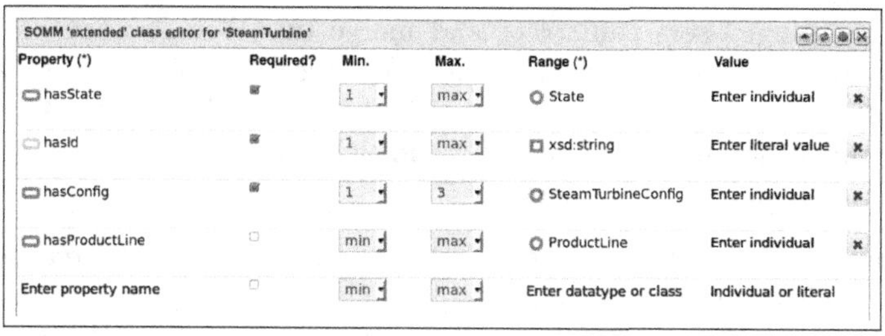

Fig. 3. SOMM editor to attach properties to classes.

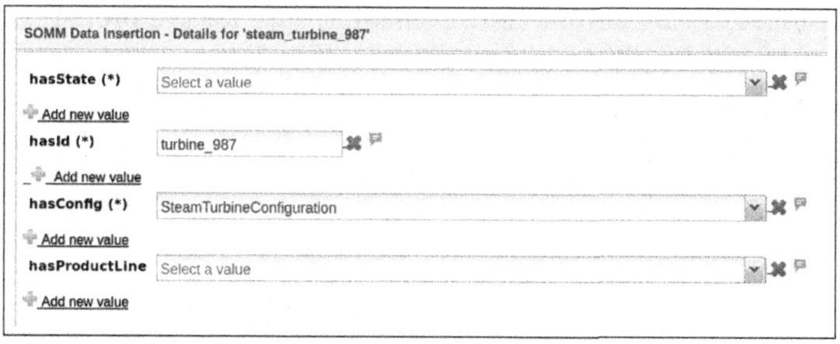

Fig. 4. Data insertion in SOMM.

Automatically generated data forms.SOMM exploits the capabilities of the 'knowledge acquisition forms' in Web-Protégé to guide engineers during data entry. The main use of data forms that we envision is ontology validation during the time of ontology development. The forms are automatically generated for each class based on its relevant mandatory and optional properties. For this, SOMM considers (i) the explicitly provided properties; (ii) the inherited properties; and (iii) the properties explicitly attached to its descendant classes. The latter were deemed useful by Siemens engineers, e.g., although *Turbine* does not have directly attached properties, the SOMM interface would suggests adding data for the properties attached to its subclass *SteamTurbine*. Figure 4 shows an example of the property fields for an instance of the class *SteamTurbine*, where required fields (i.e., those for which a value must be provided) are marked with (*).

Extended hierarchies. In addition to subsumption hierarchies, SOMM allows also for hierarchies based on arbitrary properties. These can be seen as a generalisation of partonomy hierarchies, and assume that the dependencies between classes or individuals based on the relevant property are 'tree-shaped'. Figures 5a and b show the hierarchy for the *follows* property, which determines which kinds of processes can follow other processes; for instance, *Conveying* follows *Loading* and is followed by *Testing*.

Alignment. SOMM integrates the system LogMap [16] to support model alignment and merging. Users can select and merge two Web-Protégé projects, or import and merge an ontology into the active Web-Protégé project. Although LogMap supports interactive alignment [17], it is currently used in SOMM in an automatic mode; we are planning to extend SOMM's interface to support user interaction in the alignment process.

Reasoning. SOMM relies on HermiT [10] to support standard reasoning services such as class satisfiability and ontology classification. Data validation and query answering support is currently provided on top of the IRIS reasoner [3], as described in Sect. 3.2. Figures 5c and d illustrates the supported reasoning services. The left-hand-side of the figure shows that the class *GasTurbineModes* is satisfiable and *Process* is an inferred superclass. On the right-hand-side we can see that *steam_turbine_987* violates one of the integrity constraints; indeed, as shown in Fig. 4, *steam_turbine_987* is missing data for the property *hasState*, which is mandatory for all steam turbines (see Fig. 3).

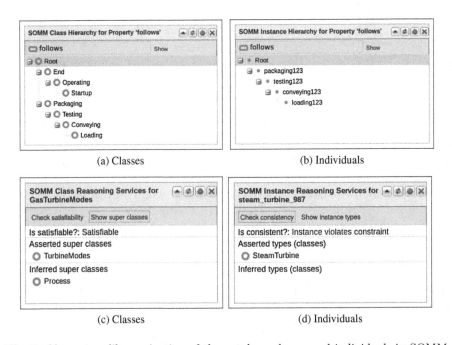

(a) Classes   (b) Individuals

(c) Classes   (d) Individuals

Fig. 5. Above: tree-like navigation of the ontology classes and individuals in SOMM. Below: reasoning services for ontology classes and individuals in SOMM

5   Evaluation

We have evaluated the practical feasibility of the data validation and query answering services provided by SOMM. For this, we have conducted two sets of experiments for the manufacturing and energy turbine scenarios, respectively. In the first experiment, we simulated the operation of a manufacturing plant using a synthetic generator that produces realistic product manufacturing data of varying size; in the second experiment, we used real anonymised turbine data.[4] All our experiments were conducted on a laptop with an Intel Core i7-4600U CPU at 2.10 GHz and 16 GB of RAM running Ubuntu 14.04 (64 bits). We allocated 15 GB to Java 8 and set up IRIS with its default configuration.

Manufacturing Experiments. In our experiments for the manufacturing use case we used the ontology, data and queries given next.

- The ontology capturing the manufacturing model illustrated in Fig. 1 from Sect. 2.1. The ontology contains 79 standard axioms and 20 constraints.
- A data generator used by Siemens engineers to simulate manufacturing of products of two types based on the aforementioned model. We used two configurations of the generator: configuration $(C1)$ simulates a situation where products were manufactured in violation of the model specifications (e.g., they used too much material of some kind); in $(C2)$, each product is manufactured according to specifications.
- A sample of three monitoring queries commonly used in practice. The first query asks for all products that use material from a given lot; the second asks for all material lots used in a given product; finally, the third one asks for the total quantity of material in lots of a specific kind.

We generated data for 6 different sizes, ranging from 50 triples to 1 million triples. For each size, we generated one dataset for each configuration of the generator. We set up configuration $C1$ so that 35 % of the manufactured products violate specification. Our experiments follow Steps 1–4 in Sect. 3.2. We checked validity of each dataset against the ontology using Steps 1–3; then, for each dataset created using $C2$ we also answered all test queries (Step 4). We repeated the experiment 5 times for each dataset and configuration (i.e., 10 times for each dataset size).

Our results are summarised in Fig. 6. Times for each data size are *wall clock* time averages (in ms). Constraint validation time (grey bar) correspond to Step 3 in Sect. 3.2. Query answering times (blue bar) measure the time for answering the use case queries (Step 4); here, only datasets satisfying the constraints (i.e., generated using $C2$) are considered. The figure also provides the average number of constraint violations in data generated according to $C1$, and the number of triples after constraint validation.

Our results demonstrate the feasibility of our ontology-based approach to model validation and query answering in realistic manufacturing scenarios. In

---

[4] We are in the process of sorting out the licenses for the ontologies and data used in our experiments; they cannot be made publicly available at this point.

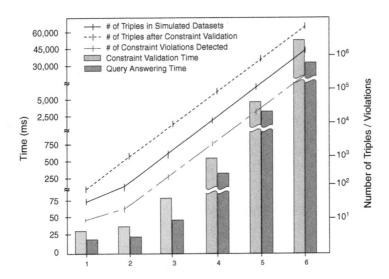

Fig. 6. Experimental results

particular, constraint validation and query answering were feasible within $87s$ on stock hardware over datasets containing over 1 million triples.

Gas Turbine Experiment. In this experiment we used the following data:

- The ontology capturing the energy plant model illustrated in Fig. 2 from Sect. 2. The ontology contains 121 standard axioms and 25 constraints.
- An anonymised dataset describing the structure of 800 real gas turbines, their sensor readings (temperature, pressure, rotor speed and position), and associated processes (e.g., expansion, compression, start up, shut down). The dataset was converted from a relational DB into RDF, and contains $25,090$ triples involving $4,076$ individuals.
- Three commonly used test queries. The first query asks for the core parts, equipment and current state of all turbines of a given type; the second asks for all components involved in a compression process; the last query asks for the temperature readings of turbines of a given type.

We followed the same steps as in the previous experiments, with very positive results. Constraint checking was completed in $2s$ and generated $27,007$ additional triples; we found $1,582$ constraint violations, which is especially interesting given that the data is real. Query answering over the valid subset took $1s$ on average.

6   Lessons Learned and Future Work

We have studied the use of ontologies to capture industrial information models in manufacturing and energy production applications.

Our study of the requirements of information models revealed that many key aspects of information models naturally correspond to integrity constraints and hence cannot be captured by standard OWL 2 ontologies. This demonstrates intrinsic limitations of OWL 2 for industrial modelling and gives a clear evidence of why constraints are essential for such modelling.

We also learned that even a rather simple form-based interface such as the one of SOMM is sufficient to capture most of the manufacturing and energy information models based on ISA and ICE standards. This was an important insight for us since at the beginning of this research project it was unclear whether designing such a simple tool to write ontologies of practical interest to our use cases would be feasible.

Finally, we have received a very positive feedback from Siemens engineers about the usability of SOMM at informal workshops organised as part of the project. This was encouraging since the development of a tool that is accessible to users without background in semantic technologies was one of the main motivations of our work.

In the future, we plan to conduct a formal user study where—with the help of SOMM—Siemens engineers will design elaborate information models and perform various tasks on these models, including validation and merging. We also plan to conduct more extensive scalability experiments. SOMM is a research prototype and, depending on the outcome these studies, we would like to deploy it in production departments.

References

1. Abele, L., Legat, C., Grimm, S., Muller, A.W.: Ontology-based of plant models. In: INDIN (2013)
2. Arancón, J., Polo, L., Berrueta, D., Lesaffre, F., Abajo, N., Campos, A.M.: Ontology- based knowledge management in the steel industry. In: The Semantic Web: Real-World Applications from Industry (2007)
3. Bishop, B., Ficsher, F.: IRIS - integrated rule inference system. In: Workshop on Advancing Reasoning on the Web (2008)
4. Calvanese, D., et al.: Optique: OBDA solution for big data. In: ESWC (Satellite Events), Revised Selected Papers (2013)
5. Classification and Product Description. http://www.eclass.eu/
6. Dantsin, E., Eiter, T., Gottlob, G., Voronkov, A.: Complexity, expressive power of logic programming. ACM Comput. Surv. **33**(3), 374–425 (2001)
7. Day-Richter, J., Harris, M.A., Haendel, M., Lewis, S.: OBO-Edit - an ontology editor for biologists. Bioinformatics **23**(16), 2198–2200 (2007)
8. Erdmann, M., Waterfeld, W.: Overview of the NeOn toolkit. In: Ontology Engineering in a Networked World (2012)
9. Forschungsunion. Fokus: Das Zukunftsprojekt Industrie 4.0, Handlungsempfehlungen zur Umsetzung. In: Bericht der Promotorengruppe KOMMUNIKATION (2012)
10. Glimm, B., Horrocks, I., Motik, B., Stoilos, G., Wang, Z.: HermiT: An OWL 2 reasoner. J. Autom. Reasoning **53**(3), 245–269 (2014)
11. Gliozzo, A., Biran, O., Patwardhan, S., McKeown, K.: Semantic technologies in IBM watson. In: Teaching NLP and CL Workshop (TNLP) at ACL (2013)

12. Grangel-González, I., Halilaj, L., Coskun, G., Auer, S., Collarana, D., Hoffmeister, M.: Towards a semantic administrative shell for industry 4.0 components. In: ICSC (2016)
13. Grimm, S., Watzke, M., Hubauer, T., Cescolini, F., Embedded $\mathcal{EL}$ + reasoning on programmable logic controllers. In: ISWC (2012)
14. Hepp, M., de Bruijn, J.: GenTax: a generic methodology for deriving OWL and RDF-S ontologies from hierarchical classifications, thesauri, and inconsistent taxonomies. In: ESWC (2007)
15. Hubauer, T., Lamparter, S., Pirker, M.: Automata-based abduction for tractable diagnosis. In: DL (2010)
16. Jiménez-Ruiz, E., Cuenca Grau, B.: LogMap: logic-based and scalable ontology matching. In: ISWC (2011)
17. Jiménez-Ruiz, E., Cuenca Grau, B., Zhou, Y., Horrocks, I.: Large-scale interactive ontology matching: algorithms and implementation. In: ECAI (2012)
18. Kagermann, H., Lukas, W.-D.: Industrie 4.0: Mit dem internet der Dinge auf dem Weg zur 4. industriellen Revolution. In: VDI Nachrichten (2011)
19. Kharlamov, E., et al.: Enabling semantic access to static, streaming distributed data with optique: demo. In: ACM DEBS (2016)
20. Kharlamov, E.: How semantic technologies can enhance data access at siemens energy. In: ISWC (2014)
21. Kharlamov, E., et al.: Ontology based access to exploration data at statoil. In: Arenas, M., et al. (eds.) ISWC 2015. LNCS, vol. 9367, pp. 93–112. Springer, Heidelberg (2015). doi:10.1007/978-3-319-25010-6_6
22. Kharlamov, E.: Ontology-based integration of streaming and static relational data with optique. In: ACM SIGMOD (2016)
23. Kharlamov, E., et al.: Optique: ontology-based data access platform. In: ISWC (P&D) (2015)
24. Kharlamov, E., et al.: Optique: towards OBDA systems for industry. In: Cimiano, P., Fernández, M., Lopez, V., Schlobach, S., Völker, J. (eds.) ESWC 2013. LNCS, vol. 7955, pp. 125–140. Springer, Heidelberg (2013). doi:10.1007/978-3-642-41242-4_11
25. Kharlamov, E., et al.: Semantic access to siemens streaming data: the optique way. In: ISWC (P&D) (2015)
26. Motik, B., Cuenca Grau, B., Horrocks, I., Wu, Z., Fokoue, A., Lutz, C.: OWL 2 web ontology language profiles (Second Edition). W3C Recommendation (2012)
27. Motik, B., Horrocks, I., Sattler, U.: Bridging the gap between OWL and relational databases. J. Web Sem. **7**(2), 41–60 (2009)
28. Nenov, Y., Piro, R., Motik, B., Horrocks, I., Wu, Z., Banerjee, J.: RDFox: a highly-scalable RDF store. In: ISWC (2015)
29. Poggi, A., Lembo, D., Calvanese, D., De Giacomo, G., Lenzerini, M., Rosati, R.: Linking data to ontologies. J. Data Semant. **10**, 237–271 (2008)
30. Qiu, R.G., Zhou, M.: Mighty MESs; state-of-the-art, future manufacturing execution systems. IEEE Robot. Automat. Mag. 11(1) (2004)
31. Richnow, J., Rossi, C., Wank, H.: Designation of wind power plants with the reference designation system for power plants - RDS-pp. VGB PowerTech. **94**, 2 (2014)
32. Ringsquandl, M., Lamparter, S., Brandt, S., Hubauer, T., Lepratti, R.: Semantic-guided feature selection for industrial automation systems. In: Arenas, M., et al. (eds.) ISWC 2015. LNCS, vol. 9367, pp. 225–240. Springer, Heidelberg (2015). doi:10.1007/978-3-319-25010-6_13
33. Shearer, R., Motik, B., Horrocks, I.: HermiT: a highly-efficient OWL reasoner. In: OWLED (2008)

34. Siemens. Modeling new perspectives: digitalization - the key to increased productivity, efficiency and flexibility (White Paper). In: DER SPIEGEL, 6 2015
35. Stetter, R.: Software im maschinenbau-laestiges anhangsel oder chance marktfuehrerschaft? In: VDMA, ITQ (2011). http://www.software-kompetenz.de/en/
36. Stolz, A., Rodriguez-Castro, B., Radinger, A., Hepp, M.: PCS2OWL: a generic approach for deriving web ontologies from product classification systems. In: Presutti, V., d'Amato, C., Gandon, F., d'Aquin, M., Staab, S., Tordai, A. (eds.) ESWC 2014. LNCS, vol. 8465, pp. 644–658. Springer, Heidelberg (2014). doi:10.1007/978-3-319-07443-6_43
37. Tao, J., Sirin, E., Bao, J., McGuinness, D.L.: Integrity constraints in OWL. In: AAAI (2010)
38. Top Quadrant. TopBraid Composer. http://www.topquadrant.com/
39. Tudorache, T., Noy, N.F., Tu, S.W., Musen, M.A.: Supporting collaborative ontologydevelopment in protégé. In: ISWC (2008)
40. Tudorache, T., Nyulas, C., Noy, N.F., Musen, M.A.: WebProtégé: a collaborative ontology editor and knowledge acquisition tool for the web. In: Semantic Web 4.1 (2013)
41. Vyatkin, V., Engineering, S.: Software engineering in industrial automation: state-of-the-art review. IEEE Trans. Ind. Inf. 9(3), 2351–2362 (2013)

Towards Analytics Aware Ontology Based Access to Static and Streaming Data

Evgeny Kharlamov[1(✉)], Yannis Kotidis[2], Theofilos Mailis[3],
ChristianNeuenstadt[4], Charalampos Nikolaou[1], Özgür Özçep[4],
Christoforos Svingos[3], Dmitriy Zheleznyakov[1], Sebastian Brandt[5],
Ian Horrocks[1], Yannis Ioannidis[3], Steffen Lamparter[5], and Ralf Möller[4]

[1] University of Oxford, Oxford, UK
evgeny.kharlamov@cs.ox.ac.uk
[2] Athens University of Economics and Business, Athens, Greece
[3] University of Athens, Athens, Greece
[4] University of Lübeck, Lübeck, Germany
[5] Siemens Corporate Technology, Munich, Germany

Abstract. Real-time analytics that requires integration and aggregation of heterogeneous and distributed streaming and static data is a typical task in many industrial scenarios such as diagnostics of turbines in Siemens. OBDA approach has a great potential to facilitate such tasks; however, it has a number of limitations in dealing with analytics that restrict its use in important industrial applications. Based on our experience with Siemens, we argue that in order to overcome those limitations OBDA should be extended and become analytics, source, and cost aware. In this work we propose such an extension. In particular, we propose an ontology, mapping, and query language for OBDA, where aggregate and other analytical functions are first class citizens. Moreover, we develop query optimisation techniques that allow to efficiently process analytical tasks over static and streaming data. We implement our approach in a system and evaluate our system with Siemens turbine data.

1 Introduction

Ontology Based Data Access (OBDA) [9] is an approach to access information stored in multiple datasources via an abstraction layer that mediates between the datasources and data consumers. This layer uses an *ontology* to provide a uniform conceptual schema that describes the problem domain of the underlying data independently of how and where the data is stored, and declarative *mappings* to specify how the ontology is related to the data by relating elements of the ontology to queries over datasources. The ontology and mappings are used to *transform* queries over ontologies, i.e., *ontological queries*, into *data queries* over datasources. As well as abstracting away from details of data storage and

This work was partially funded by the EU project Optique (FP7-ICT-318338) and the EPSRC projects MaSI[3], DBOnto, and ED[3].

P. Groth et al. (Eds.): ISWC 2016, Part II, LNCS 9982, pp. 344–362, 2016.
DOI: 10.1007/978-3-319-46547-0_31

access, the ontology and mappings provide a declarative, modular and query-independent specification of both the conceptual model and its relationship to the data sources; this simplifies development and maintenance and allows for easy integration with existing data management infrastructure.

A number of systems that at least partially implement OBDA have been recently developed; they include D2RQ [7], Mastro [10], morph-RDB [38], Ontop [39], OntoQF [33], Ultrawrap [41], Virtuoso[1], and others [8,17]. Some of them were successfully used in various applications including cultural heritage [13], governmental organisations [15], and industry [20,21]. Despite their success, OBDA systems, however, are not tailored towards analytical tasks that are naturally based on data aggregation and correlation. Moreover, they offer a limited or no support for queries that combine streaming and static data. A typical scenario that requires both analytics and access to static and streaming data is diagnostics and monitoring of turbines in Siemens.

Siemens has several service centres dedicated to diagnostics of thousands of power-generation appliances located across the globe [21]. One typical task of such a centre is to detect in real-time potential faults of a turbine caused by, e.g., an undesirable pattern in temperature's behaviour within various components of the turbine. Consider a (simplified) example of such a task:

> In a given turbine report all temperature sensors that are reliable, i.e., with the average score of validation tests at least 90 %, and whose measurements within the last 10 min were similar, i.e., Pearson correlated by at least 0.75, to measurements reported last year by a reference sensor that had been functioning in a critical mode.

This task requires to extract, aggregate, and correlate static data about the turbine's structure, streaming data produced by up to 2,000 sensors installed in different parts of the turbine, and historical operational data of the reference sensor stored in multiple datasources. Accomplishing such a task currently requires to pose a collection of hundreds of queries, the majority of which are semantically the same (they ask about temperature), but syntactically differ (they are over different schemata). Formulating and executing so many queries and then assembling the computed answers take up to 80 % of the overall diagnostic time that Siemens engineers typically have to spend [21]. The use of ODBA, however, would allow to save a lot of this time since ontologies can help to 'hide' the technical details of *how* the data is produced, represented, and stored in data sources, and to show only *what* this data is about. Thus, one would be able to formulate this diagnostic task using only one ontological query instead of a collection of hundreds data queries that today have to be written or configured by IT specialists. Clearly, this collection of queries does not disappear: the OBDA query transformation will automatically compute them from the the the high-level ontological query using the ontology and mappings.

Siemens analytical tasks as the one in the example scenario typically make heavy use of aggregation and correlation functions as well as arithmetic operations.

---

[1] http://virtuoso.openlinksw.com/.

In our running example, the aggregation function min and the comparison operator $\geq$ are used to specify what makes a sensor reliable and to define a threshold for similarity. Performing such operations only in ontological queries, or only in data queries specified in the mappings is not satisfactory. In the case of ontological queries, all relevant values should be retrieved prior to performing grouping and arithmetic operations. This can be highly inefficient, as it fails to exploit source capabilities (e.g., access to pre-computed averages), and value retrieval may be slow and/or costly, e.g., when relevant values are stored remotely. Moreover, it adds to the complexity of application queries, and thus limits the benefits of the abstraction layer. In the case of source queries, aggregation functions and comparison operators may be used in mapping queries. This is brittle and inflexible, as values such as 90 % and 0.75, which are used to define 'reliable sensor' and 'similarity', cannot be specified in the ontological query, but must be 'hard-wired' in the mappings, unless an appropriate extension to the query language or the ontology are developed. In order to address these issues, OBDA should become

> *analytics-aware* by supporting declarative representations of basic analytics operations and using these to efficiently answer higher level queries.

In practice this requires enhancing OBDA technology with ontologies, mappings, and query languages capable of capturing operations used in analytics, but also extensive modification of OBDA query preprocessing components, i.e., reasoning and query transformation, to support these enhanced languages.

Moreover, analytical tasks as in the example scenario should typically be executed continuously in data intensive and highly distributed environments of streaming and static data. Efficiency of such execution requires non-trivial query optimisation. However, optimisations in existing OBDA systems are usually limited to minimisation of the textual size of the generated queries, e.g. [40], with little support for distributed query processing, and no support for optimisation for continuous queries over sequences of numerical data and, in particular, computation of data correlation and aggregation across static and streaming data. In order to address these issues, OBDA should become

> *source and cost aware* by supporting both static and streaming data sources and offering a robust query planning component and indexing that can estimate the cost of different plans, and use such estimates to produce low-cost plans.

Note that the existence of materialised and pre-computed subqueries relevant to analytics within sources and archived historical data that should be correlated with current streaming data implies that there is a range of query plans which can differ dramatically with respect to data transfer and query execution time.

In this paper we make the first step to extend OBDA systems towards becoming analytics, source, and cost aware and thus meeting Siemens requirements for turbine diagnostics tasks. In particular, our contributions are the following:

- We proposed analytics-aware OBDA components, i.e., (*i*) ontology language $DL\text{-}Lite_{\mathcal{A}}^{\mathsf{agg}}$ that extends $DL\text{-}Lite_{\mathcal{A}}$ with aggregate functions as first class citizens, (ii) query language STARQL over ontologies that combine streaming and static data, and (iii) a mapping language relating $DL\text{-}Lite_{\mathcal{A}}^{\mathsf{agg}}$ vocabulary and STARQL constructs with relational queries over static and streaming data.

– We developed efficient query transformation techniques that allow to turn STARQL queries over $DL\text{-}Lite_{\mathcal{A}}^{\mathsf{agg}}$ ontologies, into data queries using our mappings.

– We developed source and cost aware (i) optimisation techniques for processing complex analytics on both static and streaming data, including adaptive indexing schemes and pre-computation of frequent aggregates on user queries, and (ii) elastic infrastructure that automatically distributes analytical computations and data over a computational cloud for fastest query execution.

– We implemented (i) a highly optimised engine EXASTREAM capable of handling complex streaming and static queries in real time, (ii) a dedicated STARQL2SQL$^{\oplus}$ translator that transforms STARQL queries into queries over static and streaming data, (iii) an integrated OBDA system that relies on our and third party components.

– We conducted a performance evaluation of our OBDA system with large scale Siemens simulated data using analytical tasks.

Due to space limitations we could not include all the relevant material in this paper and refer the reader to its online extended version for further details [26].

2 Analytics Aware OBDA for Static and Streaming Data

In this section we first introduce our analytics-aware ontology language $DL\text{-}Lite_{\mathcal{A}}^{\mathsf{agg}}$ (Sect. 2.1) for capturing static aspects of the domain of interest. In $DL\text{-}Lite_{\mathcal{A}}^{\mathsf{agg}}$ ontologies, aggregate functions are treated as first class citizens. Then, in Sect. 2.2 we will introduce a query language STARQL that allows to combine static conjunctive queries over $DL\text{-}Lite_{\mathcal{A}}^{\mathsf{agg}}$ with continuous diagnostic queries that involve simple combinations of time aware data attributes, time windows, and functions, e.g., correlations over streams of attribute values. Using STARQL queries one can retrieve entities, e.g., sensors, that pass two 'filters': static and continuous. In our running example a static 'filter' checks whether a sensor is reliable, while a continuous 'filter' checks whether the measurements of the sensor are Pearson correlated with the measurements of reference sensor. In Sect. 2.3 we will explain how to translate STARQL queries into data queries by mapping $DL\text{-}Lite_{\mathcal{A}}^{\mathsf{agg}}$ concepts, properties, and attributes occurring in queries to database schemata and by mapping functions and constructs of STARQL continuous 'filters' into corresponding functions and constructs over databases. Finally, in Sect. 2.4 we discuss how to optimise resulting data queries.

2.1 Ontology Language

Our ontology language, $DL\text{-}Lite_{\mathcal{A}}^{\mathsf{agg}}$, is an extension of $DL\text{-}Lite_{\mathcal{A}}$ [9] with concepts that are based on aggregation of attribute values. The semantics for such concepts adapts the closed-world semantics [32]. The main reason why we rely on this semantics is to avoid the problem of empty answers for aggregate queries under the certain answers semantics [11,30]. In $DL\text{-}Lite_{\mathcal{A}}^{\text{agg}}$ we distinguish between individuals and data values from countable sets Δ and D that intuitively correspond to the datatypes of RDF. We also distinguish between atomic roles P that denote binary relations between pairs of individuals, and attributes F that denote binary relations between individuals and data values. For simplicity of presentation we assume that D is the set of rational numbers. Let agg be an aggregate function, e.g., min, max, count, countd, sum, or avg, and let $\circ$ be a comparison predicate on rational numbers, e.g., $\geq, \leq, <, >, =,$ or $\neq$.

$DL\text{-}Lite_{\mathcal{A}}^{\text{agg}}$ Syntax. The grammar for concepts and roles in $DL\text{-}Lite_{\mathcal{A}}^{\text{agg}}$ is as follows:

$$B \to A \mid \exists R, \quad C \to B \mid \exists F, \quad E \to \circ_r(\text{agg } F), \quad R \to P \mid P^-,$$

where F, P, agg, and $\circ$ are as above, r is a rational number, A, B, C and E are atomic, basic, extended and aggregate concepts, respectively, and R is a basic role.

A $DL\text{-}Lite_{\mathcal{A}}^{\text{agg}}$ ontology $\mathcal{O}$ is a finite set of axioms. We consider two types of axioms: *aggregate* axioms of the form $E \sqsubseteq B$ and *regular* axioms that take one of the following forms: (*i*) *inclusions* of the form $C \sqsubseteq B$, $R_1 \sqsubseteq R_2$, and $F_1 \sqsubseteq F_2$, (*ii*) *functionality* axioms (funct R) and (funct F), (*iii*) or *denials* of the form $B_1 \sqcap B_2 \sqsubseteq \bot$, $R_1 \sqcap R_2 \sqsubseteq \bot$, and $F_1 \sqcap F_2 \sqsubseteq \bot$. As in $DL\text{-}Lite_{\mathcal{A}}$, a $DL\text{-}Lite_{\mathcal{A}}^{\text{agg}}$ dataset $\mathcal{D}$ is a finite set of assertions of the form: $A(a)$, $R(a,b)$, and $F(a,v)$.

We require that if (funct R) (resp., (funct F)) is in $\mathcal{O}$, then $R' \sqsubseteq R$ (resp., $F' \sqsubseteq F$) is *not* in $\mathcal{O}$ for any R' (resp., F'). This syntactic condition, as well as the fact that we do not allow concepts of the form $\exists F$ and aggregate concepts to appear on the right-hand side of inclusions ensure good computational properties of $DL\text{-}Lite_{\mathcal{A}}^{\text{agg}}$. The former is inherited from $DL\text{-}Lite_{\mathcal{A}}$, while the latter can be shown using techniques of [32].

Consider the ontology capturing the reliability of sensors as in our running example:

$$precisionScore \sqsubseteq testScore, \quad \geq_{0.9} (\text{min } testScore) \sqsubseteq Reliable, \quad (1)$$

where *Reliable* is a concept, *precisionScore* and *testScore* are attributes, and finally $\geq_{0.9} (\text{min } testScore)$ is an aggregate concept that captures individuals with one or more *testScore* values whose minimum is at least 0.9.

$DL\text{-}Lite_{\mathcal{A}}^{\text{agg}}$ Semantics. We define the semantics of $DL\text{-}Lite_{\mathcal{A}}^{\text{agg}}$ in terms of first-order interpretations over the union of the countable domains Δ and D. We assume the unique name assumption and that constants are interpreted as themselves, i.e., $a^{\mathcal{I}} = a$ for each constant a; moreover, interpretations of regular concepts, roles, and attributes are defined as usual (see [9] for details) and for aggregate concepts as follows:

$$(\circ_r(\text{agg } F))^{\mathcal{I}} = \{a \in \Delta \mid \text{agg}\{\!|v \in D \mid (a,v) \in F^{\mathcal{I}}|\!\} \circ r\}.$$

Here $\{\!\!\{\cdot\}\!\!\}$ denotes a multi-set. Similarly to [32], we say that an interpretation $\mathcal{I}$ is a *model* of $\mathcal{O} \cup \mathcal{D}$ if two conditions hold: (i) $\mathcal{I} \models \mathcal{O} \cup \mathcal{D}$, i.e., $\mathcal{I}$ is a first-order model of $\mathcal{O} \cup \mathcal{D}$ and (ii) $F^{\mathcal{I}} = \{(a, v) \mid F(a, v)$ is in the deductive closure of $\mathcal{D}$ with $\mathcal{O}\}$ for each attribute F. Here, by deductive closure of $\mathcal{D}$ with $\mathcal{O}$ we assume a dataset that can be obtained from $\mathcal{D}$ using the chasing procedure with $\mathcal{O}$, as described in [9]. One can show that for $DL\text{-}Lite_{\mathcal{A}}^{\text{agg}}$ satisfiability of $\mathcal{O} \cup \mathcal{D}$ can be checked in time polynomial in $|\mathcal{O} \cup \mathcal{D}|$.

As an example consider a dataset consisting of assertions: *precisionScore* $(s_1, 0.9)$, *testScore*$(s_2, 0.95)$, and *testScore*$(s_3, 0.5)$. Then, for every model $\mathcal{I}$ of these assertions and the axioms in Eq. (1), it holds that $(\geq_{0.9}$ (min *precisionScore*$))^{\mathcal{I}} = \{s_1\}$, $(\geq_{0.9}$ (min *testScore*$))^{\mathcal{I}} = \{s_1, s_2\}$, and thus $\{s_1, s_2\} \subseteq Reliable^{\mathcal{I}}$.

Query Answering. Let $\mathcal{Q}$ be the class of conjunctive queries over concepts, roles, and attributes, i.e., each query $q \in \mathcal{Q}$ is an expression of the form: $q(\vec{x})$:- conj$(\vec{x})$, where q is of arity k, conj is a conjunction of atoms $A(u)$, $E(v)$, $R(w, z)$, or $F(w, z)$, and u, v, w, z are from $\vec{x}$. Following the standard approach for ontologies, we adapt certain answers semantics for query answering:

$$\mathsf{cert}(q, \mathcal{O}, \mathcal{D}) = \{\vec{t} \in (\Delta \cup D)^k \mid \mathcal{I} \models \mathsf{conj}(\vec{t}) \text{ for each model } \mathcal{I} \text{ of } \mathcal{O} \cup \mathcal{D}\}.$$

Continuing with our example, consider the query: $q(x)$:- $Reliable(x)$ that asks for reliable sensors. The set of certain answers $\mathsf{cert}(q, \mathcal{O}, \mathcal{D})$ for this q over the example ontology and dataset is $\{s_1, s_2\}$.

We note that by relying on Theorem 1 of [32] and the fact that each aggregate concept behaves like a $DL\text{-}Lite$ closed predicate of [32], one can show that conjunctive query answering in $DL\text{-}Lite_{\mathcal{A}}^{\text{agg}}$ is tractable, assuming that computation of aggregate functions can be done in time polynomial in the size of the data (see more details in [26]). We also note that our aggregate concepts can be encoded as aggregate queries over attributes as soon as the latter are interpreted under the closed-world semantics. We argue, however, that in a number of applications, such as monitoring and diagnostics at Siemens [21], explicit aggregate concepts of $DL\text{-}Lite_{\mathcal{A}}^{\text{agg}}$ give us significant modelling and query formulation advantages (see more details in [26]).

2.2 Query Language

STARQL is a query language over ontologies that allows to query both streaming and static data and supports not only standard aggregates such as count, avg, etc. but also more advanced aggregation functions from our backend system such as Pearson correlation. In this section we illustrate on our running example the main language constructs and semantics of STARQL (see [26, 35] for more details on syntax and semantics of STARQL).

Each STARQL query takes as input a static $DL\text{-}Lite_{\mathcal{A}}^{\text{agg}}$ ontology and dataset as well as a set of live and historic streams. The output of the query is a stream of timestamped data assertions about objects that occur in the static input data and satisfy two kinds of filters: (*i*) a conjunctive query over the input static ontology and data and (*ii*) a diagnostic query over the input streaming data—which can be live and archived (i.e., static)— that may involve typical mathematical, statistical, and event pattern features needed in real-time diagnostic scenarios. The syntax of STARQL is inspired by the W3C standardised SPARQL query language; it also allows for nesting of queries. Moreover, STARQL has a formal semantics that combines open and closed-world reasoning and extends snapshot semantics for window operators [3] with sequencing semantics that can handle integrity constraints such as functionality assertions.

In Fig. 1 we present a STARQL query that captures the diagnostic task from our running example and uses concepts, roles, and attributes from our Siemens ontology [19, 21–25, 28] and Eq. (1). The query has three parts: declaration of the output stream (Lines 5 and 6), sub-query over the static data (Lines 8 and 9) that in the running example corresponds to '*return all temperature sensors that are reliable, i.e., with the average score of validation tests at least 90 %*' and sub-query over the streaming data (Lines 11–17) that in the running example corresponds to '*whose measurements within the last 10* min *Pearson correlate by at least 0.75 to measurements reported by a reference sensor last year*'. Moreover, in Line 1 there is declarations of the namespace that is used in the sub-queries, i.e., the URI of the Siemens ontology, and in Line 3 there is a declaration of the pulse of the streaming sub-query.

```
1  PREFIX ex : <http://www.siemens.com/onto/gasturbine/>
2
3  CREATE PULSE examplePulse WITH START = NOW, FREQUENCY = 1min
4
5  CREATE STREAM StreamOfSensorsInCriticalMode AS
6  CONSTRUCT GRAPH NOW { ?sensor a :InCriticalMode }
7
8  FROM STATIC ONTOLOGY ex:sensorOntology , DATA ex:sensorStaticData
9  WHERE { ?sensor a ex:Reliable }
10
11 FROM STREAM    sensorMeasurements              [NOW - 1min, NOW]-> 1sec
12               referenceSensorMeasurements 1year <-[NOW - 1min, NOW]-> 1sec,
13 USING PULSE   examplePulse
14 SEQUENCE BY   StandardSequencing AS MergedSequenceOfMeasurementes
15 HAVING EXISTS i IN MergedSequenceOfMeasurementes
16      (GRAPH i { ?sensor ex:hasValue ?y. ex:refSensor ex:hasValue ?z })
17      HAVING PearsonCorrelation(?y, ?z) > 0.75
```

Fig. 1. Running example query expressed in STARQL

Regarding the semantics of STARQL, it combines open and closed-world reasoning and extends snapshot semantics for window operators [3] with sequencing semantics that can handle integrity constraints such as functionality assertions. In particular, the window operator in combination with the sequencing operator provides a sequence of datasets on which temporal (state-based) reasoning can be applied. Every temporal dataset frequently produced by the window operator is converted to a sequence of (pure) datasets. The sequence strategy determines how the timestamped assertions are sequenced into datasets. In the case of the presented example in Fig. 1, the chosen sequencing method is *standard sequencing* assertions with the same timestamp are grouped into the same dataset. So, at every time point, one has a sequence of datasets on which temporal (state-based) reasoning can be applied. This is realised in STARQL by a sorted first-order logic template in which state stamped graph patterns are embedded. For evaluation of the time sequence, the graph patterns of the static WHERE clause are mixed into each state to join static and streamed data. Note that STARQL uses semantics with a real temporal dimension, where time is treated in a non-reified manner as an additional ontological dimension and not as ordinary attribute as, e.g., in SPARQLStream [8].

2.3   Mapping Language and Query Transformation

In this section we present how ontological STARQL queries, Q_{starql}, are transformed into semantically equivalent continuous queries, $Q_{\text{sql}\oplus}$, in the language $SQL^{\oplus}$. The latter language is an expressive extension of SQL with the appropriate operators for registering continuous queries against streams and updatable relations. The language's operators for handling temporal and streaming information are presented in Sect. 3.

As schematically illustrated in Eq. (2) below, during the transformation process the static conjunctive Q_{StatCQ} and streaming Q_{Stream} parts of Q_{starql}, are first independently *rewritten* using the 'rewrite' procedure that relies on the input ontology $\mathcal{O}$ into the union of static conjunctive queries Q'_{StatUCQ} and a new streaming query Q'_{Stream}, and then *unfolded* using the 'unfold' procedure that relies on the input mappings $\mathcal{M}$ into an aggregate SQL query Q''_{AggSQL} and a streaming $SQL^{\oplus}$ query Q''_{Stream} that together give an $SQL^{\oplus}$ query $Q_{\text{sql}\oplus}$, i.e., $Q_{\text{sql}\oplus} = \text{unfold}(\text{rewrite}(Q_{\text{starql}}))$:

$$Q_{\text{starql}} \approx Q_{\text{StatCQ}} \wedge Q_{\text{Stream}} \xrightarrow[\mathcal{O}]{\text{rewrite}} Q'_{\text{StatUCQ}} \wedge Q'_{\text{Stream}}$$

$$\xrightarrow[\mathcal{M}]{\text{unfold}} Q''_{\text{AggSQL}} \wedge Q''_{\text{Stream}} \approx Q_{\text{sql}\oplus}. \qquad (2)$$

In this process we use the rewriting procedure of [9], while the unfolding relies on mappings of three kinds: (*i*) *classical*: from concepts, roles, and attributes to SQL queries over relational schemas of static, streaming, or historical data, (*ii*) *aggregate*: from aggregate concepts to aggregate SQL queries over static data, and (*iii*) *streaming*: from the constructs of the streaming queries of STARQL into $SQL^{\oplus}$ queries over streaming and historical data. Our mapping language extends the one presented in [9] for the classical OBDA setting that allows only for the classical mappings.

We now illustrate our mappings as well as the whole query transformation procedure.

Transformation of Static Queries. We first show the transformation of the example static query that asks for reliable sensors. The rewriting of this query with the example ontology axioms from Eq. (1) is the following query:

$$\text{rewrite}(Reliable(x)) = Reliable(x) \vee (\geq_{0.9} (\text{min } testScore))(x).$$

In order to unfold 'rewrite($Reliable(x)$)' we need both classical and aggregate mappings. Consider four classical mappings: one for the concept '$Reliable$' and three for the attributes '$testScore$' and '$precisionScore$', where sql_i are some SQL queries:

$$Reliable(x) \leftarrow \text{sql}_1(x), \qquad\qquad testScore(x,y) \leftarrow \text{sql}_3(x,y),$$
$$precisionScore(x,y) \leftarrow \text{sql}_2(x,y), \qquad testScore(x,y) \leftarrow \text{sql}_4(x,y).$$

We define an aggregate mapping for a concept $E = \circ_r(\text{agg } F)$ as $E(x) \leftarrow \text{sql}_E(x)$, where $\text{sql}_E(x)$ is an SQL query defined as

$$\text{sql}_E(x) = \text{SELECT } x \text{ FROM } SQL_F(x,y) \text{ GROUP BY } x \text{ HAVING } \text{agg}(y) \circ r \tag{3}$$

where $SQL_F(x,y) = \text{unfold}(\text{rewrite}(F(x,y)))$, i.e., the SQL query obtained as the rewriting and unfolding of the attribute F. Thus, a mapping for our example aggregate concept $E = (\geq_{0.9} (\text{min } testScore))$ is

$$\text{sql}_E(x) = \text{SELECT } x \text{ FROM } SQL_{testScore}(x,y) \text{ GROUP BY } x \text{ HAVING } \text{min}(y) \geq 0.9$$

where $SQL_{testScore}(x,y) = \text{sql}_2(x,y) \text{ UNION } \text{sql}_3(x,y) \text{ UNION } \text{sql}_4(x,y)$.
Finally, we obtain

$$\text{unfold}(\text{rewrite}(Reliable(x))) = \text{sql}_1(x) \text{ UNION } \text{sql}_E(x).$$

Note that one can encode $DL\text{-}Lite_A^{\text{agg}}$ aggregate concepts as standard $DL\text{-}Lite_A$ concepts using mappings. We argue, however, that such an approach has practical disadvantages compared to ours as it would require to create a mapping for each aggregate concept that can be potentially used, thus overloading the system (see more details in [26]).

Transformation of Streaming Queries. The streaming part of a STARQL query may involve static concepts and roles such as $Rotor$ and $testRotor$ that are mapped into static data, and dynamic ones such as $hasValue$ that are mapped into streaming data. Mappings for the static ontological vocabulary are classical and discussed above. Mappings for the dynamic vocabulary are composed from the mappings for attributes and the mapping schemata for STARQL query clauses and constructs. The mapping schemata rely on user defined functions of $SQL^{\oplus}$ and involve windows and sequencing parameters specified in a given STARQL query which make them dependent on time-based relations and temporal states. Note that the latter kind of mappings is not supported by traditional OBDA systems.

For instance, a mapping schema for the 'GRAPH i' STARQL construct (see Line 16, Fig. 1) can be defined based on the following classical mapping that relates a dynamic attribute $ex{:}has\,Val$ to the table $Msmt$ about measurements that among others has attributes sid and $sval$ for storing sensor IDs and measurement values:

$$ex{:}has\,Val(Msmt.sid, Msmt.sval) \leftarrow \textsf{SELECT } Msmt.sid, Msmt.sval \textsf{ FROM } Msmt.$$

The actual mapping schema for 'GRAPH i' extends this mapping as following:

$$\textsf{GRAPH } i \ \{?sensor \ \ ex{:}has\,Val \ ?y\} \leftarrow \textsf{SELECT } sid \ \ as \ ?sensor, \ \ sval \ \ as \ ?y$$
$$\textsf{FROM } \textsf{Slice}(Msmt, i, r, sl, st),$$

where the left part of the schema contains an indexed graph triple pattern and the right part extends the mapping for $ex{:}has\,Val$ by applying a function $Slice$ that describes the relevant finite slice of the stream $Msmt$ from which the triples in the ith RDF graph in the sequence are produced and uses the parameters such as the window range r, the slide sl, the sequencing strategy st and the index i. (See [34] for further details.)

2.4 Query Optimisation

Since a STARQL query consists of analytical static and streaming parts, the result of its transformation by the rewrite and unfold procedures is an analytical data query that also consists of two parts and accesses information from both live streams and static data sources. A special form of static data are archived-streams that, though static in nature, accommodate temporal information that represents the evolution of a stream in time. Therefore, our analytical operations can be classified as: (i) *live-stream operations* that refer to analytical tasks involving exclusively live streams; (ii) *static-data operations* that refer to analytical tasks involving exclusively static information; (iii) *hybrid operations* that refer to analytical tasks involving live-streams and static data that usually originate from archived stream measurements. For static-data operations we rely on standard database optimisation techniques for aggregate functions. For live-stream and hybrid operations we developed a number of optimisation techniques and execution strategies.

A straightforward evaluation strategy on complex continuous queries containing *static-data operations* is for the query planner to compute the static analytical tasks ahead of the live-stream operations. The result on the static-data analysis will subsequently be used as a filter on the remaining streaming part of the query.

We will now discuss, using an example, the *Materialised Window Signatures* technique for hybrid operations. Consider the relational schema depicted in Fig. 2 which is adopted for storing archived streams and performing hybrid operations on them. The relational table **Measurements** represents the archived part of the stream and stores the temporal identifier (**Time**) of each measurement

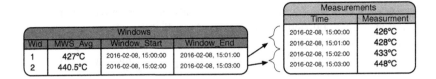

Fig. 2. Schema for storing archived streams and MWSs

and the actual values (attribute `Measurement`). The relational table `Windows` identifies the windows that have appeared up till now based on the existing window-mechanism. It contains a unique identifier for each window (`Wid`) and the attributes that determine its starting and ending points (`Window_Start`, `Window_End`). The necessary indices that will facilitate the complex analytic computations are materialised. The depicted schema is flexible to query changes since it separates the windowing mechanism —which is query dependent— from the actual measurements.

In order to accelerate analytical tasks that include hybrid operations over archived streams, we facilitate precomputation of frequently requested aggregates on each archived window. We name these precomputed summarisations as *Materialised Window Signatures (MWSs)*. These MWSs are calculated when past windows are stored in the backend and are later utilised while performing complex calculations between these windows and a live stream. The summarisation values are determined by the analytics under consideration. E.g., for the computation of the Pearson correlation, we precompute the *average* value and *standard deviation* on each archived window measurements; for the cosine similarity, we precompute the *Euclidean norm* of each archived window; for finding the absolute difference between the average values of the current and the archived windows, we precompute the *average* value, etc.

The selected MWSs are stored in the Windows relation with the use of additional columns. In Fig. 2 we see the MWS summary for the `avg` aggregate function being included in the relation as an attribute termed `MWS_Avg`. The application can easily modify the schema of this relation in order to add or drop MWSs, depending on the analytical workload.

When performing hybrid operations between the current and archived windows, some analytic operations can be directly computed based on their MWS values with no need to access the actual archived measurements. This provides significant benefits as it removes the need to perform a costly join operation between the live stream and the, potentially very large, `Measurements` relation. On the opposite, for calculations such as the Pearson correlation coefficient and the cosine similarity measures, we need to perform calculations that require the archived measurements as well, e.g., for computing cross-correlations or inner-products. Nevertheless, the MWS approach allows us to avoid recomputing some of the information on each archived window such as its *avg* value and *deviation* for the Pearson correlation coefficient, and the Euclidean norm of each archived window for the cosine similarity measure. Moreover, in case when there is a selective additional filter on the query (such as the avg value exceeds a threshold), by creating an index on the MWS attributes, we can often exclude large portions of the archived measurements from consideration, by taking advantage of the underlying index.

3   Implementation

In this section we discuss our system that implements the OBDA extensions proposed in Sect. 2. In Fig. 3 (Left), we present a general architecture of our system. On the application level one can formulate STARQL queries over analytics-aware ontologies and pass them to the query compilation module that performs query rewriting, unfolding, and optimisation. Query compilation components can access relevant information in the ontology for query rewriting, mappings for query unfolding, and source specifications for optimisation of data queries. Compiled data queries are sent to a query execution layer that performs distributed query evaluation over streaming and static data, post-processes query answers, and sends them back to applications. In the following we will discuss two main components of the system, namely, our dedicated STARQL2SQL$^\oplus$ translator that turns STARQL queries to SQL$^\oplus$ queries, and our native data-stream management system EXASTREAM that is in charge of data query optimisation and distributed query evaluation.

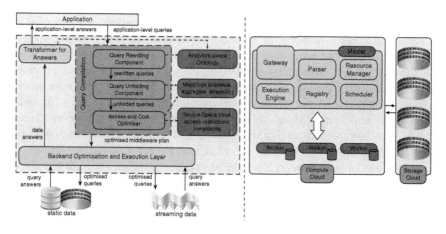

Fig. 3. (Left) General architecture. (Right) Distributed stream engine of EXASTREAM

STARQL to SQL$^\oplus$ Translator. Our translator consists of several modules for transformation of various query components and we now give some highlights on how it works. The translator starts by turning the window operator of the input STARQL query and this results in a *slidingWindowView* on the backend system that consists of columns for defining *windowID* (as in Fig. 2) and *dataGraphID* based on the incoming data tuples. Our underlying data-stream management system ExaStream already provides *user defined functions* (UDFs) that automatically create the desired streaming views, e.g., the *timeSlidingWindow* function as discussed below in the ExaStream part of the section.

The second important transformation step that we implemented is the transformation of the STARQL HAVING clause. In particular, we normalise the HAVING clause into a relational algebra normal form (RANF) and apply the described slicing technique illustrated in Sect. 2.3, where we unfold each state of the temporal sequence into slices of the *slidingWindowView*. For the rewriting and unfolding of each slice, we make use of available tools using the OBDA paradigm in the static case, i.e., the Ontop framework [39]. After unfolding, we join all states together based on their temporal relations given in the HAVING sequence.

ExaStream Data-Stream Management System. Data queries produced by the STARQL2SQL$^\oplus$ translation, are handled by ExaStream which is embedded in Exareme, a system for elastic large-scale dataflow processing in the cloud [29, 42].

ExaStream is built as a streaming extension of the SQLite database engine, taking advantage of existing Database Management technologies and optimisations. It provides the declarative language SQL$^\oplus$ for querying data streams and relations. SQL$^\oplus$ extends SQL with *UDF*s that incorporate the algorithmic logic for transforming SQLite into a *Data Stream Management Systems* (*DSMS*). E.g., the *timeSlidingWindow* operator groups tuples from the same time window and associates them with a unique window id. In contrast to other *DSMS*s, the user does not need to consider low-level details of query execution. Instead, the system's *query planner* is responsible for choosing an optimal plan depending on the query, the available stream/static data sources, and the execution environment.

ExaStream system exploits parallelism in order to accelerate the process of analytical tasks over thousands of stream and static sources. It manages an elastic cloud infrastructure and dynamically distributes queries and data (including both streams and static tables) to multiple worker nodes that process them in parallel. The architecture of ExaStream's distributed stream engine is presented in Fig. 3 (Right). One can see that queries are registered through the Asynchronous Gateway Server. Each registered query passes through the ExaStream parser and then is fed to the Scheduler module. The Scheduler places the stream and relational operators on worker nodes based on the node's load. These operators are executed by a Stream Engine instance running on each node.

4   Evaluation

The aim of our evaluation is to study how the MWS technique and query distribution to multiple workers accelerate the overall execution time of analytic queries that correlate a live stream with multiple archived stream records.

Evaluation Setting. We deployed our system to the Okeanos Cloud Infrastructure[2]. and used up to 16 virtual machines (VMs) each having a 2.66 GHz processor with 4 GB of main memory. We used streaming and static data that contains measurements produced by 100, 000 thermocouple sensors installed in 950 Siemens power generating turbines. For our experiments, we used three *test queries* calculating the similarity between the current live stream window and 100,000 archived ones. In each of the test queries we fixed the window size to 1 h which corresponds to 60 tuples of measurements per window. The first query is based on the one from our running example (see Fig. 1) which we modified so that it can correlate a live stream with a varying number of archived streams. Recall that this query evaluates window measurements similarity based on the Pearson correlation. The other two queries are variations of the first one where, instead of the Pearson correlation, they compute similarity based on either the *average* or the *minimum* values within a window. We defined such similarities between vectors (of measurements) $\vec{w}$ and $\vec{v}$ as follows: $|avg(\vec{w}) - avg(\vec{v})| < 10°C$ and $|min(\vec{w}) - min(\vec{v})| < 10°C$. The archived streams windows are stored in the Measurements relation, against which the current stream is compared.

MWS Optimisation. This set of experiments is devised to show how the MWS optimisation affects the query's response time. We executed each of the three test queries on a single VM-worker with and without the MWS optimisation. In Fig. 4 (Left) we present the results of our experiments. The reported time is the average of 15 consecutive live-stream execution cycles. The horizontal axis displays the three test queries with and without the MWS optimisation, while the vertical axis measures the time it takes to process 1 live-stream window against all the archived ones. This time is divided to the time it takes to join the live stream and the Measurements relation and the time it takes to perform the actual computations. Observe that the MWS optimisation reduces the time for the Pearson query by 8.18 %. This is attributed to the fact that some computations (such as the avg and standard deviation values) are already available in the Windows relation and are, thus, omitted. Nevertheless, the join operation between the live stream and the very large Measurements relation that takes 69.58 % of the overall query execution time can not be avoided. For the other two queries, we not only reduce the CPU overhead of the query, but the optimiser further prunes this join from the query plan as it is no longer necessary. Thus, for these queries, the benefits of the MWS technique are substantial.

Intra-query Parallelism. Since the MWS optimisation substantially accelerates query execution for the two test queries that rely on average and minimum similarities, query distribution would not offer extra benefit, and thus these queries were not used in the second experiment. For complex analytics such as the Pearson correlation that necessitates access to the archived windows, the EXASTREAM backend permits us to accelerate queries by distributing the load

---

[2] https://okeanos.grnet.gr/home/.

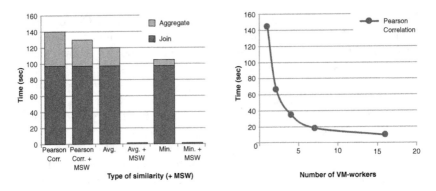

Fig. 4. (Left) Effect of MWS optimisation (Right) Effect of intra-query parallelism

among multiple worker nodes. In the second experiment we use the same setting as before for the Pearson computation without the MWS technique, but we vary this time the number of available workers from 1 to 16. In Fig. 4 (Right), one can observe a significant decrease in the overall query execution time as the number of VM-workers increases. EXASTREAM distributes the `Measurements` relation between different worker nodes. Each node computes the Pearson coefficient between its subset of archived measurements and the live stream. As the number of archived windows is much greater than the number of available workers, intra-query parallelism results is significant decrease to the time required to perform the join operation.

To conclude this section, we note that MWSs gave us significant improvements of query execution time for all test queries and parallelism would be essential in the cases where MWSs do not help in avoiding the high cost of query joins since it allows to run the join computation in parallel. Due to space limitations, we do not include an experiment examining the query execution times w.r.t. the number of archived windows. Nevertheless, based on our observations, scaling up the number of archived windows by a factor of n has about the same effect as scaling down the number of workers by $1/n$.

5   Related Work

OBDA System. Our proposed approach extends existing OBDA systems since they either assume that data is in (static) relational DBs, e.g [15,39], or streaming, e.g., [8,17], but not of both kinds. Moreover, we are different from existing solutions for unified processing of streaming and static semantic data e.g. [36], since they assume that data is natively in RDF while we assume that the data is relational and mapped to RDF.

Ontology Language. The semantic similarities of $DL\text{-}Lite_{\mathcal{A}}^{\mathrm{agg}}$ to other works have been covered in Sect. 2. Syntactically, the aggregate concepts of $DL\text{-}Lite_{\mathcal{A}}^{\mathrm{agg}}$ have counterpart concepts, named local range restrictions (denoted by $\forall F.T$)

in $DL\text{-}Lite_A$ [4]. However, for purposes of rewritability, these concepts are not allowed on the left-hand side of inclusion axioms as we have done for $DL\text{-}Lite_A^{agg}$, but only in a very restrictive semantic/syntactic way. The semantics of $DL\text{-}Lite_A^{agg}$ for aggregate concepts is very similar to the epistemic semantics proposed in [11] for evaluating conjunctive queries involving aggregate functions. A different semantics based on minimality has been considered in [30]. Concepts based on aggregates functions were considered in [5] for languages $\mathcal{ALC}$ and $\mathcal{EL}$ with concrete domains, but they did not study the problem of query answering.

Query Language. While already several approaches for RDF stream reasoning engines do exist, e.g., CSPARQL [6], RSP-QL [1] or CQELS [37], only one of them supports an ontology based data access approach, namely SPARQL-stream [8]. In comparison to this approach, which also uses a native inclusion of aggregation functions, STARQL offers more advanced user defined functions from the backend system like Pearson correlation.

Data Stream Management System. One of the leading edges in database management systems is to extend the relational model to support for continuous queries based on declarative languages analogous to SQL. Following this approach, systems such as TelegraphCQ [14], STREAM [2], and Aurora [16] take advantage of existing Database Management technologies, optimisations, and implementations developed over 30 years of research. In the era of big data and cloud computing, a different class of *DSMS* has emerged. Systems such as Storm and Flink offer an API that allows the user to submit dataflows of user defined operators. EXASTREAM unifies these two different approaches by allowing to describe in a declarative way complex dataflows of (possibly user-defined) operators. Moreover, the Materialised Window Signature summarisation, implemented in EXASTREAM, is inspired from data warehousing techniques for maintaining selected aggregates on stored datasets [18,31]. We adjusted these technique for complex analytics that blend streaming with static data.

6   Conclusion, Lessons Learned, and Future Work

We see our work as a first step towards the development of a solid theory and new full-fledged systems in the space of analytics-aware ontology-based access to data that is stored in different formats such as static relational, streaming, etc. To this end we proposed ontology, query, and mapping languages that are capable of supporting analytical tasks common for Siemens turbine diagnostics. Moreover, we developed a number of backend optimisation techniques that allow such tasks to be accomplished in reasonable time as we have demonstrated on large scale Siemens data.

The lessons we have learned so far are the encouraging evaluation results over the Siemens turbine data (presented in Sect. 4). Since our work is a part of an ongoing project that involves Siemens, we plan to continue implementation and then deployment of our solution in Siemens. This will give us an opportunity to do further performance evaluation as well as to conduct user studies.

Finally, there is a number of important further research directions that we plan to explore. On the side of analytics-aware ontologies, we plan to explore bag instead of set semantics for ontologies since bag semantics is natural and important in analytical tasks; we also plan to investigate how to support evolution of such ontologies [12,27] since OBDA systems are dynamic by its nature. On the side of analytics-aware queries, an important further direction is to align them with the terminology of the W3C RDF Data Cube Vocabulary and to provide additional optimisations after the alignment. As for query optimisation techniques, exploring approximation algorithms for fast computation of complex analytics between live and archived streams is particularly important. That is because these algorithms usually provide quality guarantees about the results and in the average case require much less computation. Thus, we intend to examine their effectiveness in combination with the MWS approach.

References

1. Aglio, D.D., Valle, E.D., Calbimonte, J.-P., Corcho, O., Semantics, R.-Q.: A unifying query model to explain heterogeneity of RDF stream processing systems. IJSWIS 10(4), 17–44 (2015)
2. Arasu, A., Babcock, B., Babu, S., Datar, M., Ito, K., Nishizawa, I., Rosenstein, J., Widom, J.: STREAM: the stanford stream data manager. In: SIGMOD (2003)
3. Arasu, A., Babu, S., Widom, J., Continuous, T.C., Language, Q.: Semantic foundations and query execution. VLDBJ 15(2), 121–142 (2006)
4. Artale, A., Ryzhikov, V., Kontchakov, R.: DL-Lite with attributes and datatypes. In: ECAI (2012)
5. Baader, F., Sattler, U.: Description logics with aggregates and concrete domains. IS 28(8), 979–1004 (2003)
6. Barbieri, D.F., Braga, D., Ceri, S., Valle, E.D., Grossniklaus, M.: C-SPARQL: a continuous query language for RDF data streams. Int. J. Seman. Comput. 4(1), 3–25 (2010)
7. Bizer, C., Seaborne, A.: D2RQ-treating non-RDF databases as virtual RDF graphs. In: ISWC (2004)
8. Calbimonte, J.-P., Corcho, Ó., Gray, A.J.G.: Enabling ontology-based access to streaming data sources. In: Patel-Schneider, P.F., Pan, Y., Hitzler, P., Mika, P., Zhang, L., Pan, J.Z., Horrocks, I., Glimm, B. (eds.) ISWC 2010. LNCS, vol. 6496, pp. 96–111. Springer, Heidelberg (2010). doi:10.1007/978-3-642-17746-0_7
9. Calvanese, D., Giacomo, G., Lembo, D., Lenzerini, M., Poggi, A., Rodriguez-Muro, M., Rosati, R.: Ontologies and databases: the *DL-Lite* approach. In: Tessaris, S., Franconi, E., Eiter, T., Gutierrez, C., Handschuh, S., Rousset, M.-C., Schmidt, R.A. (eds.) Reasoning Web 2009. LNCS, vol. 5689, pp. 255–356. Springer, Heidelberg (2009). doi:10.1007/978-3-642-03754-2_7
10. Calvanese, D., De Giacomo, G., Lembo, D., Lenzerini, M., Poggi, A., Rodriguez-Muro, M., Rosati, R., Ruzzi, M., Savo, D.F.: The MASTRO system for ontology-based data access. Seman. Web 2(1), 43–53 (2011)
11. Calvanese, D., Kharlamov, E., Nutt, W., Thorne, C.: Aggregate queries over ontologies. In: ONISW, October 2008
12. Calvanese, D., Kharlamov, E., Nutt, W., Zheleznyakov, D.: Evolution of DL-Lite knowledge bases. In: ISWC (2010)

13. Calvanese, D., Liuzzo, P., Mosca, A., Remesal, J., Rezk, M., Rull, G.: Integration, ontology-based data in EPNet: production and distribution of food during the Roman empire. Eng. Appl. AI **51**, 212–229 (2016)
14. Chandrasekaran, S., Cooper, O., Deshpande, A., Franklin, M.J., Hellerstein, J.M., Hong, W., Krishnamurthy, S., Madden, S.R., Reiss, F., Shah, M.A.: TelegraphCQ: continuous dataflow processing. In: SIGMOD (2003)
15. Civili, C., Console, M., De Giacomo, G., Lembo, D., Lenzerini, M., Lepore, L., Mancini, R., Poggi, A., Rosati, R., Ruzzi, M., Santarelli, V., Savo, D.F.: MASTRO STUDIO: managing ontology-based data access applications. In: PVLDB, vol. 6, no. 12 (2013)
16. Abadi, D., Carney, D. et al.: Aurora: a data stream management system. In: SIGMOD (2003)
17. Fischer, L., Scharrenbach, T., Bernstein, A.: Scalable linked data stream processing via network-aware workload scheduling. In: SSWKBS@ISWC (2013)
18. Gray, J., Chaudhuri, S., Bosworth, A., Layman, A., Reichart, D., Venkatrao, M., Pellow, F., Pirahesh, H., Cube, D.: A relational aggregation operator generalizing group-by, cross-tab, and sub-totals. Data Min. Knowl. Discov. **1**(1), 29–53 (1997)
19. Kharlamov, E., Brandt, S., Jimenez-Ruiz, E., Kotidis, Y., Lamparter, S., Mailis, T., Neuenstadt, C., Özçep, Ö., Pinkel, C., Svingos, C., Zheleznyakov, D., Horrocks, I., Ioannidis, Y., Möller, R.: Ontology-based integration of streaming and static relational data with optique. In: SIGMOD (2016)
20. Kharlamov, E., Hovland, D., Jiménez-Ruiz, E., Pinkel, D.L.C., Rezk, M., Skjæveland, M.G., Thorstensen, E., Xiao, G., Zheleznyakov, D., Bjørge, E., Horrocks, I.: Enabling ontology based access at an oil and gas company statoil. In: ISWC (2015)
21. Kharlamov, E., et al.: How semantic technologies can enhance data access at siemens energy. In: Mika, P., et al. (eds.) ISWC 2014. LNCS, vol. 8796, pp. 601–619. Springer, Heidelberg (2014). doi:10.1007/978-3-319-11964-9_38
22. Kharlamov, E., Brandt, S., Giese, M., Jiménez-Ruiz, E., Kotidis, Y., Lamparter, S., Mailis, T., Neuenstadt, C., Özçep, Ö.L., Pinkel, C., Soylu, A., Svingos, C., Zheleznyakov, D., Horrocks, I., Ioannidis, Y.E., Möller, R., Waaler, A.: Enabling semantic access to static, streaming distributed data with optique: demo. In: DEBS (2016)
23. Kharlamov, E., Brandt, S., Giese, M., Jiménez-Ruiz, E., Lamparter, S., Neuenstadt, C., Özçep, Ö.L., Pinkel, C., Soylu, A., Zheleznyakov, D., Roshchin, M., Watson, S., Horrocks, I.: Semantic access to siemens streaming data: the optique way. In: ISWC (P&D) (2015)
24. Kharlamov, E., Jiménez-Ruiz, E., Pinkel, C., Rezk, M., Skjæveland, M.G., Soylu, A., Xiao, G., Zheleznyakov, D., Giese, M., Horrocks, I., Waaler, A.: Optique: ontology-based data access platform. In: ISWC (P&D) (2015)
25. Kharlamov, E., Jiménez-Ruiz, E., Zheleznyakov, D., Bilidas, D., Giese, M., Haase, P., Horrocks, I., Kllapi, H., Koubarakis, M., Özçep, Ö.L., Rodriguez-Muro, M., Rosati, R., Schmidt, M., Schlatte, R., Soylu, A., Waaler, A.: Optique: towards OBDA systems for industry. In: ESWC (Selected Papers) (2013)
26. Kharlamov, E., Kotidis, Y., Mailis, T., Neuenstadt, C., Nicolaou, C., Özçep, Ö., Svingos, C., Zheleznyakov, D., Brandt, S., Horrocks, I., Ioannidis, Y., Lamparter, S., Möller. R.: Towards analytics aware ontology based access to static and streaming data (extended version). In: CoRR (2016)
27. Kharlamov, E., Zheleznyakov, D., Calvanese, D.: Capturing model-based ontology evolution at the instance level: the case of DL-Lite. J. Comput. Syst. Sci. **79**(6), 835–872 (2013)

28. Kharlamov, E. et al.: Optique 1.0: semantic access to big data: the case of norwegian petroleum directorate factpages. In: ISWC (P&D) (2013)
29. Kllapi, H., Sakkos, P., Delis, A., Gunopulos, D., Ioannidis, Y.: Elastic processing of analytical query workloads on IaaS clouds. arXiv (2015)
30. Kostylev, E.V., Reutter, J.L.: Complexity of answering counting aggregate queries over DL-Lite. J. Web Seman. **33**, 94–111 (2015)
31. Kotidis, Y., Roussopoulos, N., DynaMat: a dynamic view management system for data warehouses. In: SIGMOD (1999)
32. Lutz, C., Seylan, I., Wolter, F.: Mixing open, closed world assumption in ontology-based data access
33. Munir, K., Odeh, M., McClatchey, R.: Ontology-driven relational query formulation using the semantic and assertional capabilities of OWL-DL. KBS **35**, 144–159 (2012)
34. Neuenstadt, C., Möller, R., Özçep, Ö.L.: OBDA for temporal querying and streams with STARQL. In: HiDeSt (2015)
35. Özçep, Ö.L., Möller, R., Neuenstadt, C.: A stream-temporal query language for ontology based data access. In: Lutz, C., Thielscher, M. (eds.) KI 2014. LNCS (LNAI), vol. 8736, pp. 183–194. Springer, Heidelberg (2014). doi:10.1007/978-3-319-11206-0_18
36. Le-Phuoc, D., Dao-Tran, M., Xavier Parreira, J., Hauswirth, M.: A native and adaptive approach for unified processing of linked streams and linked data. In: Aroyo, L., Welty, C., Alani, H., Taylor, J., Bernstein, A., Kagal, L., Noy, N., Blomqvist, E. (eds.) ISWC 2011. LNCS, vol. 7031, pp. 370–388. Springer, Heidelberg (2011). doi:10.1007/978-3-642-25073-6_24
37. Le-Phuoc, D., Dao-Tran, M., Pham, M.-D., Boncz, P., Eiter, T., Fink, M.: Engines, linked stream data processing: facts and figures. In: ISWC (2012)
38. Priyatna, F., Corcho, O., Sequeda, J.: Formalisation and experiences of R2RML-based SPARQL to SQL query translation using Morph. In: WWW (2014)
39. Rodriguez-Muro, M., Kontchakov, R., Zakharyaschev, M.: Access, ontology-based data: ontop of databases. In: ISWC (2013)
40. Rodrıguez-Muro, M., Calvanese, D.: High performance query answering over DL-lite ontologies. In: KR (2012)
41. Sequeda, J., Miranker, D.P.: Ultrawrap: SPARQL execution on relational data. JWS **22**, 19–39 (2013)
42. Tsangaris, M.M., Kakaletris, G., Kllapi, H., Papanikos, G., Pentaris, F., Polydoras, P., Sitaridi, E., Stoumpos, V., Ioannidis, Y.E.: Dataflow processing and optimization on grid and cloud infrastructures. IEEE Data Eng. Bull. **32**(1), 67–74 (2009)

QuerioDALI: Question Answering Over Dynamic and Linked Knowledge Graphs

Vanessa Lopez[✉], Pierpaolo Tommasi, Spyros Kotoulas,
and Jiewen Wu

Smarter Cities Technology Centre, IBM Research, Dublin, Ireland
{vanlopez,ptommasi,Spyros.Kotoulas,
jiewen.wu}@ie.ibm.com

Abstract. We present a domain-agnostic system for Question Answering over multiple semi-structured and possibly linked datasets without the need of a training corpus. The system is motivated by an industry use-case where Enterprise Data needs to be combined with a large body of Open Data to fulfill information needs not satisfied by prescribed application data models. Our proposed Question Answering pipeline combines existing components with novel methods to perform, in turn, linguistic analysis of a query, named entity extraction, entity/graph search, fusion and ranking of possible answers. We evaluate QuerioDALI with two open-domain benchmarks and a biomedical one over Linked Open Data sources, and show that our system produces comparable results to systems that require training data and are domain-dependent. In addition, we analyze the current challenges and shortcomings.

1 Introduction

With the advent of Open Data in all aspects of enterprise and government operations, valuable information is accumulating in disparate stores. Recognizing the infeasibility of a fully integrated model, Linked Data has presented itself as a paradigm for inter-operability and (partial) integration. In many scenarios, such as Smart Cities [20], enterprise data can be enriched with relevant Linked Open Data (LOD) to help users find relevant entities, as well as answering more specific information needs. In this paper, we present a Question Answering (QA) system that exploits LOD and Knowledge Graphs (KGs) extracted from tabular enterprise data to answer user queries expressed in Natural Language (NL).

Traditionally, QA has focused on unstructured information (to extract answers from text fragments in documents). Notably, the IBM Watson system [1] won the Jeopardy challenge against human experts in 2011. In Watson, structured data is used just to find additional evidence, in the form of typing of answers, spatial-temporal constraints or semantic expansion for commonly appearing entity types (e.g., countries, US presidents).

With the growth of Linked Data, various QA systems over structured semantic data have been proposed to allow end users not familiar with formal queries to express arbitrarily complex information using NL. Complementary to QA over documents, QA over large graphs is useful to find answers to factoid questions in a scenario with

P. Groth et al. (Eds.): ISWC 2016, Part II, LNCS 9982, pp. 363–382, 2016.
DOI: 10.1007/978-3-319-46547-0_32

evolving (dynamic) semi-structured interlinked data sources, no fixed schema and no training corpora. Yet, very few systems address the fact that, often, questions can only be answered by combining information from several sources.

Driven by the QALD benchmarks [2, 3], the state of the art on QA over Linked Data has focused on answering open domain queries over the DBpedia dataset [11] and, to some extent over biomedical LOD. There are two other QA benchmarks over Freebase (which content has been recently migrated to Wikidata): WebQuestions and Free917 [17]. The Free917 benchmark includes the correct KB query and answers, while WebQuestions only provides the answers collected using crowdsourcing. These benchmarks exhibit quite different challenges: QALD involves more complex questions requiring multiple relations and operators such as comparatives and superlatives; Freebase is an order of magnitude bigger than DBpedia, and with several hundreds of possible mappings for a given keyword in a user query, a major challenge is to find the matching entities and relations in the KG. Differently from ontology-based *QA* over DBpedia, QA approaches over Freebase are based on IR techniques and weak-supervised learning. Often, QA systems are evaluated for only one of the benchmarks. None of the ontology-based QA approaches, many of them unsupervised, have been evaluated over both DBpedia and Freebase.

Structured QA is also gaining popularity among search engines, e.g., Google KG. With the rise of the Web of Data, more questions can be answered, thus reducing the sparseness typical of this kind of systems (for a question to be answered the knowledge need to be encoded in a KG). However, differently from industry systems that rely on a proprietary curated KG, this incremental growth comes at a cost: noise, heterogeneity, incomplete schema, and missing linkage across graphs. As a result, a large space of candidates and mapping combinations need to be checked to find translations to a user query in quasi real-time, particularly in cases where different KGs may contain only part of the answers.

In sum, the main contributions of our work are the following:

- A practical approach outlining the most significant design choices to support practitioners in the open-domain QA. Our implemented system, QuerioDALI, combines a novel linguistic analysis built on the top of the IBM Watson NLP pipeline to obtain a set of PAS (predicate argument structures) that link the user query terms together, and a Graph Pattern (GP) search component to translate these PAS into a logical representation that coveys the meaning of the query, obtaining and ranking answers based on evidence extracted from the KGs using semantic techniques, without the need of a training corpus.
- A novel component to merge facts (GPs) across graphs, obtained from both enterprise and Open Data. LOD is used both to extract answers or partial answers (enriching the enterprise data) and to understand the user query (through semantic expansion). It uses interlinked data to answer queries that could not be answered by a single KG.
- We perform a benchmark-based evaluation in two scenarios to show that the system can produce comparable results to state of the art systems that require training: (a) Open domain over large and heterogeneous KGs: using both QALD-5 and Free917; (b) Based on a real-world use case in the Smarter Care domain, where

relational enterprise data and biomedical LOD need to be combined to address care worker's information needs regarding a specific client: the evaluation is based on the QALD-4 biomedical task.

In this paper, we analyze the challenges of balancing precision and recall in every step of the process: from understanding the user query to obtaining ranked interpretations leading to answers. Moreover, differences in terms of how the data is structured across graphs (e.g., Freebase vs. DBpedia) raises more challenges than scalability, regarding the mismatch between the NL query and KGs. The next section elaborates on a motivating use case. Section 3 presents the main architectural components. Section 4 presents the evaluations and open challenges, followed by related work and conclusions in Sects. 5 and 6.

2   Use Case: Smarter Care

For most enterprise applications, user interfaces are based on the premise that business analysts and designers are able to capture the information that will be required by a user and display it when requested or as part of a business process. For example, a nurse doing home calls will most likely be interested in the address of the patients, their conditions and the medication that they are taking, among others. As the body of open knowledge available expands, there is a significant opportunity to satisfy the information needs of a user when it is difficult to predict these needs. Search systems can be used to look up information, but fail when they need to combine information from an enterprise system with Open Data. For example, answering the questions *What are the side-effects for all the medications of this patient?* or *Is any of the patient's medication related with insomnia?* requires retrieving the medications of a patient from the enterprise system, looking each one of them up individually on an online source and examining each list of side-effects.

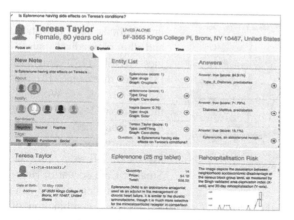

Fig. 1. BlueLENs application for care workers

Figure 1 shows BlueLENS, an example system to integrate multi-source information in the domain of care. The collected information is presented as a set of self-describing elements (in each rectangle) to allow care workers to obtain a 360-degree view of a client. Although beyond the scope of this paper, this system allows navigation and ranking of data from multiple sources.

But what happens when the user has an information need that is both specific and not covered by the existing configurations? How can we enrich the data that is already handled by this system with additional information about, among others, diseases, drugs and their interactions, openly available on the Web? The work in this paper has been driven by this real-world problem[1]. In the following Sections, some of the examples are related to this base scenario.

3 Approach for QA Over Knowledge Graphs

We propose a QA pipeline to build up the formal graph queries needed to satisfy complex information needs, expressed in NL, from both open and enterprise KGs.

In our Smarter Care scenario, the system DALI [20] is used to lift the enterprise relational data from IBM Cúram [10] into a KG with explicit semantics and linked to well-known W3C and LOD models. We distinguish between two types of data-sources depending on how the data is interfaced by the system: (1) The *QA KGs* from which answers are extracted, built from open and enterprise datasets and exposed through federated SPARQL endpoints; (2) The *Annotators*, which are the distributed LOD sources and dictionaries used to perform query expansion. For the **open-domain scenario**, the QA KGs consists of DBpedia 2015-14 and Freebase. For the **Smarter Care scenario** it contains enterprise data from Cúram about patient conditions, prescriptions and care plans, and the biomedical ontologies SIDER [13], drugbank [8] and diseasome [7]. As Annotators, we use WordNet, the lemon lexicon (http://lemon-model.net) and schema.org for both scenarios; and DBpedia for the Smarter Care scenario. However, our system can be con-

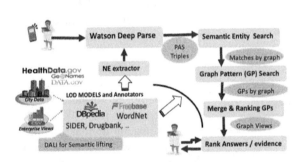

Fig. 2. QA Architecture pipeline

figured to use any ontologies or dictionaries. QuerioDALI uses a combination of off-the-shelf and novel components (see Fig. 2). In this paper we detail the following steps and components:

Step 1: **Deep NLP and Named Entity (NE) extraction.** Based on the Watson parsing pipeline and off-the-shelf NE annotators to generate domain-independent PAS triples for the query.

Step 2: **Anchoring.** Required to bridge the gap between the user vocabulary and the terminology in the KGs. Semantic expansion is used to find the candidate URIs.

---

[1] BlueLENS: https://ibm.biz/Bd465t and QuerioDALi video demos at: https://ibm.biz/BdHEwF.

Step 3: **Graph Pattern Search.** From this large pool of candidates, a novel component translates the PAS triples into GPs, which convey into a formal query (in terms of entities, relations and logical operators) that can be executed against the KGs.

Step 4: **Merging of GPs.** Merging facts and partial translations, from the same or different sources, requires finding join terms across the GPs to be joined, as well as entity co-reference if the query involves merging across graphs.

Step 5: **Ranking.** Due to ambiguity, a query can have alternative representations that can be combined in different ways, based on the query type (factoid, boolean, etc.), answers are ranked using a score that reflects the confidence of the system on the evidence.

The coverage and accuracy of each component influences the design choices and performance of the next component in the pipeline.

3.1   Step 1: Deep NLP Parsing and NER

For the linguistic processing of a question, we adopted Watson's general purpose deep parsing implemented as an Apache UIMA application. It receives as input a question and identifies syntactic, morphological and semantic elements of the question, building a dependency parse tree. The predicate arguments of a tree node are other nodes that may come from remote positions on the tree. A dependency parse tree is shown in Fig. 3.

The challenge here is generating domain-independent PAS triples in the form of <subject, predicate, object> . To do that, we use the *Watson Subtree Pattern Matching framework,* which allows expressing and executing rules over the dependency tree. We wrote 25 general-purpose rules to extract the triple patterns for the kind of NL sentences found in the QALD training sets for English [3]. However, its coverage can be extended. On Fig. 3 we see the PAS triples obtained for an example query and one example rule fired to detect the PAS with the superlative "lowest readmission" (pattern = *nounAdjSup*).

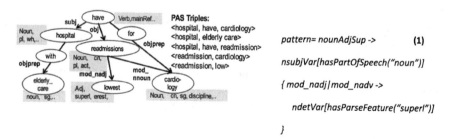

$$pattern = nounAdjSup \rightarrow \qquad (1)$$

$$nsubjVar[hasPartOfSpeech("noun")]$$

$$\{ mod\_nadj \,|\, mod\_nadv \rightarrow$$

$$ndetVar[hasParseFeature("superl")]$$

$$\}$$

Fig. 3 Dependency tree for Which hospitals with elderly care have the lowest readmissions for cardiology?

In general terms, all the rules that can be fired generate one of the following structures: (1) Basic PAS: with explicit subject-predicate-object, the subject can be a wh-term (e.g., *Who climbed Mount Everest?*); (2) Noun modifiers PAS for which the predicate is not explicitly given (e.g., *Russian mountain*); (3) PAS Tuples: those for which the predicate is associated to 3 instead of 2 modifier terms (e.g., *Who climbed Mount Everest the first?*); (4) PAS with a superlative/comparative determiner (e.g., *highest mountain*); and (5) PAS with spatial or temporal modifiers for the whole sentence (e.g., *in May 2008*).

NE extraction is crucial to find the syntactic roles in a sentence and multi-word terms, in the example "elderly care". Thus, we extend the scope of Watson's NE recognizer (based on its own domain-independent ontology) by also using LOD NE extractors to detect multi-words before parsing the sentence; in particular DBpedia Spotlight [16] and Alchemy API (www.alchemyapi.com) For each term, or NE, in the PAS we capture the following features: its syntactic role (subj, obj, pred, nadj, etc.), lemma, covered text, if it is the focus of the sentence, the Part-Of-Speech and associated features (plural/singular, verb tense, etc.). If the entity is recognized, the semantic type is also provided (city, sports team, etc.).

3.2 Step 2: Anchoring and Semantic Expansion

The PAS Triples obtained in Step 1 represent how the terms in a sentence are linked together, but it does not necessarily correspond to the way information is structured in the KGs, e.g., the linguistic predicate *designed* in *Who designed the giant dipper?* may translate to an ontological type *designer*, instead of (or in addition to) an ontological property. Thus, before we can translate each PAS triple into one or more GPs (from the same or different KGs), Step 2 returns all candidate entities for each term in the PAS, together with their type and a matching score, without forcing any heuristic on the kind of expected type.

The result is a Mapping Table per linguistic term, containing the candidate URIs for each underlying graph to be queried. The matching is based on exact and approximate index searches over the entity labels (*rdfs:label*) for instances, properties and concepts, as well as literal values. Wh-words in the PAS Triples are replaced with semantic equivalent term(s), if appropriate: person/organization for who, date for when, place/location for where.

Semantic expansion of query terms is performed using the lexicons and KGs selected for the pool of annotators, such as synonyms and hypernyms from WordNet, and other lexically related words obtained following properties such as owl:sameAs, SKOS:broader, dcterms:subject and redirects if using DBpedia. For example, the term penicillin G (in the question *What are the side effects of penicillin G?*) is annotated with the DBpedia term *Benzylpenicillin*, which has an *owl:sameAs* link to the entity in SIDER labeled *Penicillin G Potassium Injection* that links to the list of side effects that answers the question.

Lessons Learned. The entity matching problem is hard. For example, there are more than 200 entities with the exact label *Founders* in Freebase. Furthermore, noun-modifier PAS, composed of two terms with an implicit relation, may often refer to a compound term (multi-word) missed by the previous NE recognition. Thus, besides looking for URIs matching each independent term it also identifies if there are candidate matches for the compound (e.g., the PAS <order, ?, dragon> in *Who founded the order of the dragon*). While expanding the search space is needed to find the correct translation, the more accurate the matches are, the less computational effort is required to find the GPs in the next step. Thus, techniques are required to prune the space of candidate solutions:

- A syntactic score (based on the index search and string distance metrics) is used to rank and select only those matches over a minimum threshold and up to a maximum number (that can be adjusted for each KG). In addition, unconnected properties, classes with no instances, and instances with no type are removed.
- Approximate (fuzzy) matches are returned only if exact matches are not found for the term, including the lemma and plural forms. The score of lexically related matches (synonyms, hypernyms) is penalized with respect to matches to the original query term, as they can also introduce noise. Expansion is based on linguistic features (POS, Watson types), e.g., derived words from WordNet (e.g., *Russia* in *Russian mountains...*) are not obtained for entities which semantic type is *person*, such as *musician* (derived word: *music*), hypernyms are excluded in the case of proper nouns, and meronyms are only used for geographical entities.

This is the only component that requires tuning the exact and index searches according to the source to be queried. While tuning was not needed to query the biomedical ontologies, Freebase/DBpedia have some unique ways to structure data that should be considered not to miss relevant mappings: aliases (alternative labels) are presented through non-standard RDF properties. Ambiguous terms (with different URIs) in DBpedia are represented with different labels, e.g.: db:Sunflowers_ (Van_Gogh_series), as well as redirects (from db:Sunflower to db:Helianthus), while in Freebase they share the same label. Properties in DBpedia may be covered by the ontology (e.g., dbo:founder), while in Freebase they have different namespaces for different domains (fb:organization.founders, fb:formula1.founders, and so on). Finally, DBpedia types are structured in a taxonomy (requiring inference through the subsumption hierarchy), while, in Freebase, instances belong to a flat list of types (where the most relevant one is given by fb:common.topic.notable.type).

3.3   Step 3: Template-Based Graph Pattern Search

The output of this component is a set of *Graph-Views*, one per graph. Each Graph-View consists in a set of ranked GPs. We define a GP as: (i) a set of triple patterns or BGPs (Basic Graph Patterns); (ii) JOINS among them; (iii) FILTER and modifiers such as ORDER BY, COUNT, OFFSET, LIMIT (as a query can be translated into alternative GPs, OPTIONALS are covered); (iv) the variables that are the focus (the variables we seek an answer for) of the GP; and (v) a confidence score, explained in Sect. 3.5.

GPs can be readily translated to SPARQL queries. In this step, each GP is formed with BGPs that belong to the same graph and that join together provide a translation to one or more given PAS Triples in the query. The algorithm is iterative, looking for the best translations first, and expanding the search only if required:

Step 3.1.   For each PAS Triple, KGs are sorted based on their coverage. Thus, the graphs that have candidate matches for all or most of the terms in the PAS are selected first.

Step 3.2.   For each PAS Triple and each covering graph with candidate matches, the search for GPs is performed using *parametrized pattern templates*. Only GPs that produce bindings after executing the associate SPARQL query are selected.

Step 3.3.   For queries with multiple PAS Triples, Graph-Views are created by merging the GPs belonging to the same graph but covering different PAS. Two GPs are merged if they have a common join (focus term) and their joining produces a non-empty set of bindings. A merged GP will have a higher ranking than a GP that could not be merged with any other, as it provides a more complete translation (coverage) to the user query.

If the Graph-View combines GPs that translate each PAS in the query (or if there is only one PAS in the query) the query can be answered within one graph and this step already generates a complete translation. If the Graph-View contains alternative translations (i.e., different interpretations of the query), answers are ranked (Sect. 3.5). However, for some queries, Graph-Views contains only partial translations and a complete translation (answers) can only be obtained by combining GPs across graphs (Sect. 3.4).

Parametrized Pattern Templates. To execute the right pattern templates to search for the GPs providing a more meaningful representation of each PAS, while reducing the amount of queries, we use the types of the candidate matches. There are 11 direct patterns and 7 indirect patterns (corresponding to more costly SPARQL queries that are executed only if no GPs are found using the direct ones). In Table 1, we give some examples of patterns executed for a given combination of candidate types[2].

Often, properties are either implicit or difficult to match. If there are matches not only for the subject and/or object of the PAS but also for the properties, a FILTER is added to the patterns above. If the GP does not produce any binding, the query is executed again without the filter: FILTER (?prop = <property1> || ?prop = <property2> || etc.). Schema information such as domain and range is often missing so properties need to be inferred by looking at the instance-level when invalid ontological matches are found for the property. String metrics and annotators are used to calculate semantic relatedness between all possible properties for the matched entity(-ies) and the property in the query. In particular, we use WordNet taxonomy to measure their

---

[2] Due to space constraints a full list of templates is presented in https://ibm.biz/BdHEwF.

Table 1. Example of GP templates according to entity types to support factoid queries (variables are preceded by the "?" and the parameters to substitute by the matches are between <>).

Pattern 3: Direct instance - property (Focus:?o/?s) – {<instance> <prop> ?o} UNION {?s <prop> <instance>}   E.g.: **what is the population of Japan?** – {db:Japan db:populationTotal ?o}
Pattern 5: Direct type - type (Focus:?so/?os) -?so a <type1>. ?os a <type2> {?so ?p ?os} UNION{?os ?p ?so}   E.g.: **Languages in countries** – {?so a db:Language. ?os a db:Country. ?os ?px ?so}
Pattern 6.a: Superlative datatype property(Focus:?o)-Pattern 4(type-prop) + order by desc(?o) offset 0 limit 1   Pattern 6.b: Superlative object property (Focus:?o)- Pattern4/5+ order by desc(count(?o)) offset 0 limit 1   E.g.: **highest mountain in Australia** – {?s a db:Mountain. ?s ?px db:Australia. ?s db:elevation ?o} ORDER   BY DESC(?o) LIMIT 1 OFFSET 0
Pattern 7.a: Comparative instance-instance - <inst1> <prop> ?o1. <inst2> <prop> ?o2. FILTER (?o1 > ?o2)   Pattern 7.b: Comparative type-instance- ?s a <type>. <inst> <prop> ?o2. ?s <prop> ?o1. FILTER (?o1 > ?o2)   E.g.: **Is Lake Baikal bigger than the Great Bear Lake?** – db:Lake_Baikal ?prop ?o1. db:Great_Bear_Lake   ?prop ?o2. FILTER (?o1 > ?o2). FILTER (?prop = db:volume)
Pattern 10: Indirect instance-property (Focus ?o)- E.g.: <instance> ?px ?ent. ?ent <property> ?o.   E.g.: **give me the prescriptions of Teresa** - patient:Teresa ?p1 ?ent. ?ent patient:hasPrescription ?o

semantic distance. Thus, the most related ontological properties to the user query term are those over a given threshold when applying Wu and Palmer formula [21] (Formula 1) and the rest are discarded. For example, in *Which mountains are higher than Anapourna?* (pattern 7: ?s a db:Mountain. ?s db:elevation ?o1. db:Anapourna db: elevation ?o2 FILTER (?o1 > o2)), none of the properties for the instance Anapourna could be lexically matched; the right ontological property *elevation* is found among all of them due to its semantic relatedness to the query term *high*:

Formula 1. Relatedness between term C1 and C2, where C is the lowest common ancestor between C1 and C2, Ni is the length of the path from Ci to C, and N is the length of the path from C to the root.

$$Similarity(C_1, C_2) = \frac{2 \times N}{(N_1 + N_2 + (2 \times N))}$$

Merging patterns. These patterns also determine the way BGPs are combined to create a merged GP for the same graph. For instance, for two type-instance GPs (pattern 2) the join answers are given by the common focus: ?s, as in *Which Russian rivers flow into the Black Sea?* (join term: the subject *river*) the merged GP joins all BGPS in both GPs:

GP1: ?s rdf:type dbo:river.?s ?p1 db:Russia. FILTER (?p1 = db:sourceCountry)
GP2: ?s rdf:type dbo:river. ?s ?p2 db:BlackSea. FILTER (?p2 = db:mouth)

Lessons Learned. All semantic patterns are domain- and dataset-independent. Some patterns may occur more commonly in some ontologies, but no assumption is made a-priori about this, e.g., commonly Freebase requires patterns in which intermediate entities act as blank nodes. For each of the indirect pattern templates (instance-type, instance-property and instance-instance), variations were added (one for each) to provide answers to that kind of queries. Take as example: *What character does Ellen play in Finding Nemo?*, answered by merging *Pattern 9* (for PAS <Ellen, play, Finding Nemo>) and *Pattern 10* (for PAS <Ellen, play, character>) through a blank node:

Pattern 9: Indirect Instance -Instance	**Pattern 10**: Indirect Instance-Property (focus: ?o) .
?bnode ?p1 fb: Ellen DeGeneres	?bnode fb:film.performance.character ?o.
?bnode ?p2 fb:Finding Nemo	?bnode ?p3 fb:Ellen DeGeneres

3.4   Step 4: Merging Across Graphs

Partial answers from GPs across graphs that are translations to different PAS Triples need to be combined to generate complete answers. For GPs to be merged they need to have at least one variable (binding) in common, as well as one BGP with a common mapping (the subject or object) that corresponds to the Join Term. However, when merging across graphs, the Join Term may not necessarily be represented by the same URI across graphs, even if they refer to the same real world entity; the same applies to the bindings in common. Therefore, merging across federated graphs presents two challenges:

(1) Finding the join term between each pair of GPs to be merged considering the query.
(2) On-the-fly entity co-reference based on both a syntactic (similar labels) and a semantic (semantic equivalent link) merging between the join term and bindings across graphs.

Consider an example query in the biomedical scenario: *Is Eplerenone having side effects on Teresa's conditions?*, finding an answer requires combining a GP from the patients' graph representing Teresa's conditions and a GP from the SIDER ontology with the side effects for the drug *Eplerenone*. The **first step** is to find the join term for the query, which in this case is represented by different URIs and query terms in each graph. From the three PAS Triples: <Eplerenone, have, side effects> <side effects, conditions> <conditions, Teresa>, the first one is translated in SIDER and the third in the patient's graph. The second PAS, without any valid translation, links *Side Effects* (in SIDER) and *Conditions* (in the patients' graph); that is, the two terms that need to be joined to retrieve the answer.

The **second step** is to retrieve the common bindings for the join terms across graphs, which will likely have different URIs even if they represent the same real-world entity. We perform two types of merging: syntactic and semantic. The **semantic**

merging relies on the linkage across entities through equivalence relations. To find the linkage between the join variables (renamed as ?s for instances of *Conditions* and ? sjoin for instances of *Side Effects*), while avoiding pairwise comparison of entities binding each join, the following BGPs are added to the merged ones (GP1 + GP2) before executing the final SPARQL query to retrieve the answers over the federated graphs:

> **GP1**: ?s rdf:type <Condition> . <Teresa> ?p3 ?ent. ?ent p2 ?s.
> **GP2**: ?sjoin rdf:type <Side_effects> . <drug:Eplerenone> ?p1 ?sjoin.
> {?s ?rel1 ?sjoin}. UNION {?sjoin ?rel1 ?s} UNION {?sjoin ?rel1 ?same. ?s ?rel2 ? same.}
> FILTER ((?rel1 == owl.sameAs ‖ skos.closeMatch) && (?rel2 == owl.sameAs ‖ skos.closeMatch))

The answer to the query is true, as there is an evidence graph, shown in Fig. 4, where Teresa's Condition, *prediabetes*, links to one of the side effects of Eplerenone, through the DBpedia instance *Diabetes_mellitus*.

However, as equivalence links across entities are often sparse, we also perform **syntactic merging**. An index search is added to the merged GP to find those instances from each join with similar labels.

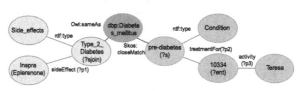

Fig. 4 Evidence graph after merging

3.5   Step 5: Ranking of Answers

The answers are the bindings of the variables representing the focus of the query over a given graph. Ranking is crucial for obtaining the answers with the best confidence among all the alternative GP representations and their combination (in the same or across graphs).

Two different scores reflect how well a GP matches a PAS in a user query: (1) At BGP level: the average score of each individual mapping in a BGP is computed. A higher score is assigned to BGPs that combine more candidate matches (e.g., if besides subject/object, the property is also match); and (2) At GP level: this is based on the pattern executed – direct patterns carry more weight than indirect ones. The rationale behind it is that the longer the distance between two candidate terms, the more likely the translation is noisy. Active and passive forms are also considered, if the same relation is valid in the two directions (from subject to object and vice-versa), e.g., *who was J.F Kennedy successor of?*, a higher score is given to the GP with the right directionality for the relationship.

In short, as show in Formula 2, the final confidence score of each answer A, denoted by *CF*(A), is a combination of the score of the GPs where the answer is computed and the scores of the BGPs in such GPs, irrespective from the graphs that they are computed from:

Formula 2. Specifically, let A be an answer in GPA and BGPA, where there are in total m GPs across all data graphs and n BGPs in GPA

$$CF(A) = \frac{CF(GP_A)}{\sum\limits_{i=1}^{m} CF(GPi)} * \frac{CF(BGP_A)}{\sum\limits_{j=1}^{n} CF(BGP_j)}$$

Note that answers can be represented by an entity (or a list), a datatype literal, a boolean a count (*how many*), or blank nodes (e.g., *What is the revenue of Apple?* is answered in Freebase by a set of answer nodes, where each one contains a set of property-values representing the amount, currency and date). For each answer (or set of), supporting evidence can be retrieved as per user request. An evidence graph is the subset of the graph containing all the bindings and mappings the query is translated into. When partial answers are obtained across graphs, co-reference across entities is performed; therefore, an answer may correspond to more than one URI from different graphs.

4 Evaluation

We evaluate the system with a set of blind questions based on two scenarios and for each question q, we compute precision (P), recall (R) and F-measure (F_1), as defined in QALD:

$$R(q) = \frac{\text{number of correct system answer for } q}{\text{number of gold standard answers for } q} \quad F_1(q) = \frac{2 * P(q) \times R(q)}{P(q) + R(q)}$$

$$P(q) = \frac{\text{number of correct system answer for } q}{\text{number of system answer for } q}$$

To evaluate the ranking when there are multiple translations, we calculate: P/R@1 by considering only the answers which confidence score is the highest, ranked in position 1; P/R@2 for the results in the first and second positions; and P/R@3 for the results from the first to the third position. The last (P/R@3), essentially considers all answers (till position 3), without ranking. Detailed results are documented online (https://goo.gl/0o0KYy).

4.1 Scenario 1: Open Domain

DBpedia QALD-5 Evaluation: We used the 2015 QALD-5 test set [3], which consists of 50 questions annotated with the corresponding SPARQL query and answers. From those 50 questions, question Q21 is missing and question Q42 is classified as out of scope for DBpedia KB. Therefore, we measure the average P, R and F_1 over 48 English queries of different complexity; QALD queries may require operations beyond triple matching such as counting or ordering. Retrieving all possible answers to a given query (total recall) over large and heterogeneous sources such as DBpedia is a challenge, even for manually created gold standards. This is due to the presence of duplicated entities (same real world entity represented with different URIs), heterogeneous properties (dateOfbirth, birthdate, etc.), and literal values that should correspond to entities, but were not mapped to. QALD queries contain UNIONs to get all answers when a question can be translated into alternative valid SPARQL queries. However, for 6 of the benchmark queries, QuerioDALI found different translations leading to an extended set of valid answers. For those queries, we have updated the set of answers in the benchmark and updated the P/R accordingly (the updated queries are documented online). Take Q4: *Which animals are critically endangered?* the SPARQL query in QALD retrieves a total count of 1613 distinct animals as answers: Select ?uri where {? uri rdf:type db:Animal. ?uri db:conservationStatus 'CR' }

However, there are 1629 answers if one considers the UNION with other valid translations: { ?uri rdf:type dbo:Animal . ?uri dbp:status res:Critically_endangered } UNION

{ ?uri dcterms:subject dbc:Critically_endangered_animals . ?uri dbo:kingdom res: Animal}

In Table 2 we present our P/R and F_1 results. The set of results obtained in the first position are more precise. The best F_1 of 0.61 is for the answers ranked first ($F_1@1$), proving that our ranking mechanism ranks the best answers first. From 48 questions, the system was unable to find answers for 10 of the queries (F_1 strictly 0). Only for 2 of the queries the answers were not ranked in the first position but the second. In the first position, 25 queries were answered with an F_1 of strictly 1, while for 11 queries either some answers were missing (4 of them with P = 1, R = (0,1)) or inaccurate answers were retrieved (6 of them with R = 1, P = (0,1)), or both (1 of them with P = R = (0,1)). For the second and third position, recall is increased but at a cost, as more inaccurate answers are also retrieved.

QALD campaigns are notably challenging, an F1 of 0.61, even if far from perfect, is a promising result. To put it into perspective, the average F_1 of QALD-4 was 0.34 while the average F_1 of QALD-5 is 0.43. The relatively low values show that the complexity of the questions is still high. Note that our P, R and F_1 above are with respect to the total number of questions. This is what QALD reports as global-measures, which considers all the questions and not just the processed questions (those for which a system is able to provide a query or an answer, even if wrong). Our system is designed to give results even if only partial results are found (e.g., in Q38 *Where did the architect of the Eiffel Tower study?* QuerioDALI finds who is the architect of the Eiffel Tower but not where he studied, giving partial results rather than no answers). Thus, QuerioDALI only fails to process 3 out of the 48 questions, reaching an F_1 - if only processed questions are considered- of 0.84.

Xser [12] has the best global F_1 reported in QALD-5: 0.63, just 0.02 over our $F_1@1$. This is well over the second ranked system with a global F_1 of 0.3. Like our system, Xser analyzes the query through a semantic parser, and then instantiates the query with respect to the KB. However, differently from ours, Xser requires training data as it relies on a structure prediction approach implemented using a Collins-style hidden perceptron.

Freebase Free917 Evaluation: We used the first 101 queries from Free917 to evaluate and manually analyse why some queries failed to reach P/R of 1; 14 of those queries did not produce any answer (out of scope using the last available Freebase data dump). Thus, we measure the average P, R and F_1 over the remaining 87 queries covering a wide range of domains. We did not use any manually crafted lexicons tailored to answer these queries [9], as addressing entity recognition and anchoring without domain adaptation is an integral part of the challenge we want to evaluate.

As shown in Table 2, the best F_1 of 0.72 is for the answers ranked first (F1@1). The results obtained in the first position are more precise, with a small drop on recall compared with position 2. There is not significant drop on precision between position 2 and 3. Interestingly this is because fewer alternative translations are generated when using Freebase instead of DBpedia. We believe F_1 is better for this benchmark because Free917 queries tend to be more tailored to the underlying KG than QALD queries. However, while they are linguistically less complex (based on a single relation without comparisons or superlatives) they can generate complex GPs due to the presence of intermediate (blank) nodes.

Table 2. Precision, Recall and F1 over QALD-5 test questions and 100 Free917 questions

	P@1	R@1	P@2	R@2	P@3	R@3	F_1@1	F_1@2	F_1@3
QALD5	0.64	0.69	0.53	0.73	0.52	0.73	0.61	0.55	0.55
Free917	0.723	0.727	0.637	0.733	0.625	0.733	0.720	0.658	0.647

The system was unable to find correct answers for 22 out of 87 queries (F_1 strictly 0). In the first position, 60 queries were answered with an F_1 of strictly 1, while for 1 query some answers were missing (P = 1, R = (0,1)) and 3 queries retrieved inaccurate answers (R = 1, P = (0,1)). Only for 1 query the answers were not ranked in the first position but the second, thus R@2 slightly increases by 0.006, but with a drop in P of 0.08.

OPEN CHALLENGES: Entity identification is a crucial step for QA over KGs. Existent tools to extract NEs while very useful are not enough to retrieve all relevant entities to translate a user query. We reflect on this challenge with a small experiment to compare the P/R of the URIs retrieved by just using DBpedia Spotlight and QuerioDALI's final selection of URIs to semantically interpret the user questions, using as a ground truth the entities in the QALD-5 SPARQL queries. We witnessed a significant jump in P/R from QuerioDALI results over just using Spotlight: the average P was improved from 0.31 to 0.61 and the average R from 0.44 to 0.71. Note that QuerioDALI uses Spotlight in Step 1, thus this shows the baseline over which QuerioDALI had to improve to answer the user questions, i.e., by considering the semantic linkage among query terms.

Next, we analyze the reasons behind the failed queries and those with inaccurate or missing answers (F_1@1 < 1). Errors may be introduced by the different components in the pipeline, in each step of the process. Thus, a query may fail for more than a reason:

Linguistic coverage. When the system fails to linguistically understand the query and correctly represent how the terms link together. This occurs if the dependency parse tree generated for the query is incorrect or if the generic rules used to create the PAS triples from the tree do not cover a particular type of sentence. Only one DBpedia query, db40: *What is the height difference between Mount Everest and K2?* and one ill-formed Freebase query (fb17:*How many beers come a can?*) failed because the rules could not capture the linguistic dependencies. These kind of errors need to be solved by extending the coverage.

NER + Anchoring. When the system fails to bridge the lexical gap between the NL expressions and the data: a total of 13 queries in DBpedia and 24 in Freebase. In some cases the anchoring can recover if a multi-word NE is not captured by the annotators – for fb27:*When was home depot founded?* it finds the entity db:date_founded combining the property and the subject - but 7 out of 24 Freebase errors are due to NEs not found. In the worst case, missing the right anchoring leads to no answers or wrong answers (db:4/13, fb:21/24), such for db8:*Is Barack Obama a democrat?* where the term *democrat* could not be mapped to db:Democratic_Party_(United_States) and it was mapped instead to the hypernym db:politician, therefore the *true* answer is based on not accurate enough evidence (P/R = 0). Inaccurate anchoring often leads to a combination of noisy answers among good answers (db:7/13, fb:2/24), or a loss in recall because relevant properties or instances were not mapped (db:2/13, fb:1/24). For example, in db11:*Who killed John Lennon?*, QuerioDALI can not find the property (db:conviction) and it returns all the instances of a person related to the two instance matches db: John_Lennon and db:Death_of_John_Lennon (only the latter is related to John Lennon's killer), obtaining a R@1 of 1 but a P@1 of 0.09.

Synonym features are important for both datasets. WordNet semantic relatedness is useful to find the right properties, such as *weigh* for *heavy* in db48:*Who is the heaviest player of the Chicago Bulls?*. Hypernyms are also needed to map *players* to db:person and find the relevant property for *heaviest* (the exact mapping for player does not lead to any valid GPs). However, the use of approximate (index) matches, aliases and alternative names (e.g., *foaf:name*) tends to be noisy. Thus, we only use them if non exact matches are found for a term. The use of manually or semi-automatically created lexicons (if training data is available) or through user feedback, could alleviate the anchoring problem.

Template coverage. When the system fails to bridge the structural gap between the NL expressions and the data. Queries fail if there is not a template to properly translate the query and bridge the structural gap, 5 queries failed in DBpedia due to this (0 in Freebase), of which 4 of them involved temporal reasoning in one way or another. For example, queries involving the temporal adjectives, such as *youngest/oldest*, that were not mapped to the relevant property db:birth-date can be partially answered by searching among the properties with datatype date, but it also obtains unrelated properties like db:death-date; Queries with *since*, such as *since 2000*, require to add a filtering pattern to find dates whose year is over 2000 (FILTER (year(?-date) > = 2000)). The query db49:*Show me everyone who was born on Halloween* is particularly challenging as there is not temporal adjective such as "same date/year as" to indicate we are looking for persons born on the same date as they day in which

Halloween is celebrated (as in the correctly answered query db14:*Which artists were born on the same date as Rachel Stevens?*)

Merging. In this scenario, merging across sources is not required, even if DBpedia uses different vocabularies, they are all interlinked as part of the same graph. However, queries with more than one PAS Triple require that the respective GPs are combined in order to obtain a complete answer. In this evaluation 18 DBpedia queries require to fuse GPs, where only one query failed (db22): the system found the right GPs but the relevant property contained a String literal, with comma separated values, that could not be linked.

Ranking. There were only 2 ranking errors in Freebase, mostly for ambiguous queries such as fb22:*Who designed the Parthenon?*, where the two translations retrieve the architect for two different instances of Parthenon - the right and expected one, the Parthenon in Athens, and a full scale replica in Tennessee. For queries evaluated in a federated manner, the quality can be improved by giving a higher rank when the union of answers from each data graph forms the final answers (i.e., most popular answers across graphs).

4.2   Scenario 2: Smarter Care

We use the QALD-4 Task 2 over biomedical data, based on the ontologies SIDER, Drugbank and Diseasome. There is no benchmark to evaluate patient-based questions that can be answered using our enterprise dataset and Open Data, but, nonetheless, QALD-4 Task2 includes questions that need to be answered by combining facts from graphs, thus, it allows us to evaluate the merging. We used the training dataset used to evaluate the SINA system [18], which also provides QA over interlinked datasets and high P/R measures, using a trained Hidden Markov Model for the domain. For one of the benchmark queries (Q5), we were able to find different translations due to duplicated entities leading to an extended set of valid answers, not present in QALD-4. The results are given in Table 3.

Table 3. Precision, Recall and F1 over all QALD-4 (biomedical) train questions

P@1	R@1	P@2	R@2	P@3	R@3	F_1@1	F_1@2	F_1@3
0.85	0.88	0.78	0.92	0.78	0.92	0.85	0.83	0.83

As expected, the results on a domain-specific set up are better than for open-domain QA, with an F_1 of 0.85 (the SINA system obtains an F_1 of 0.89). From the 25 queries in this task, 7 of them can only be answered by merging GPs across two graphs. Only for one of them QuerioDALI could not get an F_1 of 1 (*Q19*) but, as analyzed below, the reason was not because of the merging. The system can find answers for 23 out of 25 queries (92 % coverage). The number of queries with F_1 of strictly 1 are 21, thus for 84 % of the queries, perfect answers are retrieved in the first position. For two queries, the system was not able to retrieve any answer. The rest of the queries introduce noisy answers together with the good answers. We analyze the reasons behind the inaccurate results:

Linguistic coverage. For one query, Q19: *Which are the drugs whose side effects are associated with the gene TRPM6?*, the linguistic component failed to retrieve one of the PAS needed to fully understand the query. From the two PAS retrieved <drugs, associate, gene RTPM6> <side effects, associate, gene RTPM6> only the second is correct. The system can recover from ambiguous associations by using the KGs to find the right linkage, but it can hardly recover if an association is missing, in this case <drugs, side effects> .

NER + Anchoring. In domain specific scenarios, ambiguity and errors due to anchoring are notably reduced, as long as the system can rely on annotator sources with a good coverage of the domain vocabulary, in order to find both the right lexically related words and NEs (multi-words). In particular, Q14: *Give me drug references of drugs targeting Prothrombin* failed, because the system was unable to detect the multi-word *drug reference*. The system could retrieve all drugs targeting Prothrombin, but failed to find the references ($F_1 = 0$). In Q17: *Which are possible drugs against rickets?* the valid answers are ranked in the second position, this is because the relation *against* was not mapped to any ontological property and the system ranks first all drugs with side effect on rickets (from SIDER) and after all drugs with *diseasome:possibleDrugs* to rickets (the valid answers).

Templates coverage. Bridging the structural gap is particularly challenging when schema information is missing. This is the case for Q10:*Which foods does allopurinol interact with?* ($F_1@1 = 0.35$), where the *interact* maps to the ontological properties for allopurinol: *drugbank:interactionDrug* and *drugbank:foodInteraction*. The system retrieves all the values as answers, but only the latter leads to precise answers. This is because the relevant property is a datatype and the system can not infer based on the type *food*.

5   Related Work

QA approaches over Linked Data are popular because they balance expressivity with posing queries in an intuitive way, using NL. Here we look at approaches that are agnostic to the underlying KG or can scale to large open domain scenarios.

On Generating Templates, user questions are generally mapped into appropriate artifacts that can be processed over RDF datasets. One type of mapping is from questions to triples, e.g., PowerAqua [19], and FREyA [5]. FreyA leverages users' feedback for disambiguating the right relation and improve accuracy over time, while PowerAqua ranks across alternative representations. However, complex questions that require performing aggregations or comparisons are not mapped in these systems. An alternative mapping is from questions to SPARQL query templates to capture more complex structures. For instance, TBSL [4] uses 22 manually curated templates to match a parsed user question into a SPARQL template, which is then instantiated using the domain vocabulary. However, capturing the semantics of input questions can lead to too rigid templates to fix the triple structure, queries with unknown syntactic constructions or domain-independent expressions cannot be parsed (F_1 of 0.62 on 39 questions that could be processed out of the 50 in QALD-1). In Yahya et al. [14], phrases in the questions are extracted and matched to RDF entities using a large dictionary. To optimize the way of yielding SPARQL queries, an integer linear program has been used. The SINA system [18] parses questions into a list of keywords and links keywords to resources across datasets to form triple patterns using a Hidden Markov Model. Differently, our approach does not require training (as in [18]) or building indexes over all relation phrases in a paraphrase dictionary (as in [14]).

On **Merging and ranking,** only a few existing systems consolidate candidate answers among a collection of inter-connected, distributed datasets, considering the fact that sometimes answers to a question can only be provided if information from several sources are combined. Specifically, PowerAqua can query distributed sources without requiring federation, but it imposes an overhead for merging the data by performing entity co-reference (linear to the number of answers). The SINA system constructs federated SPARQL queries leveraging the built-in *owl:sameAs* property as linkage across entities.

On **Entity Recognition** against Freebase, the systems mainly use IR techniques to score answers. The Aqqu system [9] performs entity identification, template generation, relation matching and ranking. It reports an F_1 of 0.65 using Free917 without lexicon and 0.76 with a manually crafted lexicon (from which 276 questions are tested and 641 used as training). Supervised learning is used to find the mappings. Assuming all mappings are found, they only required the use of 3 templates to map the candidate instances and relations to queries. YodaQA [15] uses both Freebase and DBpedia, however, it has been designed to answer questions over text (filtering the passages containing the most clues from all NEs and noun phrases), which is why it only reaches a F_1 of 0.18 in QALD-5 [3].

6   Conclusions

Our system performs at the same level as the state of the art systems without the use of training data, both for open scenarios on the Web of Data or domain specific, particularly in the biomedical domain, where a large number of specialized ontologies are available [6].

Large scale deployment to scenarios with large evolving data and graphs, requires (besides supporting concurrency) strategies for iteratively pruning the large search space of solutions (and speed up the search for GPs by considering fewer candidates), while keeping the most promising ones based on the context of the query. In turn, (1) NL queries are converted into PAS Triples to focus on finding the entities and links that matter; relevant graphs are selected based on their coverage (not for the whole query but for a given PAS); (2) unconnected and less promising mappings are filtered out based on their syntactic relevancy (exact vs. approximate); (3) remaining mappings are assigned a confidence score based on their semantic relation to the query term (synonyms, etc.); (4) templates are then used to semantically validate the mappings according to their type, and search first for the GPs that more accurately represent the user query, extending the coverage to indirect relations only if nothing is found; (5) GPs are merged according to the entities to be joined, merging across graphs is based on semantic linkage or label similarity.

As shown in the evaluation, a major challenge is the anchoring of query terms to the ontological terms that would lead to the answer. Inevitably, the use of fuzzy search and semantic expansion to bridge the lexical gap also introduces lots of irrelevant mappings that, in some cases, may lead to GPs with inaccurate translations. Thus, ranking is crucial to compare across the possible solutions and select the translations with higher accuracy and better coverage for the query. The second challenge for both scenarios is coverage: we can incrementally add new patterns to increase the complexity of the questions, accommodating more complex queries requiring temporal reasoning, some basic statistical analysis or negations. However, covering the whole spectrum of different manners in which users express a question is an open problem. While QuerioDALI reaches F-measures comparable to systems that use training data, we believe that for these systems to overcome the lexical-structural gap and reach higher P/R over novel questions, they require a certain level of domain training. As the availability of training data is sometimes challenging, especially when dealing with sensitive information, we believe that the system should evolve into a cognitive system, with the ability to learn over time. This can be done if the system leverages users' feedback, using a hybrid approach between open factoid QA and guided explorative queries. In turn, the learning will not only improve the QA but it can also be used to augment the KGs with relevant connections.

References

1. Kalyanpur, A., Boguraev, B., Patwardhan, S., Murdock, J.W., et al.: Structured data and inference in DeepQA. IBM J. Res. Dev. **56**(3), 10 (2012)
2. Unger, C., Forascu, C., Lopez, V., Ngomo, A.N., Cabrio, E., Cimiano, P., Walter, S.: Question answering over linked data (QALD-4). In: Working Notes for CLEF 2014 (2014)
3. Unger, C., Forascu, C., Lopez, V., Ngomo, A.N., Cabrio, E., Cimiano, P., Walter, S.: Question answering over linked data (QALD-5). In: Working Notes of CLEF 2015 (2015)
4. Unger, C., Buhmann, L., Lehmann, J., Ngonga Ngomo, A.-C., Gerber, D., Cimiano, P.: Template-based question answering over rdf data. In: World-Wide Web, WWW 2012 (2012)
5. Damljanovic, D., Agatonovic, M., et al.: Improving habitability of natural language interfaces for querying ontologies with feedback and clarification dialogues. Web Semant. **19**(1), 21 (2013)
6. Weissenborn, D., Tsatsaronis, G., Schroeder, M.: Answering factoid questions in the biomedical domain. In: Workshop on Bio-Medical Semantic Indexing and QA, CLEF 2013 (2013)
7. DISEASOME. http://diseasome.kobic.re.kr/. Accessed April 2016
8. Wishart, D.S., Knox, C., Guo, A.C, Shrivastava, S., et al.: DrugBank: a comprehensive resource for in silico drug discovery and exploration. Nucleic Acids Res. **34**, D668–D672 (2006)

9. Bast, H., Haussmann, E.: More Accurate Question Answering on Freebase. In: CIKM 2015 (2015)
10. IBM Cúram. http://www-03.ibm.com/software/products/en/social-programs
11. Lehmann, J., Isele, R., Jakob, M., Jentzsch, A., et al.: DBpedia - a large-scale, multilingual knowledge base extracted from wikipedia. Semant. Web J. **6**(2), 167–195 (2015)
12. Xu, K., Feng, Y., Zhao, D.: Answering natural language questions via phrasal semantic parsing. In: CLEF 2014 Working Notes Papers (2014)
13. Kuhn, M., Letunic, I., Jensen, L.J., Bork, P.: The SIDER database of drugs and side effects. Nucleic Acids Res. **44**, 1075–1079 (2016)
14. Yahya, M., Berberich, K., Elbassuoni, S., Weikum, G.: Robust question answering over the web of linked data. In: CIKM 2013 (2013)
15. Baudis, P., Sedivy, J.: Modeling of the QA task in the YodaQA system. In: CLEF 2015 (2015)
16. Mendes, P.N., Jakob, M., Garcıa-Silva, A., Bizer, C.: Dbpedia spotlight: Shedding light on the web of documents. In: I-Semantics (2011)
17. Cai, Q., Yates, A.: Large-scale semantic parsing via schema matching and lexicon extensions. In: ACL, pp. 423–433 (2013)
18. Shekarpour, S., Auer, S., Ngonga Ngomo, A.C., Auer, S.: SINA: Semantic interpretation of user queries for question answering on interlinked data. J. Web Semant. **30**, 39–51 (2015)
19. Lopez, V., Nikolov, A., Fernandez, M., Sabou, M., Uren, V., Motta, E.: Merging and ranking answers in the semantic web. In: The Asian Semantic Web Conference, ASWC 2009 (2009)
20. Lopez, V., Stephenson, M., Kotoulas, S., Tommasi, P.: Data access linking and integration with DALI: building a safety net for an ocean of city data. In: Arenas, Marcelo, et al. (eds.) ISWC 2015. LNCS, vol. 9367, pp. 186–202. Springer, Heidelberg (2015). doi:10.1007/978-3-319-25010-6_11
21. Wu, Z., Palmer, M.: Verb semantics and lexical selection. In: ACL (1994)

Automatic Classification of Springer Nature Proceedings with Smart Topic Miner

Francesco Osborne[1(✉)], Angelo Salatino[1], Aliaksandr Birukou[2], and Enrico Motta[1]

[1] Knowledge Media Institute, The Open University, Milton Keynes MK7 6AA, UK
{francesco.osborne,angelo.salatino,enrico.motta}@open.ac.uk
[2] Springer-Verlag GmbH, Tiergartenstrasse 17, 69121 Heidelberg, Germany
aliaksandr.birukou@springer.com

Abstract. The process of classifying scholarly outputs is crucial to ensure timely access to knowledge. However, this process is typically carried out manually by expert editors, leading to high costs and slow throughput. In this paper we present Smart Topic Miner (STM), a novel solution which uses semantic web technologies to classify scholarly publications on the basis of a very large automatically generated ontology of research areas. STM was developed to support the Springer Nature Computer Science editorial team in classifying proceedings in the LNCS family. It analyses in real time a set of publications provided by an editor and produces a structured set of topics and a number of Springer Nature Classification tags, which best characterise the given input. In this paper we present the architecture of the system and report on an evaluation study conducted with a team of Springer Nature editors. The results of the evaluation, which showed that STM classifies publications with a high degree of accuracy, are very encouraging and as a result we are currently discussing the required next steps to ensure large-scale deployment within the company.

Keywords: Scholarly data · Ontology learning · Bibliographic data · Scholarly ontologies · Data mining · Conference proceedings · Metadata

1 Introduction

The process of classifying and annotating scholarly publications is crucial to enable scholars, students, companies and other stakeholders to easily discover and access this knowledge. To facilitate this classification process, a number of scholarly ontologies (e.g., SWRC[1], BIBO[2], BiDO[3], PROV-O[4], AKT[5], FABIO[6]) and bibliographic repositories in the Linked Data Cloud [1–3] have been proposed in the past decade, while at the same time the major publishing companies are starting to adopt richer data models [4, 5].

---

[1] http://ontoware.org/swrc/.
[2] http://bibliontology.com.
[3] http://purl.org/spar/bido.
[4] https://www.w3.org/TR/prov-o/.
[5] http://www.aktors.org/publications/ontology.
[6] http://purl.org/spar/fabio.

© Springer International Publishing AG 2016
P. Groth et al. (Eds.): ISWC 2016, Part II, LNCS 9982, pp. 383–399, 2016.
DOI: 10.1007/978-3-319-46547-0_33

In this paper, we present Smart Topic Miner (STM), a novel application, developed in collaboration with Springer Nature (SN), which classifies scholarly publications according to an automatically generated ontology of research areas.

STM analyses in real-time a collection of publications and returns a description of the given corpus in terms of (i) a taxonomy of research topics drawn from a large scholarly ontology and (ii) a set of Springer Nature Classification tags – see Fig. 1. This information is then used for a variety of tasks such as: (i) classifying proceedings in digital and physical libraries; (ii) enhancing semantically the metadata associated with publications and consequently improving the discoverability of the proceedings in both the Springer digital library, SpringerLink, as well as third-party sites such as Amazon.com; (iii) deciding where and when to market a specific book; and (iv) detecting novel and promising research areas that may deserve more attention from the publisher.

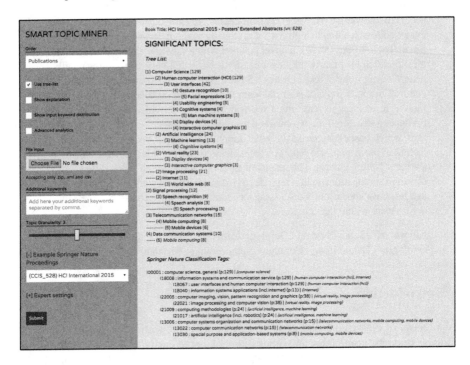

Fig. 1. The STM interface.

Traditionally, when classifying proceedings, editors choose a list of related terms and categories according to their own experience, a visual exploration of titles and abstracts, and, optionally, a list of keywords given by the curators or derived by calls for papers. However, this is a complex and time-consuming process and it is easy to miss the emergence of a new topic or assume that some topics are still popular when this is no longer the case. In addition, the keywords used in the call of papers are often a reflection of what a venue aspires to be, rather than the real contents of the proceedings.

For these reasons, there is a real need for more objective and scalable methods for identifying the research areas relevant to a proceedings book.

In this kind of scenario, it is critical for the editors to build confidence in the tool by being able to analyse the rationale behind the outcomes and understand why a certain research area or classification tag was chosen. Hence, we designed STM to produce intuitive explanations and to give the user full control over the granularity and nature of the topic characterization. Actually, one of the main advantages of adopting semantic web technologies is that they make it easier to generate a user-friendly explanation, as discussed in Sect. 2.3. Of course, the final decision of which topics and tags to associate with the proceedings still rests on SN editors.

In this paper, we describe STM in terms of its knowledge bases, algorithm and user interface. We also report the outcome of an evaluation study performed with eight Springer Nature editors with expertise in a variety of different fields, as well as a coverage study on a set of 200 proceedings. Finally, we conclude by discussing the steps required for large-scale deployment of the technology within the company.

2 Smart Topic Miner

Smart Topic Miner (STM) was designed to automatically classify proceedings and more in general any collection of articles by tagging them with a number of research areas and SN classification labels. It can be used for supporting editors in classifying new books and for quickly annotating a large number of proceedings, thus creating a comprehensive knowledge base to assist the analysis of venues, journals and topic trends. In this paper, we focus on the classification/annotation task.

STM can take as input either an XML file containing metadata about a publication or a ZIP including multiple XML files. Each XML file represents a paper in a proceedings volume published in the LNCS family of book series and contains title, abstract, the keywords provided by the authors, section title and book title. Springer books are thus usually represented as collections of XML files.

STM analyses the publication metadata and returns:

- A taxonomy (or optionally a plain list) of the most significant topics annotated with the number of relevant papers/chapters, structured according to an automatically generated ontology of research areas;
- A taxonomy of Springer Nature Classification tags;
- A number of analytics to allow the editors to further analyse the content of a proceedings volume, including the list of terms and topics associated to each paper;
- Optionally, an explanation for each topic, in term of the keyword distributions that triggered the topic recognition. For instance, the Semantic Web may have been inferred as a research area for a book by recognizing terms such as "linked data", "ontology matching", and "semantic web services".

Figure 1 shows the main interface and how the tool classified the Springer Nature book "HCI International 2015 - Posters' Extended Abstracts". Figure 2 shows the STM architecture, which consists of four main components: (1) the user interface, (2) the parser, which elaborates the input files, (3) the back-end API, and (4) the knowledge bases. Every time the user uploads a file and submits it to the system using the GUI, the parser analyses the XML files and extracts the relevant metadata. This data are sent as a JSON file to the background API via a POST query. The API analyses the data and returns the results either as a JSON or HTML file, which is in turn visualized by the interface. The API and the parser are realized in PHP and save cached data in a MariaSQL[7] database, while the front-end uses HTML5 and Javascript.

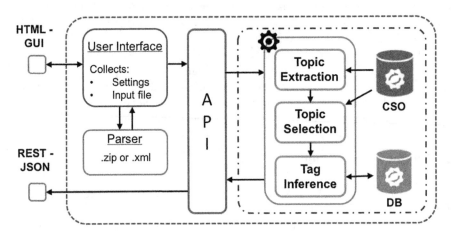

Fig. 2. The STM architecture.

In the next sections we will discuss the system in detail. In Sect. 2.1 we will elaborate on the knowledge bases, in Sect. 2.2 we will discuss the approach to infer research areas and Springer Nature Classification tags from the metadata and finally in Sect. 2.3 we will describe the user interface and the options available to the users.

2.1 Background Data

STM uses two knowledge sources: the Klink-2 Computer Science Ontology (CSO) and the Springer Nature Classification for Computer Science (SNC).

CSO was created and subsequently updated every 6 months by applying the Klink-2 algorithm [6] on the Rexplore dataset [7], which consists of about 16 million publications, mainly in the field of Computer Science. The Klink-2 algorithm combines semantic technologies, machine learning and knowledge from external sources (e.g., DBpedia, calls for papers, web pages) to automatically generate a fully populated ontology of research areas, which uses the Klink data model[8]. This model is an extension of the BIBO ontology[9] which in turn builds on SKOS[10]. It includes three semantic relations:

---

[7] https://mariadb.org/.
[8] http://technologies.kmi.open.ac.uk/rexplore/ontologies/BiboExtension.owl.
[9] http://purl.org/ontology/bibo/.
[10] http://www.w3.org/2004/02/skos/.

relatedEquivalent, which indicates that two topics can be treated as equivalent for the purpose of exploring research data – e.g., Ontology Matching and Ontology Mapping; *skos:broaderGeneric*, which indicates that a topic is a sub-area of another one – e.g., Linked Data is considered a sub-area of Semantic Web; and *contributesTo*, which indicates that the research outputs of one topic significantly contribute to research into another. For instance, research in Ontology Engineering contributes to the Semantic Web, but arguably Ontology Engineering is not a sub-area of the Semantic Web – that is, there is plenty of research in Ontology Engineering outside the context of Semantic Web research. The current version of STM uses the first two relationships.

An important characteristic of the CSO ontology is that it allows for a research topic to have multiple super-areas – i.e., the taxonomic structure is a graph rather than a tree. This is a very important difference with respect to other taxonomies of research areas notably because research topics often derive from multiple areas and can be categorized under a variety of fields. For example, it can be argued that a topic such as Inductive Logic Programming should be a sub-area of both Machine Learning and Logic Programming. Hence, a representation that forces a research area to be subsumed by only one other area fails to capture adequately the network of relationships between research topics.

The current version of the CSO ontology comprises about 17 k topics linked by 70 k semantic relationships and includes 8 levels of granularity. The main root is the topic Computer Science, however CSO includes also a number of secondary roots, such as Geometry, Semantics, Linguistics and so on. STM uses CSO for a variety of tasks, including (i) inferring a list of well-defined and human readable semantic topics from the very large distributions of terms extracted from publications, (ii) supporting the set-covering algorithm, and (iii) structuring the outcome as a taxonomy, in order to help the editors to understand the relationships between research areas.

CSO presents two main advantages over the classic manually crafted categorizations used in Computer Science, such as the well-known 2012 ACM Classification[11]. Firstly, it is able to recognize a very large number of terms which do not appear in these other classifications. In fact, it is about seventeen times larger than ACM in terms of number of concepts and about seventy times larger in terms of number of relationships. For this reason, it is able to characterize higher-level research areas by means of hundreds of sub-topics and related terms, which allows STM to effectively map specific terms from research publications to higher-level research areas. Secondly, the ontology can be easily updated to include novel research areas simply by adding the most recent publications to the dataset and running Klink-2 over again. Conversely, human crafted classifications cannot keep up with the evolution of the research domain and tend to age very quickly, especially in rapidly changing fields such as Computer Science. A more comprehensive discussion of the advantages of adopting an automatically generated ontology in the scholarly domain can be found in [6].

The Springer Nature Classification for Computer Science is a three level classification, containing 76 categories characterizing both research fields (e.g., *I23001 – Computer Applications*) and domains (e.g., *I23028 - Computer App. In Social and*

---

[11] http://www.acm.org/about/class/2012.

Behavioral Sciences). It is an internal company classification, which is used in order to categorize proceedings, books, and journals. This helps to appropriately channel the contents. For instance, users browsing the Springer Nature website can retrieve all contents on Computer Science or its sub-disciplines. These codes are also used in the metadata describing the contents for third parties (libraries, bookshops).

We integrated CSO and SNC by means of 349 relationships, so that every SNC tag is now associated to a set of related topics. For example, we mapped the *systems and data security* category to topics such as Cryptography, Security Of Data, Network Security, Computer Crime, Data Privacy and so on.

The mapping was performed in three phases. First, we used Klink-2 to generate automatically a number of *relatedEquivalent* and *skos:broaderGeneric* relationships between the SN label and the topics. Then, we manually cleaned these links and created additional ones by analysing the 158 topics at the first two levels of the CSO ontology. Finally, these links were revised by a Springer editor with extensive experience in using SNC for classifying conference proceedings.

```
function STM (metadata, CSO, SNC)
    Result: topics, tags
    /* Topic extraction                                          */
    metadata ← extractKeywordsFromText(metadata);
    topics_ini ← inferTopicsFromKeywords(metadata);
    /* Topic selection, via greedy set-covering algorithm        */
    foreach level in KCS do
        topics_in_level ← getTopicsInLevel(topics_ini, level, CSO);
        while count(topics[level]) < limit for level do
            weights ← computePubWeight(metadata, topics[level]);
            /* selects the topic covering the publications with
               highest weight                                    */
            topics[level] ← selectTopic(topics_in_level, weights);
        end
    end
    /* Tag inference                                              */
    tags ← inferTags(topics, CSO, SNC);
    return topics, tags
end
```

Algorithm 1. The STM algorithm

2.2 The STM Approach

The STM approach for generating topics and tags associated to a set of publications consists of three phases:

- *Topic extraction*, in which the metadata of the publications are analysed and each publication is mapped to a list of semantic topics in the CSO ontology;
- *Topic selection*, in which a greedy set-covering algorithm is used to reduce the topics to a user-friendly number, usually 10–20;
- *Tag inference*, in which the selected topics are used to infer a number of SNC tags, using the mapping between CSO ontology and SNC.

Figure 3 illustrates the steps of STM for inferring significant topics. The first panel shows the keywords provided by the authors, the second one shows the set of enriched keywords that include also the keywords extracted from titles and abstracts, the third one shows the output taxonomy.

In the next sections we will discuss the details of each step.

2.2.1 Topic Extraction

In the first step, STM extracts the title, the abstract, the list of keywords and the chapter name from the XML denoting each publication. It analyses the text and extracts frequent keyphrases and the terms that coincide with the topic labels in the CSO ontology. The publication ID is then associated to a set of keywords which include these terms, the original keywords, and optionally some keywords suggested by the editor.

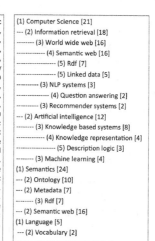

| linked data:3, relational constraints:1, semantical regularizations:1, question answering:1, graph traversal:1, non-aggregation questions:1, implicit information:1, knowledge base completion: 1, dbpedia:1, recommender system:1, relation extraction:1, weakly supervised:1, baidu encyclopedia:1, svm:1, path ranking:1, medical events:1, competitor mining:1, description logics:1, multi-strategy learning: 1, distant supervision:1, relation reasoning: 1, non-standard reasoning services:1, concept similarity measures:1, semantic data:1, medical guidelines:1, rdf:1, prolog:1, preference profile:1, similarity measure:1, ontology development:1, knowledge representation:1, graph simplification:1, rdf visualization:1, semantic web application:1, triple ranking:1, sparql-rank:1, rank-join operator:1, "shaowei" (稍微 'a little'):1, minimal degree adverb:1, a little:1, rdf native storage:1, top-k join:1, news analysis: 1, meta-data extraction:1, database integration:1, , elderly nursing care:1 [...] | semantic:24, rdf:7, applications:5, semantic web:5, knowledge base:4, linked data:4, ontology:4, ontologies:4, language:3, knowledge bases:3, algorithms:2, integration:2, architecture:2, semantics:2, knowledge management:2, query answering:2, recommendation:2, question answering system:2, semantic similarity:2, question answering:2, vocabulary:2, svm:1, graph traversal:1, information needs:1, natural language:1, path ranking:1, baidu encyclopedia:1, non-aggregation questions: 1, support vector machine:1, implicit information:1, construction:1, knowledge base completion:1, classifiers:1, relational constraints:1, semantical regularizations:1, support vector machine (svm):1, machine learning:1, support vector:1, facts:1, logic programming:1, multi-strategy learning:1, distant supervision:1, competitor mining:1, lossy compression:1, comprehensive evaluation:1, relation reasoning:1, websites:1, competition:1, decision support:1, learning algorithm:1, learning process:1 [...] | (1) Computer Science [21]
--- (2) Information retrieval [18]
-------- (3) World wide web [16]
------------ (4) Semantic web [16]
---------------- (5) Rdf [7]
---------------- (5) Linked data [5]
--------- (3) NLP systems [3]
------------ (3) Recommender systems [2]
--- (2) Artificial intelligence [12]
-------- (3) Knowledge based systems [8]
------------ (4) Knowledge representation [4]
---------------- (5) Description logic [3]
------- (3) Machine learning [4]
(1) Semantics [24]
--- (2) Ontology [10]
--- (2) Metadata [7]
------- (3) Rdf [7]
--- (2) Semantic web [16]
(1) Language [5]
--- (2) Vocabulary [2] |

Fig. 3. Example of author keywords, enriched keywords and topics from CSO.

In this phase, the proceedings can also be represented as a distribution of keywords, as shown in the second panel of Fig. 3. However, this representation is usually very noisy: many terms are redundant and consist of different labels for the same topics and the keyword distribution contains a long tail of terms associated with a single paper. The editors who tried STM (see Sect. 3.1) usually considered this representation unfriendly and very time consuming to browse.

For this reason, STM uses the CSO ontology to infer a list of semantic topics from these keywords. To do so it normalizes the terms, by eliminating plural, genitive forms and common affixes and postfixes [8], and then it identifies the terms with the same label as the ontology concepts and associates to each publication tagged with them also all the relevant super areas. For example, a publication associated with the term SPARQL will be tagged with higher-level topics such as RDF, Linked Data, Semantic Web, World Wide Web, and Computer Science. Finally, it generates the topic distribution of all input publications.

The keywords for which it was not possible to find a related concept in the ontology are not included in the topic distribution, unless the user checks the "Include keywords not in the ontology" checkbox in the GUI (see Sect. 2.3).

The drawback of this method is that an erroneous semantic connection in the ontology can sometimes lead to inferring a wrong topic and the error will then be propagated to the higher-level topics. For example, if the ontology were to state that Genetic Algorithms is a sub area of Genetics, the resulting high-level topics may include Biology, even if the proceedings do not address Natural Sciences at all. Although Klink-2 is actually able to infer semantic relationships with very high precision (>90 %, see [6]), incorrect links may still be present. However, the probability of having multiple incorrect links to the same node is quite low and the probability of multiple errors regarding all nodes in the path to the roots is extremely low. We thus addressed this problem by discarding from the topic distribution any research area which is not supported by at least n direct sub-topics. This prevents isolated errors in the lower levels of the ontology from easily propagating to the upper nodes. However, as n grows, the result set becomes smaller, since many high-level topics may be discarded. Hence, in a realistic setting it makes sense to adopt either $n = 1$, equivalent to switching off this functionality, or $n = 2$, potentially sacrificing recall for precision. We labelled this functionality 'robust mode'. Editors preferred $n = 2$ as default, but they also have the option of turning it off in the user interface.

The output of this process is a large set of topics associated with the relevant papers.

2.2.2   Topic Selection

The list of topics returned in the previous step is richer and more human-friendly than the term distribution, but in most cases will still suffer from prolixity, being composed of a very large number of topics. For this reason, we apply a greedy covering-set algorithm with the aim of selecting a smaller set of topics that could be easily handled by Springer Nature editors.

Since we want a comprehensive representation of the corpus given as input, which will include both high level fields and very granular research areas, we run the algorithm separately on the set of topics at each level of the ontology. The level of a topic is computed as 1 + the shortest path from the Computer Science root to the topic in question. For example, high-level fields, such as Human Computer Interaction and Artificial Intelligence, are at level 2 in the ontology, while more specific areas, such as Gesture Recognition and Speech Analysis, are at level 4. The maximum number of topics to be selected at any level depends on the granularity preferred by the user (see Sect. 2.3). The keywords that were not mapped to the ontology are considered in a level of their own.

The greedy covering-set algorithm assigns an initial weight equal to 1 to each paper, and at each iteration selects the topic which covered the publications with the highest total weight and reduces the weight of every covered paper (by a 0.5 factor in the prototype).

We chose this solution because the simplest version of the greedy set-covering algorithm [9], which selects at each iteration the category which covers the largest number of uncovered items, did not work well in this domain. In fact, the proceedings of a conference tend to be related to a number of topics that are often at the intersection of two or more high level topics. For example, in a Semantic Web conference the topics

Artificial Intelligence and Ontology will probably cover a very similar set of publications. Hence, an algorithm that simply selects the topic that covers the larger number of uncovered publications may discard one of them. In addition, when a prominent research area has multiple super topics, the algorithm may exclude all its super topics but one. Our implementation solves this problem, by allowing topics associated to already covered publications to be chosen when they appear in enough papers to be significant.

The output of the set-covering algorithm depends on the maximum number of topics for each level and the robust mode factor, and can be further filtered by defining a minimum number of publications that a topic should cover to be taken into consideration. The user can control these settings by switching the 'granularity level' in the GUI between 1 and 5. Each granularity level is associated to a number of settings that will yield a more succinct or richer topic characterization. A granularity level of 1 will result in very few high level topics, while a granularity of 5 will result in a very long and comprehensive list of the topics in the proceedings.

In some cases, an unusual input, such as a book with few chapters or associated keywords, may produce very few topics when using the normal granularity settings. For this reason, STM uses by default a mechanism for adjusting the settings to the input. It checks that the output meets some minimal requisites in terms of number of topics and number of covered publications and, if this is not the case, it automatically changes the granularity settings and re-runs the topic selection process. This modality can be disabled by changing manually the granularity or deselecting the 'automatic settings' checkbox in the user interface.

The result of the summarization process can either be represented as a plain list or a taxonomy of topics. The second solution makes it easier to understand the context of each topic and why each topic was inferred; it is therefore used as default. In both cases the topics are associated to the number of papers they cover and, optionally, they are annotated with the weight computed by the set-covering algorithm.

2.2.3   Tag Inference

In the final step, STM uses the mapping between the SNC and CSO to infer the SNC tags. It does so by inferring each tag that subsumes one of the selected topics according to the previously discussed mapping. For example, if the Cryptography topic was yielded by the previous step, STM will infer the tags 'I15033 - Data Encryption' (at the third level of SN Computer Science Classification), 'I15009 - Data Structures, Cryptology and Information Theory' (second level) and 'I00001 - Computer Science, general' (root). It then associates to each tag the total number of publications covered by the associated topics, so as to help the editor to assess how representative it is.

2.3   User Interface

Figure 1 shows the user interface of STM. Using the pane on the left, the user can upload the metadata, input some additional keywords and customise different settings, while the pane on the right displays the output of the process.

The interface was iteratively improved according to the feedback of experienced Springer Nature editors. In particular, the editors explained that they need a flexible tool for investigating the proceedings and for producing different kinds of annotations, rather than an automatic pipeline for annotating books. Hence, STM offers two kinds of options: those for investigating the output and those for modifying it according to the editor's needs.

The editors can control the outcome by changing the granularity of topics/tags, the metric used to order them (e.g., number of covered papers, the weight assigned by the set-covering algorithm) and the visualization style (tree list or plain list). The most used setting is the *granularity* value, which goes from 1 to 5 (default is 3) and, as discussed in Sect. 2.2.2, allows users to choose how comprehensive should be the classification. It is mostly used to 'zoom' into the topic taxonomy, especially when the editor suspects that some significant topics may have been left out by the default visualization. In addition, the editors can choose to allow in the classification also frequent keywords that could not be mapped to the CSO ontology. This functionality allows STM to take in consideration also terms outside the Computer Science field or terms that are not strictly research areas, but may be important for assessing the content of a book, such as "commercial applications" or "empirical evaluation". The output becomes noisier, but potentially more informative.

The main tool for exploring a proceedings book and assessing the quality of the classification is the advanced analytics functionality, which shows (1) the title of each paper/chapter, (2) the list of keywords and the percentage of publications which are not covered by the produced classification, and (3) the title and ID of each paper and its associated list of keywords and topics. Figure 4 shows a detail of its output. The list of uncovered terms is particularly useful since it reveals how complete is the representation yielded by STM. The advanced analytics functionality is often used when editors find out that a topic that they would have normally assigned to a conference does not appear in the output. In many occasions, the resulting analysis lead to the discovery that a topic, which used to be prominent, was not so popular anymore in the conference under analysis or that some topics mentioned in the call for papers were almost absent in the proceedings.

Similarly, sometimes editors find some topics in the output that seem inconsistent with their previous experience of the conference and need a way to assess them. For this reason, we included the *show explanation* checkbox, which displays near each topic the full list of terms used to infer it and how many papers they cover. For example, this functionality could show that the topic Semantic Web was inferred because the system found the terms "linked data", "semantic similarity", "RDF" and so on. During the tests conducted in Springer Nature, the editors used often this functionality and in most cases they discovered that the proceedings actually contained a number of terms that suggested the emergence of that topic. They were able to further confirm this intuition by examining the related papers with the *advanced analytics* functionality. In addition, the user can also inspect the full keyword distribution extracted from the text of the proceedings by checking the *show input keyword distribution* checkbox.

Finally, the users can configure more complex settings by means of the 'expert setting' menu, which allows them to switch on and off: (1) the order of the topics

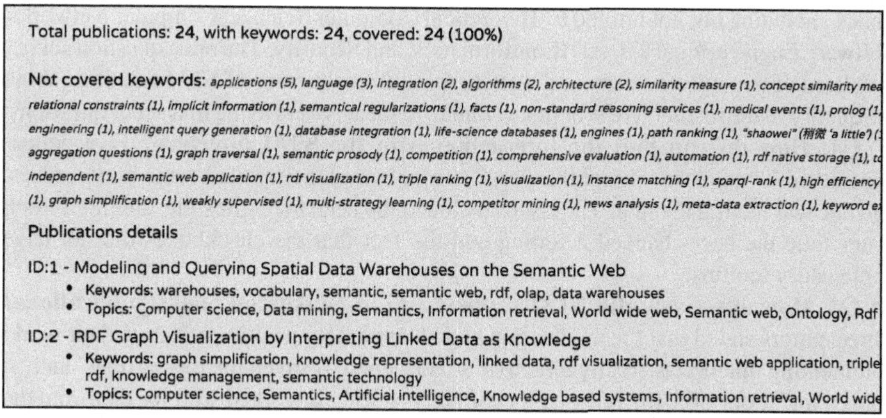

Total publications: **24**, with keywords: **24**, covered: **24 (100%)**

Not covered keywords: applications (5), language (3), integration (2), algorithms (2), architecture (2), similarity measure (1), concept similarity mea relational constraints (1), implicit information (1), semantical regularizations (1), facts (1), non-standard reasoning services (1), medical events (1), prolog (1 engineering (1), intelligent query generation (1), database integration (1), life-science databases (1), engines (1), path ranking (1), "shaowei" (稍微 'a little') (aggregation questions (1), graph traversal (1), semantic prosody (1), competition (1), comprehensive evaluation (1), automation (1), rdf native storage (1), t independent (1), semantic web application (1), rdf visualization (1), triple ranking (1), visualization (1), instance matching (1), sparql-rank (1), high efficiency (1), graph simplification (1), weakly supervised (1), multi-strategy learning (1), competitor mining (1), news analysis (1), meta-data extraction (1), keyword e

Publications details

ID:1 - Modeling and Querying Spatial Data Warehouses on the Semantic Web
- Keywords: warehouses, vocabulary, semantic, semantic web, rdf, data warehouses
- Topics: Computer science, Data mining, Semantics, Information retrieval, World wide web, Semantic web, Ontology, Rdf

ID:2 - RDF Graph Visualization by Interpreting Linked Data as Knowledge
- Keywords: graph simplification, knowledge representation, linked data, rdf visualization, semantic web application, triple rdf, knowledge management, semantic technology
- Topics: Computer science, Semantics, Artificial intelligence, Knowledge based systems, Information retrieval, World wide

Fig. 4. Fragment of the *advanced analytics* section.

according to the ontology level, (2) the text-mining from titles, abstracts and SN metadata, (3) the robust mode, (4) the automating setting, and (5) the suggestion of SNC tags.

3   Evaluation

3.1   User Study

We conducted a qualitative study on STM to assess the quality of its output, the clarity of the explanations, the impact on the editor workflows and the usability of the user interface[12]. To this end, we organized individual sessions with eight experienced SN editors. We introduced STM and its main functionalities for about 15 min and then asked them to use the application for classifying a number of proceedings in their fields of expertize for about 45 min. Every session was recorded to further analyse their interactions with the GUI, as well as their reactions and feedbacks. After the hands-on session the editors filled a three-parts survey about their experience. The first part assessed the editor background and expertise, the second part included 8 open questions, and the third part was a standard SUS questionnaire to assess the usability of the application. A demo version of STM used in this evaluation is available at http://rexplore.kmi.open.ac.uk/STM_demo. The reader can try it by using the 'Example Springer Nature Proceedings' option, which allows testing the application by using six default SN proceedings covering a variety of distinct fields.

On average, the users had 13 years of experience as editors, with seven out of eight of them having at least 5 years. All of them stated to have extensive knowledge of the main topic classifications in their fields and seven an extensive knowledge of Springer Nature Classification. Four of them considered themselves also experts at working with digital proceedings. The expertise of the editors covered a variety of Computer Science

---

[12] The data collected for the evaluation and the publication coverage study are available at http://technologies.kmi.open.ac.uk/rexplore/iswc2016/stm/.

topics, including but not limited to Theoretical Computer Science, Computer Networks, Software Engineering, HCI, AI, Bioinformatics, and Security. The open question survey consisted of five questions about the strengths and weaknesses of STM and three about the quality of the results. We will first summarize the answers to the first set of questions.

Q1. How do you find the interaction with the STM interface? Five editors considered it "good and straightforward" to use, two of them found some minor issues, and one was neutral about it. The issues included the need to re-click the 'submit' button every time the user changed a setting and the fact that the checkboxes did not have explanatory tooltips.

Q2. How effectively did STM support you in classifying books/publications? Three editors stated that the application had an extremely positive effect on their work, commenting that it was "really effective", "very good, it saved me lots of time" and "it helps a lot". Four of them assessed it positively, stating it worked well for them and the result looked correct, and one was neutral. When asked to assess the accuracy of the results the estimates varied between 75 % and 90 %.

Q3. What were the most useful features of STM? The most useful features included: the ability to produce taxonomies of semantic topics (7 editors), the mapping to the SNC tags (5), the ability to explore topics at different granularities (2) and the speed of the analysis (1).

Q4. What are the main weaknesses of STM? The main issues suggested by the editors were: the scope is limited to the Computer Science field (2 editors), the occasional noisy results when examining books with very few chapters/keywords (2), and the wrong capitalization of some topics (1). Two editors also commented that they would like to use STM on the full text of publications, while at the moment the system can only process SN metadata.

Q5. Can you think of any additional features to be included in STM? The suggested features were: being able to produce analytics about the evolution of a venue or a journal in terms of significant topics (4 editors); allowing users to find the most significant proceedings for a certain topic (3); improving the SNC (1); and mapping the topic ontology also to the ACM classification (1).

We mapped the three remaining questions on a 1 to 5 scale, where 1 is the most negative assessment and 5 the most positive. Figure 5 shows the quality of SNC tags and topics, the usability (according to the SUS statement "I thought the system was easy to use"), and the willingness of the users to use the application regularly for their work. These features obtained a similar average score: quality of SNC tags 4.0 ± 0.8, quality of topics 3.9 ± 0.8, usability 4.0 ± 0.5, and willingness to use it regularly 4.0 ± 0.8. Interestingly, the quality of the topics was considered generally higher by the three editors working exclusively with proceedings (4.7 ± 0.6).

The SUS questionnaire confirmed the good opinion of the editors, scoring a 77/100 (the average system scores a 68), which places STM in the 80 % percentile rank in term of usability.

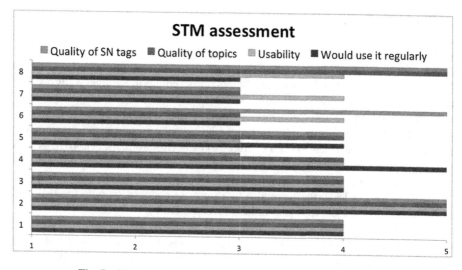

Fig. 5. STM performance according to the editors (labelled 1–8).

3.2 Assessing Publication Coverage

Editors need a topic summarization that is succinct but covers most of the publications. We thus performed a study about how the semantic topics produced by STM compare with keywords in terms of coverage.

To this end, we selected a dataset of 200 SN proceedings and generated for each of them three sets of categories: (1) the keywords defined by the authors, originally associated with each publication, (2) the enriched set of keywords, which also included additional terms extracted from abstracts, titles, and SN metadata (as discussed in Sect. 2.2.1), and (3) the semantic topics produced by STM. Then, for each proceedings book, we computed the average number of papers covered by each member of the first n-th category, using the three sets. We used the average coverage rather than the total coverage, since the latter grows monotonically with the number of descriptors and thus the top level categories (e.g., Computer Science), that often cover most of the papers, would obscure the more fine-grained ones.

Table 1 shows the average result across the proceedings. The topics produced by STM performed significantly better than the enriched keywords (the Wilcoxon test for two correlated distributions yielded $p < 0.0007$), which in turn outperformed the author keywords ($p < 0.0007$). Hence, we can conclude that while extracting keywords from text allows for a more representative set of categories, adding semantic to this representation produces a much more complete set of categories.

Table 1. Average number of papers covered by the first n descriptors.

	1	2	3	4	5	10	20	30
Author keywords	2.83	2.42	2.18	2.04	1.92	1.57	1.33	1.25
Enrich. keywords	8.27	6.95	6.11	5.56	5.14	3.95	2.92	2.44
Topics	25.26	21.08	18.83	17.19	15.93	12.03	8.62	6.93

4 Plans for Large Scale Deployment

While the tool was very well received in Springer Nature, making it part of the daily workflow of the editors requires additional steps. Before outlining these, let us take a closer look at the context of the Computer Science proceedings in Springer Nature. Every year Springer Nature publishes about 1,200 proceedings volumes. Almost 800 of these are published in Computer Science, more specifically in the Lecture Notes in Computer Science (LNCS) series family. This includes LNCS itself, its subseries Lecture Notes in Artificial Intelligence (LNAI) and Lecture Notes in Bioinformatics (LNBI), as well as more recently launched series, such as Lecture Notes in Business Information Processing (LNBIP), and Communications in Computer and Information Science (CCIS). Last but not least, there are also two series in cooperation with the IFIP and ICST/EAI societies (IFIP-AICT and LNICST, respectively).

In order to deploy STM at such a large scale (classifying 800 proceedings/year) we have to connect STM with existing production systems. In an ideal case, STM could receive inputs already from the conference submission system used by the conference. This could happen during the preparation of the material for publication by the conference chairs. In practice, however, the diversity of the submission systems and the lack of commonly accepted standards used beyond the Semantic Web community makes it difficult to expect each of the existing conference management systems to adopt a standard for exporting the data about abstracts, titles, and keywords required by STM. Therefore, we are going to explore the integration of STM with the Springer Nature's own submission system, OCS (Online Conference Service[13]).

For the proceedings not using OCS, the data required for STM will be provided during the production process, after the metadata about individual papers have been finalized and before the proceedings are published. Depending on how STM performs and the overall metadata strategy of Springer Nature, STM might be used for annotating already published contents (roughly 10,000 proceedings volumes). In combination with the data already available at the Springer LOD portal,[14] this would allow editors to analyse the evolution of conference topics.

We are also looking into how STM could be used to improve the existing Springer Nature Classification. One possible way of approaching this problem would be to set up periodic updates of the SNC based on the most recently published material. During such updates new categories could be added, corresponding to the emerging topics, while the categories corresponding to disappearing topics could be deprecated.

Finally, we also plan to expand the scope of the research area ontology to fields other than Computer Science, to support the classification of books and proceedings in other domains as well.

---

[13] https://ocs.springer.com/ocs/.
[14] http://lod.springer.com.

5   Related Work

STM identifies research areas from a corpus of metadata by using an automatically generated ontology of topics. In this regard, it can be considered a classic name-entity linking approach. In particular, many historical approaches focus on linking entities to general knowledge bases, such as Wikipedia or DBpedia. For example, Mihalcea and Csomai [10] and Bunescu and Pasca [11], introduced some of the first approaches for mapping text to Wikipedia pages. Since then, we saw the creation of a number of systems for name-entity linking which exploited DBpedia or Wikipedia, including DBPedia Spotlight [12], Microsoft Entity Linking[15], BabelFly [13], Illinois Wikifier [14], KORE [15], AGDISTIS [8] and many others. DBpedia Spotlight is also used by the Klink-2 [6], the algorithm which generated the CSO ontology, for linking keywords to DBpedia entities and informing the identification of semantic relationships between research topics. However, using directly DBpedia as source for research areas presents some issues, since the research fields taxonomy in DBpedia is quite coarse-grained: it does not contain some of the most recent or specific topics and lacks a number of links between them. Another alternative is the Machine Aided Indexer[16], a rule-based document indexer that can map the full text of a document to a taxonomy of topics. However, this method requires the manual definition of a number of rules for the mapping.

Similarly to STM, a number of methods for topic detection extract topics from a corpus of documents. The best-known technique is the Latent Dirichlet Allocation (LDA) [16], which considers each document as a distribution of topics and each topic as a distribution over words. This approach is applicable to any kind of documents and has been influential in the topic detection community in the last decade. For this reason, we saw the emergence of a number of solutions tailored to the scholarly domain. For example, He et al. [17] introduced an approach which makes use of the citation graph while the Author-Conference-Topic (ACT) model used by AMiner [18] exploits also information about authors, conferences and journals. However, LDA and similar methods are a good solution mainly in scenarios where a very large numbers of documents need to be classified, there is no good domain knowledge, fuzzy classification is acceptable and it is not important for users to understand the rationale of a classification or customise the output. None of these tenets apply to our case.

A number of digital libraries (e.g., ACM, Springer Nature, Scopus[17]) and academic search engines (e.g., Microsoft Academic Search[18]) rely on taxonomies of topics for supporting the classification of research publications. STM uses a similar solution by adopting the CSO ontology. Indeed, ontologies of research topics have proved to be very useful to enrich semantically a number of analytics models [7], as well as supporting trend detection [19] and community detection [20, 21].

---

[15] https://www.microsoft.com/cognitive-services/en-us/entity-linking-intelligence-service.
[16] http://www.dataharmony.com/services-view/mai/.
[17] https://www.elsevier.com/solutions/scopus.
[18] http://academic.research.microsoft.com/.

6   Conclusions

In this paper, we have presented Smart Topic Miner, a novel Semantic Web application designed to assist Springer Nature editors in classifying conference proceedings. The evaluation, performed with a number of experienced Springer Nature editors, showed that STM produces accurate and useful results. In particular, the semantic model on which STM builds was considered very helpful since it allows the editors to obtain a more concise representation, which can be easily analysed. A key lesson learned during the STM development regards the critical value of producing human-friendly explanations and the value of an explicit semantic representation for supporting this task.

We are planning to integrate the STM tool into Springer Nature workflows, in particular those used for publishing Computer Science proceedings. The use of such controlled topic vocabulary will improve discoverability and navigation of the contents of Springer Nature proceedings, as well as enable new applications. In addition, STM could be extended to indicate the emergence of new topics, as well as the fading of some traditional ones. Finally, we also plan to explore the possibility of using STM for directly supporting authors in defining the set of topics which best describe their paper.

Acknowledgements.   We would like to thank the Springer Nature editors for assisting us in the evaluation of STM.

References

1. Möller, K., Heath, T., Handschuh, S., Domingue, J.: Recipes for semantic web dog food – the ESWC and ISWC metadata projects. In: Aberer, K., et al. (eds.) ASWC 2007 and ISWC 2007. LNCS, vol. 4825, pp. 802–815. Springer, Heidelberg (2007)
2. Latif, A., Afzal, M.T., Helic, D., Tochtermann, K., Maurer, H.A.: Discovery and construction of authors' profile from linked data (A case study for Open Digital Journal). In: LDOW 2010 (2010)
3. Glaser, H., Millard, I.: Knowledge-enabled research support: RKBExplorer.com. In: Proceedings of Web Science (2009)
4. Bryl, V., Birukou, A., Eckert, K., Kessler, M.: What's in the proceedings? Combining publisher's and researcher's perspectives. In: SePublica 2014 (2014)
5. Hammond, T., Pasin, M.: The nature.com ontologies portal. In: 5th Workshop on Linked Science 2015, Colocated with International Semantic Web Conference 2015, Bethlehem, USA (2015)
6. Osborne, F., Motta, E.M.: Klink-2: integrating multiple web sources to generate semantic topic networks. In: Arenas, M., et al. (eds.) ISWC 2015. LNCS, vol. 9366, pp. 375–391. Springer, Heidelberg (2015). doi:10.1007/978-3-319-25007-6_22
7. Osborne, F., Motta, E., Mulholland, P.: Exploring scholarly data with rexplore. In: Alani, H., et al. (eds.) ISWC 2013, Part I. LNCS, vol. 8218, pp. 460–477. Springer, Heidelberg (2013)
8. Usbeck, R., Ngonga Ngomo, A.-C., Röder, M., Gerber, D., Coelho, S.A., Auer, S., Both, A.: AGDISTIS - graph-based disambiguation of named entities using linked data. In: Mika, P., et al. (eds.) ISWC 2014, Part I. LNCS, vol. 8796, pp. 457–471. Springer, Heidelberg (2014)
9. Chvatal, V.: A greedy heuristic for the set-covering problem. Math. Oper. Res. **4**, 233–235 (1979)

10. Mihalcea, R., Csomai, A.: Wikify!: linking documents to encyclopedic knowledge. In: Proceedings of the Sixteenth ACM Conference on Information and Knowledge Management, pp. 233–242. ACM, New York (2007)

11. Bunescu, R.C., Pasca, M.: Using encyclopedic knowledge for named entity disambiguation. In: EACL, vol. 6, pp. 9–16 (2006)

12. Mendes, P.N., Jakob, M., García-Silva, A., Bizer, C.: DBpedia spotlight: shedding light on the web of documents. In: Proceedings of the 7th International Conference on Semantic Systems, pp. 1–8. ACM, New York (2011)

13. Moro, A., Raganato, A., Navigli, R.: Entity linking meets word sense disambiguation: a unified approach. Trans. Assoc. Comput. Linguist. **2**, 231–244 (2014)

14. Cheng, X., Roth, D.: Relational inference for wikification. Urbana **51**, 61801 (2013)

15. Hoffart, J., Seufert, S., Nguyen, D.B., Theobald, M., Weikum, G.: KORE: keyphrase overlap relatedness for entity disambiguation. In: Proceedings of the 21st ACM International Conference on Information and Knowledge Management, pp. 545–554. ACM, New York (2012)

16. Blei, D.M., Ng, A.Y., Jordan, M.I.: Latent dirichlet allocation. J. Mach. Learn. Res. **3**, 993–1022 (2003)

17. He, Q., Chen, B., Pei, J., Qiu, B., Mitra, P., Giles, L.: Detecting topic evolution in scientific literature: how can citations help? In: Proceedings of the 18th ACM Conference on Information and Knowledge Management, pp. 957–966. ACM, New York (2009)

18. Tang, J., Zhang, J., Yao, L., Li, J., Zhang, L., Su, Z.: Arnetminer: extraction and mining of academic social networks. In: Proceedings of the 14th ACM SIGKDD International Conference on Knowledge Discovery and Data Mining, pp. 990–998. ACM, New York (2008)

19. Decker, S.L., Aleman-Meza, B., Cameron, D., Arpinar, I.B.: Detection of bursty and emerging trends towards identification of researchers at the early stage of trends (2007). http://athenaeum.libs.uga.edu/handle/10724/9958

20. Erétéo, G., Gandon, F., Buffa, M.: SemtagP: semantic community detection in folksonomies. In: Proceedings of the 2011 IEEE/WIC/ACM International Conferences on Web Intelligence and Intelligent Agent Technology (2011)

21. Osborne, F., Scavo, G., Motta, E.: Identifying diachronic topic-based research communities by clustering shared research trajectories. In: Presutti, V., d'Amato, C., Gandon, F., d'Aquin, M., Staab, S., Tordai, A. (eds.) ESWC 2014. LNCS, vol. 8465, pp. 114–129. Springer, Heidelberg (2014)

Semantic Technologies for Data Analysis in Health Care

Robert Piro[1]([✉]), Yavor Nenov[1], Boris Motik[1], Ian Horrocks[1], Peter Hendler[3],
Scott Kimberly[2], and Michael Rossman[2]

[1] University of Oxford, Oxford, UK
robert.piro@cs.ox.ac.uk
[2] Kaiser Permanente, Oakland, USA
[3] IHTSDO, Copenhagen, Denmark

Abstract. A fruitful application of Semantic Technologies in the field of healthcare data analysis has emerged from the collaboration between Oxford and Kaiser Permanente a US healthcare provider (HMO). US HMOs have to annually deliver measurement results on their quality of care to US authorities. One of these sets of measurements is defined in a specification called HEDIS which is infamous amongst data analysts for its complexity. Traditional solutions with either SAS-programs or SQL-queries lead to involved solutions whose maintenance and validation is difficult and binds considerable amount of resources. In this paper we present the project in which we have applied Semantic Technologies to compute the most difficult part of the HEDIS measures. We show that we arrive at a clean, structured and legible encoding of HEDIS in the rule language of the RDF-triple store RDFox. We use RDFox's reasoning capabilities and SPARQL queries to compute and extract the results. The results of a whole Kaiser Permanente regional branch could be computed in competitive time by RDFox on readily available commodity hardware. Further development and deployment of the project results are envisaged in Kaiser Permanente.

1 Introduction

Modern healthcare critically depends on data analysis, particularly in the context of quality assurance. In the US, the National Committee for Quality Assurance (NCQA)[1] specifies a wide range of quality measures; these include, e.g., the proportion of diabetic patients having regular eye examinations, because diabetes can cause retinal damage and eventually blindness. Health Maintenance Organisations (HMOs) are required to demonstrate satisfactory performance w.r.t. NCQA measures if they wish to participate in government funded healthcare schemes such as Medicare that cover more than 48 million patients in the US and represent a substantial share of the healthcare market.

Relevant quality measures can depend on many factors, and their computation may require complex analysis of the data. Moreover, data may be derived

---

[1] http://ncqa.org/.

P. Groth et al. (Eds.): ISWC 2016, Part II, LNCS 9982, pp. 400–417, 2016.
DOI: 10.1007/978-3-319-46547-0_34

from multiple sources and have heterogeneous structure. Currently, a combination of SAS programs and SQL queries is used to compute quality measures. The resulting software systems are complex, inefficient, and difficult to validate and maintain—a critical issue given that quality measures are regularly revised and augmented.

Semantic technologies offer a possible solution to this problem: RDF can be used to integrate data from heterogeneous sources, ontologies can provide flexible and adaptable schemas, and declarative rules can be used to capture relevant quality measures. A triple store could then be used to apply the rules to the data, with SPARQL queries[2] being used to return the results.

To test this hypothesis, the Knowledge Representation and Reasoning (KRR) group at the University of Oxford, together with the US HMO Kaiser Permanente[3] (KP), undertook a joint project in which they used declarative rules to capture a particularly complex set of quality measures relating to diabetes care, and used these rules with the RDFox [5] triple store in order to compute the corresponding quality measures for the 466,000 patients in KP's Georgia region. The results were extremely encouraging: firstly, only 174 rules were required, compared to the roughly 3,000 lines of complex and hard to maintain SQL code of their previously used solution, which has since been replaced by a vendor product. Secondly, RDFox was easily able to handle the relevant patient data (which amounted to approximately 1.6 billion triples), and computed the quality measures via application of the rules in approximately 30 min. The KP data analyst in charge of quality assurance confirmed that this was fast in comparison to their existing solution, and was also impressed with the small number of iteration cycles needed to check the correctness of our results—a consequence of the relative legibility of the declarative rules.

2   Motivation

The NCQA maintains and publishes the Healthcare Effectiveness Data and Information Set (HEDIS)[4], which uses (relatively) precise natural language to define sets of measures concerning the performance of HMOs in areas such as cancer screening, immunisation and Comprehensive Diabetes Care (CDC). The measures are usually expressed as a percentage of a population of interest and are designed to facilitate performance comparisons across multiple HMOs. In the case of CDC, the quality measures concern diabetic patients in the age range of 18 to 75; one measure, for example, is the percentage of the patients who received an eye exam during the relevant reporting period.

To compute the quality measures, the data first needs to be aggregated from various patient data systems. This typically involves the invention of one or more ad hoc schemas into which the data is cast. Such schemas are designed to facilitate analysis rather than to accurately model the domain, and so they are

---

[2] http://www.w3.org/TR/sparql11-query/.

[3] http://www.kaiserpermanente.org.

[4] http://www.ncqa.org/HEDISQualityMeasurement.aspx.

difficult to maintain and are prone to inconsistent interpretation by the members of the data analysis team.

Computation of NCQA measures over the aggregated data is typically done via SAS-programs or a sequence of SQL-queries. This process is also complex and error prone; for example, as already mentioned, computation of the CDC measures uses roughly 3,000 lines of SQL code. As a result, existing systems are costly, unreliable, and difficult to maintain.

2.1   Overview of Project

The aim of the project with Kaiser Permanente was to evaluate the effectiveness of Semantic Technologies for computing NCQA quality measures. The power of Semantic Technologies lies in the clearly defined declarative formalisms with which complex relationships can be expressed. One such formalism are Datalog-like rule languages that are supported by many triple stores and that can be used, e.g., to perform OWL 2 RL reasoning [4]. Rules can express relationships via intuitive if-then-statements such as

```
[?Pat, aux:countedFor, aux:measureEyeExam] :-
          [?Pat, rdf:type, aux:diabeticPatient], [?Pat, aux:has, aux:EyeExam] .
```

which says that '*if* the patient has diabetes and an eye exam, *then* the patient is to be counted for the measure Eye Exam'. These statements are succinct and relatively close to natural language.

In this project we used the RDFox triple store with its RDFox-Datalog rule language. Our goal was to investigate whether RDFox-Datalog is expressive enough to encode HEDIS specifications, if RDFox could handle datasets of the necessary size and provide competitive performance w.r.t. existing solutions, and if the resulting semantic technology solution could overcome some of the shortcomings of existing solutions. We decided to implement HEDIS CDC, as its variety of logical connexions between the data is rich, and it is particularly difficult to implement with traditional methods such as SQL and SAS; HEDIS CDC therefore makes an impressive use case for data analysts who conduct HEDIS measurements in HMOs.

The project is split into three tasks. The first task is to create a coherent and extensible data model into which the relevant patient data can be transformed. We used a data model that is close to human conceptualisation as this makes the data easier to understand. The data model makes it also easier to develop and maintain the rules that capture HEDIS measures (see Sect. 3). The second task is the development of such rules. We implemented the HEDIS CDC specification as a Datalog ontology (rule set), but had to augment the rules with SPARQL-queries to fully compute the measures (see Sect. 4). Finally, the approach was evaluated on data provided by the Kaiser Permanente Georgia region. We translated this data into RDF-triples according to our data model, computed the HEDIS CDC measures using RDFox, and compared our results with those computed using existing systems (see Sect. 5).

3   Healthcare Data Modelling in OWL

In this section we describe the conceptual model that we developed to describe the clinical and administrative data in KP. When designing our model we tried to satisfy the following three requirements. Firstly, the model had to be as close as possible to domain expert conceptualisation so as to facilitate the faithful representation of domain knowledge. Secondly, the model had to be sufficiently flexible to uniformly capture the diverse business processes that take place in a typical healthcare organisation. Thirdly, the model had to be readily amenable to Semantic Web technologies. We identified a healthcare data modelling standard from the field of healthcare informatics, called HL7 RIM, that satisfies the first two requirements. To satisfy the third requirement, we used the methodology behind the HL7 RIM to build a conceptual model in OWL. In the following two sections we give a short overview of the HL7 RIM standard and a description of how it was used to build an OWL ontology describing our data.

3.1   The HMO Data Model

Healthcare data modelling is an important topic in the field of healthcare informatics, and a number of standards have been developed to facilitate the exchange of clinical and administrative data between HMOs and other third party organisations. One such standard is the Reference Information Model (RIM)[5], which is issued by the international organisation for standardisation Health Level Seven International (HL7)[6].

Fig. 1. Core concepts in the RIM, using the standard RIM colour scheme: Act: red, Participation: blue, Role: yellow, Entity: green (Color figure online)

The RIM standard models a wide range of healthcare business processes, including clinical processes, such as clinical visits and laboratory tests, as well as administrative processes of HMOs, such as patient enrolment and insurance plan authoring. All business processes in the RIM are uniformly represented using the notions *entity*, *role*, *participation*, and *act*. Each act is characterised by the participation of entities each of which fulfils a particular role (see Fig. 1). Acts are used to represent business processes, and participations are used to describe the different parties involved in an act, such as the performer of an act and the subject of an act. Entities are used to describe physical things, such as persons and organisations, while roles describe the different competencies of entities, such as employee and patient, in the case of a person, and insurer, in

---

[5] http://www.hl7.org/implement/standards/rim.cfm.
[6] http://www.hl7.org/.

the case of an organisation. For further details on the RIM standard, please refer
to [2,9].

The scope of the RIM data model far exceeds the needs of this project, so we
used its underlying design principles to build a simplified data model that better
suits our needs. We modelled the different types of entities, roles, participations
and acts as OWL classes that are specialisations of the classes Entity, Role,
Participation, and Act, respectively. Similarly, we modelled the relationships
between these notions as object properties whose domains and ranges are as
specified in Fig. 1.

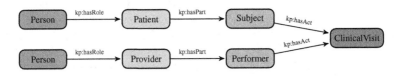

Fig. 2. The model of a clinical visit

Consider for example the part of our conceptual model depicted in Fig. 2,
which describes the clinical visits of a patient to their health care provider. The
clinical visit is modelled as the class ClinicalVisit, which is a subclass of Act.
Similarly, the two entities involved in a clinical visit are modelled as members
of the class Person, which is a subclass of Entity. One person in the role of a
Provider participates in the clinical visit as a Performer, while the other person
in the role of a Patient participates in the clinical visit as a Subject.

Our model also describes properties relevant to clinical visits, such as diagnoses and clinical procedures. In healthcare informatics these concepts are represented by *codes* from standard vocabularies such as ICD-9 [7], which describes
diagnosis, CPT, which describes clinical procedures, and SNOMED-CT, which
describes clinical terms in general. For example, ICD-9 assigns the code 250.60
to the diagnosis 'diabetes with neurological manifestations', and the code 250.70
to the diagnosis 'diabetes with peripheral circulatory disorders'. We model the
ICD-9 concepts using the class ICD9Term, and we connect its instances to the
ClinicalVisits in which they occur using the object property kp:hasDiag (see
Fig. 3). In healthcare informatics, broader clinical concepts are often modelled
as collections of codes, which are commonly referred to as *value sets*. For example, HEDIS defines the term 'Diabetes Diagnosis' as a value set that contains
amongst others the ICD-9 codes 250.60 and 250.70. We model value sets using the
class ValueSet, and the associations between codes and value sets are realised
using the object property kp:hasValueSet. Hence, in our model, the instance
of ICD9Term representing the ICD-9 code 250.60 is connected via the object
property kp:hasValueSet to the instance of ValueSet that represents 'Diabetes
Diagnosis'.

Finally, for each class in the model we introduce datatype properties that are
used to specify relevant values. For example, every Person has a specified name,

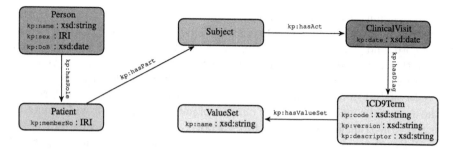

Fig. 3. Extended upper row of Fig. 2 showing how we capture health record data according to the developed schema

sex, and date of birth, every Patient has a member number with the HMO, every ClinicalVisit has a date, every ICD-9 term has a code, a version and a description, and every value set has a name (see Fig. 3).

3.2 Translating KP Data into RDF

We shall refer to the data provided by the Kaiser Permanente regional branch as *raw data*. This is the same data that serves as input to the NCQA approved vendor product, so it was already appropriately aggregated and curated. The data is obtained from KP's internal relational database and consists of delimited text files each of which represents a relational table. For example the file that stores clinical visits looks as follows.

```
VID   MBR    SERV-DT         DIAG-1        DIAG-22   PROVNBR
101   M4711  2013-09-10 ...  250.70  ...   NULL      P8736
```

Each line in this file specifies the visit ID as primary key, the patient's member number, the service date, up to twenty two ICD-9 diagnosis codes, and finally a provider number.

Before translating KP's raw data into RDF-triples, we first had to establish a naming scheme that assigns IRIs to the different objects participating in our model. We chose a naming scheme that allows us to easily map IRIs to the objects in the raw data that they represent. To this end we used IRIs that capture both the types and the identities of the encoded object. This was particularly important when we had to correct formatting errors in our translation, and in the recapitulation stage of the project in which we had to justify our results with the raw data.

For most types of objects the assignment of IRIs was relatively straightforward. For example, we encoded the patient with member number 'M4711' using the IRI <http://www.kp.org/Patient/M4711> and the provider with provider number 'P8736' using the IRI <http://www.kp.org/Provider/P8736>. Similarly, we encoded the ICD-9 code '250.70' using the IRI <http://www.kp.org/ICD9Term/250.70>. Note that, as mentioned above, each IRI captures both the type and the identity of the encoded object.

The assignment of IRIs to clinical visits was slightly more involved. In the initial translation we assumed that each clinical visit is described in a single database record, and thus we used the primary key VID of the record to identify each visit. In the recapitulation stage of the project, however, it became clear that this assumption was wrong, as clinical visit may reside in multiple database records. The data analyst in KP clarified that the identity of a visit is uniquely determined by the date of the visit, the member number of patient, and the provider number. To correct the translation, we therefore encoded a clinical visit using IRIs of the form <http://www.kp.org/Visit/UID>, where UID encodes the slash-separated values of the date, the member number, and the provider number. So, for example, for the clinical visit listed in the record above, UID is equal to 2013-09-10/M4711/P8736. Finally, since there is a one-to-one correspondence between a visit and its subject and a visit and its performer, we use a visit's UID in the IRIs of its subject and performer. Hence, the subject and the performer of the visit in our example are encoded using the IRIs <http://www.kp.org/Subject/UID> and <http://www.kp.org/Performer/UID>, respectively.

Having assigned IRIs to the different objects, we can now easily translate each record of a clinical visit into RDF-triples by simply referring to the data model described in Fig. 3. Some of the triples encoding the clinical visit record in the previous example are given below.

```
<http://www.kp.org/Patient/M4711>rdf:type<http://www.kp.org/Patient>.
<http://www.kp.org/Patient/M4711>kp:hasPart<http://www.kp.org/Subject/UID>.
<http://www.kp.org/Subject/UID>kp:hasAct<http://www.kp.org/Visit/UID>.
<http://www.kp.org/Visit/UID>kp:date"2013-09-10"^^xsd:date.
<http://www.kp.org/ICD9Term/250.70>rdf:type<http://www.kp.org/ICD9Term>.
<http://www.kp.org/Visit/UID>kp:hasDiag<http://www.kp.org/ICD9Term/250.70>.
               . . .
```

Since the translation of each clinical visit for a given patient uses the same patient IRI, the patient's entire medical history is connected in a contiguous RDF-graph.

Note that our choice of naming scheme allows us to translate each database record independently of other records. As a result, in addition to being relatively simple, the translation could also be easily executed concurrently, as it maintains no global state. Observe, however, that the record-by-record translation results in repetition of triples. For example, the triple asserting that <http://www.kp.org/Patient/M4711> is a member of the class <http://www.kp.org/Patient> will be generated once for every database record that mentions that patient. Similarly, there may be repetitions involving each provider and each diagnosis. As we will see in Sect. 5, this redundancy increases the number of triples by a factor of 5.5.

Finally, we also had to add to our RDF-graph triples related to the HEDIS specification. Firstly, as discussed in the previous section, the HEDIS specification defines a number of value sets. The membership of codes to value sets is naturally encoded using assertions for the object property kp:hasValueSet. Additionally, the HEDIS specification refers to the begin and end dates of the current measurement year, which in our case is the year 2013, as well as to the measurement period, which in our case consists of the years 2012 and 2013. Instead of explicitly referring to these dates and years in our rules, we exploit the following triple encoding of the relevant information.

```
kp:HEDIS kp:measuredPeriod 2012.   kp:HEDIS kp:beginDate "2013-01-01"^^xsd:date.
kp:HEDIS kp:measuredPeriod 2013.   kp:HEDIS kp:endDate "2013-12-31"^^xsd:date.
```

4 Encoding HEDIS CDC and Its Challenges

This section describes how we encoded the HEDIS specification in RDFox-Datalog. The resulting RDFox-Datalog ontology defines the different patient classes stipulated by the HEDIS specification, e.g. 'patient with eye exam'; we then use RDFox to compute class membership for all patients. Simple SPARQL counting queries determine the number of patients in each class. These numbers are used to calculate the percentage of the population of interest, which is then reported to the NCQA.

As we show in the following, capturing the HEDIS specification involved the use of recursive datalog rules, and hence went beyond what could be achieved via SPARQL query answering alone. We also needed value manipulation, stratified negation, and stratified aggregation, which are not commonly supported reasoning features. In standard materialisation-based triple stores, these features can be simulated by iteratively answering full SPARQL queries, adding the query results to the store, and applying the rules with respect to the enriched data. Since RDFox supports BIND and FILTER constructs in rule bodies, we had to simulate only stratified negation and aggregation.

4.1 Encoding Basic Concepts

HEDIS CDC is specified using natural language. The following extract (which we will refer to as *extract 1*) is drawn from the chapter that defines which patients are diabetic.

> [Diabetics are those patients] who met any of the following criteria during the measurement year [2013] or the year prior to the measurement year [2012] (count services that occur over both years):
> – At least two outpatient visits (Outpatient Value Set), observation visits (Observation Value Set) or nonacute inpatient visits (Nonacute Inpatient Value Set) on different dates of service, with a diagnosis of diabetes (Diabetes Value Set). Visit types need not be the same for the two visits.
> – ...

We first encode some basic concepts, starting with the notion of a *diabetes diagnosis*. Extract 1 specifies that a clinical visit has a diabetes diagnosis if it has a code in the value set named "Diabetes". The following rule classifies such clinical visits by deriving a triple of the form [?CV, rdf:type, aux:diabetesDiagnosis] where ?CV is an instance of ClinicalVisit and the prefix aux indicates a derived property or class.

```
[?CV, rdf:type, aux:diabetesDiagnosis] :-[?CV, kp:hasDiag, ?ICD9],
                [?ICD9, kp:hasValueSet, ?VS],[?VS, kp:name, "Diabetes"] .
```

We add similar rules to classify outpatient visits, observation visits and non-acute inpatient visits, which derive triples of the form [?CV, rdf:type, aux:outpatient], etc.

We can now associate each patient with their "admissible visits" using triples of the form [?Pat, aux:admissibleVisit, ?CV]. According to Extract 1, a patient's visit counts as admissible if it has a diabetes diagnosis and is also either an outpatient, a non-acute inpatient, or an observation visit. For outpatient visits we thus use the following rule:

```
[?Pat, aux:admissibleVisit, ?CV] :-[?Pat, aux:patientHasAct, ?CV],
        [?CV, rdf:type, aux:outpatient], [?CV, rdf:type, aux:diabetesDiagnosis] .
```

For non-acute inpatient visits and the observation visits we use analogous rules.

Abstractions such as aux:admissibleVisit help to keep subsequent rules shorter and more easily legible. The declarative nature of Datalog allows to introduce such abstractions without the explicit creation of tables. The flexible RDF-schema is simply extended and no data needs to be copied into the new schema.

Finally, we need to ensure that there are at least two admissible visits on different dates in the relevant measurement period. We achieve this with SPARQL BIND and FILTER constructs which RDFox supports in rules bodies. Note that these features can also be simulated by interrupting the reasoning process and computing the relevant values using SPARQL queries, as in the case of stratified negation (see Sect. 4.3).

```
[?Pat, rdf:type, aux:diabeticPatient]:-
        [?Pat, aux:admissibleVisit, ?CV0], [?Pat, aux:admissibleVisit, ?CV1],
        [?CV0, kp:date, ?date0],              [?CV1, kp:date, ?date1],
        BIND( YEAR(?date0) AS ?y0 ),          BIND( YEAR(?date1) AS ?y1 ),
        [kp:HEDIS, kp:measurePeriod, ?y0], [kp:HEDIS, kp:measurePeriod, ?y1]
        FILTER ( ?date0 != ?date1 ) .
```

This rule says that a patient is a diabetic patient if they had two admissible visits in the years ?y0 and ?y1 (computed using BIND), each of which is either the measurement year or the year prior to that, and that the visits occurred on different dates (established using FILTER). The measurement year and the year prior to that are retrieved from the dataset as described in Sect. 3.2. Note that this rule is also non-treeshaped, and cannot be expressed in OWL 2 or its fragments. The non-treeshapedness is unavoidable since we need to compare for each patient the dates of each pair of admissible visits.

4.2   Recursion

We were able to encode all notions discussed so far by using only non-recursive Datalog rules, which means that we could also compute these notions using (large and complex) SPARQL queries. However, as we show next, HEDIS CDC also contains notions that require genuine recursion, and thus cannot be computed using SPARQL queries alone.

A period in which a patient is insured with a HMO is called an enrolment. Patients often have multiple consecutive enrolments within the measurement year, which is due to changes in circumstances such as retirement, change of workplace or switching between health plan packages. A patient may also have gaps in their enrolment history, because they switched HMOs or were uninsured. However, the NCQA requirements on HMOs apply only to patients who have a *continuous enrolment* with the HMO which, according to the HEDIS CDC specification, is when they have:

> no more than one gap in enrolment of up to 45 days during the measurement year. [... Patients must be insured with the HMO on] December 31 of the measurement year.

To determine whether a patient has a continuous enrolment we proceed as follows. First, we identify as *connected* all enrolment acts that are connected to the end date of the measurement period via a sequence of enrolment acts without any gaps. Second, we identify as *gap-connected* all enrolment acts that are connected to the end date of a measurement period via a sequence of enrolment acts with one gap. Finally, we identify that a patient has a continuous enrolment if they have a (gap-)connected enrolment act containing the begin date of the measurement period.

The notion of connected enrolment act has the following recursive definition. An enrolment act is connected (1) if its period contains the end date of the measurement year, or (2) if it is directly succeeded by a connected enrolment. Case (1) of the definition is handled by the following rule.

```
[?Enr, rdf:type, aux:connEnr] :-
            [kp:HEDIS, kp:endDate, ?anchor], [?Pat, aux:hasEnr, ?Enr],
            [?Enr, kp:beginDT, ?dateB],       [?Enr, kp:endDT, ?dateE],
            FILTER ( ?dateB <= ?anchor && ?dateE >= ?anchor ) .
```

For the recursive case (2), we need to identify the connecting successor. This involves date manipulation because we have to compute the date of the previous day. Unfortunately, SPARQL BIND does not provide arithmetic on the data type xsd:date and instead we had to compute the previous day during data translation which was stored using the data value property kp:beginDT-1.

```
[?Enr, rdf:type, aux:connEnr]:-
            [?Pat, aux:hasEnr, ?Enr],    [?Pat, aux:hasEnr, ?SuccEnr],
            [?Enr, kp:beginDT, ?dateB], [?SuccEnr, rdf:type, aux:connEnr],
            [?Enr, kp:endDT, ?dateE],    [?SuccEnr, kp:beginDT-1, ?prev],
            FILTER ( ?dateB <= ?prev && ?dateE >= ?prev ) .
```

The notion of gap-connected enrolment can again be defined recursively. An enrolment act is gap-connected (1) if it has a gap of at most 45 days to a connected enrolment act, or (2) if it is directly succeeded by a gap-connected enrolment. Similarly to before, during our translation we precomputed the date that is 46 days earlier than the start date of an enrolment act and stored it using the data property kp:beginDT-46.

Note that we compare all pairs of enrolments of each patient, which is quadratic in the number of patient's enrolments. To reduce the workload, we restricted aux:connEnr to enrolments whose period intersects the measurement year. We measured the outdegree of aux:connEnr, which was maximally 6 and thus manageable.

Finally, we determine if a patient was continuously enrolled using two simple rules that identify all patients having a connected or a gap-connected enrolment act whose period contains the begin date of the measurement period. We are thus able to encode this HEDIS section using just 6 recursive Datalog rules. This compares to 500 lines of heavily commented and involved SQL-code previously used by the Kaiser Permanente.

4.3   Stratified Negation

Stratified negation is a feature that is not commonly supported by RDF-triple stores but that can be very useful when conclusions need to be drawn based on the lack of some information. We next give an example of a HEDIS CDC measure whose computation requires negation, and we describe how it was computed before RDFox was extended to handle stratified negation.

HbA1c is a special type of haemoglobin, whose level is used as an indicator for average blood glucose levels over three months and whose healthy level is below 7 %. HMOs are required to pursue good levels of HbA1c, but only in patients without severe health issues, such as by-pass operations, etc. Concretely, the measure for HbA1c control is computed as $\frac{\text{\#patients in HbA1c denom. with HbA1c}<\%7}{\text{\#patients in the HbA1c denominator}}$, where the HbA1c denominator contains those patients in the population of interest that have no exclusions. For example, HEDIS CDC states in the definition of the HbA1c denominator:

> Exclude members [from the pop. of interest] who meet any of the following criteria:
> - IVD [Ischemic Vascular Disease]. Members who met at least one of the following criteria during both the measurement year and the year prior to the measurement year. Criteria need not be the same across both years.
>   - At least one outpatient visit (Outpatient Value Set) with an IVD diagnosis (IVD Value Set).
>   - At least one acute inpatient encounter (Acute Inpatient Value Set) with an IVD diagnosis (IVD Value Set).

From what we have seen earlier, it is not difficult to imagine how to write rules which identify patients with IVD and with other excluded properties. All final rules computing the excluded patients have the head [?Pat, rdf:type, aux:HasExclusion] and thus mark a patient excluded from the HbA1c denominator for the respective reporting year. Yet, computing the HbA1c denominator requires selecting all patients from the population of interest who do *not* have an exclusion and thus it requires negation.

Stratified negation [1,6] is a well established extension of recursive Data-log and is sufficient for our purposes. However, RDFox did not support strat-ified negation at the time, so we applied a well-known work-around [3,8] that uses the FILTER NOT EXISTS construct in SPARQL. After populating the class aux:Denominator with the population of interest and the class aux:HasExclusion with the part of the population that needs to be excluded, we halt RDFox's reasoning process and execute the following query.

```
SELECT ?Pat rdf:type aux:HbA1cDenom WHERE {
          ?Pat rdf:type aux:Denominator .
          FILTER NOT EXISTS { ?Pat rdf:type aux:HasExclusion }.}
```

We save the answers as triples into a file, which we then load back into RDFox.

This solution, however, is not optimal in a setting where transparency and proximity to the natural language specification is a major selling point. RDFox has since been extended to support stratified negation. The query can now be expressed as the following rule that can be listed and evaluated together with the other axioms:

```
[?Pat, rdf:type, aux:HbA1cDenom]:-
     [?Pat, rdf:type, aux:Denominator], not [?Pat, rdf:type, aux:HasExclusion].
```

4.4   Aggregates

Aggregate functions collapse multiple inputs into one single value, like 'max', 'count', 'average' but also 'list' or 'set'. The HEDIS CDC specification requires for measurement results always the latest and 'best' reading, if more than one measurement was taken on the same date. For example HEDIS CDC requires to

[...] identify the most recent BP reading taken during an outpatient visit (Outpatient Value Set) or a non-acute inpatient encounter (Nonacute Inpa-tient Value Set) during the measurement year. The member is numerator compliant if the BP is <140/80 mmHg. The member is not compliant if [...] the systolic or diastolic level is missing. If there are multiple BPs on the same date of service, use the lowest systolic and lowest diastolic BP on that date as the representative BP.

We first use a rule to classify a clinical visits which has both systolic and diastolic measurements as instances of aux:HasCompleteBP. Amongst these the latest, i.e. the date maximal, measurement has to be determined. Since RDFox had no support for aggregate functions at the time, we used a workaround which incidentally shows the connexion between aggregates such as max, min etc. and negation: We first mark all those visits which have a later visit using the rule

```
[?CV0, rdf:type, aux:HasLaterVisit]:-
     [?Pat, aux:patientHasAct, ?CV0],     [?Pat, aux:patientHasAct, ?CV1],
     [?CV0, rdf:type, aux:HasCompleteBP], [?CV1, rdf:type, aux:HasCompleteBP],
     [?CV0, kp:date, ?date0],             [?CV1, kp:date, ?date1],
     FILTER ( ?date0 < ?date1 ).
```

and we use, as done in Sect. 4.3, a SPARQL query to determine all those clinical visits that do not have a later visit:

```
SELECT ?CV rdf:type aux:latestBP WHERE {
              ?CV rdf:type aux:completeBP .
              FILTER NOT EXISTS {?CV rdf:type aux:HasLaterVisit}.}
```

The answers of the this query are added to the running store in RDFox. We can then populate the class aux:latestBP-140-80 with clinical visits from the class aux:latestBP. Note that our work-around computes aux:hasLaterVisit using quadratically many rule instantiations in the number of the blood pressure measurements per patient per year, and that a native implementation of the aggregate function max can achieve the same in linear time by iterating through all measurements and retaining the binding with the latest date.

5   Performance and Evaluation

We evaluated our approach on a commodity server provided by the Kaiser Permanente data centre. The server was security compliant according to the sensitive nature of the data. All tests were performed on this server and none of the provided data has been transferred outside of the security compliant infrastructure. The server runs Linux RedHat, has 8 Intel Xenon E5-2680 CPUs, clocked at 2.7 GHz and has 64 GB RAM. In what follows we shall first discuss data translation then the computation of HEDIS CDC using RDFox and finally the reconciliation of the results.

5.1   Data Translation

Kaiser Permanente provided the data in several files listed in Table 1. A multi-threaded Scala application translated the data into an RDF-graph. As discussed in Sect. 3.2, the application produced many duplicate RDF-triples expanding the number of triples by a factor of 5.5 from 293 M triples to 1.6 G triples. The translation took 47 min and produced 8 GZip files amounting to 8.8 GB.

Table 1. Files provided by Kaiser Permanente regional branch in Georgia

Content	Records	Size	Content	Records	Size	RDF-graph
Providers	113 k	6.8 MB	Prescriptions	8.9 M	892 MB	Total triples: 1.6 G triples
Members	466 k	84 MB	Labs	28.3 M	1.4 GB	Unique triples: 293 M triples
Enrolment	3.3 M	332 MB	Visits	54 M	8.6 GB	Translation time: 45 min (8 CPUs)

5.2   Computing HEDIS CDC

RDFox imported the 1.6 G RDF-triples in 11 min using 8 threads (Table 2, first row), which, due to duplicate elimination, resulted in a store containing 293 M unique triples. RDFox's importation process comprises reading, parsing, resolving the IRIs in an internal dictionary, eliminating duplicates and populating the store and its index structures.

The Datalog encoding of the HEDIS CDC specification consists of 174 rules of which approximately 65 % can be expressed in OWL 2 RL. Many of the OWL 2 RL expressible rules contain at most two body atoms whilst longer rules tend to be not tree-shaped and are thus not OWL 2 RL expressible, as the examples in Sect. 4 show.

The evaluation of larger rules, such as those for computing continuous enrolments, incur a high work-load, which leads to unacceptable computation times when applied to the whole RDF-graph. We therefore apply the full HEDIS Datalog ontology on a much smaller subgraph, which we compute using Datalog reasoning, and which contains all the data for the population of interest. To this end, we first identify the patients defined in extract 1 by evaluating the relevant rules on the full RDF-graph. These are simple rules, which RDFox can evaluate efficiently. Next, we use rules to mark all relevant triples connected to the identified patients. Finally, using a SPARQL query, we load the marked triples into a new store, on which we evaluate the full HEDIS Datalog ontology.

This strategy considerably reduces total computation time from 1 h 45 min to 30 min (sum of times in Table 2). Computing and extracting the subgraph on the full RDF-graph takes 13 min (795 s) using all 8 CPUs (Table 2, second row). Just before the subgraph extraction, the memory consumption peaks at 28 % at which RDFox uses 53 Bytes per RDF-triple. In dropping the store that contains the full RDF-graph, we release 18.1 GB of RAM. RDFox imports the subgraph in 32.4 s and consumes 2.5 % of the available RAM (Table 2, third row). The subgraph contains 14,000 patients of which almost all belong to the population of interest. This effectively reduces the original RDF-graph from 293 M triples to 23.4 M triples or to 8 % of its original size. Evaluating the full HEDIS Datalog ontology on the subgraph as well as running the counting queries then takes 4.5 min (258 s) on 8 CPUs (Fig. 2, fourth row). We could not properly compare our performance to the vendor product's performance. The vendor product not only computes HEDIS CDC but all HEDIS measure sets in approximately 8 h. The vendor product generally acts as black box and it is not possible to separate all different stages of computation from outside. Loading and initialising the database takes the vendor product 1 h. Then it executes a 4 h long pre-processing step which also includes the computation of the expensive continuous enrolment. The following stage contains an 18 min phase which can be associated with HEDIS CDC. Lastly the computation times of the vendor product were achieved on the more powerful licensed production server which has 16 CPUs but was not at our disposal.

5.3   Reconciliation of Results

For each category, we compared the membership numbers output by RDFox with those output by the vendor solution, and we found differences. The results are reported in Table 3. The row 'RDFox' reports the number of patients computed by RDFox. 'RDFox+' reports the number of excess patients which were not

Table 2. Computing HEDIS CDC with RDFox

	Patients	RDF-triples	RAM	% of 64G B	Time
Import	466 k	1.6 G (293 M)	17.8 GB	28 %	661 s
Extract 1 and extraction	466 k	367 M	18.1 GB	28 %	795 s
Import subgraph	14 k	23.4 M	1.6 GB	2.5 %	32.4 s
CDC numerators/counting	14 k	32.0 M	1.6 GB	2.5 %	258 s

included in the results of the vendor solution. Analogously 'Vendor+' shows the number of patients computed by the vendor that were not included in the RDFox results. 'CDC denominator' reports the population of interest, whilst each other category is a subset of the CDC denominator. The RDFox excess of 4 reported for the CDC denominator propagates through all other categories. We therefore indicate with the second summand for each category, how many of the excess patient were contained in the excess of the CDC denominator.

All results that were computed by RDFox were approved by the HEDIS data analyst. We shall shortly explain why discrepancies still remain. For each derived triple, RDFox can provide a proof tree that shows the rule instances and the RDF-triples in the RDF input graph which contributed to its derivation. Using these RDFox explanations and the information encoded in the IRIs (see Sect. 3.2), we can easily look up the records in the raw data and find the diagnosis codes in question. We can thus argue the correctness of RDFox's deviation directly. We showed, for instance, that RDFox's CDC denominator excess is actually correct and should also be output by the vendor solution. It is however much more difficult to argue why an excess in the Vendor product occurs. The vendor product only gives hints as to why it counts a patient into a certain category. For example it prints out the relevant visit date which is meant to help looking up the triggering visit. However in the case of Nephrological Attention, these visit dates of patients in the vendor excess could not be found in the raw data and it was not possible to explain the origin of these dates. The lack of explanations is a clear and typical short coming of traditional solutions.

Table 3. Computed numbers by RDFox and deviations

Results	CDC Denom	LDL-C		BP			Eye exam	Neph attent	HbA1c			HbA1c <7 %	
		Lab	<100 mg/dl	<140/80	<140/90				Lab	<8 %	>9 %	Denom	Lab
RDFox	14402	13217	7952	8963	11442	5430	13204	13474	9465	3132	8939	3702
RDFox+	0 + 4	0 + 3	0 + 2	3 + 3	0 + 4	0	2 + 3	0 + 3	0 + 3	0 + 1	1 + 3	0 + 2
Vendor+	0	0	0	0	0	1692	230	0	0	0	13	5

Since the HbA1c <7 % denominator uses negative information, the roles were reversed. We could show using RDFox explanations that all 13 patients in the vendor product's excess had an explicit exclusion and should not be counted. This excess propagates into HbA1c <7 % Lab. The minor RDFox excess in BP

140/80 could be traced to us interpreting a rule in a different way, which was subsequently approved by the HEDIS help-desk. The large discrepancy in Eye Exam was due to data that was not delivered by Georgia region.

6 Lessons Learnt

The project was very successful and we have learnt useful lessons in particular with regards to encoding and representing the data. However, the project also revealed some limitations of Semantic Technologies and suggested several ways in which they could be adapted to better fit data analysis applications of this kind.

Expressivity of the Ontology Language. The project revealed that OWL 2 alone is insufficient to compute the HEDIS measures. As we saw in Sect. 4.1, non-treeshaped expressions are necessary to determine the diabetic population. However, OWL 2, and consequently its tractable fragment OWL 2 RL, prohibits such expressions in order to ensure decidability. Furthermore, OWL 2 supports neither stratified negation nor stratified aggregation. As witnessed in Sects. 4.3 and 4.4 the absence of such constructs necessitates the introduction of non-declarative workarounds that make the behaviour of the system as a whole more difficult to understand. Finally, as we saw in Sect. 4.2, value manipulation during reasoning is an important language feature for data analysis applications. Although unrestricted value manipulation endangers the termination, non-recursive value manipulation preserves the termination guarantee and, and as seen in this project, is sufficient to encode the HEDIS specification.

Use of RDF as Data Format. RDF restricts the user to triples which correspond to unary and binary predicates. Hence rule bodies feature a large number of atoms, as n-ary relations have to be reified. Within rules this amounts to a named parameter perspective, since predicate names appear as constant in every rule body atom. It is therefore helpful to have meaningful predicate names which also indicate whether or not a triple is derived, as this makes rules legible and comprehensible. We also applied naming conventions that indicate which class of the data model an individual instantiates and from which data it originates. Both allowed us to debug and judge the correctness of the HEDIS encoding much faster. The resulting large number of joins that need to be performed in order to evaluate rules with many body atoms is not prohibitive in practical applications as RDF-triple stores are optimised for computing these large numbers of joins.

Due to the flexible schema of RDF, the data has a fully normalised representation. In particular, our data does not contain any null values and, for example in the case of clinical visits allows a variable number of diagnosis per clinical visit. The flexibility of RDF also helped us in the recapitulation stage of the project, in which, as discussed in Sect. 3.2, it became apparent that the data pertaining to a given clinical visit might be spread over multiple records in the raw data. This led to the misclassification of certain patients. The solution was simply to modify the assignment of IRIs to clinical visits in the data translation phase, which effectively merged database records referring to the same visit. Due to the flexibility of RDF, we did not have to change the conceptual schema or the way in which we compute the HEDIS measures.

The RIM modelling standard. Successful deployment of (semantic) technology also requires addressing 'soft' issues such as user expectations and familiarity. In this project it was crucial to win the support of the domain experts, who are the future users. This can be achieved by exploiting modelling standards in the respective field as we did with HL7 RIM. Following these conceptualisations makes it easier to argue clarity and intuitiveness of Semantic Technologies which are their major selling points.

We also used the RIM modelling standard to structure the types of clinical processes that occur in our project, which allowed us to uniformly represent healthcare data regardless of whether it describes visits, prescriptions or lab results. Due to the uniformity, the data model could be more easily memorised which facilitated rule authoring as it was not necessary to frequently refer to the data model documentation

7   Conclusion

In this paper we described the project conducted in collaboration with Kaiser Permanente to investigate the benefits of using Semantic Technologies in data analysis. Using the RIM modelling standard, we developed a schema ontology that mirrors how domain experts conceptualise business processes in healthcare, and we translated the raw data into an RDF-graph following this schema. With this data model in hand we encoded in RDFox-Datalog the HEDIS CDC specification which is renowned for its complexity. The declarative nature of RDFox-Datalog allowed us to succinctly express HEDIS CDC in a rule ontology which is close to the language of the specification. During the development of the rule ontology RDF proved to be a flexible data format that keeps the vocabulary explicit and thus confers its legibility to the rules.

The process of evaluating the rules on the patients and reconciling the results exceeded our and the HEDIS data analyst's expectations. The data of 466 000 patients fit easily into memory and the results were computed on modest resources within 30 min using the highly efficient triple store RDFox. Due to RDFox's good scalability we are confident that we could significantly reduce this time on a machine with more threads such as the vendor licensed production server.

The HEDIS data analyst noted that we had very few discrepancies from the outset and appreciated the ease with which changes and amendments could be done, not least because the data model and the rules provided comprehensible context. The explanation facilities in RDFox allowed us to easily trace discrepancies into the raw data. This reduced the number of development cycles of our application and we even discovered problems with the vendor solution. All results computed by our solution using RDFox were approved by the HEDIS analyst.

With this project we have successfully demonstrated the advantages of Semantic Technologies over traditional solutions in the context of data analysis in healthcare, and we are planning a further project with Kaiser Permanente in which the approach will be extended to all of HEDIS and all of their regions. The project also shows a possible avenue for applications of Semantic Technologies in encoding regulatory corpora in the field of data analysis in general, and demonstrates that it has the potential to cut development and maintenance costs in business settings.

Acknowledgments. The project was funded by the DBOnto Platform Grant, the MaSI[3] Fellowship, and Kaiser Permanente. Thanks are particularly due to Alan Abilla, Andy Amster, Patrick Courneya, Paul Glenn, Peter Hendler, Joseph Jentzsch, Scott Kimberly, and Mike Sutten, without whom this project would not have been possible.

References

1. Apt, K.R., Blair, H.A., Walker, A.: Towards a theory of declarative knowledge. In: Foundations of Deductive Databases and Logic Programming. pp. 89–148 (1988)
2. Benson, T.: Principles of Health Interoperability HL7 and SNOMED. Springer, New York (2010)
3. Chaussecourte, P., Glimm, B., Horrocks, I., Motik, B., Pierre, L.: The energy management adviser at EDF. In: Alani, H., et al. (eds.) ISWC 2013. LNCS, vol. 8219, pp. 49–64. Springer, Heidelberg (2013)
4. Krötzsch, M.: OWL 2 profiles: an introduction to lightweight ontology languages. In: Eiter, T., Krennwallner, T. (eds.) Reasoning Web 2012. LNCS, vol. 7487, pp. 112–183. Springer, Heidelberg (2012)
5. Nenov, Y., Piro, R., Motik, B., Horrocks, I., Wu, Z., Banerjee, J.: RDFox: a highly-scalable RDF store. In: Arenas, M., et al. (eds.) ISWC 2015. LNCS, vol. 9367, pp. 3–20. Springer, Heidelberg (2015)
6. Ross, K.A.: Modular stratification and magic sets for datalog programs with negation. In: Proceedings of the ACM Symposium on Principles of Database Systems, pp. 161–171 (1990)
7. Slee, V.N.: The international classification of diseases ninth revision (ICD-9). Ann. Intern. Med. **88**(3), 424–426 (1978). http://dx.doi.org/10.7326/0003-4819-88-3-424
8. Tao, J., Sirin, E., Bao, J., McGuinness, D.L.: Integrity constraints in OWL. In: Proceedings of the 24th AAAI Conference, AAAI 2010, Atlanta, GA, USA (2010). http://www.aaai.org/ocs/index.php/AAAI/AAAI10/paper/view/1931
9. Vizenor, L., Smith, B.: Speech acts and medical records: the ontological nexus. In: Proceedings of the International Joint Meeting EuroMISE 2004 (EuroMISE, Prague, CZ) (2004)

Building and Exploring an Enterprise Knowledge Graph for Investment Analysis

Tong Ruan[1(✉)], Lijuan Xue[1], Haofen Wang[1], Fanghuai Hu[2],
Liang Zhao[1], and Jun Ding[2]

[1] East China University of Science and Technology, Shanghai 200237, China
{ruantong,whfcarter}@ecust.edu.cn, {xuelijuanjsj,tracy_zl1993}@163.com
[2] Shanghai Hi-knowledge Information Technology Corporation,
Shanghai 200082, China
{hufh,dingjun}@hiekn.com

Abstract. Full-fledged enterprise information can be a great weapon in investment analysis. However, enterprise information is scattered in different databases and websites. The information from a single source is incomplete and also suffers from noise. It is not an easy task to integrate and utilize information from diverse sources in real business scenarios. In this paper, we present an approach to build knowledge graphs (KGs) by exploiting semantic technologies to reconcile the data from diverse sources incrementally. We build a national-wide enterprise KG which incorporates information about 40,000,000 enterprises in China. We also provide querying about enterprises and data visualization capabilities as well as novel investment analysis scenarios, including finding an enterprise's real controllers, innovative enterprise analysis, enterprise path discovery and so on. The KG and its applications are currently used by two securities companies in their investment banking and investment consulting businesses.

Keywords: Knowledge graphs · D2R · Information extraction · Data fusion · Investment analysis

1 Introduction

Full-fledged information about enterprises is useful in different areas, including analysis of regional industry distribution for the government, competitor intelligence for corporation executives, credit analysis for banks and so on. While applications of enterprise information may be similar or different, a large-scale enterprise knowledge base (KB) is always in great demand. The challenges of building such an enterprise KB are enormous. For example, the information is

This work was partially supported by "Action Plan for Innovation on Science and Technology" Projects of Shanghai (project No: 16511101000), and Software and Integrated Circuit Industry Development Special Funds of Shanghai Economic and Information Commission (project NO: 140304).

P. Groth et al. (Eds.): ISWC 2016, Part II, LNCS 9982, pp. 418–436, 2016.
DOI: 10.1007/978-3-319-46547-0_35

scattered in different databases and websites with noise. Besides, it is not realistic or necessary to get all data sources on hands at the beginning of the project. New data sources along with new pieces of data should be added incrementally on demand.

Recently, semantic technologies based on RDF representation, graph databases as well as other related technologies have been key enablers to build and explore KGs. In particular, a big KG could be constructed by integrating a few separate KGs with schema alignment and instancing matching mechanisms, and new properties or concepts as well as new instances can be easily added to the fused KG. Similarly, the ever growing KG can support more intelligent applications.

In this paper, we build a national-wide Enterprise Knowledge Graph, called EKG, which incorporates information about 40,000,000 enterprises in China. Based on the EKG, we provide querying about enterprises and data visualization capabilities as well as novel investment analysis applications, including finding an enterprise's real controllers, innovative enterprise analysis, enterprise path discovery and so on. We encounter business challenges as well as technology challenges in constructing and deploying the KG. The major business challenge is how to provide deep and useful analysis services without violating the privacies of a company and its employees. The technology challenges include constructing problems such as transforming the databases to RDF (D2R), representing and querying difficulties when meta properties and n-ary relations are involved, and performance issues since currently the KG contains more than one billion triples.

The product name of the EKG is "Magic Mirror". We expect that the "mirror" can reflect every important aspect of a company. The product is sold as a service, and currently we target the service to securities companies. Most securities companies in China provide investment bank services and investment consulting services. With the opening of "New Three Board", small innovation companies can be listed on the "New Three Board" with the endorsement of securities companies. Therefore the securities companies serve customers from big enterprises to small and medium-sized enterprises. There are about 40M companies in China. It is difficult for the securities companies themselves to gather authentic and full-fledged company information of their customers and potential customers. We collect company information from different sources for them and represent it in easy-to-use graphs. The "Magic Mirror" targets to help the securities companies to know and to approach their target companies better and quicker.

The rest of the paper is organized as follows. Section 2 describes the challenges that we are facing. Section 3 gives a brief overview of our approach. Section 4 presents our approach for building a national-wide EKG and how we addressed these challenges. Section 5 shows how the EKG is being used in practice. Section 6 lists the related work, and we conclude the paper in Sect. 7.

2 Challenges

There are business challenges as well as technology challenges in constructing and deploying the EKG. The business challenges include:

- **Data Privacy** One of our data sources is from China's State Administration for Industry and Commerce[1] (CSAIC), a government agency which pays serious attention to privacy issues. In such cases, we have to be very careful to balance the information requirements from EKG users with the privacy constraints.
- **Killer Services on the Graph** Since our EKG is complex and big, users will be overloaded with too much information if we deliver the raw information directly to users.

To address the privacy problem, first we transform the original data into the rank form or the ratio form instead of using real accurate values. Secondly, we obscure critical nodes (e.g., person related information) which should not be shown when visualizing the EKG as a graph. Thirdly, we provide UI interfaces which only allow particular types of queries. We deliver services which directly meet business requirements of users. For example, the service *finding an enterprise's real controllers* tells the investors from investment banks who is the real owner of a company, and the service *enterprise path discovery* provides hints on how the investors could reach the enterprises they want to invest in.

Technology challenges arise from the diversity and the scale of the data sources. At the first stage of our project, we mainly utilize relational databases (RDBs) from CSAIC. Secondly, we supplement the EKG with bidding information from Chinese Government Procurement Network[2] (CGPN) and stock information from Eastern Wealth Network[3] (EWN). Then the EKG is fused with the patent information extracted from the Patent Search and Analysis Network of State Intellectual Property Office[4] (PASN-SIPO) in another project. At last, the competitor relations and acquisition events are added to the EKG. This information is extracted from encyclopedia sites, namely Wikipedia, Baidu Baike and Hudong Baike. The following challenges are encountered during the above process:

- **Data Model** In addition to the basic data types (e.g., integer, float, and string), we need to store some complex data types including sequential data (e.g., phone number list of an enterprise), range-type data (e.g., effective operating time of an enterprise), map-type data (e.g., the annual sales of an enterprise should be stored in a map structure in which the key is the particular year and the value is sales). Furthermore, the relations in EKG are not only binary relations, there exist "property of relations" and "n-ary

---

[1] http://www.saic.gov.cn/.

[2] http://www.ccgp.gov.cn/.

[3] http://www.eastmoney.com/.

[4] http://www.pss-system.gov.cn/.

relations". The former are called *meta property* or *property graph* in the literature[5], and the latter sometimes referred as *event*. For the meta property example in our database, if a person is employed by a company, there is an additional property "entry time" which relates to the "employ" relation. For the event example, the investment relationship contains investors, companies, investment time, investment amount, investment ratio and so on. Since relations in RDF are binary, there are many discussions on how to represent events and meta properties, such as the W3C Working Group Note "Defining N-ary Relations on the Semantic Web"[6]. However, we do not find existing mature solutions on representing and querying meta properties and events in an efficient way.

- **D2R Mapping** At first, we use existing D2R tools (e.g., D2RQ[7]) to map RDBs from CSAIC into RDF. However, we encountered the following challenges that we can not solve directly with existing tools: (a) Mapping of meta property. As mentioned above, there exist meta properties in our applications, and metafacts are facts related to meta properties. (b) Data in the same column of RDBs map to different classes in RDF (Referred as *conditional class mapping* later). For example, patent applicants could be natural persons as well as companies. (c) Data in the same RDB tables may map to different classes having subClass relations (Referred as *conditional taxonomy mapping* later). For example, "Listed Company" is a subCalss of "Company", a record in the company table can be mapped to an instance of a parent class "Company" or an instance of a child class "Listed company".

- **Information Extraction** The "competitive", "acquisition" and other relations are extracted from encyclopedia sites in our paper. Since the related information might not only be contained in semi-structured sources like infoboxes, lists and tables, but also be mentioned in free texts. It is not an easy task to extract them from various types of data with high accuracy. However, entity extraction becomes difficult when there are abbreviations of company names in encyclopedic sites. The same company may be written in various variations or abbreviations. For example, abbreviations like "中铝 (Chalco)", "中铝公司 (Chinalco)", and "中国铝业 (Chinalco)" can all represent the company "中国铝业股份有限公司 (Aluminum Corporation of China Limited)". Moreover, Several company names may share the same abbreviation. For example, "大连万达集团股份有限公司 (Dalian Wanda Group Co., Ltd)" and 万达信息股份有限公司 (Wonders Information Co., Ltd)" can all be abbreviated as "万达 (Wanda)".

- **Query Performance** We encounter performance bottlenecks since the number of triples of our EGK has reached billions. Furthermore, there are more complex query patterns when the EKG usage scenarios increases: (a) Query all instances of a class which is a superClass in class hierarchy. For example, each patent in the patent KG has an International Patent Classification

---

[5] http://www.w3.org/TR/rdf11-concepts/.

[6] http://www.w3.org/TR/swbp-n-aryRelations/.

[7] http://d2rq.org/.

(IPC) code as its class. The IPC is a hierarchical patent classification system used to classify the content of patents. When users query the KG on an IPC code, we should find all the subClasses of the IPC code recursively, then find the patents belong to the subClasses. (b) Query all properties of an instance. The problem arises since different properties of the same instance may store as different triples in graph store. The operation usually has many I/O operations if the number of properties becomes large. For example, there are more than 100 properties of companies in our EKG, if each property of an instance is stored in a different property table (vertical partitioning method in [1]), the operation requires read operation for more than 100 times. (c) The queries on meta properties and n-ary relations. There are queries containing join operations between meta properties and ordinary properties, combined with filtering and sorting criteria. For example, find investment events where investment ratio is bigger than 10 % and investment amount is bigger than ten million. If we use the triple store method in [8] to store properties and meta properties separately, when join operations are required on these properties, we need to load all triples to the memory.

In order to solve the above challenges, we carefully select the most suitable methods and algorithms, and adapt them to our problems.

1. The following steps are designed to do D2R mapping. First, we split the original tables into atomic tables and complex tables, then we use D2RQ tools to handle mappings on atomic tables. At last, we develop programs to process ad hoc mappings on complex tables.
2. We use multi-strategy learning methods in [16] to extract competitor relations and acquisition events from various data sources of encyclopedic sites. We use HTML wrappers to extract information from semi-structured sources firstly, then we use Hearst patterns to extract information from free texts. The extracted data are fed as seeds to distantly supervise the learning process to extract more data from free texts.
3. We adopt a graph-based entity linking algorithms in [2] to accomplish the task of entity linking. First we calculate the similarity between mentions and entities in KB to find candidate entities, then we construct an undirected graph to complete disambiguation.
4. Since we do not find enough features to support meta properties or events in existing graph databases, we design our own storage structure to fully optimize the performance of miscellaneous queries in EKG. We use a hybrid storage solution composed of multiple kinds of databases. For large-scale data, we use NoSQL database namely Mongodb as the underlying storage. NoSQL can store mass data and has good scalability. For high-frequency query data, we use a memory database to store data, which can greatly accelerate query speed. We also partition the main data table to reduce the number of records in a single data table. We store meta properties of a relationship in the same table so that the ordinary property and meta property can be fetched in one operation. We also build indexes on n-ary relations and meta properties.

3   Approach Overview

3.1   Problem Definition

Description of the Target KG. Our goal is to build a national-wide enterprise KG. Firstly we give a quasi-definition of what a KG is in our paper.

As shown in Fig. 1, a knowledge graph G consists of schema graph G_s, data graph G_d and the relations R between G_s and G_d, denoted as $G = <G_s, G_d, R>$.

The schema graph $G_s = <N_s, P_s, E_s>$, where N_s is a set of nodes representing classes (concepts); P_s is a set of nodes representing properties, and P_s contains *rdfs:subClassOf*, *rdfs:equivalentClass*, and other user defined properties such as *applicant*; and E_s is a set of edges representing the relationships between classes in the graph G_s. $E_s \subseteq N_s \times P_s \times N_s$. For example, the domain of the property *applicant* in Fig. 1 is *patent* and the range of the property is *company* or *person*. There are two particular situations: (a) The subject of P_s is a relation instead of a class, we call the P_s *meta property*. For example, if a person is employed by a company, there is a meta property "entry time". (b) If P_s links to more than one subject or object, we call it *n-ary relation*.

The data graph $G_d = <N_d, P_d, E_d>$, where N_d is a set of nodes representing instances and literals; P_d is a set of nodes representing properties; and E_d is a set of edges representing the relationships between nodes in the graph G_d. Actually, each edge (Subject, Predicate, Object) stands for a fact. N_d includes two disjoint parts, namely instances (N_i) and literals. If the object of a triple is an instance, we call the property *relation* or *object property* in our paper. Otherwise, we call the property *datatype property*.

The relations R between schema graph G_s and data graph G_d are the relations which link the instances in the data graph to classes in the schema graph by the *rdf:type* property. $R = \{(instance, rdf:type, class) | instance \in N_i, class \in N_s\}$.

Figure 1 shows a small part of our EKG. The graph has nodes representing companies, stocks, persons, patents, investment events and data extracted from

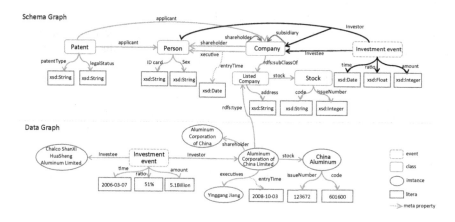

Fig. 1. Part of our enterprise knowledge graph

multiple sources, including stock code, entry time and issue number. The figure does not show a wide array of other properties extracted from data sources. Note that the graph includes edges that represent the executives, shareholders, subsidiary and so on.

Data Sources and Related Tasks to Construct the EKG. We have five major sources for constructing our EKG:

1. The KG is mainly based on structured enterprise information from CSAIC. It contains the information of 40,000,000 companies, 60,000,000 people, 8,000,000 pieces of litigation, and 1,000,000 sources of credit information. The information about a company contains company executives, registration number, address and so on. The information about a person contains entry time, ID card number, position and so on. The information about a piece of litigation contains complaint, case number, trial date and so on. A source of credit information contains performance status, court and so on. We transform RDBs into the RDF to build a basic KG[8].

2. PSAN-SIPO, one of the largest and most successful patent websites, contains a large amount of patent information. We extract 5,000,000 pieces of patent information from PSAN-SIPO including patent applicant, application number, patent name and so on to build a patent KG[9]. The basic KG and the patent KG are linked with companies and persons to form EKG.

3. We extract 3,000,000 pieces of enterprise bidding information from CGPN including investor, investee, invest time and so on. Stock information (e.g., stock code, issue number and so on) of listed companies is extract from EWN. Then the EKG is fused with the above extracted information.

4. Competitive relations as well as acquisition events extracted from encyclopedic sites are added to the EKG through the company name and person name.

An Example. Here we take "中国铝业股份有限公司 (Aluminum Corporation of China Limited)" as an example to show how the EKG can be constructed from different sources, as shown in Fig. 2. Firstly, the databases from CSAIC contain information about the directors and shareholders as well as general managers of "中国铝业股份有限公司 (Aluminum Corporation of China Limited)". We transform RDBs into RDF to form the basic KG, and we get triples such as < "中国铝业股份有限公司 (Aluminum Corporation of China Limited)", director, "熊维平 (Weiping Xiong)"> and < "中国铝业股份有限公司 (Aluminum Corporation of China Limited)", general manager, "罗建川 (Jianchuan Luo)"> from the transformed results.

Secondly, we can utilize the PSAN-SIPO website to investigate a company's technology advantages. We extract attributes and the values of attributes from

---

[8] http://ent.hiekn.com:28080/.

[9] http://kechuang365.com/charts.html.

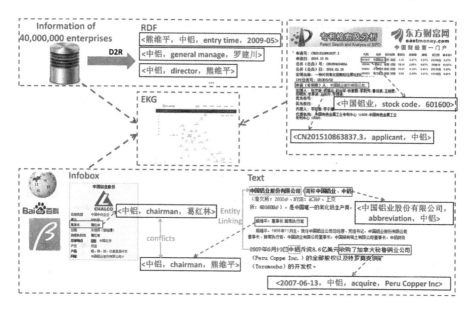

Fig. 2. Building EGK from multiple sources :*Aluminum Corporation of China Limited* Example

the website, and also convert the attribution value pairs into triples to form a patent KG. The patent KG may contain triples such as < CN201510863837.3, applicant, "中国铝业股份有限公司 (Aluminum Corporation of China Limited)" >. Data fusion algorithms are expected to link the two KGs with companies and persons.

Finally, we extract attribute value pairs of stock from EWN and also convert them into triples such as < "中国铝业 (Chinalco)", stock code, 601600 >. We can extract corporate executives from the infobox of Baidu Baike (e.g., < "中铝(Chalco)", chairman, "葛红林 (Honglin Ge)" >) and Wikipedia (e.g., < "中铝(Chalco)", chairman, "熊维平 (Weiping Xiong)" >), acquisition events from free texts of encyclopedia sites (e.g., < 2007-06-13, "中铝 (Chalco)", acquire, Peru Coppe Inc >). We find that information extracted from different sources has the problem of inconsistency. We determine which data source is correct according to the pages' update times. Information extracted from EWN and encyclopedia sites is linked to the KG with companies and persons. As a result, the information of "中国铝业股份有限公司 (Aluminum Corporation of China Limited)" is obtained, and builds a KG of "中国铝业股份有限公司 (Aluminum Corporation of China Limited)".

3.2   Data-Driven KG Constructing Process

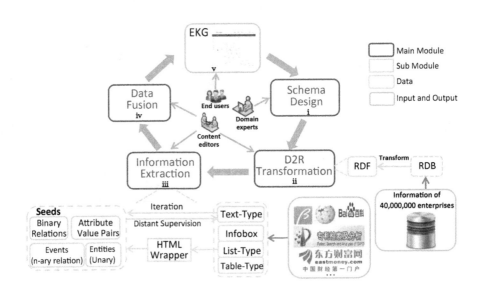

Fig. 3. Data-driven KG constructing process

As shown in Fig. 3, there are five major steps, namely *Schema Design, D2R Transformation, Information Extraction, Data Fusion with Instance Matching, Storage Design and Query Optimization*. When the EKG is built, we provide *Usage Scenarios* to securities companies on the EKG. The whole process is data-driven and iterative. In particular, whether the *D2R Transformation* step or the *Information Extraction* step is initiated is based on the type of data source. Whether the new iteration begins depends on the input of new data sources. Besides, if there are multiple sources in one iteration, we always use more structured data in the first place. There are two iterations in our example. For the first iteration there are two separate projects. One is an enterprise KG transformed from CSAIC; the other one is a patent KG extracted from PASN-SIPO. The two KGs serve different users. For the second iteration, we use data fusion algorithms to link the two KGs with companies and persons. There are also other data sources which could supplement the KG. We use specific HTML wrappers to extract information from semi-structured sources. Then we use Hearst patterns and distant supervision to extract more information from free texts. At last, we use *Instance Matching* algorithms to check whether instance pairs can be aligned.

4   Building Knowledge Graphs

4.1   Schema Design

While most general KGs such as DBpedia and YAGO are built in a bottom-up manner to ensure wide coverage of cross-domain data, we adopt a top-down approach in EKG construction to ensure the data quality and stricter schema. While methods exist to automatically extract schema-level knowledge such as taxonomies and class definitions from Web sites and databases, this approach is quite useful when the domain schema is complex. We manually design or extend the schema of the EKG since the schema is subject change when new data sources are added.

At the first iteration, the EKG includes four basic concepts, namely "Company", "Person", "Credit" and "Litigation". Major relations include "subsidiary", "shareholder", and "executive". The concepts in patent KG only include "Patent". Major relation is "applicant". At the second iteration, we add "ListedCompany", "Stock", "Bidding" and "Investment" to the EKG. Besides properties of the newly added concepts such as "stock code" and "issue number", there are also new relations between the existing EKG, for example "acquisition" and "competitive" as well as "subClassof" relation between "Company" and "ListedCompany", as shown in Fig. 1.

4.2   D2R Transformation

We take three steps to transform RDBs to RDF, namely *table splitting, basic D2R transformation by D2RQ* and *post processing*. The original data tables from CSAIC are integrated from multiple databases of provincial bureaus. These tables do not follow the basic design principles of databases (e.g., BNCF[10]). Some tables may contain multiple entities and relations. Furthermore, the table may contain n-ary relations or the same table column may refers to different entity types, as mentioned in Sect. 2. In order to make the tables easier to understand and to handle, we split the original tables virtually into smaller ones, namely atomic entity tables, atomic relation tables, complex entity tables (e.g., tables require *conditional class mapping* mentioned in Sect. 2) and complex relation tables (e.g., relation table with meta properties). An atomic entity table corresponds to a class, and an atomic relation table corresponds to relation instances where the domain and the range are two classes. We use D2RQ to transform the atomic entity tables and the atomic relation tables into RDF. We also write special programs to deal with complex relation tables which may require *meta property mapping, conditional taxonomy mapping* and so on, which have been mentioned in Sect. 2.

- *Table Splitting:* As shown in Fig. 4, the original table *Person Information* also contains enterprise information. We divide the table into *Person_P*, *Enterprise_E* and *Person Enterprise_PE*. The *Enterprise_E* table is furthered

---

[10] BCNF: Boyce and Codd Normal Form.

merged with the original *Enterprise Information* table, because the two tables share the similar information about enterprises.

- *Basic D2R Transformation by D2RQ:* We write a customized mapping file in D2RQ to map fields related to atomic entity tables and atomic relation tables into RDF format. We map table names into classes, columns of the table into properties, and cell values of each record as the corresponding property values for a given entity.
- *Post Processing:* (a) Meta property Mapping. The program gives a self-increasing ID annotation to the fact which has meta properties. The meta properties will then be properties of this n-ary relation identify by this ID. Thus we get some new triples(e.g., <ID, meta property, value>). (b) *Conditional taxonomy mapping.* Our program determines whether the entity maps to the subClass according to whether the entity appears in the table related to the subClass. For example, if a company exists in the relation table of company and stock, it implies that the company is a listed company, so we add a triple <registration number, rdf:type, Listed Company> to illustrate the fact, otherwise we add a triple <registration number, rdf:type, Company>. (c) *Conditional class mapping.* In our example, in the entity table of applicant, there is a column called "applicant type" which indicates whether the applicant is a or a natural person. Our program uses this field as the condition of the mapping choice.

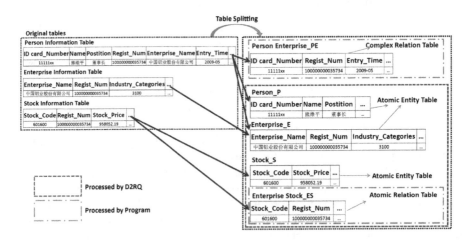

Fig. 4. An example of table splitting

4.3   Information Extraction

The EKG extracts information from various data sources, including HTML websites like PSAN-SIPO, EWN, and CGPN, and encyclopedic sites like Wikipedia,

Baidu Baike, and Hudong Baike. Furthermore, while most information extraction research work focuses on extracting one particular kind of target such as entities or relations between entities, we have to extract different types of entities(e.g., companies), binary relations(e.g., competitors), and attribute value pairs(e.g., CEO of a company). Our tasks also include event(n-ary relation) extraction (e.g., company acquiring) and synonym extraction (e.g., abbreviation of companies). The extraction strategy varies according to the data sources and extraction targets. We adopt a multi-strategy learning method to extract multiple types of data from various data sources. The whole process is as follows:

1. Entities and attribute value pairs of patent, stock and bidding information are extracted from PSAN-SIPO, EWN and CGPN respectively by using HTML wrappers.
2. Attribute value pairs (e.g., the chairman of an enterprise) of enterprises are extracted from infoboxes of encyclopedic sites by using HTML wrappers. Information extracted from different sources has the problem of inconsistency. We evaluate this information according to the pages' update time to determine which data source is correct. This extracted information can supplement the null value of databases.
3. Binary relations, events and synonyms identification on free texts require seeds annotation in sentences to learn patterns. These patterns are further used in other sentences to extract information. The quality of the extracted information heavily depends on the number of annotated sentences, whereas manual annotation costs too much human effort. Thus, for binary relation, event and synonym, we define a set of Hearst patterns to extract data from free texts of encyclopedic sites. For example, leveraging the Hearst pattern "X收购 (acquire) Y" can extract triples such as < 中国铝业 (Chinalco), acquire, 永晖焦煤股份有限公司 (Winsway Coking Coal Holdings)>, and < 中铝 (Chalco), acquire, 秘鲁铜业公司 (Peru Coppe Inc.)> from free texts. Then the extracted triples are fed as seeds automatically label free texts. This kind of distant supervision can significantly reduce the effort of manually labeling sentences. We first collect sentences that contain seeds and label these sentences. Then we generate extraction patterns from the annotated sentences. A good pattern should be generated by several sentences, thus we compute the support of each pattern. The pattern with a score greater than the threshold is selected as the extraction pattern. Finally, we use the generated extraction patterns to extract new information from other free texts. The newly extracted information is added to the seeds for bootstrapping. The whole process is iterative until no new information can be extracted.

In the process of information extraction, there are many abbreviations of company names in encyclopedic sites, as mentioned in Sect. 2. Here we use entity linking algorithms to link a company mentioned in text to companies in the basic EKG. We adopt a graph-based method to accomplish the task of entity linking in two steps: (a) *candidate detection*, that is, finding candidate entities in the KB that are referred by each mention. We first normalize company names, including company names extracted from multiple data sources and company names in the KB. More specifically, for a company name, if it contains any suffix (e.g., "股份有限公司 (Corp.)", "有限公司 (Co.,Ltd)", "集团 (Group)", and so on), the suffix is deleted. The purpose of this step is to be able to calculate the similarity between the core word of the mention and the core word of the entity in KB. Then, we use *Context Similarity* to calculate the similarity between the mentions and the entities in KB which are normalized to find candidates. Context similarity is to compute cosine similarity between the sentence containing the mention and the textual description (the first section of the Wikipedia article) of the entity in the KB. (b) *disambiguation*, selecting the most possible candidate to link. Here, we use the disambiguation algorithm proposed in the literature [2].

4.4 Data Fusion with Instance Matching

Information from different sources should be fused into EKG. For example, if the value of the "applicant" property of a patent is a company name, it should link to the instance of the company in the EKG. The problem is simple for instance matching of companies. In China, as requested by China's State Administration for Industry and Commerce, the full company name should be unique. The company names on patent and bidding web sites are also full names. Therefore it is very easy to link the patent KG with the basic KG. However, the problem is tough for instance matching of persons. While there are personal ID numbers for every person, there are no such IDs in the patent data sources. Currently, we use a simple heuristic rule to match the person in the patent KG to that person in the basic KG. If the name of the patent inventor and the applicant equals the name of the person and company in the basic KG respectively, we say the patent inventor matches the person name in the basic KG.

4.5 Storage Design and Query Optimization

We design our own triple store on the top of existing NoSQL databases. In particular, we use MongoDB as main storage for its large install bases, good query performance, mass data storage, and scalability with clustering support. We implement varies data types on top of MongoDB including List type, Range type and Map type. The corresponding query interfaces for each type are also implemented. For example, the Map type implements interfaces such as accessing all values, a key of specific values and the maximum or minimal value in the map.

Query performance is improved in varies ways: (a) Design a storage structure which supports efficient querying on meta properties and n-ary relations. We store meta properties and their values in different columns of the same table as original SPO triples. In the similar way, we store n-ary relations in the same row of a table. The consequence of such design is promising. The property and meta property are be retrieved together with one operation. Furthermore, when filtering and sorting operations were performed on n-ary relations, query can be completed on the database level by using the indexes we built. There are no additional in-memory operation required. (b) Use in-memory database Redis to store the data which are heavily accessed. Redis supports abundant data structures, which is very useful in our application context. The data stored in Redis includes the schema definition table and the class hierarchy relation table. (c) Construct sufficient indexes. Besides commonly used indexes such as SP, SO, PS and so on, we also build indexes on meta properties and n-ary relations based on application requirements. For example, the indexes on investment amount and investment ratio. (d) Data sharding. We partition triples into different tables according to the data type of the property value. For example, basic types are divided into Integer, Float, DateTime, String and Text, and each is stored in separate table.

5   Deployment and Usage Scenarios

The EKG platform is deployed on the Internet with access control. There are about two hundred million entities, one billion attribute value pairs, and two hundred million relations in EKG. It takes an hour to extract entities, three hours to extract attribute value pairs, and three hours to extract relations from various sources. It is rebuilt once a month to incorporate newly added enterprise data.

We sell the whole solution as services instead of software. Securities companies have customized the EKG portal and have integrated it into their own applications. We provide general querying and graph visualization capabilities, including browsing shareholders, subsidiaries, patents and executives of a particular company. We also provide in-depth analyzing services dedicated to investing requirements in securities companies, including finding an enterprise's real controllers, innovative enterprise analysis, enterprise path discovery, multidimensional relationships discovery of enterprises and so on, as shown in Fig. 5.

A typical example to query an entity is shown in Fig. 5. A recorded demo about this example can be accessed from YouTube[11]. Firstly users can select the types of entities in our EKG, including Company, Person, Patent, Litigation and so on. Users can further select the number of relation levels which shown in the results. The level is limited to be less than three, otherwise the graph would become too big. Users may further design the filtering criteria based on what kinds of nodes and relations are included. The result of the query is shown in the right panel of Fig. 5. The target entity and its relationships as well as related entities are shown in different shapes and colors depending on the types of relations and entities.

– *Finding an enterprise's real controllers.* When a securities company wants to approach a new customer, they should know the real decision makers of the potential customer. The person who owns the biggest equity share is the real decision maker. However, a person may also control a company indirectly. For example, he or she can control a company by holding equity shares of companies which in turn are shareholders of the target company. We develop an algorithm to traverse the KG to find the real controllers of a company

---

[11] https://youtu.be/y3ZCMNrisGM.

on the EKG. The shareholders of a company can be roughly divided into two types: enterprises and natural persons. The real controllers here refer to natural persons. To find the persons, we calculate the equity shares of all shareholders of the target company recursively until the shareholders are natural persons. When the shareholders are natural persons, we multiply all the equity shares on the recursion path as the nature persons' equity share, and add equity shares to the same person. Ultimately, the persons who have the largest equity shares are chosen as real controllers of the target company. The results of querying includes the information of real controllers and control paths. We find many different investment structures that a person can use to control a company. Figure 6(a) is a simple linear structure. A person P controls a company T through company A plus company B. Figure 6(b) is a typical triangle structure. It shows that a person P controls a company T by simultaneously investing in the company T and a shareholder of the company T. Figure 6(c) is more complex. The person P controls the target company T in three ways, namely from company A, from company B, and from company C plus company B.

- *Innovative enterprise analysis.* Securities companies want to find new and innovative companies that are worthy of investment. We connect the notion of innovativeness with the number of patents a company holds. In general, securities companies provide the field they are interested in, for example, robotics or remote healthcare, and the EKG system returns a list of companies which owns the largest number of patents in this field. The fusion of the patent KG into basic KG gives customers benefits that they can use to further investigate other information about the target company.

- *Enterprise path discovery.* Securities companies would like to know whether there are paths to reach their new customers, and they also want to know whether their potential customers have paths to their competitors, namely other securities companies. We use path discovery to find the path between any two companies or persons. As shown in Fig. 7, in practice, we found most targeting companies can be approached by our securities company in less than four hops, since both securities companies and their targeting companies are fairly big and their investors are very famous.

- *Multidimensional relationships discovery.* Given two companies, there might be varies relationships between them (e.g., competitive relationship, patent transferring relationship, acquisition relationship or investment relationship). Currently, we provide visualized graphs to help securities companies investigate different relationships between companies. Securities companies use these graphs to find new customers they are interested in. For example, if they find there are difficulties in investing in their target companies, they can approach the competitors of their target companies.

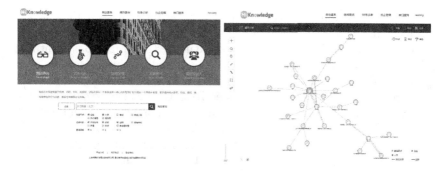

Fig. 5. Screenshot of EKG query interface showing results of a query on the *Long Credit Data (Beijing) CO.,Ltd*

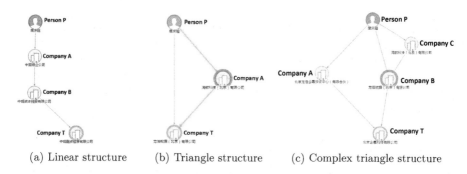

(a) Linear structure   (b) Triangle structure   (c) Complex triangle structure

Fig. 6. Different structures for a person to control an enterprise

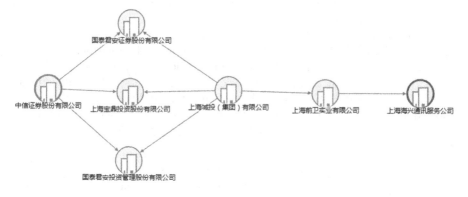

Fig. 7. Enterprise path discovery

6 Related Work

6.1 Knowledge Graphs and Their Applications

Knowledge graphs have attracted more and more attention from both academia and industry. Linked Open Datasets such as DBpedia and YAGO can be viewed as cross-domain knowledge graphs. Several Internet search engine companies have made an effort to build knowledge graphs in order to improve their search engine capabilities, including Google Knowledge Graph, Baidu Intimate, and Sogou Cubic Know. Nguyen et al. [12] analyzed the fitness for use of two encyclopedic datasets, namely DBpedia and Freebase, in music recommendations. Pirrò et al. [13] provided a tool called RECAP to generate explanations of relatedness on entities in encyclopedic datasets such as DBpedia and Freebase.

In contrast to the general-purposed KGs such as DBpedia and YAGO, Szekely et al. [18] presented a system called *DIG* and discussed how it can be used to build a knowledge graph for combating human trafficking. We also present our solution on building marine-oriented knowledge graphs in [15]. The notion of enterprise knowledge graph is also used by IBM [7] and other information technology companies when they adopt knowledge graph technology to particular enterprises.

While all the above work shows the benefit of semantic technologies, the large-scale enterprise knowledge graph built in this paper have a wider business perspective since EKG can further be used in more business scenarios such as competitor intelligence and credit analysis. The incrementally construction model proposed in this paper reduces the investment risk in early stages of the KG project.

6.2   Technologies in Constructing and Storing KGs

D2R. In the process of transforming databases to RDF formats, we use the D2RQ tool. While the tool in general is to provide a virtual RDF layer on top of the relational database [3,17], it can also export RDF triples from RDB. D2RQ mainly consists of the D2R Server, D2RQ Engine and D2RQ Mapping. D2RQ Engine [5] use D2RQ Mapping to convert the data in relational database into RDF format. The D2RQ Engine does not convert the relational database into real RDF data, but uses D2RQ Mapping file to map the database into virtual RDF format. In this paper we use D2RQ to export RDF triples from simple tables generated from the table-splitting step, and then post-process the exported RDF triples for metafacts, n-ary relations and other complex D2R mapping situations.

Information Extraction. Information extraction has been studied for a considerable amount of time. *Wrapper induction* is a sort of information extraction, which extracts knowledge from semi-structured data. Dalvi et al. [4] presented a generic framework to learn wrappers across websites. Gentile et al. [6] presented a methodology called multi-strategy learning, which combines text mining with wrapper induction to extract knowledge from lists, tables, and web pages. Distant supervision is an effective method to leverage redundancies among different sources, which has been used in [11] [14]. In this paper, we combine multi-strategy learning with distant supervision to extract information from various data sources. Entity Linking is the task of linking named entity mentions in

text with their referent entities in a specific KB. Alhelbawy et al. [2] proposed
a collective disambiguation method using a graph model. He zhengyan et al. [9]
proposed an entity disambiguation model based on Deep Neural Networks. [10]
provides an overview of recent instance linking approaches. In this paper, we
use entity linking algorithms proposed by Alhelbawy et al. to help determine
abbreviated company names in the text of encyclopedic websites.

Graph Database. As we have mentioned, our EKG requires representation and
querying on meta properties and n-ary relations. We have evaluated a few graph
databases including Jena[12], Blazegraph[13], sesame[14], AllegroGraph[15] and Hexastore[16]. Portion of graph databases such as Hexastore support named graph or
4-ary relation. Weiss et al. [19] also built an all permutation indexing of (S,P,O,C)
to speed the structural query with data source constraints in Hexastore. However, these databases do not directly support more than 4-ary relations which
frequently take place in our usage scenarios.

7   Conclusion and Future Work

We find real-world enterprise information could be represented by the KG in a
very natural way, and the KG also provides an easy way to integrate new data
sources even after the basic KG has been built. The application requirements can
conveniently be transformed to graph traversing and graph mining algorithms.
The analyzing results visualized in the graph can be easily understood by non-IT
customers.

We sell the EKG as services, and it can be easily integrated into different
applications. For example, one of our customers, China Securities, who is one of
the top ten securities companies in China, integrates the EKG into their customer
relation management system. Thus the enterprise information provided by the
EKG can not only be used by the investment banking sector, but also be used by
other sectors such as asset management sector or investment consulting sector.

In the future, we plan to add more data sources to the KG, such as tax
and invoice information per month. With the applications of such information,
investors can analyze the status of enterprises' business operations. Furthermore,
we will also try to monitor the change of shareholders as well as share ratios. As
we know, if investors want to control a company, they may not directly buy the
share of the target company; instead, they can buy shares of companies who are
shareholders of the company. In that case, we could develop interesting applications such as "Control intention recognition" to warn the current controller of
the company.

---

[12] http://jena.apache.org/.
[13] https://www.blazegraph.com/.
[14] http://rdf4j.org/.
[15] http://franz.com/.
[16] https://www.npmjs.com/package/hexastore.

References

1. Abadi, D.J., Marcus, A., Madden, S.R., Hollenbach, K.: Scalable semantic web data management using vertical partitioning. In: Proceedings of the 33rd International Conference on Very Large Data Bases, pp. 411–422. VLDB Endowment (2007)
2. Alhelbawy, A., Gaizauskas, R.J.: Graph ranking for collective named entity disambiguation. ACL **2**, 75–80 (2014)
3. Bing, B.H.L.: Semantic pattern mapping between rdbms and linked data based on open source software. New Technol. Lib. Inf. Serv. V **25**(9), 34–39 (2011)
4. Dalvi, N., Kumar, R., Soliman, M.: Automatic wrappers for large scale web extraction. Proc. VLDB Endowment **4**(4), 219–230 (2011)
5. Eisenberg, V., Kanza, Y.: D2rq/update: updating relational data via virtual rdf. In: Proceedings of WWW, pp. 497–498. ACM (2012)
6. Gentile, A.L., Zhang, Z., Ciravegna, F.: Web scale information extraction with lodie. In: 2013 AAAI Fall Symposium Series, pp. 197–212 (2013)
7. Guy, I.: Mining and analyzing the enterprise knowledge graph. In: Proceedings of the 22nd International Conference on World Wide Web, pp. 497–498. ACM (2013)
8. Harris, S., Gibbins, N.: 3store: Efficient bulk rdf storage (2003)
9. He, Z., Liu, S., Li, M., Zhou, M., Zhang, L., Wang, H.: Learning entity representation for entity disambiguation. ACL **2**, 30–34 (2013)
10. Heflin, J., Song, D.: Ontology instance linking: Towards interlinked knowledge graphs. Proc. AAAI **2016**, 4163–4169 (2016)
11. Mintz, M., Bills, S., Snow, R., Jurafsky, D.: Distant supervision for relation extraction without labeled data. In: Proceedings of the Joint Conference of the 47th Annual Meeting of the ACL, the 4th International Joint Conference on Natural Language Processing of the AFNLP: Observation of strains. Infect Dis Ther. 3(1), 35–43. : vol. 2, pp. 1003–1011. ACL (2011)
12. Nguyen, Phuong, T., Tomeo, Paolo, Noia, Tommaso, Sciascio, Eugenio: Content-based recommendations via DBpedia and freebase: a case study in the music domain. In: Arenas, M., et al. (eds.) ISWC 2015. LNCS, vol. 9366, pp. 605–621. Springer, Heidelberg (2015). doi:10.1007/978-3-319-25007-6_35
13. Pirrò, G.: Explaining and suggesting relatedness in knowledge graphs. In: Arenas, M., et al. (eds.) ISWC 2015. LNCS, vol. 9366, pp. 622–639. Springer, Heidelberg (2015). doi:10.1007/978-3-319-25007-6_36
14. Roth, B., Barth, T., Wiegand, M., Singh, M., Klakow, D.: Effective slot filling based on shallow distant supervision methods (2014). arXiv preprint arXiv:1401.1158
15. Ruan, T., Wang, H., Hu, F., Ding, J., Lu, K.: Building and exploring marine oriented knowledge graph for zhoushan library. In: Proceedings of ISWC 2014 (2014)
16. Ruan, T., Xue, L., Wang, H., Pan, J.Z.: Bootstrapping Yahoo! finance by wikipedia for competitor mining. In: Qi, G., Kozaki, K., Pan, J.Z., Yu, S. (eds.) JIST 2015. LNCS, vol. 9544, pp. 108–126. Springer, Heidelberg (2016). doi:10.1007/978-3-319-31676-5_8
17. Sahoo, S.S., Halb, W., Hellmann, S., Idehen, K., Thibodeau, T., Jr., Auer, S., Sequeda, J., Ezzat, A.: A survey of current approaches for mapping of relational databases to rdf. In: W3C RDB2RDF Incubator Group Report (2009)
18. Szekely, P., et al.: Building and using a knowledge graph to combat human trafficking. In: Arenas, M., et al. (eds.) ISWC 2015. LNCS, vol. 9367, pp. 205–221. Springer, Heidelberg (2015). doi:10.1007/978-3-319-25010-6_12
19. Weiss, C., Karras, P., Bernstein, A.: Hexastore: sextuple indexing for semantic web data management. Proc. VLDB Endowment **1**(1), 1008–1019 (2008)

Extending SPARQL for Data Analytic Tasks

Julian Dolby[1]([⊠]), Achille Fokoue[1], Mariano Rodriguez Muro[1],
Kavitha Srinivas[1], and Wen Sun[2]

[1] IBM Thomas J. Watson Research Center, Yorktown, USA
{dolby,achille,mrodrig,ksrinivs}@us.ibm.com
[2] IBM Research China, Beijing, China
sunwenbj@cn.ibm.com

Abstract. SPARQL has many nice features for accessing data integrated across different data sources, which is an important step in any data analysis task. We report on the use of SPARQL for two real data analytic use cases from the healthcare and life sciences domains, which exposed certain weaknesses in the current specification of SPARQL, specifically when the data being integrated is most conveniently accessed via RESTful services and in formats beyond RDF, such as XML. We therefore extended SPARQL with *generalized service*, constructs for accessing services beyond the SPARQL endpoints supported by `service`. For efficiency, our constructs support posting data, which is also not supported by `service`. We provide an open source implementation of this SPARQL endpoint in an RDF store called Quetzal, and evaluate its use in the two data analytic scenarios over real datasets.

1 Introduction

SPARQL, due to its minimal schema requirements and declarative graph queries, is ideally suited for the painful data integration steps that precedes any data analysis. However, when applying SPARQL to two data analytic use cases in the healthcare and life sciences domain, we uncovered roadblocks in its use: in our use cases, the data is most conveniently accessed via RESTful services and in formats beyond RDF, such as XML. The current SPARQL specification provides no construct to access such data; the existing federation construct, `service`, handles only static queries to other RDF endpoints, so we extend `service` to send data to and retrieve data from RESTful services. These services take and return solution bindings as do other SPARQL constructs, and so integrate seamlessly with SPARQL syntax, and preserve its declarative nature and opportunities for query optimization.[1]

The first use case is in the domain of drug safety, in particular Drug-Drug Interaction (DDI) predictions. Drug-Drug Interactions are a major cause of preventable adverse drug reactions (ADRs), causing a significant burden on the patients' health and the healthcare system [6]. It is widely known that clinical

---

[1] This is similar to what RDBMS do with foreign functions that are declared to be functional.

© Springer International Publishing AG 2016
P. Groth et al. (Eds.): ISWC 2016, Part II, LNCS 9982, pp. 437–452, 2016.
DOI: 10.1007/978-3-319-46547-0_36

studies cannot sufficiently and accurately identify DDIs for new drugs before they are made available on the market. As a result, the only practical way to explore the large space of unknown DDIs is through in-silico prediction of drug-drug interactions. We have built a system called Tiresias [7] that takes in various sources of drug-related data and knowledge as inputs, and provides DDI predictions as outputs. After semantic integration of the input data, we computed several similarity measures between all the drugs, and used the resulting similarity metrics as features in a large-scale logistic regression model to predict potential DDIs. In order to gather comprehensive information about drugs and their relevant associated bio-entities, Tiresias originally relied on Drugbank for the data integration task of combining information from multiple sources such as Uniprot. Unfortunately, data integration performed by a data source such as Drugbank is static, and it becomes out-of-date after the release of a specific version. A second problem was that numerous tools (R, scala, Java) were used to create the pipeline for data integration in the system, making it difficult to manage the different pieces of the pipeline.

The second use case investigates the potential health consequences of rising air pollution in China, especially high PM2.5 levels in many major cities in China. A critical question is whether PM2.5 levels have any correlation with the onset of some particular diseases. To provide an answer to this question, we looked for a potential correlation between diagnosis for particular diseases in Electronic Medical Record (EMR) data of a hospital in Beijing and elevated PM2.5 levels available in public sources such as StateAir[2]. The second use case requires flexibility in being able to specify different time windows for the data, to examine possible correlations between air pollution and disease, which is not difficult to manage across different tabular data and statistical software, but it does tend to be static.

In the above two use cases, to address the staleness and management issues, we tried to leverage SPARQL to perform a more dynamic and complete integration of information coming directly from the most authoritative sources for the entities of interest (e.g., Uniprot for proteins and genes, Drugbank for direct properties of drugs and StateAir for PM2.5 levels). A difficulty using SPARQL is its lack of means to integrate computation, such as the similarity metrics computed for drug pairs in the first use case or the statistical correlations between pollution and diseases in the second case. Our extension allows these computation to be expressed as RESTful services and hence integrated cleanly into our overall queries.

First, as is evident in our two use cases, the data being integrated is often most conveniently accessed via RESTful services and in formats beyond RDF, such as XML, which the current SPARQL specification does not fully support. We therefore generalize SPARQL's existing federation construct, service, to refer to RESTful services, using HTTP POST or GET methods to post solution bindings to a service. We also specify general mechanisms to extract tuples from

---

[2] StateAir U.S. Department of State Air Quality Monitoring Program: http://www. stateair.net/web/mission/1/.

data returned by the web service. This allows a SPARQL query to inject data from anywhere on the web directly into the query computation, which is a very useful ability in integrating disparate data. Our use of GET and POST allow RESTful services to be integrated unchanged. We define both service and table functions for web data ingestion, the distinction being that service functions send solution bindings singly and table functions post entire solutions. The latter is especially useful for more efficient federated computations, as we show in our use cases. Our service functions use RESTful services, as opposed to foreign function calls because these functions can provide language neutrality, statelessness, and insulation from operational details. Last, we added the construct of row numbers to SPARQL to allow batching of computations to external web services. We show how all these extensions operate in our use cases to dynamically inject data from a web service (e.g., from Uniprot) or include sophisticated computations on the data (e.g. computation of chemical similarity between drugs) into a query.

Overall, our core contributions in this paper are as follows:

1. We describe two use cases from the healthcare and life sciences domain, where we encountered problems initially in using SPARQL for data analytic tasks.
2. We introduce a minimal set of three extensions to SPARQL to broaden the scope of its use in data integration tasks.
3. We define the semantics of each extension formally, and describe extensions to the SPARQL algebra for them.
4. We provide a open source reference implementation of these extensions in a github project (Quetzal-RDF)[3] over relational and non-relational backends.
5. We discuss important lessons learnt that could be generalized to other applications.

The rest of this paper is organized into the following sections. Section 2 describes a simplified version of two real use cases for SPARQL as a data preparation tool, and we use them as examples for the extensions described in the paper. Section 3 presents our extensions to SPARQL, with a corresponding Sect. 4 to define the algebra. Section 5 shows the use of these extensions in the 2 real use cases and discusses lessons learnt. Section 6 describes related work.

2 Motivating Examples

We present two motivating examples from health care and life sciences; the first involved drug-drug interactions and the second involved potential health effects of pollution.

2.1 Drug Drug Interactions

Our first motivating example is a simplified version of the DDI problem. For simplicity we describe the steps to determine drug similarity based on chemical structure and their biological functions in Fig. 1:

---

[3] https://github.com/Quetzal-RDF/quetzal.

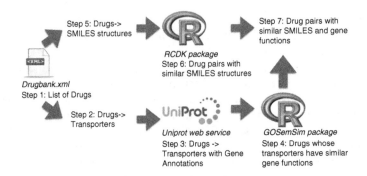

Fig. 1. Workflow for the DDI example

1. A list of drugs is obtained from Drugbank
2. Their transporter proteins are obtained from Drugbank
3. Transporter gene functions are obtained from Uniprot.[4]
4. Pairs with similar transport gene functions are found with a complex algorithm in an R package called GoSemSim [18].
5. Drug chemical structures in the standard SMILES format are obtained from Drugbank.
6. Pairs with similar structures are found with an R version of another complex algorithm implemented in the Chemistry Development Kit (CDK) [15].
7. These two sets of pairs are intersected to get similar drugs.

These steps map directly to the steps in our example query in Fig. 2.[5] We divide the query into functions corresponding to each portion of the work flow. These functions are called with `bind`. Within the `select`, the functions are shown in the same order as the list above; however, the last step of intersecting the two sets of similar drugs is implicit in the join of `?drugName`, `?similarDrug` and `?similarityValue` triples obtained from steps 4 and 6.

The implementations of these steps illustrate aspects of our extensions, and will be given in Sect. 3.1. Each function is assumed to be published at the URL specified by the function. The `getDrug` function (step 1) illustrates a simple use of *service functions*, applying XPath to extract the list of drugs from the DrugBank XML file. The functions `getDrugChemicalStructure` (step 5) and `getDrugTransporters` (step 2) illustrate *table functions*, in which a service is called for all drugs at once using an HTTP POST to compute corresponding `?smiles` and `?transporter` values respectively. The function `getProteinGeneFunctions` (step 3) is once again a service, but we will show later how we batch requests to make it more efficient. The details are involved, and are described in Sect. 3.1.

---

[4] While gene functions for these proteins are specified in Drugbank as well, the most comprehensive and up-to-date annotation of gene functions is in UniProt.

[5] Note that translation of the workflow to the query can be done in multiple ways, we choose this one for illustration.

```
prefix drug: <http://www.drugbank.ca>
prefix xs: <http://www.w3.org/2001/XMLSchema>

function drug:getDrugNames ( -> ?drugName)
function drug:getDrugTransporters (?drugName -> ?drugName ?transporter)
function drug:getProteinGeneFunctions (?protein -> ?protein ?geneFunction)
function drug:getProteinGeneFunctionSimilarity
                  (?drugName ?geneFunction -> ?drugName ?similarDrug ?similarityValue)
function drug:getDrugChemicalStructure (?drugName -> ?drugName ?smile)
function drug:getDrugChemicalSimilarities
                  (?drugName ?smile -> ?drugName ?similarDrug ?similarityValue)
select drugName, ?similarDrug where {
  BIND( drug:getDrugNames() AS (?drug))
  BIND( drug:getDrugTransporters(?drugName)
                                     AS (?drugName ?transporter ) )
  BIND( drug:getProteinGeneFunctions(?transporter) AS (?transporter ?geneFunction) )
  BIND( drug:getProteinGeneFunctionSimilarity(?drugName ?geneFunction
                                     AS (?drugName ?similarDrug ?similarityValue) )
  BIND( drug:getDrugChemicalStructure(?drugName) AS (?drugName ?smile) )
  BIND( drug:getDrugChemicalSimilarities(?drugName ?smile)
                                     AS (?drugName ?similarDrug ?similarityValue) )
}
```

Fig. 2. Simplified Example Query

The other two steps, functions `getDrugChemicalSimilarities` (step 6) and
`getProteinGeneFunctionSimilarities` (step 4) are also *table functions*: they
are each invoked once on the entire set of tuples of drug names and SMILES (or
gene functions) for a complex similarity computation.

2.2   Air Pollution and Health

Another example is due to rising air pollution levels, especially high PM2.5 levels
in many major cities in China, where there is increasing attention to the impact
of air pollution on people's health. A critical problem is to determine whether
PM2.5 levels have any correlation with the onset of particular diseases. To study
this problem, a possible workflow may contain the following steps as shown in
Fig. 3:

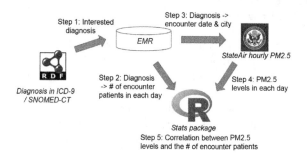

Fig. 3. Workflow for the air pollution and health example

1. Get a set of diagnosis codes of interest.
2. The total number of the encountered patients with the diagnosis in each day is queried in the EMR data.
3. The date of the encounter and city of patients are obtained from the electronic medical record (EMR) data.
4. PM2.5 levels of the treated dates are found in public sources like StateAir
5. Correlations between PM2.5 levels and the number of encountered patients are calculated with the R statistics package.

Similar to the previous example, the above steps can be supported by using our proposed extensions. Particular diagnosis codes or disease codes can be selected from open data repositories. Then EMR datasets are queried to retrieve the target patients' historical encounter information and the PM2.5 levels for each encounter date. At step 5, R's statistics package can be used to calculate the Person's correlation between the PM2.5 levels and the patient encounter.

3   Extensions

In this section, we first motivate our language extensions based on our example use case. Then we present definitions of our functions.

3.1   Constructs

The first step of our DDI use case is to gather the drugs of interest, which we obtain from Drugbank, a standard drug information repository that is conveniently accessed as XML. That motivates our first extension, *service functions*, which provide the ability to get sets of solution bindings from an HTTP call; the result is expected to be in XML and is interpreted in a manner analogous to XMLTABLE in SQL using XPath. The functions are analogous to service calls in SPARQL; the function executes once for each solution binding and the URL can be computed as an expression.

```
function drug:getDrugNames
service drug:getDrugNames [] -> "//row" : "./drug"
```

In this case, the getDrugNames function invokes the service bound to the getDrugNames URL, and expects an XML result where the rows are obtained by applying the XPath expression //row, which gets every XML element with that name. Within each row, the ./drug XPath expression is applied to get a column value. These functions are called using a BIND syntax, one call for each current solution binding; in this case, the result is a set of solution bindings, each having the name ?drug bound to a drug name returned from Drugbank.

```
bind(drug:getDrugNames() as (?drug))
```

Once the drug names have been obtained, the next steps for determining DDIs are to compute the *transporters* for each drug and its drug structure, called SMILES. These illustrate the need for the next extension: the *service function* interface used above is not suitable, since it would result in one call for each drug, which becomes expensive for large numbers of drugs. What we need is an interface that efficiently posts a set of solution bindings as a table, which we call *table functions*.

```
function drug:getDrugTransporters
table drug:getDrugTransporters
[?drug->?drug ?transporter] -> "//row" : "./drug" "./transporter"

function fn:getSMILES ( ?drug -> ?smiles )
table fn:getSMILES [ "funcData" -> post data ]
 -> "//x:row" :: "./x:drug" "xs:string" "./x:smiles" "xs:string"
```

When these functions are called, the ?drug element of each binding in the current solution is used to construct an XML document that is posted to the given URL. The result is a solution in which each binding has ?drug and ?transporter (or ?smiles respectively) elements. The key distinction with service functions is that the call is made once per solution rather than once per solution binding.

The motivation for our last extension comes from the fact that many services provide the capability to *batch* service requests. Continuing with our example, once we have the transporters for each drug, one way of looking for potential DDI is based on genes, using the gene annotations that have been associated with each transporter. These annotations are maintained in UniProt, where each transporter is an entity, and we query these annotations for each transporter. The standard UniProt website provides a query interface in which a UniProt identifier can be specified in the URL, and data about that entity returned in a variety of formats. The original service function from the example was defined as follows:

```
function drug:getProteinGeneFunctions (?protein -> ?gene ?id ?type)
service CONCAT("http://www.uniprot.org/uniprot/", CONCAT(?protein, ".xml"))
   [ ] -> "/up:uniprot/up:entry/up:dbReference" ::
   "./@id" "xs:string"
    "../accession[1]" "xs:string"
   "./@type" "xs:string"
```

But this interface is not suited to querying a large number of identifiers at once, and issuing hundreds of queries to a website will likely result in requests being simply denied. Fortunately, the European Bioinformatics Institute (EBI), part of the European Molecular Biology Laboratory (EBML), provides a flexible interface to freely-available life sciences data, including UniProt, that does allow querying identifiers in a batch[6]. The interface for calling this service is below; note that the URL itself is an expression in which the parameter ?ids becomes part of the URL.

---

[6] http://www.ebi.ac.uk/Tools/dbfetch/dbfetch.

```
function drug:getProteinGeneFunctionsGroup ( ?ids -> ?gene ?id ?type)
service CONCAT("http://www.ebi.ac.uk/...&id=", ?ids)
  [ ] xs-> "/uniprot/entry/dbReference" ::
    "./@id" "xs:string"
    "../accession[1]" "xs:string"
    "./@type" "xs:string"
```

To use this interface, we construct URLs that encode a collection of identifiers; the identifiers are encoded as a +-separated string in the query URL. We can easily express this with existing constructs in SPARQL; specifically the `group_concat` function and a group by clause. However, to group the transporters into chunks, we added one additional extension: the function `rowNumber` which is analogous to the SQL function. The `rowNumber` function simplifies grouping URLs into chunks to batch to a service request. To maintain the interface we had before, we encapsulate this mechanism in a SPARQL function as follows. Functions will be defined later, but essentially they work as in programming languages: ?transporter bindings are the parameter, and ?gene and ?id are results. Now the original query does not have to change.

```
function drug:getProteinGeneFunctions (?transporter -> ?gene ?id)
{ select (group_concat(?transporter; separator='+') AS ?ids) where
  {
    { select distinct ?transporter where
      {
        BIND( fn:getDrugBankNames() AS ( ?drug ) )

        BIND( fn:getTransporters( ?drug )
          AS ( ?drug ?transporter ) )
      }
    group by (xsd:int(rowNumber() / 30)) }
  }

  BIND( drug:getProteinGeneFunctionsGroup( ?ids )
    AS ( ?geneFunction ?t2 ?type ) )

}
```

As we show in the query fragment above, since the length of URLs is limited, we need a series of `ids` strings, each of which has up to 30 transporter ids. This set of `ids` strings then are used in a series of requests of the EBI Web service. The call to getProteinGeneFunctions needs to be a get operation due to the nature of the dbfetch URL, but it uses the groups of transporters computed above to minimize the number of get transactions it makes.

Thus, our extensions allow us to flexibly access services using both GET and POST mechanisms, and give us the power of SPARQL to construct complex URLs.

3.2   Functions

We introduce a *function* and we add a form of BIND to represent calls.

Definition 1 (Function call). *We extend BIND in SPARQL to express function calls: we allow BIND to specify the function uri and the input and output variables:*

$$\text{BIND } \{output~var\}^* \text{ AS uri}(\{input~var\}^*)$$

Definition 2 (Function). *A function encapsulates an operation; it is identified by a uri and specifies sets of input and output variables. They are specified by a function declaration as follows:*

`function` *uri* (*input var* * -> *output var* *)

There are multiple types of function within this interface; each type of function extends the definition above with an implementation specification. For function declarations as above that do not have a definition, the body of the function must be obtainable using the uri that defines the function. This extension to SPARQL is analogous to 'include' in languages such as C or 'import' in Java, but it is more powerful because it incorporates URIs as a mechanism for providing function bodies rather than specifying files. This allows SPARQL service providers to provide, in addition to a general SPARQL endpoint, specific functions that provide building blocks from which users can construct more complex queries of their data.

Definition 2.1 *(SPARQL Function).* *A SPARQL function encapsulates a SPARQL Pattern. When these functions are called, the input and output variables of the function definition are substituted by the formal parameters. Other variables in the pattern are private to each call, so there can be no name collisions between them and the surrounding query. Its definition simply specifies a pattern in addition to the declaration above.*

`{` *pattern* `}`

While the notion of input and output variables from the function interface is kept in SPARQL functions, there is in fact no semantic distinction between them. Both input and output variables are substituted with their formal parameters at call sites, other pattern variables are renamed to unique temporaries, and the resulting SPARQL included in the overall query.

Definition 2.2 *(Service Function).* *A Service function denotes calls to an HTTP service. The semantics is that the service is invoked for each distinct set of bindings for all the variables in the input vars; each solution mapping is extended by adding all variables in the output vars corresponding to its bindings of the input vars. This means that a service with no input vars is invoked once and its output extends all solution mappings. These functions extend the declaration above as follows:*

```
service uri (get | post) [(param -> expr)*] -> (xml | json) row : (col-
umn type)*
```

This uri *is what is used to invoke the HTTP service; it is not the uri that
defines the function. This* uri *can be a uri literal or an expression that returns
a string. The parameters are bound to the specified expressions, and passed to
the service call using a GET or POST request as specified. Only input variables
may be used in these expressions.*

*'xml' or 'json' must be specified. If 'xml' is specified, the results is expected to
be an XML document which is parsed into result rows using the* row *XQuery or
XPath expression; each row is parsed for each* column *using the corresponding
XQuery or XPath expression, and each such column is given the type denoted
by the corresponding* type *XQuery/XPath expression. If 'json' is specified, the
same process expects JSONPath [1] expressions[7].*

*In general, each column expression may return multiple values. In this case,
we define the result as a set of rows corresponding the cross product of all of
the column results. Each solution mapping is thus replicated and extended with
each binding in the cross product. Note that this cross-product behavior is also
exhibited by* XMLTABLE *as defined by SQL/XML.*

Definition 2.3 (Table Function). *A Table function denotes calls to an HTTP
service. The semantics is that the service is invoked for each group with a table
consisting of all sets of bindings for the input parameters. Any variables in
the scope of any table call that are not used as input parameters must be in a
group by clause. Note that query parameters can be defined using grouping functions like sum over the input variables. Every solution binding in each group
is extended with the bindings of output variables generated by the query. The
output is processed either with XQuery/XPath or JSONPath, just as for service
functions. These functions extend the declaration above as follows:*

```
table uri [(param -> expr)*] -> (xml | json) row : (column type)*
```

4   Translation

We define our extensions by augmenting the algebraic translation of SPARQL
given in §18 of the specification [9]. We then present the algebra for service and
table functions as an extension to the existing algebra. We then define the basic
support for invoking HTTP services.

Algebra Extensions. We define our extensions by extending the translation of
Graph Patterns, which is what §18 calls all the patterns of SPARQL. The overall
algorithm is in §18.2.2.6, and we extend it to handle service and table functions.
In this algorithm Ω represents the entire current solution, i.e. a collection of

---

[7] Systems that fully support SQL XML such as Oracle or DB2 can support XQuery,
   but we realize that in other implementations such as PostgreSQL, only XPath is
   supported. Allowing XPath accommodates more backend engines.

tuples; μ denotes a single tuple, called a solution binding. The new algorithm is Algorithm 1, and we present just the additional pieces. Service functions (lines 5–11) call the service for each solution binding μ and combine the results. Table functions (lines 11–13) call the service once with current solution Ω. Join must be used rather than extend for service functions because the call might return zero or more results, and following SQL/XML, we define that using join semantics. We use $\{\mu\}$ to denote the singleton multiset containing solely the binding μ.

Algorithm 1. if the form is *Group Graph Patterns*

1: $FS \leftarrow \emptyset$
2: $G \leftarrow$ the empty pattern
3: **for all** $E \in GroupGraphPattern$ **do**
4:     **if** E is BIND($f(i_1, \ldots, i_n)$ AS $(o_1, \ldots o_m)$) **then**
5:         **if** f is a service function **then**
6:             $G' \leftarrow \emptyset$
7:             **for all** $\mu \in G$ **do**
8:                 $G' \leftarrow G' \cup JOIN(\{\mu\}, CALL(uri))$
9:             **end for**
10:            $G \leftarrow G'$
11:        **else if** f is a table function **then**
12:            $G \leftarrow JOIN(G, CALL(uri))$
13:        **end if**
14:    **end if**
15:    Cases from §18.2.2.6
16: **end for**
17: **return** G

HTTP Calls. Service and table functions are based on HTTP services, as is the existing service pattern in SPARQL. For service functions, our HTTP operation allows computing query parameters based on the current solution binding; in the case of table functions, the expression must be a grouping function because the entire solution Ω is passed to the HTTP query. When the call returns data, the specified row definition is used to partition the result into rows; each such row is then parsed for each column value and type. We use the follow state of function f:

$$
\begin{aligned}
uri(f) &\equiv \text{the URI expression of} f \\
params(f) &\equiv \text{the} inputvar, expression \text{ pairs of } f \\
out(f) &\equiv \text{the} outputvars \text{ of } f \\
row(f) &\equiv \text{the row XQuery/XPath/JSONPath of} f \\
cols(f) &\equiv \text{the columns XQuery/XPath/JSONPath of} f \\
types(f) &\equiv \text{the column type XQuery/XPath/JSONPath of} f
\end{aligned}
$$

The function **process** is used to abstract over whether XQuery, XPath or JSON-Path is used to handle the result.

Algorithm 2. HTTP calls

1: **procedure** CALL(f)
2:     $req \leftarrow new\ HttpRequest$
3:     $req.uri \leftarrow uri(f)$
4:     **for all** $\langle param, expr \rangle \in params(f)$ **do**
5:         $req[param] \leftarrow expr$
6:     **end for**
7:     $\Omega \leftarrow \emptyset$
8:     $\Omega_{raw} \leftarrow req.send()$
9:     **for all** $row_{raw} \in process(row(f), \Omega_{raw})$ **do**
10:         $\mu \leftarrow \emptyset$
11:         **for all** $var, colExpr, typeExpr \in out(f), col(f), type(f)$ **do**
12:             $type \leftarrow process(row, typeExpr)$
13:             $val \leftarrow process(row, colExpr)$
14:             $\mu[var] \leftarrow type(val)$
15:         **end for**
16:         $\Omega \leftarrow \Omega \cup \{\mu\}$
17:     **end for**
18:     **return** Ω
19: **end procedure**

---

5   Evaluation and Lessons Learnt

This section presents our evaluation and a summary of lessons learnt.

5.1   Evaluation

In the overall drug-drug interaction use case, evaluations with real data reported in [8] show that Tiresias achieves very good DDI prediction quality with an average F-Score of 0.74 (vs. 0.65 for the baseline) and area under PR curve of 0.82 (vs. 0.78 for the baseline) using standard 10-fold cross validation. Furthermore, a retrospective analysis demonstrated the effectiveness of Tiresias to predict correct, but yet unknown DDIs. Up to 68 % of all DDIs found after 2011 were correctly predicted using only DDIs known in 2011 as positive examples in training. For the simplified DDI prediction use case described in Sect. 2, all our experiments were conducted on MacBook laptops with 8 G memory. The SPARQL extensions were implemented in Quetzal-RDF and our tests were conducted on the storage backends of Apache Spark and DB2. Our first use case had the following characteristics:

- There are 541 approved drugs in Drugbank with 113 unique transporters in Uniprot (drugs to transporters is a many to many relationship, with 941 drug-transporter pairs), and 541 unique chemical fingerprints.
- Getting data from UniProt for 113 unique transporters means 113 separate GET requests. For efficiency, we batched the transporter IDs into batches of 10 or 30 depending on the storage layer (10 for Spark, 30 for DB2). There

is a tradeoff between the size of the data we get back from Uniprot when transporters are batched into a single request (each transporter returns about 33 K) which can place significant burden on the storage layer to parse the data and the efficiency that is gained by batching the requests. After processing the Uniprot data, we had 1916 Gene Ontology (GO) functions for 113 transporters, for a total of 17384 rows, and we finished processing the drugs to transporters and transporters to GO functions computations in 51 s in DB2, and 2.4 min in Spark.

- Drug-drug chemical fingerprint similarity computation is $O(n^2)$, so when results are streamed back as XML from the REST service, it can place significant demands on resources needed to parse the 12.5 Mb data file in storage layers, and join it with existing data from DrugBank. For the Apache Spark backend, the query to fetch the drugs from Drugbank, fetch the appropriate chemical fingerprints and compute the chemical similarity for each drug pair took 9.5 min, with a significant amount of time being used by the join code that joined the pairs coming back with the drugs computed in prior steps. In DB2, the bottleneck was in parsing the 12.5 Mb file and not in the joins, so we needed to be more careful in the query to reduce the amount of data we shipped back. We did so by altering the SPARQL query to add *rowNumber* to the rows we send to the service, which reduces the amount of data we need to ship back, and that resulted in DB2 handling the query in 9 min. In general, the limitations of the storage layer can become a bottleneck for the sorts of queries that can be processed using the approach advocated in the paper (e.g., this will not work if terabytes of data need to be shipped between machines), but clearly the approach can work for a number of real workloads like this one.

- Drug similarity computation for GO annotations is another bottleneck similar to the one discussed above for chemical similarity, but we faced a hurdle here in terms of the ability of R code to handle this computation efficiently. The drug pair similarity computation required an average of 4–5 s in the GOSemSim R library, and there was no mechanism to batch the computation in the R code as we did for the chemical similarity computations. Performing the drug-pairs similarity for 541 drugs is therefore not feasible without parallelization of the computation on a cluster. Being able to extend our approach to farm out a specific workload to a cluster would be useful in future work.

In addition, we evaluated the air pollution use case with a real EMR dataset from a tertiary hospital in Beijing. The dataset contains about 0.7 million outpatient encounters of 6506 patients during Feb. 2015 to Nov. 2015. The PM2.5 levels of the same period were also collected from the StateAir website. By testing a number of different diagnosis codes and time windows for aggregation, we found that there is a correlation between the weekly average PM2.5 levels and the total number of the encounter patients with three specific diagnosis codes related to the respiratory system. The Pearson's correlation coefficient is 0.303, and the p-value is 0.043.

5.2   Comparison with Alternative Solutions

We compared the proposed approach in this paper to alternative solutions. In the air pollution use case, according to the data scientists working on the clinical data analysis of the EMR dataset, two alternative solutions are typically used to handle the problem. Some data scientists perform the steps in Sect. 2.2 separately with different tools, and rely on glue code or manual steps to connect the different steps. A problem with this approach, which was also encountered in the DDI prediction use case (where some similarity computations happen in R, java and scala, for example), is that users have to leverage multiple tools and programming languages to perform the analysis, which ends up being difficult to manage. Alternatively, data scientists tend to use existing software to manage pipelines (e.g., in Spark or SAS or Python ML libraries). While this solution solves the management problem, it is a procedural approach to managing pipelines and ties one to a specific set of tools, compared to a more declarative one proposed in this paper. In many cases, the declarative one as advocated in this paper can allow more flexibility, and quick experimentation needed in the data preparation and curation steps of the data analytic process.

5.3   Lessons Learnt

We have learnt important lessons on applying SPARQL for data preparation in data analytic use cases, which can be translated into the following key requirements:

- We need flexibility in SPARQL in being able to connect to different RESTful services, which we provided here by generalizing `service`.
- We need flexibility in SPARQL to batch requests from a set of query parameters specified in a table into a single URL, which we provided using `rowNumber`.
- We need the ability in SPARQL to compute over a table of data for at least 2 reasons. First, REST is by definition stateless, and the computation is often a type of aggregation over the data, which requires posting an entire table to a service. Second, it is simply more efficient to send data from a table when one is shipping data that does not fit within a URL. We propose `table functions` to handle this requirement.
- We need to be able to farm out workloads to external data sources when the computation requires parallelization as in the use case computing drug pair GO similarities, which is something we plan for future work.

6   Related Work

Mechanisms for integrating non-RDF data and SPARQL break into two broad categories. R2RML [4] and Direct Mapping [3] map relational data into a virtual RDF, providing the ability to ask SPARQL queries directly on the data, and hence integrate with RDF data using a SPARQL endpoint. We extend this approach to other forms of data. Furthermore, if these endpoints implemented our extended version of service, they could be posted data, opening up a wider variety of federated computing. The second category is tools to create RDF data, such as CSV2RDF [5,16] for CSV and spreadsheet data, and XSPARQL [2] for XML. These tools create RDF data that then can be queried as a separate step; hence it is not integrated into the language as is our approach. Finally, our approach opens up data available only as web services (e.g., the Google Maps API within a SPARQL query).

SPARQL federation resembles traditional federated SQL [10,14]. SPARQL 1.1 Service Descriptors [17] describe datasets, including simple statistics. Systems like DARQ [12], SIHJoin [11] and FedX [13] use queries to compute statistics at runtime. FedX also minimizes data transmission by restricting queries sent to end-points using embedded bindings using the VALUES keyword, to pass pre-computed results. However, the ability to federate is currently severely restricted since one can only transmit a limited amount of data, since it must be encoded in the URL.

7   Conclusions and Future Work

We have shown a few simple extensions to SPARQL that can greatly enhance its utility for query diverse sources of data: *functions* that allow modularizing queries, and *service* and *table* functions for integrating data from RESTful services. We define these constructs, and show how they were useful in providing a declarative approach to the problem of data integration that precedes any data analysis.

While we provide a general mechanism for accessing services that use XML and JSON formats, the same constructs can natrually handle other formats such as Protobuf, which can be more efficient. We intend to support such formats in the future. And service calls taking data from a large table are a natural basis for concurrent calls to the service; we intend to explore constructs for managing such parallelism.

References

1. JSONPath. https://www.npmjs.com/package/JSONPath
2. XSPARQL specification. http://www.w3.org/Submission/xsparql-language-specification/
3. Arenas, M., Sequeda, J., Prud'hommeaux, E., Bertails, A.: A direct mapping of relational data to RDF. W3C Recommendation, W3C, September 2012
4. Das, S., Cyganiak, R., Sundara, S.: R2RML: RDB to RDF mapping language. W3C Recommendation, W3C, September 2012
5. Ermilov, I., Auer, S., Stadler, C.: CSV2RDF: user-driven CSV to RDF mass conversion framework. In: Proceedings of the ISEM 2013, September 2013
6. Flockhart, D.A., Honig, P., Yasuda, S.U., Rosebraugh, C.: Preventable adverse drug reactions: a focus on drug interactions. In: Centers for Education & Research on Therapeutics

7. Fokoue, A., Hassanzadeh, O., Sadoghi, M., Zhang, P.: Predicting drug-drug interactions through similarity-based link prediction over web data. In: Proceedings of the 25th International Conference on World Wide Web, WWW 2016. ACM (2016)

8. Fokoue, A., Sadoghi, M., Hassanzadeh, O., Zhang, P.: Predicting drug-drug interactions through large-scale similarity-based link prediction. http://researcher.watson.ibm.com/researcher/files/us-achille/adrTechreport.pdf

9. Harris, S., Seaborne, A.: SPARQL 1.1 query language. W3C Recommendation, W3C, March 2013

10. Kossmann, D.: The state of the art in distributed query processing. ACM Comput. Surv. (CSUR) **32**(4), 422–469 (2000)

11. Ladwig, G., Tran, T.: SIHJoin: querying remote and local linked data. In: Antoniou, G., Grobelnik, M., Simperl, E., Parsia, B., Plexousakis, D., Leenheer, P., Pan, J. (eds.) ESWC 2011. LNCS, vol. 6643, pp. 139–153. Springer, Heidelberg (2011). doi:10.1007/978-3-642-21034-1_10

12. Quilitz, B., Leser, U.: Querying distributed RDF data sources with SPARQL. In: Bechhofer, S., Hauswirth, M., Hoffmann, J., Koubarakis, M. (eds.) ESWC 2008. LNCS, vol. 5021, pp. 524–538. Springer, Heidelberg (2008). doi:10.1007/978-3-540-68234-9_39

13. Schwarte, A., Haase, P., Hose, K., Schenkel, R., Schmidt, M.: FedX: a federation layer for distributed query processing on linked open data. In: Antoniou, G., Grobelnik, M., Simperl, E., Parsia, B., Plexousakis, D., Leenheer, P., Pan, J. (eds.) ESWC 2011. LNCS, vol. 6644, pp. 481–486. Springer, Heidelberg (2011). doi:10.1007/978-3-642-21064-8_39

14. Sheth, A.P., Larson, J.A.: Federated database systems for managing distributed, heterogeneous, and autonomous databases. ACM Comput. Surv. (CSUR) **22**(3), 183–236 (1990)

15. Steinbeck, C., Han, Y., Kuhn, S., Horlacher, O., Luttmann, E., Willighagen, E.: The Chemistry Development Kit (CDK): an open-source java library for chemo- and bioinformatics. J. Chem. Inf. Comput. Sci. **43**(2), 493–500 (2003)

16. Tandy, J., Herman, I., Kellogg, G.: Generating RDF from tabular data on the web. W3C Proposed Recommendation, W3C, November 2015

17. Williams, G.: SPARQL 1.1 service description. W3C Recommendation, W3C (2013)

18. Yu, G., Li, F., Qin, Y., Bo, X., Wu, Y., Wang, S.: GOSemSim: an R package for measuring semantic similarity among GO terms and gene products. Bioinformatics **26**(7), 976–978 (2010)

Author Index

Aberer, Karl II-140
Alani, Harith II-257
Allik, Alo II-3
Almeida, João Paulo A. I-53
Alse, Suresh I-446
Alshukaili, Duhai I-3
Ambite, José Luis I-549, II-12
Angles, Renzo I-20
Arcan, Mihael II-241
Asprino, Luigi II-168
Atencia, Manuel I-360
Auer, Sören II-38

Bakke, Eirik II-65
Bakry, Menna I-218
Balduini, Marco II-140
Beek, Wouter I-184
Bereta, Konstantina I-37
Bernstein, Abraham I-252
Biega, Joanna II-177
Biemann, Chris II-56
Bienvenu, Meghyn I-532
Birukou, Aliaksandr II-383
Bloem, Peter I-184
Boncz, Peter I-463
Bosque-Gil, Julia II-47
Božić, Bojan II-195
Brambilla, Marco II-140
Brandt, Sebastian II-344
Brasileiro, Freddy I-53
Brennan, Rob II-195
Buitelaar, Paul II-241
Burel, Grégoire II-257
Burns, Gully A. II-12

Calbimonte, Jean-Paul II-140
Cao, Ermei II-104
Carral, David I-70
Carvalho, Victorio A. I-53
Chapellier, Cyril I-532
Charron, Bruno II-273
Chekol, Melisachew Wudage I-86
Chen, Jiaqiang I-102
Cheng, Gong I-119

Christen, Victor I-135
Clarke, Barry II-30
Cochez, Michael I-151
Cohn, Anthony II-30
Cudré-Mauroux, Philippe II-220
Curioni, Giulio II-30

d'Aquin, Mathieu I-581
Daniele, Laura II-21
De Giacomo, Giuseppe I-167
de Melo, Gerard I-102
de Rooij, Steven I-184
de Vries, Gerben Klaas Dirk II-186
Decker, Stefan I-151
Dehghanzadeh, Soheila I-252
Dell'Aglio, Daniele I-252, II-140
Della Valle, Emanuele I-252, II-140
den Hartog, Frank II-21
Dengel, Andreas I-218
Di Sciascio, Eugenio II-131
Difallah, Djellel Eddine II-220
Dimitrova, Vania II-30
Ding, Jun II-418
Dirschl, Christian II-195
Dolby, Julian II-437
Domingue, John I-566
Dong, Guozhu II-113
Dragisic, Zlatan I-200
Dragoni, Mauro II-241
Drosatos, George I-566
Du, Heshan II-30

Egami, Shusaku II-291
Eisenmann, Harald II-308
Eldesouky, Bahaa I-218
Ermilov, Ivan II-38
Euzenat, Jérôme I-360

Fang, Zhijia II-47
Faralli, Stefano II-56
Faria, Daniel I-200
Fazekas, György II-3, II-229
Feier, Cristina I-70
Felder, Victor II-220

Feng, Zhiyong I-632
Fernandes, Alvaro A.A. I-3
Fokoue, Achille II-437
Freudenberg, Markus II-195

Gad-Elrab, Mohamed H. I-234
Galstyan, Aram I-649
Gangemi, Aldo II-150, II-168
Gao, Shen I-252
García, Roberto II-65
Gavankar, Chetana I-271
Genevès, Pierre II-80
Gentile, Anna Lisa II-150
Ghasemi-Gol, Majid I-649
Gil, Rosa II-65
Gimeno, Juan Manuel II-65
Gkotsis, George I-566
Glimm, Birte II-159
Gonçalves, Rafael S. II-159
Gracia, Jorge II-47
Gramegna, Filippo II-131
Grau, Bernardo Cuenca II-325
Graux, Damien II-80
Grimm, Stephan II-325
Groß, Anika I-135
Guclu, Isa I-289
Guizzardi, Giancarlo I-53
Gutierrez, Claudio I-20

Haag, Florian II-122
Hall, Wendy I-515
Hariman, Charles Darwis I-102
Hartig, Olaf I-305
Hendler, Peter II-400
Hennig, Christian II-308
Hermjakob, Ulf II-12
Hernández, Daniel II-88
Hirate, Yu II-273
Hitzler, Pascal I-70, II-113
Hoffart, Johannes II-177
Höffner, Konrad I-325
Hogan, Aidan II-88
Horrocks, Ian I-480, II-325, II-344, II-400
Hose, Katja I-341
Hu, Fanghuai II-418
Hu, Wei II-104

Ibragimov, Dilshod I-341
Ieva, Saverio II-131
Inants, Armen I-360

Ioannidis, Yannis II-344
Ivanova, Valentina I-200

Jachiet, Louis II-80
Jiménez-Ruiz, Ernesto I-200, II-325
Joshi, Amit Krishna II-113

Kaldoudi, Eleni I-566
Kämpgen, Benedikt II-308
Karger, David R. II-65
Kawamura, Takahiro II-291
Kharlamov, Evgeny II-325, II-344
Kimberly, Scott II-400
Knoblock, Craig A. I-446, I-549, I-649
Kollingbaum, Martin J. I-289
Kontokostas, Dimitris II-195
Kotidis, Yannis II-344
Kotoulas, Spyros II-363
Koubarakis, Manolis I-37
Krötzsch, Markus I-376
Kuzey, Erdal II-177

Lambrix, Patrick I-200
Lamparter, Steffen II-325, II-344
Layaïda, Nabil II-80
Lehmann, Jens I-325, II-38
Li, Haoxuan II-104
Li, Yuan-Fang I-271, I-289
Link, Vincent II-122
Liu, Daxin I-119
Lodi, Giorgia II-168
Lohmann, Steffen II-122
Lopez, Vanessa II-363
Loseto, Giuseppe II-131

Magee, Derek II-30
Mailis, Theofilos II-344
Martin, Michael II-38
Matentzoglu, Nicolas II-159
Mauri, Andrea II-140
Maus, Heiko I-218
Megdiche, Imen I-393
Mehdi, Gulnar II-325
Menard, Colette I-532
Mileo, Alessandra I-252
Möller, Ralf II-344
Motik, Boris I-480, II-400
Motta, Enrico I-581, II-383
Muro, Mariano Rodriguez II-437

Nakatsuji, Makoto I-411
Nenov, Yavor I-480, II-325, II-400
Neuenstadt, Christian II-344
Neumaier, Sebastian I-428
Nikolaou, Charalampos II-344
Nuzzolese, Andrea Giovanni II-150

Ohsuga, Akihiko II-291
Oriol, Xavier I-167
Osborne, Francesco II-383
Özçep, Özgür II-344
Özsu, M. Tamer I-305

Pafili, Kalliopi I-566
Pan, Jeff Z. I-289
Panchenko, Alexander II-56
Parreira, Josiane Xavier I-428
Parsia, Bijan II-159
Paton, Norman W. I-3
Paulheim, Heiko I-498, II-186
Pedersen, Torben Bach I-341
Peroni, Silvio II-168
Pesquita, Catia I-200
Pham, Minh I-446
Pham, Minh-Duc I-463
Piccolo, Lara S.G. II-257
Piro, Robert II-400
Pirrò, Giuseppe I-86
Polleres, Axel I-428
Ponzetto, Simone P. II-56
Portokallidis, Nick I-566
Potter, Anthony I-480
Presutti, Valentina II-150, II-168
Prud'hommeaux, Eric I-151
Purcell, David II-273

Qian, Xinqi II-104
Qu, Yuzhong I-119, II-104

Rahm, Erhard I-135
Ramakrishnan, Ganesh I-271
Rao, Guozheng I-632
Rebele, Thomas II-177
Reeves, Helen II-30
Rettinger, Achim I-598, I-615
Rezk, Martin II-273
Ringsquandl, Martin II-325
Ristoski, Petar I-498, II-186
Riveros, Cristian II-88
Rizzo, Giuseppe II-204

Roes, Jasper II-21
Rojas, Carlos II-88
Rosati, Riccardo I-167
Roshchin, Mikhail II-325
Rossman, Michael II-400
Roumeliotis, Stefanos I-566
Ruan, Tong II-47, II-418
Ruta, Michele II-131

Salatino, Angelo II-383
Sandler, Mark B. II-3, II-229
Savo, Domenico Fabio I-167
Schlobach, Stefan I-184
Scioscia, Floriano II-131
Siow, Eugene I-515
Solanki, Monika II-21, II-195
Srinivas, Kavitha II-437
Steigmiller, Andreas II-159
Stepanova, Daria I-234
Stirling, Ross II-30
Suchanek, Fabian M. I-532, II-177
Sun, Wen II-437
Sun, Zequn II-104
Svingos, Christoforos II-344
Sygkounas, Efstratios II-204
Szekely, Pedro I-446, I-549, I-649

Taheriyan, Mohsen I-549
Tandon, Niket I-102
Teste, Olivier I-393
Third, Allan I-566
Thost, Veronika I-376
Tiddi, Ilaria I-581
Tiropanis, Thanassis I-515
Tommasi, Pierpaolo II-363
Tonon, Alberto II-220
Trojahn, Cassia I-393
Troncy, Raphaël II-204

Umbrich, Jürgen I-428
Urbani, Jacopo I-234
Usbeck, Ricardo I-325

van Harmelen, Frank I-184
Viehl, Alexander II-308

Wang, Haofen II-47, II-418
Wang, Xin I-632
Weikum, Gerhard I-234, II-177
Wilmering, Thomas II-229

Wu, Jiewen   II-363
Wu, Wenrui   I-632

Xue, Lijuan   II-418
Xue, Lingkun   II-104

Zerega, Enzo   II-88
Zhang, Ji   I-598, I-615

Zhang, Lei   I-598, I-615
Zhang, Xiaowang   I-632
Zhao, Liang   II-418
Zheleznyakov, Dmitriy   II-344
Zhu, Linhong   I-549
Zimányi, Esteban   I-341

Printed in the United States
By Bookmasters